COMPSTAT

Proceedings in
Computational Statistics

11th Symposium held
in Vienna,
Austria, 1994

Edited by
R. Dutter and W. Grossmann

With 78 Figures

Springer-Verlag
Berlin Heidelberg GmbH

Professor Dr. Rudolf Dutter
Department for Statistics and
Probability Theory
University of Technology Vienna
Wiedner Hauptstraße 8-10
A-1040 Vienna, Austria

Professor Dr. Wilfried Grossmann
Department of Statistics,
OR and Computer Science
University Vienna
Universitätsstraße 5
A-1010 Vienna, Austria

ISBN 978-3-7908-0793-6   ISBN 978-3-642-52463-9 (eBook)
DOI 10.1007/978-3-642-52463-9

© Springer-Verlag Berlin Heidelberg 1994
Originally published by Physica-Verlag Heidelberg in 1994

2201/2202-543210 - Printed on acid-free paper

Preface

This book assembles papers which were presented at the biennial symposium in Computational Statistics held under the auspices of the International Association for Statistical Computing (IASC), a section of ISI, the International Statistical Institute. This symposium named COMPSTAT '94 was organized by the Statistical Institutes of the University of Vienna and the University of Technology of Vienna, Austria.

The series of COMPSTAT Symposia started 1974 in Vienna. Meanwhile they took place every other year in Berlin (Germany, 1976), Leiden (The Netherlands, 1978), Edinburgh (Great Britain, 1980), Toulouse (France, 1982), Prague (Czechoslovakia, 1984), Rom (Italy, 1986), Copenhagen (Denmark, 1988), Dubrovnik (Yugoslavia, 1990) and Neuchâtel (Switzerland, 1992). This year we are celebrating the 20*th* anniversary in Vienna, Austria.

It has obviously been observed a movement from "traditional" computational statistics with emphasis on methods which produce results quickly and reliably, to computationally intensive methods like resampling procedures, Bayesian methods, dynamic graphics, to very recent areas like neural networks, accentuation on spatial statistics, huge data sets, analysis strategies, etc.

For the organization of the symposium, new guidelines worked out by the IASC in written form were in effect this time. The goal was to refresh somehow the spirit of the start of COMPSTAT '74, keep the tradition of the series and ensure a certain continuity in the sequence of biannual meetings.

This year, twelve speakers were invited to introduce, expose and show the boarder lines of special topics. About 60 papers were selected as "contributed" ones, and approximately 130 as short communications and posters; they were submitted by researchers from more than 30 countries all over the world.

The present volume contains the "invited papers" as well as the "contributed papers", and brings together a wide range of topics in the field of computational statistics. They may roughly be classified into three main topic streams:

(A) **Computation and Algorithms** including I *Treatment of "Huge" Data Sets*, II *Numerical Analysis Useful in Statistics*, III *Parallel Statistical Computing*, V *Computational Aspects in Optimization*, XI — *in Statistics in Neural Networks*, XVIII *Metadata and Statistical Information Systems*, XX *Statistical Software, Evaluation and Comparison*, XXI *Using the Computer in Teaching Statistics*,

(B) **Statistical Inference and Design** including IV *Selection Procedures*, VIII *Computational Aspects in Bayesian Statistics*, IX — *in Sequential Analysis including Quality Control*, X — *in Reliability and Survival*, XII — *in Robust Statistics*, XIV — *in Time Series Analysis*, XVII *Resampling Methods*, XXII *Experimental Designs*, XXIV *Statistical Inference*,

(C) **Statistical Modelling** including VI *Computational Aspects in Spatial Statistics and GIS*, VII — *in Discrimination and Classification*, XIII — *in Multivariate Analysis*, XXIII *Model Fitting*, XXV *Applications*.

Invited speakers are: Peter J. Huber (Keynote Speaker on *Huge Data Sets*), Philippe C. Besse (on *Models for Multivariate Data Analysis*), Antonio Ciampi (on *Classification and Discrimination: The RECPAM Approach*), Dianne Cook (on *Some Dynamic Graphics for Spatial Data (with Multiple Attributes) in a GIS*), Håkan Ekblom (on *What can Numerical Analysis do for Statistics?*), Randall L. Eubank (on *Confidence Bands for Nonparametric Function Estimation*), Arnoldo Frigessi (on *Assessing Convergence of Markov Chain Monte Carlo Methods*), Guido Giani (on *Construction of Decision Procedures for Selecting Populations*), Kurt Hornik (on *Neural Networks: More Than 'Statistics for Amateurs'?*), Marie Hušková (on *Miscellaneous Procedures Connected with the Change Point Problem*), R. Douglas Martin (on *Smoothing and Robust Wavelet Analysis*), Albert Prat (on *Application of a Multiagent Distributed Architecture for Time Series Forecasting*).

All contributed papers were reviewed. The criteria used were originality, accuracy and that the subject should be concerned with statistics and computation. The reviewing was mainly done by the members of the Scientific Programme Committee (who also had the painful task to select these papers out of about 200 received): Tomas Aluja-Banet (Spain), Yadolah Dodge (Switzerland), Rudolf Dutter (chairman, Austria), Ludovic Lebart (France), Georg Pflug (Austria), Dieter Rasch (The Netherlands), Edward J. Wegman (USA), Andrew Westlake (England), Peter Sint (consultative member, Austria). Further referees were Wilfried Grossmann and Harald Strelec (both Austria).

Other presented papers as "short communications", "posters" and "software descriptions" are printed in further volumes published by the local organizers. They have also published the given Tutorials "Visualizing Data" (by W.S. Cleveland, U.S.A), "Statistical Meta Data" (by Bo Sundgren, Sweden) and "Stochastic Models in Financial Mathematics" (by W. Schachermayer, Austria).

The Local Organizing Committee consisted of Wilfried Grossmann (chairman), Harald Strelec, K.A. Fröschl, Herbert Guttmann, Marcus Hudec and Roland Rainer. They also managed presentations and demonstrations of non commercial and commercial software, exhibitions of software and books. Finally, the social programme and the local conference arrangements in one building of the University of Technology has to be acknowledged.

Final remarks concerning the layout of this volume are: Usually, the name of an author presenting a paper is printed in bold-face. Only these are entered in the Address List of Authors. However, all authors are listed in the Author Index.

Vienna, May 1994   The Editors

Contents

XII

XXV   Applications   521

XXVI   Late Paper   529

Author Index   548

Address List of Authors   550

Part I

Treatment of "Huge" Data Sets

Huge Data Sets

Peter J. Huber

Universität Bayreuth, Mathematik 7

Abstract

We identify and discuss some of the problems encountered in the data analysis of large data sets, and we suggest some strategies for overcoming them.

1 Introduction: The Himalaya Expedition Metaphor

This paper is on data analysis – not merely data processing – of larger than usual data sets. The logistic problems and the hardships of such an undertaking resemble that of a Himalaya expedition. Such an expedition requires months or even years of planning and preparation at home. Then, after moving the equipment and personnel to the theater of operations, a base camp will be established, followed by several more advanced, smaller camps higher up. Those camps provide safety and rest, they serve as bases for local exploration, and they store provisions for the ultimate attack on the summit.

In a major data analysis, a sequence of derived data sets plays a role very similar to the expedition camps just mentioned. In the beginning of the analysis, there are only the raw data (the "home"). They should be treated as read-only and can be kept on some back-up storage device during most of the actual analysis. As a first step of almost any analysis, the raw data must be transformed into a cleaned and streamlined base set (the "base camp"). This base set must be easily available all of the time. From the base set, several new sets will be derived by extracting subsets, by calculating summaries or by applying various transformations. Some derived sets will be saved as such (the "camps"), some may be kept in virtual form, to be re-generated on demand from a saved set. The derived sets tend to decrease in size, but just because of this, one will occasionally have to travel all the way back to the base set to fetch additional information.

The analysis of a large data set almost never is a one-shot affair, many people are involved, and it may be worthwhile to prepare a data analysis system specifically geared toward the data in question. Also, there is almost always a question of transition from research to routine, and making selected parts of the data and of the analysis methods available to less sophisticated workers, or returning to the metaphor: the transition from "expedition" to "trekking tour". The classic paper by Student (1927) on aspects of this transition is still worth reading.

2   On the rawness of raw data

Any data analysis ought to begin with the raw data, or at least, with the rawest form of the data that is accessible. The raw data, however, only rarely are in a form suitable for the actual analysis. They must be cleaned from gross errors, in particular from gross systematic errors. We have found that some of the hardest errors to detect by traditional methods are unsuspected gaps in the data collection (we usually discovered them serendipitously in the course of graphical checking). Then the data must be cast into a uniform format, permitting fast random access. In practice, this means that the cleaned and streamlined base set will consist of one or more binary direct access files. We have found that many programmers, in a misguided attempt to save 10%-20% of space, organize data sequentially in such a way that finding and unpacking individual items in a large data set becomes intolerably slow, even after indexing.

If data collection extends over years, methods of collection and formats of coding tend to change. Perhaps some variables are added or dropped half-way through. Perhaps there are external changes, e.g. a new system of ZIP code numbers as in Germany. Even with electronically collected data we have regularly been hit by unbelievable features, bugs or coding quirks, often undocumented and detected only in the course of a careful data analysis. I remember an instance where a fixed size record did not provide enough space and a clever programmer managed to stuff additional bits of information into the unused eighth position of 7-bit ASCII characters! A surprisingly frequent problem is that leading bits get lost due to integer overflow, perhaps because somebody misjudged the range of a variable or miscounted the number of bits to be allotted to some hardware register. Fortunately, with some effort and ingenuity, the missing bits can often be restored by continuity or monotonicity considerations, but if such a feature goes undetected, serious bias may result in data summaries.

3   On the structure of statistical data sets

Traditionally and conveniently, statistical data sets are logically (but not necessarily physically) organized as matrices, that is, as families of values $x(i,j)$ accessible through a case number i and a variable number j. Triply and quadruply indexed sets are rare, but occur occasionally. The values can be of any type: continuous, ordered or unordered categorical, labels (strings), identifiers (linking cases or variables across the same or different data sets). As a rule, the cases have all the same structure, that is, for fixed j, $x(i,j)$ has the same type for all i, whereas the variables often have different structures. The reason behind this of course is the love of statisticians for column-wise summaries. But the larger the data sets, the more often the rule is violated. For example, so-called header or calibration records with a different structure may be interspersed between rows of regular data. Large and huge data sets usually consist of a family of such matrices, interlinked through identifiers or pointers. For example, in a retrospective study of environmental hazards, we may have originally unrelated patient data sets and pollution data sets that can be linked through geographical references.

Tiny and small sets tend to be homogeneous; often, homogeneity of the data is a hypothesis to be tested. Large and huge sets almost never are homogeneous: they separate into distinct subgroups, and they have to be large so that there are enough observations pertaining to each subgroup. If large sets were just more of the same, we would not have to take that many observations! As a rule, it does not make sense to treat such sets as random samples from a multivariate distribution, and as a consequence, density estimates may not make sense either. If stochastic models are applicable at all, they will have a more complicated structure. A typical feature of large sets is that many objects are observed by many observers, both observed and observers being located in space and time. An example is furnished by radar data: Several hundred airplanes are simultaneously observed by a dozen or so radar stations. During each antenna turn, that is, every five seconds or so, each radar produces one observational record per plane, totalling about 50 MB of data per hour. Possible questions of a data analytic nature include: piecing together of flight paths, near collisions, crowding of airspace (how likely it is to occur when and where), determination of the statistical accuracy of individual radars, and so on. With such data, the traditional column-wise summaries are simply meaningless. Scatterplots make sense, especially after suitable coordinate transformations, but they may look literally like a can of worms.

4   A classification of size

Size of data sets is relative and in our mind is connected with ease of handling: smaller sets are easier to handle than large ones, and a set is huge if its size causes aggravation. A few years ago I proposed the following crude classification: a **tiny** set is one you might write to the blackboard, a **small** set fits on a few printed pages, a **medium** set fills a floppy disk, a **large** set fills a tape, and a **huge** set requires many tapes (or similar storage devices). The sizes increase in steps of roughly 100: while a tiny set comprises about 100 bytes, a small set is 10^4 bytes, a medium set 10^6, a large set 10^8, and a huge set 10^{10} or more bytes (Huber 1992).

Clearly, a classification based on ease of handling does not remain fixed over time. For example, the size of medium sets creeps upward together with the storage capacity of diskettes, and huge sets get ever larger. Though, scalability is limited; a main issue of this paper is to discuss bottlenecks and obstacles to scaling.

5   Interactive and batch

What can and should be done interactively and what in batch? Interactivity is indicated when a number of reasonably straightforward decisions have to be made sequentially in relatively quick succession, each step being based on the results of the preceding step. With large sets, all three of these conditions may be violated. The decision or its implementation may not be straightforward, since large sets tend to be more complex than smaller ones. The steps may involve the transformation and generation of large derived sets and may require too much computer time for interactivity. But there is a much more embarrassing obstacle: we may be unable to create a rational basis for making the next decision because of a lack of methods for direct presentation and visualization of large data sets! In other words, so long as we are operating on large sets, we may have to base our decisions on *a priori* knowledge and on knowledge gained from exploring smaller data sets, themselves extracted or otherwise condensed from a large set on *a priori* grounds. The practical strategy will therefore be to prototype such steps interactively with smaller, but hopefully representative sets, and then to run them in batch.

6   Stepping to large and huge sets

Over the years a large canon of data analytical tools has been built up for tiny, small and medium sets. Thus, Tukey's EDA (1977) describes methods for pencil-and-paper exploration of tiny to small sets. Many of those methods, e.g. stem-and-leaf displays, do not scale up to the next higher data set size. Most of the tools embodied in computer packages are directed towards small to medium sets. Their spectrum ranges from graphical exploration by scatterplot matrices, to regression analysis, to contingency tables and correspondence analysis, to cluster analysis, and so on. Comprehensive textbooks on data analysis of medium sized sets do not yet exist. For larger sets there is not even an established canon of tools going beyond those proven to work for medium sets. The discussions seem to be over-burdened with admittedly important questions of data entry and checking, and with the production and dissemination of summaries, cf. for example the recent survey by Dekker (1994) on computer methods in census data processing. Imputation of missing values is among the few general tools directed toward large and huge sets – its main purpose is to facilitate the computation of summaries.

With regard to data analysis we thus must primarily rely on the body of tools developed and tested with medium size (and smaller) sets. Let us call them medium procedures for simplicity. We are faced with the problem of stepping them up by a factor 100 to large sets and by a factor 10'000 to huge sets. These are steps of an awesome size. Basically, there five possible approaches, and it is obvious that all of them have their proper place:

1. Take a medium procedure and scale it up (see Section 10). This works reasonably well with analytical algorithms, except that one tends to run into space or time limits. For example, 32-bit addresses do not suffice to address every byte in a huge set, and an algorithm requiring $O(n^2)$ or more operations is beyond the horizon of any computer in existence. Scaling up does not work at all with data presentation and visualization, because we hit human limitations just above medium sized sets.

2. Apply a medium procedure to selected subsets of medium size. Here, the selection may apply either to variables or to cases. The problem is that the subsets will comprise only a tiny fraction of the entire set, and it is far from evident how one can make sure that the "interesting" or "relevant" parts are included, in particular with highly structured data. With census data, 1% random subsets sometimes are made publicly available, but with many other kinds of data random subsets are useless.

3. Apply a medium procedure to selected summaries of the data. With census data, such summaries often are the only data made publicly available.

4. Operate on smaller sets derived from the data in a more general fashion than by subsetting or summarizing.

5. Devise new data analytic procedures specifically geared toward large and huge sets.

The preparation of smaller sets in (2)-(4) intrinsically operates on the entire set; it usually involves data base type operations. To give an example of approach (4), suppose that one wants to investigate migration patterns by comparing census data separated by a few years in time. We must match up the individuals in the two censuses, determine whether they have moved and from where to where, and then prepare a contingency matrix counting the number of individuals having moved from location i to location j. Typically, one will prepare the contingency table through off-line batch at comparative leisure, but one will then want to analyze it interactively.

The preparation of smaller sets is relatively unproblematic, so long as it can be done on *a priori* grounds (i.e. on external criteria, not based on the data under scrutiny, or at least not on the whole data set). But at present, it is not at all clear whether it remains possible to identify interesting parts in a *data-based* way. On one hand, because of presentation and visualization problems our customary interactive data exploration tools do not work with large and huge sets, and on the other hand, non-visual splitting tools (such as clustering algorithms) seem to run into computational complexity barriers. A detailed discussion of these issues is in order.

7   Data presentation and visualization

Scatterplots were introduced into data analysis very early on because of their conceptual and technical simplicity. With a single pixel per point they provide the tightest possible graphical coding of all individual elements of a two-dimensional data set. Scatterplot matrices and various enhancements by lines, symbols, labels, colors, and so on, followed somewhat later. The dots in scatterplots are small, so different parts of the data do not obscure each other, at least in moderately sized data sets. Only lately we realized that scatterplots have crucially important properties in connection with high interaction graphics: if in linked graphs a group of dots is highlighted by using a different color or symbol, it is easy to follow that group through other representations and to see for example whether the group tends to stay together or to disperse. With increasing data size, interactive subset operations become ever more important, but many users seem to have trouble because of the abstractness of the notions involved. Scatterplots help since there is a visually intuitive correspondence between groups of dots and subsets. I believe these are the deeper reasons why enhanced scatterplots and scatterplot matrices have become the workhorses of exploration graphics with medium sized data.

Immediately above medium size however, scatterplots run into problems: First, larger data sets tend to be more highly structured, and if there are too many competing structures, a picture of the full, raw data can become confusing indeed. Second, with 10'000 or more points, a scatterplot may show little more than a big, amorphous blob. In some situations one can get around the second problem by summarization: summarize highly populated regions by density estimates and use individual dots only in more sparsely populated parts of the picture. While it is possible to build high-interaction features into such density estimates, it becomes harder and less intuitive to mark a subset and to follow it across linked plots. Unfortunately, the smoothing process inherent in density estimation complicates the first problem by fusing competing structures. After looking at some plots of radar data sets, I am tempted to express it crassly: If a large data set is a can of worms, then a density estimate mashes it into pulp, making it difficult to disentangle the worms. Fortunately, the coin has two sides: if on one side complex data structure may render direct visualization infeasible, on the other side it makes is easier to split and summarize the data on *a priori* grounds.

The open and possibly unsolvable problem is to create graphical summaries that highlight rather than hide unusual features of large data sets, and that in addition can be interacted with. There is a most unfortunate circularity: we should look at the data in order to decide which summary to compute, but we can do so only by looking at a summary.

Those size problems are not limited to over-abundance of points (or "cases"). If there are 30 or more variables, scatterplot matrices become over-burdened with too many small squares, most of them being uninformative or uninteresting. Here, we need machine assistance with selecting interesting variables and with combining and summarizing several variables into one. Thereby, we again may run into computational complexity problems. But otherwise, this situation seems to be under better control: Promising and seemingly quite successful approaches are based on principal component and correspondence analysis ideas, followed by inspection of the structures visible in the space spanned by the leading eigenvectors.

8   On computer performance

A factor 10^8 between tiny and huge sets is a tremendous range. On the other hand, the performance of computing devices has increased even more stupendously in the course of our century, from less than one operation per second on a mechanical calculator to more than 10^9 on today's supercomputers.

High performance computers customarily are rated in megaflops or gigaflops. A Mflop is a million floating point operations per second, often measured by the LINPACK linear algebra benchmark; 1 Gflop = 1000 Mflops. A lowly 486/25 PC does just 1 Mflop; most of today's PCs and workstations rate below 10 Mflops. The fastest single-processor machines are some very recent superworkstations standing at 100 Mflops.

For operations on the whole of a large or huge data set those 100 Mflops may at present be the highest sustainable rate. By how much we shall ever be able to go beyond that rate in the future is anybody's guess (mine is: at best another factor 100).

The general problem is that with huge data sets one tends to run into memory or data flow bottlenecks, wasting the gigaflops nominally available at national supercomputer centers.

We are approaching physical limits pretty fast in any case. Over the past years there has been much noise about massively parallel teraflop machines (10^{12} floating point operations per second), supposedly being just around the corner, and even petaflop machines sometime after the year 2001, see Bell (1992) for a critical appraisal. There is no doubt that special tasks and special algorithms can take advantage of such a performance. The obvious difficulty is that light moves merely 0.3 mm in 10^{-12} seconds, and just-in-time management of the data flow becomes a headache. One cannot say how much of a headache it will be in our case before somebody has actually implemented a full-scale general purpose data analysis system on a massively parallel machine and used it with real-life data. Such an experiment would be an exciting research project indeed.

9   Bottlenecks

It is the hallmark of a well-balanced system that several bottlenecks will be hit all together if one tries to push it beyond its limits. In this sense, present-day 486-based PCs are excellently balanced for doing interactive data analysis with medium sized data sets. We shall now try to single out the potential bottlenecks and to discuss them separately. It will appear that memory size is the most critical among them.

9.1   Disk size

Disk size rarely is a critical issue (but we once ran into it in the 1980's when we tried to deal with a 38 MB data set on a workstation with a 70 MB disk!). For comfortable work one should have free disk space corresponding to about 8-10 times the size of the raw data. It is difficult even to begin a data analysis unless free disk space amounts to 2-3 times the size of the raw data set: if one cannot keep both a copy of the raw and of the streamlined base set on disk, it becomes nearly impossible to construct the latter. Afterwards, there must be enough disk space to store the base set, plus space for 4 or 5 derived sets, plus some swap space.

9.2   Memory size

Paging or swapping works reasonably well large programs, but certainly not with large data arrays in the midst of operations, where one may encounter a 10 to 100-fold slowdown in our experience. In order to utilize the full performance of the CPU, there must be enough memory to store four copies of the largest array one intends to work with. This way one can keep the arguments and the result of a two-argument array operator (such as a matrix product) in memory, with some space to spare for smaller data items. A factor of 8-10 is preferable, otherwise the arguments may still have to be swapped too often. In other words, if one plans to do any non-trivial operations on large arrays, one may need almost as much free memory as free disk space.

9.3 CPU speed

Unless there are rigid time constraints (such as in real-time prediction problems), CPU speed rarely develops into a bottleneck with batch operations, one simply has to wait longer for the results. The range from short batch jobs (about 5 minutes) to long ones (a month or more of running time is fairly common) amounts to scale factors 10'000 or more. One of the longest batch jobs of a data analytic nature has been reported by Gonnet et al. (1992) with the noteworthy words: "Exhaustive matching [of the entire protein sequence database] required only 405 days of CPU time and was obtained in the background [...] from up to six workstations running in parallel for only 19 weeks".

Interactive operations are much more constrained, since the human side of the feedback loop gets broken whenever the machine response time exceeds the order of the combined human think and reaction time. With high interaction graphics this is a fraction of a second, otherwise a few seconds; occasionally, 10-30 seconds may be acceptable. These times are intimately connected with computational complexity, to be discussed in the next section.

10 Scalability of algorithms and computational complexity

If the size of the data is increased by a factor 100, we need equipment on an at least 100-fold bigger scale. We have to increase disk and memory size by a factor 100, and CPU and bus speeds by an even higher factor, in order to perform a given task in comparable time. Unfortunately, not all statistical computations grow linearly with the size of the data.

By now, we have acquired a fair amount of experience with interactive analysis of medium sized data sets on workstations and PCs. As already mentioned, the standard 486 PCs have a performance level just about adequate for interactive analysis of medium sized sets dealing with them in their entirety: most of the "elementary" operations have response times in the order of seconds, and computational complexity makes itself felt in a very direct and instructive fashion.

If we allow 10 bytes per data item, a representative medium sized data set (1 MB) might consist of $n=10^5$ elements either arranged in a matrix with $r=5000$ rows and $c=20$ columns, or linearly in a long vector.

Most of the basic tasks of high interaction graphics (apart from some slower preparatory ones) typically involve $O(r)$ operations; they ought, and can be, performed in a fraction of a second on a PC.

Arithmetic with the entire data set, $O(n)$ operations, also is quite fast, taking at most a few seconds.

Subjectively, anything using $O(n \log(n))$ or $O(nc)$ operations is sluggish in comparison (it may take tens of seconds). The former corresponds to Fast Fourier transforms, sorting and other data base operations, the latter to matrix operations, e.g. multiplying a (r,c)-matrix by a (c,c)-matrix, a singular value decomposition of an (r,c)-matrix, a multiple regression, and so on.

Tasks involving $O(nr)$ operations, as for instance clustering algorithms, take hours and better are done off-line as a batch job.

We note that with larger data sets the number of rows tends to grow much faster than the number of columns; for the sake of the argument, we shall assume $c = O(n^{1/3})$. Some random order-of-magnitude calculations show:

- With large data sets (100 MB), interactive operations on the entire set remain just about viable with a superworkstation in the 100 Mflop range and 1 GB memory; though, $O(nc)$ tasks will become annoyingly slow.

- With huge data sets (10 GB), no machine in existence will permit interactive work with the entire set. On the other hand, from the point of view of CPU performance alone, $O(n \log(n))$ and $O(nc)$ batch tasks remain viable even on a PC. The limiting factors will be disk speed and data swapping.

- With medium sets ($n=10^5$), tasks with $O(n^2)$ complexity still are eminently feasible; they take about a day on a PC. Such tasks become annoyingly slow with large sets ($n=10^7$) on any machine, and the so-called computer-intensive methods of statistics are simply out of the question. With huge sets ($n=10^9$), we will for instance think twice before attempting to cluster the whole set, which is $O(n^{5/3})$ or so. The practical complexity limit for computations with huge sets may be around $O(n^{1.5})$ now.

11   Conclusions

It always helps to have a faster machine if one wants to analyze larger data sets. However, insufficient memory size can be a worse bottleneck than insufficient CPU speed. For large and huge sets one can safely forget about so-called computer-intensive methods.

Direct interactive analysis of large sets (100 MB) is just about possible now (on 100 Mflop superworkstations with 1 GB memory) but is impeded by the difficulty of direct interactive presentation and visualization of large data sets. This may force us into batch operations even before excessive interactive response times will do so.

At least at present, the recommended strategy to play around the missing visualization techniques is to prepare medium size (1 MB) derived sets, which are easier to visualize and to grasp as a whole, and easier to work with through established techniques of high interaction graphics. Precisely because of the dearth of visualization techniques, most selection, summarization or other derivation of smaller sets must be done on external criteria, that is based on *a priori* knowledge, ordinarily using data-base type procedures in batch-mode. Well-programmed data base operations typically have a computational complexity of $O(n \log(n))$ and thus extend to huge sets even on relatively modest equipment.

12   Bibliography

Bell, Gordon (1992). Ultracomputers: A Teraflop Before Its Time. *Comm. of the ACM*, Vol. 35, No.8, 27-47.

Dekker, A. L. (1994). Computer methods in population census data processing. *Int. Statistical Review*, **62**, 55-70.

Gonnet, G. H., Cohen, M. A., Benner, S. A. (1992). Exhaustive matching of the entire protein sequence database. *Science*, **256**, 1443-1445.

Huber, Peter J. (1992). Issues in Computational Data Analysis. In: *Computational Statistics*, Vol. 2. Ed. Y.Dodge and J. Whittaker, Physica Verlag.

Student (1927). Errors of routine analysis. *Biometrika* 19, 151-164.

Tukey, J. W. (1977). *Exploratory Data Analysis*. Addison-Wesley, Reading, MA.

Assessing Convergence of Markov Chain Monte Carlo Methods

Arnoldo Frigessi

Laboratorio di Statistica, Università di Venezia

ABSTRACT

Markov chain Monte Carlo (MCMC) methods have led to remarkable improvements in Bayesian statistics. MCMC methods are iterative stochastic algorithms which in the limit allow to sample from a specified multivariate target distribution. However, in practice the algorithm must be stopped after a finite number of iterations. The stopping time should be such to guarantee that the distribution of the obtained sample is close enough to the target distribution. Hence one is lead to estimate the speed of convergence of MCMC algorithms. In this paper we briefly survey results useful to monitor and predict the convergence, especially in terms of the number of variables n. This is related to the question of estimating the computational complexity of approximated sampling.

Key words: Computational complexity; Markov chain Monte Carlo; Gibbs Sampler; rate of convergence; second largest eigenvalue.

1 INTRODUCTION

Markov chain Monte Carlo (MCMC) methods are a class of iterative stochastic algorithms for sampling from multivariate distributions. They are particularly useful when the correlation structure is complex, such as is the case for graphical models, belief networks or for Markov random fields (MRF) in general. The first reference to MCMC in the statistical literature is probably Hastings (1970). The first Bayesian application is in the area of image analysis, see the pioneering work of Grenander (1983) and Geman &

---

*Address for correspondence: Laboratorio di Statistica, Università di Venezia, Ca' Foscari, 30123 Venezia, Italy

Geman (1984). Today the literature is quite vast, and we mention the recent issue of the Journal of the Royal Statistical Society with the discussed papers by Besag & Green (1993), Gilks, Clayton, Spiegelhalter, Best, McNeil (1993) and Smith & Roberts (1993). Also the papers by Gelman & Rubin (1992) and Geyer (1992), and references therein, are interesting. Synonyms for MCMC are (sometimes improperly used) Glauber dynamics, stochastic relaxation methods, dynamical Monte Carlo, Metropolis algorithms etc.

Consider a vector $X = (X_1, ..., X_n)$ of n random variables each taking value in a set S. In this paper we will assume S to be discrete, with $|S|$ values. The space of all possible configurations (or images) is denoted by Ω and has cardinality $|S|^n$. Let $\pi(\cdot)$ be a probability distribution over Ω. Of course sampling from π can be performed by subdividing the unit interval $[0, 1]$ into $|\Omega|$ subintervals, each of length $\pi(\omega), \omega \in \Omega$. This procedure becomes easily unfeasible if $|\Omega|$ is too large. This is the typical situation in image analysis, where often at least 100×100 pixels (which is n) are considered. Furthermore, in certain situations, an explicit expression for π may not be available, making the above enumeration impossible. (In graphical models for instance, one operates with a set of consistent conditional distributions of the type $\pi(x_i|x_j, j \neq i)$.)

MCMC algorithms are simply irreducible, aperiodic discrete time Markov chains $(X(0) = x_0, X(1), \ldots, X(t), \ldots)$ with state space Ω and transition matrix $P(x, y)$, which have π as invariant distribution. Accordingly, if it is possible to generate in practice a trajectory of the Markov chain, for large t, $X(t)$ may be used as a sample from the distribution π. Of course this leads immediately to the question of how to estimate the rate of convergence to equilibrium of the chain. Before we turn to this point, let us introduce a quite ample class of MCMC, the Hasting methods. Until 5 years ago, the only popular MCMC methods were the so called Gibbs Sampler and the Metropolis dynamics. Nowadays we are eventually aware of the usefulness of MCMC alternative methods.

Next we describe how the Hasting method operates. We have to choose a transition matrix $Q(\omega, \omega')$ governing transitions on Ω. We are free to choose Q, but it needs to be irreducible. Assume $X(t)$ is given. The first step is to produce a candidate state x^* for $X(t + 1)$. This is done by sampling from $Q(X(t), \cdot)$. In order to decide if the move to x^* is accepted a coin is flipped:

$$X(t + 1) = \begin{cases} x^* & \text{with probability } \alpha \\ X(t) & \text{with probability } 1 - \alpha, \end{cases}$$

where α is defined as

$$\alpha = \alpha(X(t), x^*) = \min\{\frac{\pi(x^*)Q(x^*, X(t))}{\pi(X(t))Q(X(t), x^*)}, 1\}.$$

We have thus defined a Markov chain with transition probability $P(\cdot, \cdot)$, that depends on Q and α and is irreducible and aperiodic. The chain also leaves

π invariant as implied by the following property: $P(x, y)$ is reversible relative to π, that is

$$\pi(x)\,P(x,y) = \pi(y)\,P(y,x), \text{ for all } x, y \in \Omega.$$

Reversibility of the MCMC chain is useful mainly because of technical convenience: defining the scalar product $< f, g >= \sum_\omega f(\omega)g(\omega)\pi(\omega)$ for two $l^2(\Omega)$ functions, the kernel operator $Pf(x) = \sum_\omega f(\omega)P(x,\omega)$ is selfadjoint on l^2 and the eigenvalues satisfy

$$1 > \lambda_2 \geq \ldots \geq \lambda_{|\Omega|} > -1.$$

The choice of the acceptance probability α is not unique, as is mentioned in Hastings (1970). For instance one could also take

$$\frac{\pi(x^*)}{\pi(x^*) + \pi(X(t))}$$

without breaking reversibility. Although one can show that asymptotically this choice is worse than the first one we mentioned, the comparison for finite values of t is still open.

An important special case of the Hastings scheme is the Metropolis algorithm, which corresponds to a symmetric choice of the proposal transition matrix Q, i.e. $Q(x,y) = Q(y,x)$. Here the new state x^* is accepted either if $\pi(x^*)$ is larger than $\pi(X(t))$ or otherwise with probability $\pi(x^*)/\pi(X(t))$.

Also the Gibbs Sampler fits into the Hastings class, by choosing a proposal transition matrix defined as follows: first pick up a component i at random in $\{1, 2, ..., n\}$ and then update only the i-th component of $X(t)$ by sampling from the conditional probability $\pi(X_i = \cdot | X_j(t), j \neq i)$. The Hastings acceptance rate α turns out to be 1. Both the absence of rejections and the fact that conditionally equilibrium is imposed on each component at every step, induced researchers to believe that it was the 'fastest' method. This is now known to be in general false, see for instance Frigessi, Hwang, Sheu & di Stefano (1993).

In order to measure the rate of convergence of the Markov chain $X(t)$ to the invariant distribution π we consider for example the *variation distance*

$$d\{\mu, \pi\} = \frac{1}{2} \sum_{\omega \in \Omega} |\mu(\omega) - \pi(\omega)|$$

between probabilities μ and π. (The 1/2 in front is convenient.) Considering the sample $X(t)$ as drawn from the target distribution π is wrong and the sampling error after t steps can be measured by

$$\max_{x_0 \in \Omega} d\{P(X(t) = \cdot \mid X(0) = x_0), \pi(\cdot)\}, \tag{1}$$

which tends to zero as $t \to \infty$. Let $\epsilon > 0$ be an arbitrary desired accuracy. The *stopping time* t^* for approximate convergence to π with accuracy ϵ is defined as

$$t^* := \min \left[t : \max_{x_0 \in \Omega} \mathrm{d} \left\{ P(X(s) = \cdot \mid X(0) = x_0), \, \pi(\cdot) \right\} \leq \epsilon, \text{ for all } s \geq t \right].$$
$$(2)$$

How big is t^*? In section 3 we will report on some techniques that are useful to estimate t^*, especially in terms of the number of variables n. This is the issue of computational complexity: if MCMC methods take the same computational time as exact calculation for sampling, which is of order $|\Omega|$, they would be useless. In section 3 we concentrate on *a priori* estimates on t^*. In the next section we discuss some alternative ideas useful to spot convergence of MCMC methods.

This paper cannot treat in its full completeness the topic of convergence of MCMC methods. The literature is large, and new papers on this topic can be found in the main statistics and probability journals. We have tried to describe some of the important features of convergence related in particular to the situation of 'huge data sets'- one of the leading topics of the Wien Compstat conference. Huge data sets means in our context that n is extraordinary large.

2 ON–LINE MONITORING

Very important in practice are techniques to *monitor* the convergence on line, while the MCMC chain is running, trying to recognize when it has reached equilibrium.

There is a general simple idea: find a statistics $W_t = W(X(t))$ whose limiting behavior is known. Generally 4, 5 one–dimensional statistics are used, all converging to constant values as $t \to \infty$. This type of convergence is much easier to detect as the weak convergence of the chain $X(t)$. The MCMC is run and when all the test statistics are close enough to their limiting values the chain is stopped. Can we say something on the distance (1) at that point? Not in general. As far as I know, there is no example of computable statistics whose convergence happens (in some sense) at the same time (or provably slower) as the weak convergence of the chain.

Ritter & Tanner (1992) propose the so called Gibbs Stopper, which essentially is a statistic of the desired kind, since its convegence implies weak convergence of the chain. The idea behind their method consists in computing an estimate of π along the MCMC trajectory. For instance use

$$\hat{\pi}(\omega) = \frac{1}{T} \sum_{t=1}^{T} \cdot \mathbf{1}_{\{X(t)=\omega\}}$$

for all $\omega \in \Omega$, and define $W_T(\omega) = \hat{\pi}(\omega)/\pi(\omega)$. Clearly this statistics tends to

1 as T tends to infinity. However one has to check the convergence for each $\omega \in \Omega$, and this implies enumeration of all Ω. Hence this statistics cannot be computed in general exactly and in full, really this is computationally as hard as sampling from π. We will come back in section 3 on the question of computability (...we still have to show that the Hasting method itself is computable in practice).

An important contribution is the paper by Roberts (1993), where a monotonically convergent statistics is proposed. Monotonicity allows a very convenient monitoring of convergence. The type of convergence that is monitored is however not the same as (1): a different distance is introduced. But this is in principle of course allowed, if the type of convergence is still strong enough to ensure that $X(t)$ has almost distribution π. Additionally the choice in Roberts (1993) also permits to compute useful a priori bounds.

We believe that there may be hopes to obtain computable and reliable statistics by using the techniques of coupling and of stochastic domination. In order to explain how coupling may help, consider this unrealistic algorithm: together with the original chain $X(t)$, start a second MCMC chain on Ω, say $Y(t)$, that runs using the same random numbers as $X(t)$. This is coupling. Additionally assume that the starting state $Y(0)$ is sampled from π, and here we are clearly cheating, since were we able to do this the problem would be solved. However, let's hypothetically continue the argument. Because of common random numbers, once the two chains meet $(X(t) = Y(t)$, for some t) they proceed together. But since $Y(t)$ has always distribution π, the original chain has reached equilibrium at the time of meeting.

Can we make this idea practicable? A possible way could be by means of stochastic domination. The point is to *code* the MCMC chain into a new Markov chain, with a drastically smaller sample space, but whose relaxation time is not faster than that one of the original chain. This may mean that the graph on which this new chain moves has loops, or waiting times in certain states, needed to 'wait' for the original chain. The coded chain is coupled to $X(t)$. Furthermore, if its sample space is small enough one can check (for instance in the spirit of the Gibbs Stopper) if the coded chain has reached its equilibrium, which would imply that also $X(t)$ has. The paper Frigessi & den Hollander (1994) applies coupling and stochastic domination to a Gibbs Sampler chain for a quite simple distribution π, with the aim of devising a priori bounds. However the same construction maybe could be used for on-line monitoring, too.

On-line monitoring of the convergence of a statistic of a Markov chain is a related topic. We refer to Sokal (1989) for a non-statistical view of time series methods in monitoring, to the recent paper Geyer (1992) for updated informations and to Mykland, Tierney & Yu (1992) for an intersting application of regeneration times.

3 A PRIORI BOUNDS

There are many ways to bound the distance (1). Here is a simple one, from Kemeny and Snell (1960), Theorem 4.1.5.: assume there is a power κ such that the off-diagonal entries of P^κ are strictly positive; then

$$|P(X(t \cdot \kappa) = x|x_0) - \pi(x)| \le \{1 - \inf_{\omega,\omega'} P^\kappa(\omega,\omega')\}^t. \tag{3}$$

For any ϵ, from $\{1 - 2\inf_{\omega,\omega'} P^\kappa(\omega,\omega')\}^t < \epsilon$, we obtain the stopping time

$$t^* = \frac{\log(\epsilon)}{\log(1 - \min_{\omega,\omega'} P^\kappa(\omega,\omega'))}.$$

Given that a bound from below on $\min_{\omega,\omega'} P^\kappa(\omega,\omega')$ is computable, this gives an operative stopping time.

In order to show that things are not so easy, we introduce a specific π, widely used as a nontrivial example of spatial process, the Ising model. Consider a finite lattice $\Lambda \subset Z^2$, and a random vector $X = (X_1, ..., X_n)$, where $n = |\Lambda|$, whose components are associated to the sites of the lattice. Choose $S = \{-1, +1\}$, so that $|\Omega| = 2^n$. Let

$$\pi(x) = \frac{1}{Z_\beta} \exp(\beta \sum_{<i,j>} x_i x_j), \beta \ge 0$$

be the target distribution, where the sum is over all nearest neighbours sites. We chose for instance free boundary conditions, which means that sites at the boundary of Λ have less than 4 neighbours. It can be easily seen that β controls the correlations of the model, in the sense that $cor(X_i, X_j)$ increases if β becomes larger. Z_β is the normalizing constant, that cannot be computed explicitly for large values of n. When we mentioned the need for computable methods, we meant for instance that those methods should not require the knowledge of Z_β. In fact, looking back to the acceptance probability α of the Hastings scheme, one can notice that indeed it is computable, since the normalizing constants in the fraction $\pi(x^*)/\pi(X(t))$ simplify.

Consider the Gibbs Sampler applied to the Ising model with raster scan. This means that sites are ordered, for instant spiralwise from the center of Λ in order to reduce the so called boundary effect. At each step one site variable is updated, say x_i. Denote by x^i the configuration that equals x except in site i, where it has a value x_i'. Then

$$P(x, x^i) = (1 + e^{-2\beta x_i' \sum_{<i,j>} x_j})^{-1} \tag{4}$$

is the single site updating transition probability. For this site visiting schedule $\kappa = |\Lambda|$, and $\min_{\omega,\omega'} P^\kappa(\omega,\omega') \ge (1 + e^{+8\beta})^{-|\Lambda|}$. Bounding the r.h.s. of (3) from above and putting this equal to some ϵ gives

$$t^* = O\left((1 + e^{8\beta})^{|\Lambda|}\right),$$

for $|\Lambda| \to \infty$. Recall that full enumeration of Ω takes a time of order $2^{|\Lambda|}$, which is *smaller* than the above t^*. So it would seem that this MCMC is totally useless. This is not true however, it is the bound in (3) that is too weak (the bound on the transition probability is tight). We can do in fact much better using the spectral decomposition theory.

This is not difficult by Jordan decomposition. If P^t is the t-th power of the transition matrix P, one can show that $P^t = R_1 J(t) R_2$, for some matrices R_1, R_2 that do not depend on t, where the matrix $J(t)$ has a block diagonal form. One block is formed just by a 1, by ergodicity. Denoting by $1, \nu_2, \nu_3, ..., \nu_{|\Omega|}$ the algebraic multiplicities of the corresponding eigenvalues λ_i, the other blocks are $\nu_i \times \nu_i$ upper triangular matrices, with λ_i^t's on the diagonal, $t\lambda_i^{t-1}$'s just over the diagonal and further terms of the type $\binom{t}{j} \lambda_i^{t-j}$. Hence the norm $||P^t - A_\pi||$, where A_π denotes the matrix having all rows equal to π, is bounded from above by

$$\sum_{k=2}^{|\Omega|} |\lambda_k|^t t^{\nu_k - 1},$$

being matrix norms in our case equivalent. For large t the leading term is ρ_2^t, where $\rho_2 = \max(|\lambda_2|, |\lambda_{|\Omega|}|)$ is the second largest eigenvalue in absolute value of the transition matrix P.

A more elegant technique for bounding the sampling error (1) after t steps comes from Diaconis & Strook (1991):

$$\max_{x_0 \in \Omega} d(P(X(t) = \cdot \,|\, x_0), \pi(\cdot)) \le \frac{\rho_2^t}{2 \min_{x \in \Omega} \left\{ \pi(x)^{\frac{1}{2}} \right\}}. \tag{5}$$

It is not possible to think $\rho_2 = \lambda_2$ in general. One has to take into account also the last eigenvalue, which can sometimes be very close to -1. This happens for MCMC chains that behave somehow almost periodically. In fact there are MCMC method that are particularly fast for approximating expectations of the type $E_\pi(f(X))$ (for some function f), that have $\rho_2 = -\lambda_{|\Omega|}$, see Frigessi, Hwang & Younes (1993). These algorithms make positive use of their almost periodic behaviour. But also often $\rho_2 = \lambda_2$: in Frigessi *et al.* (1993) it is shown that this is true for the Gibbs Sampler and for all other single-site updating reversible MCMC algorithms provided the correlation among variables X_i w.r.t. the distribution π is suitably large. (This is really a bit sloppy here, but it can be made precise in terms of β for MRFs.)

Hence one is lead to bound the second largest eigenvalue λ_2 of the transition matrix P. This is not at all easy; and it cannot be done numerically because of the prohibitive dimension of the matrix. One way to handle the problem is again by stochastic domination. This has been proposed by Hastings (1970), later by Sokal (1989). Recently Diaconis & Saloff–Coste

(1994) have invented a new domination technique. Before we describe them both, we mention that the type of convergence in (5), often denominated geometric, is not guaranteed if the state space Ω is continuous. See the paper by Yu (1994) that gives a good reference list.

We recall the minmax characterization of eigenvalues for selfadjoint, operators. Let

$$\mathcal{E}(f,f) = \frac{1}{2} \sum_{x \in \Omega} \sum_{y \in \Omega} \{f(x) - f(y)\}^2 \, \pi(x) \, P(x,y)$$

be the Dirichlet form and let

$$\text{var}\,(f) = \sum_{x \in \Omega} \{f(x) - E_\pi[f(X)]\}^2 \, \pi(x)$$

be the (equilibrium) variance of $f(X)$, for any function $f \in l^2(\Omega)$. Then the second largest eigenvalue of P satisfies

$$\lambda_2(P) = 1 - \inf_f \left\{ \frac{\mathcal{E}(f,f)}{\text{var}\,(f)} \right\}, \tag{6}$$

where the inf is taken over all non–constant functions.

Assume that there is a MCMC chain reversible with respect to π, with transition matrix M, whose second largest eigenvalue $\lambda_2(M)$ we know. If, for all $\omega, \omega' \in \Omega$, it is $P(\omega,\omega') \geq M(\omega,\omega')$, with strict inequality sometimes, then

$$\lambda_2(P) < \lambda_2(M). \tag{7}$$

The proof is trivial, just apply the assumed bound on the definition of $\mathcal{E}(f,f)$ and apply the minmax characterization. Intuitively, the chain that stochastically 'moves' more will converge faster. The problem is to find a non trivial M. This is hard, and we are not aware of any good bound obtained by this mean for non-uniform target distributions.

A more promising type of domination is proposed by Diaconis & Saloff–Coste (1994). Denote by $\mathcal{M}$ the Dirichlet form associated with the matrix M. Assume one can compute a constant a such that $\mathcal{M} \leq a\mathcal{E}$ uniformly for non–constant f. Then,

$$\lambda_2(P) \leq 1 - \frac{1}{a}(1 - \lambda_2(M)). \tag{8}$$

For the proof: since $\mathcal{M} \leq a\mathcal{E}$, by the minmax characterization

$$\lambda_2(M) \geq 1 - a\inf_f \left\{ \frac{\mathcal{E}(f,f)}{\text{var}\,(f)} \right\}.$$

But this last inf is exactly $1 - \lambda_2(P)$, and (8) easily follows. (Really, M does not need to have the same invariant distribution π, but one needs also to compute a bound on the ratio between the two distributions.) Note that just by taking a large enough, one can always bound $\mathcal{M}$ with $a\mathcal{E}$ from above: however the bound on λ_2 becomes worse.

If $a < 1$ then (7) it follows from (8). However (8) is stronger than (7), since $M(\omega, \omega') \leq AP(\omega, \omega')$, with $A > 1$ implies $\mathcal{M} < \mathcal{E}$ and hence (7). The viceversa is not true, since $\mathcal{M} > a\mathcal{E}$ does not imply a bound on the transition probabilities in general. There is another very important advantage of (8). Given any MCMC chain P, reversible w.r.t. π, one can always take the control chain $M(\omega, \omega') = \pi(\omega')$ for all ω. In this case $\lambda_2(M) = 0$ and the bound reduces to

$$\lambda_2(P) \leq 1 - \frac{1}{a}. \tag{9}$$

This had already been found by Diaconis & Stroock (1992). Both in this case and more generally, guidelines are available in order to obtain an appropriate factor a. This techniques have been successful not only when the target distribution is uniform. We describe the way proposed by Jerrum & Sinclair (1989) to obtain a possible a. The calculation in Diaconis & Stroock (1992) is slightly tighter, although in practice one tends to relax it in any case to the expression of Jerrum & Sinclair. Our presentation here follows in part also Frigessi, Martinelli & Stander (1993).

Consider the graph G with nodes Ω. We put an edge $e = (\omega_-, \omega_+)$ if $P(\omega_-, \omega_+) > 0$. For each pair of distinct $x, y \in \Omega$ we choose a path $\gamma_{x,y}$ in the graph G from x to y. Edges should not appear more than once in a given path. Irreducibility of P guarantees that such paths exist. Let Γ denote the collection of such paths, one for each pair x, y. The constant a that we are constructing will depend on the specific choice of Γ. The tightness of the bound will depend on the criteria used in constructing paths.

Let us choose $M(\omega, \omega') = \pi(\omega')$ for all ω. In this case $\mathcal{M}(f, f) = \text{var}(f)$. The first step is to work on the square $\{f(x) - f(y)\}^2$, adding and subtracting the contribution from each node visited along the path from x to y, thus obtaining

$$f(x) - f(y) = \sum_{e \in \gamma_{x,y}} (f(e_-) - f(e_+)), \tag{10}$$

where the sum is over edges $e = (e_-, e_+)$ along the chosen path joining x to y. Applying Cauchy-Schwartz, and since the maximum number of edges along a path is n (when x and y differ in every site), we obtain

$$\{f(x) - f(y)\}^2 \leq |\gamma_{x,y}| \sum_{e \in \gamma_{x,y}} \{f(e_-) - f(e_+)\}^2 \leq n \sum_{e \in \gamma_{x,y}} \{f(e_-) - f(e_+)\}^2. \tag{11}$$

Summing up,

$$\text{var}(f) \leq \frac{n}{2} \sum_x \sum_y \left[\sum_{e \in \gamma_{x,y}} \{f(e_-) - f(e_+)\}^2 \right] \pi(x)\,\pi(y)$$

$$= \frac{n}{2} \sum_e \{f(e_-) - f(e_+)\}^2 \sum_{\gamma_{x,y} \ni e} \pi(x)\, \pi(y), \tag{12}$$

by changing the order of summation. We now multiply and divide by $\pi(e_-)P(e_-, e_+)$ and get

$$\operatorname{var}(f) \le n\, \mathcal{E}(f, f) \left\{ \max_e \sum_{\gamma_{x,y} \ni e} \frac{\pi(x)\, \pi(y)}{\pi(e_-)\, P(e_-, e_+)} \right\}, \tag{13}$$

where the maximization is over all edges in the graph. Dividing and rearranging terms gives

$$\lambda_2(P) \le 1 - \frac{1}{n\, \max_e \rho(e; \Gamma)} \tag{14}$$

where we abbreviated

$$\rho(e; \Gamma) = \sum_{\gamma_{x,y} \ni e} \frac{\pi(x)\, \pi(y)}{\pi(e_-)\, P(e_-, e_+)}.$$

for every fixed choice of the set Γ.

We have thus found a way to obtain the coefficient a. To get a value for it, one has to choose a specific Γ and compute the $\max_e \rho(e; \Gamma)$. This is generally however also hard. In the literature we are aware of essentially two choices of the set of paths Γ: in both cases exact calculations of $\max_e \rho(e; \Gamma)$ are not available, but only asymptotic first order approximations. If one is interested in asymptotics for large n (the number of random variables, also denoted as $|\Lambda|$ in the Ising model example), the Γ–set proposed in Jerrum & Sinclair is appropriate. If the asymptotic of interest is in terms of the correlation of the distribution π (more precisely in terms of β), then the choice in Holley & Stroock (1988) has to be preferred. Let us first describe the set of paths that allows an estimation of the second largest eigenvalue when $n = |\Lambda| \to \infty$. We assume for simplicity that the Hastings method we are considering updates a single component at each step. This is natural for the Gibbs Sampler, and quite common for general Hastings updates, when the proposal matrix Q selects a component $i \in \{1, ..., n\}$ and proposes a new value for the ith component x_i.

Let us now choose the path $\gamma_{x,y}$ following Frigessi, Martinelli & Stander (1993). Given $\omega \in \Omega$, recall that the n components are ordered from 1 to n. (In the case of lattice based random vectors X, the sites are order in some arbitrary fashion.) Let $x, y \in \Omega$ be arbitrary and $i_1 < \cdots < i_l$ be the components (sites) at which x and y differ. Then for $1 \le k \le l$, let the kth edge of the path $\gamma_{x,y}$ from x to y correspond to a transition in which the value taken by the i_kth component is changed from x_{i_k} to y_{i_k}. The path simply formalizes the idea that starting from x we correct its components in order making them equal to the corresponding ones in y. The set Γ_n of paths has thus been fixed. One should notice that it has been defined without using any property of the target dsitribution π.

We now describe another set of paths that allows to estimate the second largest eigenvalue when $n = |\Lambda|$ is fixed, but the correlation between X_i's increases. Define the set $\Gamma_{x,y}$ of all possible paths from x to y. For every pair x, y, consider the number

$$\delta_{x,y} = \max_{\gamma_{x,y} \in \Gamma_{x,y}} \min_{x_k} \pi(x_k), \qquad (15)$$

where the min is over all states x_k along the path $\gamma_{x,y}$. We now chose as the path $\gamma_{x,y}$ from x to y one for which

$$\delta_{x,y} = \min_{k \in \gamma_{x,y}} \pi(x_k).$$

Hence we chose to include in the set of paths those ones that along theit trajectory visit the most favourable less probable configuration. This set Γ_β is defined in terms of π.

By means of Γ_n we are only able to estimate $\rho(e; \Gamma_n)$ in terms of $n = |\Lambda|$, while using Γ_β the quantity $\rho(e; \Gamma_\beta)$ can be expanded only in terms of parameters of the distribution π. This two approaches have been taken in the recent papers by Frigessi, Martinelli & Stander (1993) and Ingrassia (1994).

We are going now to report on the type of results obtained in these two papers. The aim is to illustrate the main features of the results and here we will not pursuit generality. On the contrary we will assume π to be the Ising model distribution on a square lattice with $\sqrt{n}$ sites on every side and we will consider the Gibbs Sampler with randomly chosen component to be updated at every step. However the reader should be aware that all results are proven under much more general assumptions, see the papers in question for details.

In Frigessi, Martinelli & Stander (1993) it is proven that

$$\max_e \rho(e; \Gamma_n) \leq nc \exp(4\beta \sqrt{n}),$$

where c is a positive constant, from which it follows that

$$\lambda_2 \leq 1 - \frac{c}{n^2} \exp\left(-4\beta \sqrt{n}\right).$$

Combining this upper bound on λ_2 with (5), we get an upper bound on t^*:

$$t^* \leq O\{\exp(\beta C \sqrt{n}),$$

for some constant $C > 0$. We can clearly see here, that there is an advantage in using MCMC methods over enumerating all the states, that takes a time exponential in n, not in its square root. The stopping time $t^* = \exp(\beta C \sqrt{n})$ is however not polynomial in n. We just mention that in Frigessi, Martinelli & Stander (1993) conditions are given for t^* being of order $n \log n$, i.e.

polynomial in n. The conditions regard π and the decay of the correlation between site variables that drift apart; this can be precisely stated in terms of phase transition. For the Ising model it can be shown that the polynomial behaviour of t^* is guaranteed if $\beta < 0.44...$ only.

The choice of paths used in Ingrassia (1994), while providing useful bounds on the second largest eigenvalue in terms of β, fails to yield such tight bounds in terms of n. It is proven that

$$\lambda_2 \leq 1 - cZ_\beta \exp(-m\,\beta)$$

for a constant m that does not depend on β but depends on n. It is very interesting that asymptotically in β this is the behavior of the second largest eigenvalue also of the Metropolis algorithm. We mention that lower bounds on λ_2 can also be obtained with a similar technique.

We end this section reminding the reader that applying asymptotic results (in n or β) in practice can be often inappropriate. since all constants that are irrelevant in the limit may even govern the finite scale behaviour. On this aspect almost nothing is known rigorously. See the book of Diaconis (1988) for more informations.

Along with the search for reliable stopping criteria, in the last years researchers have concentrated on new algorithms, with the aim of speeding up the convergence. Just to point to a successful class of algorithm, we mention *auxiliary variable methods*, see Swendsen & Wang (1987). Many ot the new methods seem very promising. Some of them are adapted to quite specific situations, others are more general purpose. But we have to say that quite rarely there is a rigorous comparative analysis of rates of convergence which is needed to declare a method better than others.

One of the last birth is the so called *Simulated Tempering* algorithm, originally proposed by Marinari & Parisi (1993) in the physical literature and later independently developed by Geyer (1993) and called $(MC)^3$. This is a Hastings type algorithm on a probability space which is larger than Ω, like in the case of the auxiliary variable approach. Intuitively this algorithm should behave very well, and some initial practice with it supports optimism. I tend to be more sceptic. Further work on estimating and comparing rates of convergence has to be done.

References

BESAG, J. & GREEN, P. J. (1993). Spatial statistics and Bayesian computation (with discussion). *J. R. Statist. Soc.* B 55, 25–37.

DIACONIS, P. (1988). *Group Representations in Probability and Statistics.* IMS, Hayward, California.

DIACONIS, P. & SALOFF-COSTE, L. (1994). Comparison theorems for reversible Markov chains. *Ann. Appl. Probab.* **3**, 696–730.

DIACONIS, P. & STROOCK, D. (1991). Geometric bounds for eigenvalues of Markov chains. *Ann. Appl. Probab.* **1**, 36–61.

FRIGESSI, A. & DEN HOLLANDER, F. (1993). A dynamical phase transition in a caricature of a spin glass. To appear on *Journa of Statistical Physics*, 1994.

FRIGESSI, A., HWANG C.-R. & YOUNES, L. (1992). Optimal spectral structure of reversible stochastic matrices, Monte Carlo methods and the simulation of Markov random fields. *Ann. Appl. Probab.* **2**, 610–28.

FRIGESSI, A., HWANG C.-R., SHEU, S.-J. & DI STEFANO, P. (1993). Convergence rates of the Gibbs sampler, the Metropolis algorithm and other single-site updating dynamics. *J. R. Statist. Soc.* B **55**, 205–19.

FRIGESSI, A., MARTINELLI F. & STANDER, J. (1993). Computational complexity of Markov chain Monte Carlo methods. *Quaderno IAC*, **32**. GELMAN, A. & RUBIN, D. B. (1992). Inference from iterative simulation

using multiple sequences (with discussion). *Statistical Science* **7**, 457–511.

GEMAN, D. & GEMAN, S. (1984). Stochastic relaxation, Gibbs distributions and the Bayesian restoration of images. *IEEE Transactions on Pattern Analysis and Machine Intelligence* **PAMI-6**, 721–41.

GEYER, C. J. (1992). Practical Markov chain Monte Carlo (with discussion). *Statistical Science* **7**, 473–511.

GEYER, C. J. (1993). Markov chain Monte Carlo maximum likelihood. *Preprint.*

GILKS, W. R., CLAYTON, D. G., SPIEGELHALTER, D. J., BEST, N. G., MCNEIL, A. J., SHARPLES, L. D. & KIRBY, A. J. (1993). Modelling complexity: Applications of Gibbs sampling in medicine (with discussion). *J. R. Statist. Soc.* B **55**, 39–52.

GRENANDER, U. (1983). Tutorial in Pattern Theory. *Division of Applied Mathematics, Brown University.*

HASTINGS, W. K. (1970). Monte Carlo sampling methods using Markov chains and their applications. *Biometrika* **57**, 97–109.

INGRASSIA, S. (1994). On the rate of convergence of the Metropolis algorithm and Gibbs sampler. To appear in *Annals of Applied Probability*, 1994.

JERRUM, M. & SINCLAIR, A. (1989). Approximating the permanent. *SIAM J. Comput.* **18**, 1149–78.

KEMENY, J. G. & SNELL, J. L. (1960). Finite Markov Chains. Springer Verlag, Heidelberg.

MARINARI, E., PARISI, G. (1992) Simulated Tempering: a new Monte Carlo scheme. *Europhysics Letters* **19**, 451–458.

MYKLAND, P., TIERNEY, L. , YU, B. (1992), Regeneration in Markov chain samplers. *Technical Report Univ. of Minnesota* **585**.

RITTER , C. & TANNER, M. A. (1992). Facilitating the Gibbs Sampler: The Gibbs Stoopper and the Griddy-Gibbs sampler. *J. Am. Stat. Ass.*, **87**, 861–868.

ROBERTS, G. O. (1992). Convergence diagnostic of the Gibbs Sampler. *Bayesian Statistics 4*, Bernardo et al. editor, Proceedings of the 4th Valencia International Meeting, 775–782.

SMITH, A. F. M. & ROBERTS, G. O. (1993). Bayesian computation via the Gibbs sampler and related Markov chain Monte Carlo methods. *J. R. Statist. Soc.* B **55**, 3–23.

SOKAL, A. D. (1989). Monte Carlo methods in statistical mechanics: fondations and new algorithms. *Cours de Troisième Cycle de la Physique en Suisse Romande.* Lausanne.

Swendsen, R. H. & Wang, J.-S. (1987). Nonuniversal critical dynamics in Monte Carlo simulation. *Phys. Rev. Lett.* **58**, 86–8.

YU, R. H. (1993). Density estimation in the L^∞ norm for dependent data with applications to the Gibbs Sampler. *Ann. Stat.* **21**, 711–735.

Part II

Numerical Analysis Useful in Statistics

What can Numerical Analysis do for Statistics?

Håkan Ekblom

Department of Mathematics, Luleå University

Abstract. Basic numerical concepts, like cancellation, instability and condition number, are described and discussed. The implication on statistical computation is exemplified on computing variances and regression coefficients.

Keywords. Numerical analysis, cancellation, instability, condition number, perturbation theory, variance computation, algorithms for regression

1 Introduction

With the intensive use of computers in statistics today, where often very large problems are solved by tedious computation, it is natural to ask the question what support can be given from numerical analysis. It turns out that the numerical viewpoint can be useful also in what is usually considered a simple calculation. As an example of this, we will look at the problem how to compute variances. Other examples which show in what way numerical analysis can help statistics are taken from regression.

2 Basic numerical concepts

2.1 Floating point arithmetic

Today the overwelming amount of statistical computing is carried out on computers using floating-point arithmetic. This is often described as having a certain number of digits of accuracy. E.g. the standard IEEE single precision arithmetic [26] is said to give about 7 (decimal) digits of accuracy. This characterization is valid when numbers are represented in the computer. It is obvious that numbers like π cannot be exactly represented on a binary computer, but neither can 0.1 nor 0.6 .

The relative error due to rounding an exact value into a machine number is limited by the machine accuracy u, an important characteristic of the computer at hand. For the IEEE single precision arithmetic, u is close to 10^{-7}. Also multiplying or adding two machine numbers and storing the result in memory gives a relative error less than the machine accuracy in magnitude.

32

Of course we can instead work in higher precision arithmetic, e.g. the IEEE double precision arithmetic, for which u is close to 10^{-16}. Although using higher precision arithmetic may lead to the required accuracy, it often happens that the problems with numerical instabilty will instead appear at a later stage of the computation.

2.2 Numerical instability

To study what may happen in a slightly more complex situation, we will now consider computing function values.

Example 1 The functions

$$f_1(x) = (x - 1)^6 \tag{1}$$

and

$$f_2(x) = x^6 - 6x^5 + 15x^4 - 20x^3 + 15x^2 - 6x + 1 \tag{2}$$

are mathematically equivalent. However, when plotting them we get the results of Figures 1 and 2.

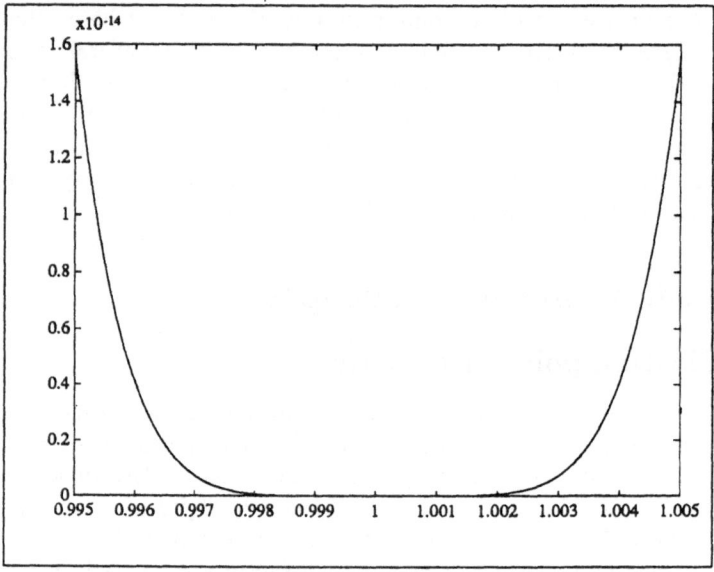

Figure 1: Picture of (1)

We see from the plots that the two expressions behave totally different numerically. The low precision in the f_2 values are due to cancellation, which occurs when subtracting almost equal numbers. The computation using (2) is said to be instable.

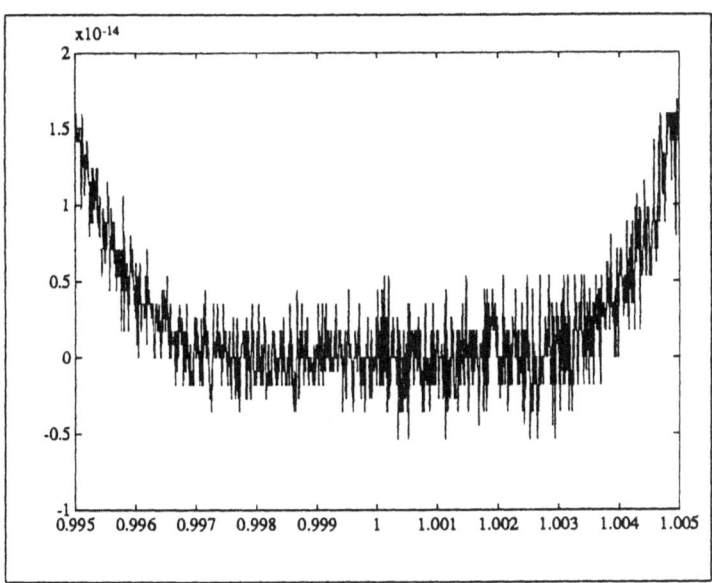

Figure 2: Picture of (2)

Figure 2 gives a good picture of the difference between numerical analysis and mathematics. "Mathematical properties" of the function, like being monotone in an interval, being non-negative, having a unique minimum, and so on, have totally disappeared. However, in a standard code for the exponential function, e.g., it is reasonable to require the function to be monotonely increasing. Thus the code designer has to try to arrange for that in some way.

We give two more examples of numerical instability.

Example 2 The Taylor expansion for the exponential function,

$$e^x = 1 + x + \frac{x^2}{2!} + \frac{x^3}{3!} + \dots$$

is convergent for any x. To use the expansion in a program, which adds every term which influences the sum, often leads to good results. E.g. in IEEE single precision arithmetic we get $e^{10} = 22026.469$, which has a relative error of only $1.3 \cdot 10^{-7}$. However, if we compute e^{-10} we get -0.000065 which of course is totally wrong. The reason is catastrophic cancellation, stemming from computing a value near zero via large terms. In this case it is easy to find a better algorithm, namely to use $e^{-x} = 1/e^x$. Computing e^{-10} this way gives a value with the same relative error as for e^{10}.

Example 3 Consider the sequence

$$x_0 = 1, \quad x_1 = 1/3, \quad x_{n+1} = (13x_n - 4x_{n-1})/3 \ . \tag{3}$$

% It is easy to check that the solution is $x_n = 3^{-n}$, but computing in IEEE single precision gives e.g. $x_{15} = 3.07$ and $x_{30} = 3.3_{10}9$. The reason is that the general solution to the difference equation is

$$x_n = c_1\, 3^{-n} + c_2\, 4^n.$$

Our start values correspond to $c_1 = 1$ and $c_2 = 0$. However, due to rounding errors, c_2 will not stay at zero, and finally the second term will take over.

2.3 Condition number

It is not only an unstable algorithm which may cause an inexact result. The problem itself may be ill-conditioned, i.e. very sensitive to errors in given data. For function evaluation, a condition number is derived in the following way.

Let x be the exact argument to f, but unfortunately only an approximation $\tilde{x} = x + h$ to x is available. How will this affect the function value? The mean value theorem gives us immediately

$$|f(\tilde{x}) - f(x)| \approx |f'(x)h| \tag{4}$$

as an approximation of the absolute error in f. However, most often the relative error is asked for. We have

$$\frac{|f(\tilde{x}) - f(x)|}{|f(x)|} \approx \frac{cond_f(x)|h|}{|x|} \tag{5}$$

where $cond_f(x)$, the condition number of f at x, is defined as

$$cond_f(x) = \frac{|x||f'(x)|}{|f(x)|}. \tag{6}$$

Example 4 Let $f(x) = \sqrt{x}$. We get $cond_f(x) = 0.5$ for all values of x, i.e. the relative error is always halved. Thus computing square roots is a very well-conditioned problem.

Example 5 Let $f(x) = e^x$. We get $cond_f(x) = |x|$.

Example 6 Let $f(x) = (x-1)^6$. We get $cond_f(1.01) = 606$, $cond_f(1.1) = 66$ and $cond_f(3) = 9$.

Note that in Example 1, f_1 and f_2 have the same condition number since they solve the same problem, but they differ in numerical stability.

3   Computing variances

Although it may seem to be a trivial task, computing variances in an efficient and robust way involves a lot of numerical consideration. Chan, Golub and LeVeque [5] give an excellent overview, which is summarized below. For a sample $x_1, \ldots, x_n$, the variance is given as S/n or $S/(n-1)$, where

$$S = \sum_{i=1}^{n} (x_i - \bar{x})^2 . \tag{7}$$

We note that if we use (7), data has to be passed twice. To avoid this, the following one-pass algorithm is often given as an alternative in text-books:

$$S = \sum_{i=1}^{n} x_i^2 - \frac{1}{n} \left(\sum_{i=1}^{n} x_i \right)^2 . \tag{8}$$

However, if $\bar{x}$ is large and S is small, the subtraction in (8) may cause catastrophic cancellation. It may even give a negative value to S, a result which is at least a very clear warning.

The problem to find S has the following condition number [5]

$$c = \sqrt{1 + \frac{n\bar{x}^2}{S}} . \tag{9}$$

Thus if we use a computer with machine constant u, the relative errors due to rounding the x values may give a relative error of S which is limited by cu. The reflects the sensitivity of the problem and is not connected to any specific algorithm being used. But error bounds can be derived for the computations (7) and (8), and they will depend on the condition number. Their expressions are $nu + n^2 c^2 u^2$ and $nc^2 u$, respectively.

Example 7 Let $n = 100$, $\bar{x} \approx 10$ and $S \approx 1$. It gives $c \approx 100$. Thus, if we use a computer with $u \approx 10^{-7}$, the relative error in S due to rounding the x-values is limited by 10^{-5}, approximately. Using (7) for computing S gives a maximum relative error which is of the same size. In contrast, the upper bound of the relative error for the text-book algorithm (8) is around 10^{-1}!

An easy way to lower the condition number is to shift the data, $x_i \leftarrow x_i - d$, with some suitable value of d. The condition number will now be

$$c_d = \sqrt{1 + n\frac{(\bar{x} - d)^2}{S}} . \tag{10}$$

If d is within one standard deviation of the mean, i.e. $|\bar{x} - d| \leq \sqrt{S/n}$, we will have $c_d \leq \sqrt{2}$. In Example 7, such a shift would change the error bound of the text-book algorithm (8) to $2 \cdot 10^{-5}$.

Of course, $\bar{x}$ is the ideal value of d. Using this shift in (2) gives

$$S = \sum_{i=1}^{n} (x_i - \bar{x})^2 - \frac{1}{n} \left(\sum_{i=1}^{n} (x_i - \bar{x}) \right)^2. \tag{11}$$

Theoretically the second term is zero, but in practice it is instead a good estimate of the computing error of the first term. The rounding errors when computing the first term are cancelled out to a high extent by the rounding errors of the second term. Thus formula (11) is actually usable ("corrected two-pass algorithm") — its error bound is $nu + n^3 c^2 u^3$.

Other ways to compute S include the updating algorithms, which can be generalized to handle two samples. If the means and variances of two subsamples are computed separately, they can easily be combined to give the variance of the full sample. If the subsamples are of equal size, this idea leads to the "pairwise algorithm", which is very well suited for parallel processing.

4 Linear least squares regression

In the standard linear regression model we have

$$y = X\beta + \epsilon \tag{12}$$

where

y is the n-vector of observations
X is the $n \times p$ design matrix
β is the p-vector of unknown regression coefficients
ϵ are the errors distributed according to $N(0, \sigma)$.

The least squares estimate b of β is obtained by solving the overdetermined system of equations $Xb = y$ through minimizing $\|y - Xb\|_2$, the L_2-norm of the residual vector.

4.1 Perturbation theory

The condition number for the matrix X is defined as

$$cond(X) = \frac{\sigma_1}{\sigma_p} \tag{13}$$

where σ_1 and σ_p are the largest and smallest singular values of X, respectively. Condition numbers and singular values are easily found using MATLAB or similar systems. A large value of $cond(X)$ means that the solution is very sensitive to errors in X (and also in y). Note that for square matrices the value of the determinant is no good measure of condition.

Example 8 The determinant of the matrix

$$X = \begin{Bmatrix} 1 & -1 & -1 & -1 & -1 & -1 & -1 & -1 \\ 0 & 1 & -1 & -1 & -1 & -1 & -1 & -1 \\ 0 & 0 & 1 & -1 & -1 & -1 & -1 & -1 \\ 0 & 0 & 0 & 1 & -1 & -1 & -1 & -1 \\ 0 & 0 & 0 & 0 & 1 & -1 & -1 & -1 \\ 0 & 0 & 0 & 0 & 0 & 1 & -1 & -1 \\ 0 & 0 & 0 & 0 & 0 & 0 & 1 & -1 \\ 0 & 0 & 0 & 0 & 0 & 0 & 0 & 1 \end{Bmatrix}$$

is one, but the condition number is 1024! Actually, changing the element a_{81} to $-1/64$ makes the matrix singular. In contrast, the determinant of

$$X = \begin{Bmatrix} 0.1 & 0 & 0 \\ 0 & 0.1 & 0 \\ 0 & 0 & 0.1 \end{Bmatrix}$$

is 0.001, but $cond(X) = 1$.

We can use the condition number of X to judge if a linear regression problem is well-conditioned or not.

Example 9 Assume that we like to regress the oil consumption (litres) in a house on the variation in temperature (F). We made our observations each Wednesday during 7 weeks, and got the values in Table 1.

Week	oil cons(y)	min(x$_1$)	max(x$_2$)	range(x$_3$)
1	505	28.8	33.7	4.9
2	456	29.7	34.6	4.9
3	526	24.0	29.6	5.6
4	433	30.2	34.9	4.7
5	488	29.1	34.0	4.9
6	417	31.5	36.3	4.8
7	397	34.8	40.2	5.4

Table 1: Weekly oil consumption, the minimum and maximum temperatures(F) and the temperature ranges.

First note that we cannot regress on x_1, x_2 and x_3, since X would be rank deficient with $\sigma_3 = 0$ and thus $cond(X) = \infty$. If we regress on x_1 and x_2, we are not far from linear dependence since the range is fairly constant. This is reflected in the condition number of X, which is 107. This means that the uncertainty of the temperature values will have quite a strong influence.

Now if we instead switch to centigrades, i.e. replace x_1 and x_2 by $(x_1 - 32)/9$ and $(x_2 - 32)/9$, respectively, we get $cond(X) = 1.2$ and thus a well-conditioned problem. We can also get a similar effect by subtracting each column by its mean.

4.2 Numerical stability

Although it is sometimes possible to reformulate the problem, like in Example 9, there is often a need to solve fairly ill-conditioned problems. Still you want to carry out the computation with good accuracy, so we will take a look at some different possibilities.

4.2.1 Normal equations

A well-known method to solve $Xb = y$ is via the normal equations

$$X^T X b = X^T y \tag{14}$$

To solve this system causes no problems, since $X^T X$ has the attractive properties of being symmetric and positive definite. However, forming (14) is numerically very dangerous, since

$$cond(X^T X) = cond(X)^2 \tag{15}$$

The following illuminating example is found in [8].

Example 10 Let

$$X = \left\{ \begin{array}{ccc} 1 & 1 & 1 \\ \epsilon & 0 & 0 \\ 0 & \epsilon & 0 \\ 0 & 0 & \epsilon \end{array} \right\}$$

with $\epsilon = 0.0001$. We get

$$X^T X = \left\{ \begin{array}{ccc} 1 + \epsilon^2 & 1 & 1 \\ 1 & 1 + \epsilon^2 & 1 \\ 1 & 1 & 1 + \epsilon^2 \end{array} \right\}.$$

With a machine epsilon around 10^{-7} (as in IEEE single precision arithmetic), $1 + \epsilon^2$ will be rounded to 1, making $X^T X$ singular.

4.2.2 QR-factorization

There exist alternatives to the normal equations (14) which are superior from numerical point of view. One is QR-factorization, where the design matrix is factorized into $X = QR$. Here Q is an orthogonal $n \times n$-matrix and R is a "right triangular" $n \times p$-matrix, with rows $p + 1$ to n equal to zero. Since multiplying a vector by an orthogonal matrix does not change its 2-norm value, we have

$$\|Xb - y\|_2 = \|Q^T(QRb - y)\|_2 = \|Rb - z\|_2 \tag{16}$$

where $z = Q^T y$.

Thus once the factorization is done, what remains to carry out is a matrix-vector-multiplication and solving a triangular system. If X is rank deficient, one or more diagonal elements in R will be zero, which calls for special treatment. However, we will postpone the discussion how to treat rank deficiency to the next section.

4.2.3 Singular Value Decomposition

Another important possibility to better up the numerical stability (compared to solving the normal equations), is to use SVD, Singular Value Decomposition. Here a matrix is factorized into $X = USV^T$ where U is $n \times n$ orthogonal, V^T is $p \times p$ orthogonal and S is $n \times p$ with the only non-zero elements in the uppermost diagonal, namely the singular values. Similarly to QR-factorization, once the SVD-factors are found, the solution to the least-squares problem is trivial:

$$\|Xb - y\|_2 = \|U^T(USV^Tb - y)\|_2 = \|Sc - z\|_2 \tag{17}$$

where $z = U^Ty$ and $c = V^Tb$.

If X has full rank, it is a trivial task to solve for c ($c_i = z_i/\sigma_i$ for all i) and then compute $b = Vc$.

One attractive feature of SVD is the elegant way cases where a matrix is rank deficient (or nearly so) can be handled. First we note that even if the columns of X are linearly dependent, rounding errors will in practice make all singular values deviate from zero. A fruitful way to define the numerical rank of a matrix is then to count the number of singular values above a treshold $\epsilon > 0$.

Returning to the least-squares problem, we can reach a solution the following way. If e.g. $\sigma_1 > \sigma_2 > \ldots > \sigma_{p-1} > \epsilon > \sigma_p$, we let

$$c_i = \begin{cases} z_i/\sigma_i & (i = 1, 2, \ldots, p-1) \\ 0 & (i = p) \end{cases}.$$

Now $b = Vc$ will give the minimal solution, i.e. out of all the solutions to the degenerate problem, the one for which $\|b\|_2$ is smallest is selected.

We summarize an example given in [20].

Example 11 U.S. Census data from 1900-1970 is used to predict the 1980 population with the model

$$y(t) = b_1 + b_2t + b_3t^2 .$$

The least squares estimate, using a single precision SVD routine, is

$$b_1 = -0.372 \cdot 10^5 , \quad b_2 = 0.368 \cdot 10^2 , \quad b_3 = -0.905 \cdot 10^{-2} .$$

The corresponding double precision calculation gives

$$b_1 = 0.373 \cdot 10^5 , \quad b_2 = -0.402 \cdot 10^2 , \quad b_3 = 0.108 \cdot 10^{-1}$$

The discrepancy is caused by the ill-conditioning of the problem, since we have the following singular values of the design matrix:

$$\sigma_1 = 1.1 \cdot 10^7 \; , \quad \sigma_2 = 6.5 \cdot 10^{-1} \; , \quad \sigma_3 = 3.5 \cdot 10^{-4}$$

and condition number $3.1 \cdot 10^{10}$.

Now if we use the treshold $\epsilon = 10^{-3}$, the matrix will be considered rank deficient. The minimal solution computed as indicated above gives in single precision

$$b_1 = -0.167 \cdot 10^{-2} \; , \quad b_2 = -0.162 \cdot 10^1 \; , \quad b_3 = -0.871 \cdot 10^{-3}$$

and in double precision

$$b_1 = -0.166 \cdot 10^{-2} \; , \quad b_2 = -0.162 \cdot 10^1 \; , \quad b_3 = -0.869 \cdot 10^{-3} \; .$$

Of course in this simple example we know how to avoid such an ill-conditioned matrix, namely by changing the base, e.g. to

$$y(t) = b_1 + b_2 \frac{t - 1935}{35} + b_3 (\frac{t - 1935}{35})^2 \; .$$

SVD has found many applications over the last decades. We quote the preface of [16] :

> "The habits of the field [numerical linear algebra] - our reliance on orthogonal matrices, our appreciation of problem sensivity, our careful consideration of roundoff - have spilled over into many areas of research. A prime example of this is the increased use of the SVD as an analytical tool by many statisticians and control engineers. People in these areas are reformulating numerous concepts in the 'language' of the SVD and as a result are finding it much easier to implement their ideas in the presence of roundoff error and inexact data."

4.3   Efficiency

Although numerical accuracy has high priority, it is also of great value to have a fast algorithm. Solving the least squares problem if $n \gg p$ via QR-factorization, as described in the previous section, approximately doubles the number of operations as compared to using the normal equations. In this case you have to pay in computing time for increased numerical accuracy, but that is not always the case. For example, to solve a square system of equations $Ax = b$ by first computing $B = A^{-1}$ and then $x = Bb$, is both less accurate and less efficient than solving $Ax = b$ by Gaussian elimination with pivoting. This follows a general rule, saying that there is seldom any need to perform matrix inversion, since most often a system of equations can be solved instead. A typical example is the expression

$$y = c^T A^{-1} d \; ,$$

where c and d are n-vectors and A an $n \times n$-matrix. The value of y should be computed the following way:

$$Ax = d$$
$$y = c^T x.$$

When leaving the traditional computer architectures for some kind of parallel computing, other factors have to be considered in designing good algorithms. The book by Golub and van Loan [16] gives a good overview for matrix computations. There is also supporting software like BLAS and LAPACK [9]. BLAS (Basic Linear Algebra Subprograms) is a standard set of fundamental routines for linear algebra computations which are both efficient and highly accurate. LAPACK a collection of transportable Fortran routines for solving the most common computational problems in linear algebra, such as systems of linear equations, linear least-squares problems, eigenvalue problems and singular value problems. It is integrated into the NAG library, but individual routines are most easily obtained by electronic mail through netlib[10].

LAPACK is supported by Level 2 BLAS for matrix-vector operations and Level 3 BLAS [11] for matrix-matrix operations. However, in practice it may sometimes help to take the special properties of the problem or the computer achitecture into account ([12] gives an example). Golub and van Loan state that

> "... the efficiency of a vector-pipeline matrix computation depends upon the vector length, the vector stride, the vector touch, and data re-use properties of the algorithm. Optimizing with respect to all these attributes is very complicated and something of an art."

5 Other types of regression

Although the dangers of the least squares method - its high sensitivity to a few "wild points" - have been known since very long, other possibilities have not really come into focus until the last three decades. Among numerical analysts, the "natural" alternative criterion was to minimize the L_p-norm with different values of p [25]. In parallel, Peter Huber introduced M-estimators and a basis for robust statistics [18]. This of course also gave a need for new algorithms.

There are many algorithms to compute M-estimators for linear models. Two main approaches are Newton-like methods [15], which have the advantage of giving fast convergence near the solution, and Iteratively Reweighted

Least Squares (IRLS), which is easier to implement. For L_p-regression the two approaches give the same search directions [28]. In some cases, e.g. for Huber regression, IRLS is globally convergent without linesearch [19, 4]. For the matrix updating to be made in each iteration, stress can be put on computational efficiency (as e.g. in [1]) or numerical stability [24].

Among the alternatives to least squares, L_1-regression has a special position, being used even prior to L_2 [14]. Once it was clear that L_1-minimization can be formulated as a linear programming problem, it was possible to solve quite large problems. The simplex method, based on moving from one corner to the next in the feasable region, was for a long time generally accepted as the optimal way to solve linear programming problems. It is a finite algorithm, since the number of corners is finite. The simplex algorithm was refined to take the special character of L_1-problems into account (e.g. [2]).

It was a big surprise when interior point methods [27] were presented as being superior to the simplex method, at least for some types of large linear programming problems. The name stems from moving in the interior of the simplex during the iterations. Interestingly, these methods are essentially IRLS methods. Of course, interior point methods for linear programming can also be adapted to solving L_1-problems.

Another promising approach is the L_1-algorithm suggested by Madsen and Nielsen [23]. It is a continuation method which makes use of the fact that the L_1 estimate is a limiting case of Huber estimates. Also Coleman and Li [6] have recently designed an L_1-algorithm with good properties.

Some M-estimators proposed give a convex object function to minimize (e.g. the Huber, Fair, and L_p estimators) while others do not (e.g. the Tukey, Hampel, and Welsh estimators). This has a strong influence on the algorithm design. For linear models, the convex object function will have a unique minimum, and the effort can be focused on computing it efficiently and with good accuracy. In contrast, a non-convex object function has many local minima, and the principal goal is to get close to the global minimum. Here good starting points are most desirable. One strategy is first to compute the Huber solution and use it as a start value to the non-convex function minimization (as is done e.g. in [1]).

The problem to minimize non-convex functions is of course even more pronounced if a non-linear model is used. Algorithms for non-linear least squares is today a thoroughly studied subject within numerical analysis. As an alternative to the classical approach, i.e. finding a search direction and then making a line-search, trust-region methods are nowadays often used (e.g. [13]). A common algorithmic approach, when using a robust criterion, is to adapt the structure of the corresponding non-linear least-squares algorithm.

Another recent computational problem, which has attracted many researchers, is to find the LMS (Least Median of Squares) solution (e.g. [17], [3]). The object function to minimize has a large amount of local minima, an optimization problem well suited for parallel computing (e.g. [29]).

6   Concluding remarks

As indicated above, many new numerical methods are immediately useful in statistics. However, the time lag until these are really used by statisticians can certainly be shortened. While a scientist doing computational work may be willing to plug a piece of software into her program, a statistician will instead meet an improved numerical method as part of some statistical package. That type of software is not always updated numerically as fast as it should be.

Modelling in regression is in the borderland between numerical analysis and statistics. As has been pointed out [22], statisticians are good at taking care of the stochastic part while numerical analysts are more open-minded in their modelling, using constraints, neural networks, wavelets and so on. So also from this point of view, the two disciplines can learn from each other.

Ideas and methods from numerical analysis can be applied in many other areas than those exemplified above. Quadrature rules are used when approximating probabilities and percentage points, eigenvalues are computed in factor analysis, etc. A good overview is found in the book by Kennedy and Gentle [21].

It is an interesting fact that what is sometimes described as "the core" of numerical analysis, namely solving differential equations, seems to be of limited interest to statistics. Cox [7] states:

> "It seems likely that statistics and applied probability have made insufficient use of developments in numerical analysis associated more with classical applied mathematics, in particular in the solution of large systems of ordinary and partial differential equations, integral equations and integro-differential equations and for the extraction of 'useful' information from integral transforms."

In summary, numerical analysis makes it easier to understand the peculiarities stemming from floating point calculation. Furthermore, the perturbation theory can reveal ill-conditioning and may help in reformulating the problem. But above all, numerical analysis enables stable and efficient algorithms to be designed, as a base for good statistical software.

7   Acknowledgements

The author would like to thank Jaromir Antoch for many helpful suggestions and stimulating discussions concerning the content of the paper, Inge Söderqvist for constructive comments to an early version of this paper and Ove Edlund for invaluable help to form the authors first LATEX document.

References

[1] Antoch, J. and Ekblom, H. (1994): *Recursive Robust Regression - Computational Aspects and Comparison.* To be published in Computational Statistics and Data Analysis.

[2] Barrodale, I. and Roberts, F.D.K. (1973): *An improved algorithm for discrete l_1 linear approximation.* SIAM J.Num.Anal. **10**, 839-848.

[3] Boček, P. and Lachout, P. (1994): *Linear programming approach to LMS-estimation.* To be published in Computational Statistics and Data Analysis.

[4] Byrd, R.H. and Pyne, D.A (1979): *Convergence of the IRLS Algorithm for Robust Regression.* Tech.Report No 313, Dept. of Math. Sciences, The John Hopkins University.

[5] Chan, T., Golub, G. and Leveque, R. (1983): *Algorithms for Computing the Sample Variance: Analysis and Recommendations.* The American Statistician, **37**(3).

[6] Coleman, T.F. and Li, Y. (1992): *A global and quadratically-convergent affine scaling method for linear l_1 problems,* Math. Programming **56**, 189-222.

[7] Cox, D. (1992): *The Role of the Computer in Statistics.* In Proceedings of COMPSTAT 92, Y.Dodge and J.Whittaker, eds., Physica-Verlag.

[8] Dahlquist, G. and Björk, Å. (1974): *Numerical Methods.* Prentice-Hall, Englewood Cliffs, N.J.

[9] Dongarra, J.J., Demmel, J.W. and Ostrouchov, S. (1992): *LAPACK: A Linear Algebra Library for High Performance Computers.* In Proceedings of COMPSTAT 92, Y.Dodge and J.Whittaker, eds., Physica-Verlag.

[10] Dongarra, J.J and Grosse, E. (1987): *Distribution of mathematical software via electronic mail.* Comm ACM, **30**(5), 403-407.

[11] Dongarra, J.J. et al. (1990): *A set of level 3 basic linear algebra subprograms.* ACM Trans. Math. Soft., **16**(1), 1-17.

[12] Edlund, O. (1994): *A Study of Possible Speed-up when using a Vector Processor.* Short communication, COMPSTAT 94.

[13] Ekblom, H. and Madsen, K. (1989): *Algorithms for non-linear Huber estimation,* BIT **29**, 60-76.

[14] Fairbrother, R.W. (1987): *The Historical Developement of the L_1 and L_∞ Estimation Procedures.* In Dodge (ed): Statistical Data Analysis Based on the L_1-Norm and Related Methods, North-Holland.

[15] Fletcher, R. (1987): *Practical Methods of Optimization*. 2nd Edition, John Wiley.

[16] Golub, G. and van Loan, C. (1989): *Matrix Computations*. John Hopkins University Press, Baltimore, MD, 2nd edition, 1989.

[17] Hawkins, D.M. (1993): *A feasable set algorithm for least median of squares regression*. Computational Statistics and Data Analysis **16**, 81-101.

[18] Huber, P.J. (1964): *Robust Estimation of a Location Parameter*. Annals of Mathematical Statistics **35**, 73-101.

[19] Huber, P.J. (1981): *Robust Statistics*. John Wiley, New York.

[20] Kahaner, D., Moler, C. and Nash, S. (1989): *Numerical Methods and Software*. Prentice Hall, New Jersey.

[21] Kennedy, W.J. and Gentle, J.E. (1980): *Statistical Computing*. Marcel Dekker, New York and Basel.

[22] Prof. Georg Lindgren, University of Lund, Sweden, personal communication.

[23] Madsen, K. and Nielsen, H.B. (1993): *A finite smooting algorithm for linear l_1 estimation*. SIAM J. Optimization, 3(2), 223-235.

[24] O'Leary, D. (1990): *Robust Regression Computation Using Iteratively Reweighted Least Squares*. SIAM J. Matrix Anal. Appl., 11(3), 466-480.

[25] Rice, J.R. and White, J.S. (1964): *Norms for Smoothing and Estimation*. SIAM Review **6**, 243-256.

[26] Stevenson, D. (1981): *A Proposed Standard for Binary Floating Point Arithmetic*. Computer **14**, 51-62.

[27] Strang, G. (1987): *Karmarkar's Algorithm and Its Place in Applied Mathematics*. The Mathematical Intelligencer 9(2).

[28] Watson, G.A. (1980): *Approximation Theory and Numerical Methods*. John Wiley, New York.

[29] Xu, C.-W and Shiue, W.-K. (1993): *Parallel algorithms for least median of squares regression*. Computational Statistics and Data Analysis 16(3), 349-362.

Computation of Statistics Integrals using Subregion Adaptive Numerical Integration

Alan Genz

Department of Mathematics, Washington State University

Keywords. Multivariate distribution, adaptive integration, Bayesian

1 Introduction

Many types of statistical analysis problems require evaluation of multidimensional integrals in the form

$$I(f) = \int_R f(\theta)p(\theta)d\theta,$$

where $\theta = (\theta_1, \theta_2, ..., \theta_m)^t$, the integration region R is infinite and $p(\theta)$ is a probability density function. Two important examples are the multivariate normal density, and the multivariate t density. In these cases it is assumed that $R = (-\infty, b_1] \times (-\infty, b_2] \times ...(-\infty, b_m]$, and $f = 1$, so that $P(\mathbf{b}) = I(1)$ is a probability distribution function. A larger but less specific class of problems arises in Bayesian analysis, where it will be assumed that $p(\theta)$ has a global maximum at a unique θ in R. With these problems p is usually unnormalized, so that I(1) is needed, and other typical f's are the vector $\mathbf{f} = \theta$ (for expected values of the variables), and the matrix $F = \theta\theta^t$ (for the covariance matrix for θ). Good review references for statistics integration problems are the papers by Naylor and Smith (1982 and 1988) and Shaw (1988) and the book by Tanner (1993).

One type of numerical integration method uses good deterministic integration rules combined with adaptive algorithms. The most successful of these algorithms are the *subregion adaptive* algorithms (see Genz, 1991, for an overview), These algorithms dynamically subdivide the initial integration region in order to adapt to the local irregularities of the integrand. The purpose of this paper is to describe how subregion adaptive algorithms can be used effectively for the solution of statistics problems of the type described at the beginning of this introduction. This paper will summarize and extend some earlier work by the author (Genz 1992, 1993, Genz and Kass, 1991, 1993).

---

[1]Department of Mathematics, Washington State University, Pullman, Washington 99164-3113 USA; email: acg@eecs.wsu.edu. This work was supported by a grant from the US National Science Foundation.

2 Adaptive Integration and Transformations

Subregion adaptive integration methods are based on the assumption that the integrand in the problem of interest can be accurately approximated locally by a low degree multivariate polynomial. The basic strategy for a subregion adaptive algorithm is to dynamically subdivide the initial integration region R into smaller and smaller subregions that are concentrated in the parts of R where the integrand is more irregular. The hope is that at some stage in this process the region R is sufficiently well-partitioned that the combined integrated polynomial approximations for all of the subregions provide an accurate approximation to the initial integral. One of the most effective adaptive algorithms was originally developed by van Dooren and de Ridder (1976). This original algorithm has been modified and improved and is now implemented and available for vectors of integrals (the DCUHRE Fortran subroutine developed by Berntsen, Espelid, and Genz, 1991a, 1991b).

Subregion adaptive integration algorithms require repeated subdivision of the integration region. The subdivision steps for these algorithms repeatedly cut selected subregions along directions determined by the coordinate axes. When the integration region is infinite, the point selected along the axis where the current subregion is to be cut is not clearly defined. Any strategy for choosing these cutting points implies some transformation of an infinite interval to a finite one, so it is convenient to initially use an appropriately chosen transformation from the initial infinite integration region to a finite region. Then the adaptive algorithm can be used directly on the transformed integrand over the finite integration region.

The choice of a good transformation, is often problem dependent. But many statistics problems are formulated in ways that suggest obvious choices for transformations. And in many of these problems, a good transformation actually simplifies the integral. In the next two sections transformations from infinite regions to the unit hypercube $[0, 1]^m$ are described. These transformations allow subregion adaptive integration to be used for the numerical computation of the multivariate normal and t distribution functions. The final section describes how generalizations of the transformations used for the multivariate normal and t can used for more general integration problems from Bayesian analysis.

3 The Multivariate Normal Distribution

The cumulative multivariate normal distribution function is defined by

$$P(\mathbf{b}) = \frac{1}{\sqrt{|\Sigma|(2\pi)^m}} \int_{-\infty}^{b_1} \int_{-\infty}^{b_2} \cdots \int_{-\infty}^{b_m} e^{-\frac{1}{2}\theta'\Sigma^{-1}\theta} d\theta,$$

with all b_i's finite and Σ a covariance matrix. A sequence of three transformations can be used to transform $P(\mathbf{b})$ into an integral over a unit hyper-cube.

This sequence begins with the transformation $\theta = Cy$, where CC^t is the Cholesky decomposition of Σ. Because $\theta^t\Sigma^{-1}\theta = y^t C^t C^{-t} C^{-1} Cy = y^t y$, $d\theta = |C|dy$ and $|C| = |\Sigma|^{\frac{1}{2}}$,

$$P(\mathbf{b}) = \frac{1}{\sqrt{(2\pi)^m}} \int_{-\infty}^{b_1'} e^{-\frac{y_1^2}{2}} \int_{-\infty}^{b_2'(y_1)} e^{-\frac{y_2^2}{2}} ... \int_{-\infty}^{b_m'(y_1,y_2,...,y_{m-1})} e^{-\frac{y_m^2}{2}} dy,$$

with $b_i'(y_1, ..., y_{i-1}) = (b_i - \sum_{j=1}^{i-1} c_{i,j} y_j)/c_{i,i}$. If each of the y_i's are now transformed separately using $y_i = \Phi^{-1}(z_i)$, $P(\mathbf{b})$ becomes

$$P(\mathbf{b}) = \int_0^{e_1} \int_0^{e_2(z_1)} ... \int_0^{e_m(z_1,z_2,...,z_{m-1})} dz,$$

with $e_i(z_1, z_2, ..., z_{i-1}) = \Phi((b_i - \sum_{j=1}^{i-1} c_{i,j} \Phi^{-1}(z_j))/c_{i,i})$. This integral can be put into constant limit form using $z_i = w_i e_i(z_1, z_2, ..., z_{i-1})$, so that

$$P(\mathbf{b}) = e_1 \int_0^1 e_2(w_1) \int_0^1 e_3(w_1, w_2) ... \int_0^1 e_m(w_1, w_2, ..., w_{m-1}) dw,$$

with $e_i(w_1, ..., w_{i-1}) = \Phi((b_i - \sum_{j=1}^{i-1} c_{i,j} \Phi^{-1}(e_j(w_1, ..., w_{j-1}) w_j))/c_{i,i})$. The inner integral over w_m can determined using the one variable cumulative normal distribution function Φ, and because e_m has no dependence on w_m, this sequence of transformations has reduced the number of integration variables by one.

Tests with large numbers of randomly chosen covariance matrices and randomly chosen b vectors have shown that subregion adaptive integration can be used to compute $P(\mathbf{b})$ for m as large as 10 in a few seconds or less of workstation time, and that this is currently the most efficient way to compute $P(\mathbf{b})$ (see Genz, 1992 and 1993).

4   The Multivariate t Distribution

The cumulative multivariate t distribution function is defined by

$$P(\mathbf{b}) = \frac{K_\nu^{(m)}}{\sqrt{|\Sigma|}} \int_{-\infty}^{b_1} \int_{-\infty}^{b_2} ... \int_{-\infty}^{b_m} \left(1 + \frac{\theta^t \Sigma^{-1} \theta}{\nu}\right)^{-\frac{\nu+m}{2}} d\theta,$$

with $K_\nu^{(m)} = \frac{\Gamma(\frac{\nu+m}{2})}{\Gamma(\frac{\nu}{2})\sqrt{(\nu\pi)^m}}$, all b_i's finite and Σ a covariance matrix..

A sequence of four transformations can be used to transform $P(\mathbf{b})$ into an integral over a unit hyper-cube. This sequence also begins with a Cholesky decomposition transformation $\theta = Cy$. Then

$$P(\mathbf{b}) = K_\nu^{(m)} \int_{-\infty}^{b_1'} \int_{-\infty}^{b_2'(y_1)} ... \int_{-\infty}^{b_m'(y_1,...,y_{m-1})} \left(1 + \frac{y^t y}{\nu}\right)^{-\frac{\nu+m}{2}} dy,$$

with $b'_i(y_1, ..., y_{i-1}) = (b_i - \sum_{j=1}^{i-1} c_{i,j} y_j)/c_{i,i}$. Next let $y_i = u_i(\frac{\nu + \sum_{j=1}^{i-1} y_j^2}{\nu + i - 1})^{\frac{1}{2}}$.

An equivalent expression is $y_i(u_1, u_2, ..., u_i) = u_i \prod_{j=1}^{i-1} (\frac{\nu + j - 1 + u_j^2}{\nu + j})^{\frac{1}{2}}$ Using the definition of $K_\nu^{(m)}$ and some algebra, it can be shown that

$$P(\mathbf{b}) = K_\nu^{(1)} \int_{-\infty}^{\hat{b}_1} (1 + \frac{u_1^2}{\nu})^{-\frac{\nu+1}{2}} ... K_{\nu+m-1}^{(1)} \int_{-\infty}^{\hat{b}_m(u_1,...,u_{m-1})} (1 + \frac{u_m^2}{\nu + m - 1})^{-\frac{\nu+m}{2}} du,$$

with $\hat{b}_i(u_1, u_2, ..., u_{i-1}) = \prod_{j=1}^{i-1} (\frac{\nu+j}{\nu+j-1+u_j^2})^{\frac{1}{2}} (b_i - \sum_{j=1}^{i-1} c_{i,j} y_j(u_1, ..., u_j))/c_{i,i}$.

Now let $u_i = t_{\nu+i-1}^{-1}(z_i)$, where $t_\nu(u) = K_\nu^{(1)} \int_{-\infty}^{u} (1 + y^2)^{-\frac{\nu+1}{2}} dy$, so

$$P(\mathbf{b}) = \int_0^{e_1} \int_0^{e_2(z_1)} ... \int_0^{e_m(z_1,...,z_{m-1})} dz,$$

with $e_i(z_1, ..., z_{i-1}) = t_{\nu+i-1}(\hat{b}_i(t_\nu^{-1}(z_1), ..., t_{\nu+i-2}^{-1}(z_{i-1})))$. Finally, using $z_i = e_i(z_1, ..., z_{i-1})w_i$,

$$P(\mathbf{b}) = e_1 \int_0^1 e_2(w_1) \int_0^1 e_3(w_1, w_2) ... \int_0^1 e_m(w_1, w_2, ..., w_{m-1}) d\mathbf{w}.$$

Subregion adaptive integration can now be used to approximately compute the transformed $P(\mathbf{b})$. Tests with randomly generated covariance matrices and randomly generated **b** vectors have shown that this is an efficient method for computing $P(\mathbf{b})$ for m as large as 10.

5   Transformations for Bayesian Analysis

A natural transformation for general Bayesian integration problems is based on assuming (Chen, 1985) that the integrand $p(\theta)$ is unimodal and that $log(p)$ is locally quadratic near the mode. This means $p(\theta) \simeq Ae^{-\frac{1}{2}(\theta - \mu)'\Sigma^{-1}(\theta - \mu)}$ for some constant A. Here, the integration region R is assumed to be $[-\infty, \infty]^m$, so that an initial transformation of the original integral might also be necessary for some problems. In most problems of interest, multivariate optimization techniques can be applied to the function $log(p(\theta))$ to obtain the *mode* μ (the point where the maximum of $log(p(\theta))$ is attained) and the matrix Σ (the inverse of the Hessian matrix for $-log(p(\theta))$ at μ).

Once Σ and μ are available, the Cholesky decomposition CC^t of Σ can be found. Then, the use of the transformation $\theta = \mu + Cy$, followed by inverse normal transformations $y_i = \Phi^{-1}(z_i)$ on the y components yields

$$\int_{-\infty}^{\infty} ... \int_{-\infty}^{\infty} f(\theta)p(\theta)d\theta = (\sqrt{2\pi})^m |C| \int_0^1 ... \int_0^1 f(\mu + Cy(z))g(y(z))dz,$$

where $y(z) = (\Phi^{-1}(z_1), ..., \Phi^{-1}(z_m))^t$ and $g(y) = e^{\frac{1}{2}y'y} p(\mu + Cy)$. Now, subregion adaptive integration can be applied to the transformed integral

50

over the unit hypercube. As long as the stated assumption of approximate normality is correct, this method should be reasonably efficient because the transformed integrand should be approximately constant.

If $p(\theta)$ has tail behavior that is not accurately modeled by the multivariate normal distribution, there are several schemes that might improve on the modal approximation transformation. One method is to use the "split-t" transformation first described by Geweke (1989, 1991). With this technique, the transformation $z_i = \Phi(y_i)$ is modified by splitting at the origin and using alternative functions in each direction that have (i) the form of univariate t distribution function and (ii) scale factors different than 1 (the standard deviations in the normal case). The function $p(\theta)$ is investigated along the directions indicated by the Cholesky decomposition of Σ, in order to determine the appropriate degrees of freedom and scale factors. Genz and Kass (1993) describe an effective numerical algorithm for determining good values for these parameters and show how the complete transformation (multivariate normal combined with split-t transformations) can be used to provide accurate results for several Bayesian integration problems.

A more direct approach when $p(\theta)$ does not have normal tail behavior is to use a multivariate t model $p(\theta) \simeq A(1 + (\theta - \mu)^t \Sigma^{-1}(\theta - \mu)/\nu)^{-\frac{m+\nu}{2}}$ for $p(\theta)$. Multivariate numerical optimization is used to find the *mode* μ and the Hessian matrix H at μ for $log(p(\theta))$. In this case $\Sigma = -\frac{\nu}{\nu+m}H^{-1}$, so an appropriate value for ν also needs to be estimated numerically.

Once Σ, μ and ν are available, the Cholesky decomposition CC^t of Σ is found. Then, the use of the transformation $\theta = \mu + Cy$, followed by inverse t transformations $y_i = t^{-1}_{\nu+i-1}(z_i)$, yields

$$\int_{-\infty}^{\infty} \cdots \int_{-\infty}^{\infty} f(\theta)p(\theta)d\theta = \frac{|C|}{K_\nu^{(m)}} \int_0^1 \cdots \int_0^1 f(\mu + Cy(z))g(y(z))dz,$$

with $y(z) = (t^{-1}_\nu(z_1), ..., t^{-1}_{\nu+m-1}(z_m))^t$ and $g(y) = (1 + y^t y/\nu)^{\frac{m+\nu}{2}} p(\mu + Cy)$. And now subregion adaptive integration can be applied to the transformed integral over the unit hypercube. This transformation method has not yet been carefully tested on Bayesian integration problems.

References

Berntsen, J., Espelid, T.O. and Genz, A. (1991a) 'An Adaptive Algorithm for the Approximate Calculation of Multiple Integrals', *ACM Trans. Math. Softw.* 17, pp. 437-451.

Berntsen, J., Espelid, T.O. and Genz, A. (1991b) 'An Adaptive Multiple Integration Routine for a Vector of Integrals', *ACM Trans. Math. Softw.* 17, pp. 452-456.

Chen, C.F. (1985) 'On Asymptotic Normality of Limiting Density Functions with Bayesian Implications', *J. Royal Statist. Soc.*, 47, pp. 540-546.

van Dooren, P. and de Ridder, L. (1976) 'An Adaptive Algorithm for Numerical Integration over an N-Dimensional Rectangular Region', *J. Comp. Appl. Math.* **2**, pp. 207-217.

Genz, A. (1991) 'Subregion Adaptive Algorithms for Multiple Integrals', in N. Flournoy and R. K. Tsutakawa (eds.) *Statistical Numerical Integration*, Contemporary Mathematics **115**, American Mathematical Society, Providence, Rhode Island, pp. 23-31.

Genz, A. (1992) Numerical Computation of Multivariate Normal Probabilities, *J. Comp. Graph. Stat.* **1**, pp. 141-150.

Genz, A. (1993) 'A Comparison of Methods for Numerical Computation of Multivariate Normal Probabilities', *Computing Science and Statistics*, **25**, pp. 400-405.

Genz, A. and Kass, R. (1991) 'An Application of Subregion Adaptive Numerical Integration to a Bayesian Inference Problem', *Computing Science and Statistics* **23**, pp. 441-444.

Genz, A. and Kass, R. (1993) 'Subregion Adaptive Integration of Functions Having a Dominant Peak', Carnegie Mellon University Statistics Department Technical Report No. 586, submitted.

Geweke, J. (1989) 'Bayesian Inference in Econometric Models Using Monte Carlo Integration', *Econometrica* **57**, pp. 1317-1340.

Geweke, J. (1991) 'Generic, Algorithmic Approaches to Monte Carlo Integration in Bayesian Inference', in *Statistical Multiple Integration*, N. Flournoy and R. K. Tsutakawa (Eds.), Contemporary Mathematics Series Volume 115, American Mathematical Society, Providence, Rhode Island, pp. 117-135.

Naylor, J. C. and Smith, A. F. M. (1982) 'Applications of a Method for the Efficient Computation of Posterior Distributions', *Appl. Stat.*, **31**, pp. 214-225.

Naylor, J. C. and Smith, A. F. M. (1988) 'Econometric Illustrations of Novel Numerical Integration Strategies for Bayesian Inference', *J. Economet.*, **38**, pp. 103-125.

Shaw, J. E. H. (1988) 'Aspects of Numerical Integration and Summarisation', in J. M. Bernado, M. H. Degroot, D. V. Lindley and A. F. M. Smith (eds.) *Bayesian Statistics 3*, Oxford University Press, Oxford, pp. 411-428.

Tanner, M. (1993), *Tools for Statistical Inference* 2nd Edition, Springer-Verlag, New York.

Universal Generators for Correlation Induction

Wolfgang Hörmann, Gerhard Derflinger

Department of Statistics, University of Economics and Business Administration

Keywords. Simulation, non-uniform random number generation, rejection method, universal generators, correlation induction

1 Introduction

Compared with algorithms specialized for a single distribution universal (also called automatic or black-box) algorithms for continuous distributions were relatively seldom discussed. But they have important advantages for the user: One algorithm coded and tested only once can do the same or even more than a whole library of standard routines. It is only necessary to have a program available that can evaluate the density of the distribution up to a multiplicative factor. Black box algorithms suggested in literature fall into two groups. Simple and short algorithms with almost no setup (eg. [4]) but only moderate speed and very fast table-methods which need a long and complicated setup (eg. [2] and [1]). In [7] and [6] we introduced new universal generators based on transformed density rejection which lie between these two groups. The suggested algorithms are quite simple and need only a moderate setup time whereas the marginal execution time for standard distributions (like the gamma- or the beta-family) is about the same as for specialized algorithms. In [8] we demonstrated how transformed density rejection can be used to construct very fast table methods as well. In this paper we show that transformed density rejection is well suited to construct universal algorithms suitable for correlation induction which is important for variance reduction in simulation.

2 Transformed density rejection

Rejection is the most flexible method for generating non-uniform random variates. To generate random variates with density function $f(x)$ we need a "hat" function $h(x)$ and a real number α with $f(x) \leq \alpha h(x)$. As the first step a random variate X with density $h(x)$ and a uniform ($U(0,1)$) random number U is generated. If $\alpha h(X)U \leq f(X)$ then X is accepted as a random variate of the desired distribution otherwise X is rejected and we have to try again. It is obvious that α is the expected number of replications to get one random number and that $1/\alpha$ is the probability of acceptance. Therefore $h(x)$ should be chosen in a way that α is close to one and that the generation of random numbers with density $h(x)$ is simple.

Transformed density rejection uses a new and general method for constructing hat functions. To do so we need a transformation $T(x)$ with the property that $T(f(x))$ is concave on the support of $f(x)$ (i.e. $\{x|f(x) > 0\}^-$), and arbitrary but fixed design points x_i, $i = 0, \cdots, n - 1$. Then we define $lin(x; x_0, \ldots, x_{n-1}) = \min_i\{f(x_i) + f'(x_i)(x - x_i)\}$ which is obviously greater or equal $T(f(x))$. Thus $T^{-1}(lin(x; x_0, \ldots, x_{n-1})) = h(x)$ is a hat function for $f(x)$. The lines, connecting neighbouring points of contact, transformed with T^{-1} can be used as simple squeezes to avoid the evaluation of $f(x)$ in most cases. Figure 1 illustrates transformed density rejection for the normal distribution with $x_0 = -1.665$, $x_1 = 0$, $x_2 = 1.665$ and $T(x) = -1/\sqrt{x}$; [left hand side: $T(f(x))$ (thick line) and $lin(x)$ (thin line); right hand side: $f(x)$ (thick line) and $T^{-1}(lin(x))$ (thin line); the dashed lines are the squeezes.]

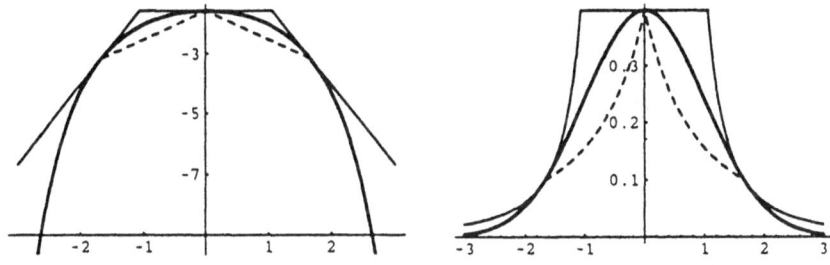

Fig. 1

If we take $T(x) = \log(x)$ (this special case was investigated in [5]) or $T(x) = -1/\sqrt{x}$ the generation of random variates with density $h(x)$ can be easily done by inversion. For the case $T(x) = \log(x)$ transformed density rejection is applicable to the well-known class of log-concave distributions. This class consists of distributions with unimodal densities that have subexponential tails and is defined by the property $\log(f(x))$ is concave or equivalently $f(x)f''(x) - f'(x)^2 \leq 0$ on the support of f. Examples are the normal the gamma (with $a \geq 1$) and the beta (with $a, b \geq 1$) distributions. For $T(x) = -1/\sqrt{x}$ we call the densities T-concave. This class of distributions consists of unimodal densities with subquadratic tails and is defined by $f(x)f''(x) - 1.5f'(x)^2 \leq 0$. Obviously T-concave is more general than log-concave. The most prominent distribution which is T-concave but not log-concave is the t-family with $\nu \geq 1$.

3 Universal algorithms

It is not difficult to use the transformed density rejection method to construct a hat function and thus a rejection algorithm for an arbitrary T-concave density that is – up to a multiplicative factor – computable by a given subprogram (black box). First we give a sketch of the universal algorithm for given design points x_i which are sorted in ascending order. (C-code is available on request.)

Algorithm UNIV

1: [Set-up]

1.0 Compute $y_i = T(f(x_i))$, $a_i = T(f(x_i))'$ and $as_i = (y_{i+1} - y_i)/(x_{i+1} - x_i)$ for $i = 0 \ldots n - 1$. Compute the first coordinates of the points of intersection b_i between tangent i and $i - 1$. b_0 and b_n are the left and right border of the domain and can be ∞.

1.1 (We define $F(x) = \int T^{-1}(x)\, dx$) Compute $lf_i = F(y_i + a_i(b_i - x_i))/a_i$ and $area_i = F(y_i + a_i(b_{i+1} - x_i))/a_i - lf_i$, the area between the x-axis and the hat-function between b_i and b_{i+1}. Compute the cumulated areas $areac_i = \sum_{j=0}^{i} area_j$.

2.1 Generate two uniform random numbers U and V.

2.3 Set $U \to U * areac_{n-1}$. Let I be the index with $areac_{I-1} \leq U \leq areac_I$. ($I$ is a random deviate from the discrete distribution with probabilities proportional to $area_I$. If n is large use a guide table for acceleration). Set $U \leftarrow U - areac_{I-1} + lf_I$. U is a uniform random number in the interval $(lf_I, lf_I + area_I)$.

2.4 (Generation of a variate from the dominating distribution)
Compute $X = x_I - y_I/a_I + F^{-1}(a_I U)/a_I$
and $lx = T^{-1}(y_I + a_I(X - x_I))$.

2.5 (Squeeze acceptance)
If $X \geq x_I$ and $I < n - 1$ and $V * lx \leq T^{-1}(y_I + as_I(X - x_I))$ return X.
If $X < x_I$ and $I > 0$ and $V * lx \leq T^{-1}(y_I + as_{I-1}(X - x_I))$ return X.

2.6 (Final acceptance)
If $V * lx \leq f(x)$ return X, else go to 2.1.

It is not necessary to know the derivative of the density function since we can replace the tangent of $T(f(x))$ in the point x_i by the "pseudo-tangent" through the point $(x_i, T(f(x_i + \Delta)))$ with slope $(T(f(x_i + \Delta)) - T(f(x_i)))/\Delta$ which is always greater than $T(f(x))$ if we chose $\Delta > 0$ for x_i on the left-hand side of the mode and $\Delta < 0$ for x_i on the right-hand side.

To use Algorithm UNIV we have to chose T, n and the x_i. Our tests showed that $T = -1/\sqrt{x}$ (then $T^{-1}(x) = 1/x^2$ and $F(x) = -1/x$) is not only more general but also considerably faster than the logarithm. By taking the number of touching points n low ($n = 3$ is a good choice) or large (for example $n = 33$ or $n = 65$) we can switch between faster setup but slower marginal execution time and short marginal execution time but slow setup. The question how to select the points of contact to minimize the expected number of iterations or the expected number of evaluations of f is not so simple. For $n = 3$, f with unbounded support and $x_1 =$ the mode, Theorem 1 of [6] gives the answer: Take x_0 and x_2 such that the equation $f(x_i) = T^{-1}(T(f(x_1)) - F(T(f(x_1)))/f(x_1))$ (i.e. $f(x_i) = f(x_1)/4$ for $T(x) = -1/\sqrt{x}$) is fulfilled. x_i that are close to the solution can be easily found by numerical search. For n large the problem is more difficult: After several attempts we obtained the best results with a formula that is asymptotically optimal in both respects: Choose the location of the points in a way that

$$\frac{(\text{length of interval}) \times \sqrt[3]{T(f(x))''}}{T(f(x))} = \text{constant}.$$

This can be established by integrating $\sqrt[3]{T(f(x))''}/T(f(x))$ numerically. Then, by interpolation, the inverse function of this indefinite integral is approximately evaluated for equidistant arguments. Points chosen by this method are very close to optimal. For details see [8]. In that paper we also describe an algorithm (UNIVD) based on the same idea as UNIV but utilizing the idea of decomposition. UNIVD is more complicated than UNIV but the expected number of uniforms required is reduced almost to one.

4 Correlation induction

In many simulation experiments variance reduction can be obtained by inducing positive (common random numbers) or negative (antithetic variates) correlation between the random deviates generated (see simulation text books e.g. [3]). The highest (or lowest) correlation possible can always be obtained with the inversion method which is very slow for most distributions. In [9] and [10] it is demonstrated that it is also possible to obtain correlation induction with rejection methods. Transformed density rejection is especially well suited to install correlation induction facilities as it is so close to the inversion method. In algorithm UNIV monotonicity is already at hand and it is only necessary to establish synchronization by using two random number streams: The first one for the two random numbers necessary for the first acceptance-rejection experiment, the second stream if the first pair was rejected. As the decomposition logic destroys monotonicity in Algorithm UNIVD we changed algorithm UNIV following the guidelines of [10] but using only two random numbers from the first stream instead of four for one random deviate. The argument *ianti* (either 1 or -1) controls the orientation of the induced correlation: For positive correlation take *ianti* = 1 (or -1) for both sequences. For negative correlation take the opposite sign for the second sequence. It is necessary to have to independent streams of random numbers available, usually by starting the same generator with two different starting values.

Algorithm UNIVCI
(Add or replace the following steps in algorithm UNIV)

2.0 Generate two uniform random numbers U, V from the first stream. Go to step 2.2.

2.1 Generate two uniform random numbers U, V from the second stream.

2.2 If ianti= -1 set $U \to 1 - U$.

But there are more reasons than simplicity to implement correlation induction for transformed density rejection: In many simulation applications, especially when comparing the performance of different systems, it is important to induce correlation between random numbers of different distributions. This is best done by a universal algorithm that uses the same algorithm for both cases. A second advantage is the high acceptance probability of transformed rejection if we take n large. For $n = 33$ our experiments with several distributions showed that the correlation induced by UNIVCI is almost the same as for inversion, in no case it was more than 0.015 away. As an example Table 1 shows the negative correlation obtained with UNIVCI between beta and gamma distributions compared with the correlation induced by inversion. The strongest negative correlation varies due to the different shapes of the distributions whereas the difference of the correlations induced by the two methods remains close to 0.01.

Table 1 Negative correlation induced by UNIVCI and by inversion

	Beta			
	100,100	10,100	2,100	2,2
Gamma 2	-0.932/-0.946	-0.887/-0.897	-0.925/-0.932	-0.925/-0.932
Gamma 10	-0.979/-0.989	-0.950/-0.961	-0.883/-0.892	-0.970/-0.979
Gamma 100	-0.990/-0.999	-0.974/-0.984	-0.920/-0.934	-0.979/-0.990
Beta 2,2	-0.980/-0.992	-0.973/-0.983	-0.928/-0.937	-1.000/-1.000
Beta 2,100	-0.961/-0.971	-0.892/-0.901	-0.800/-0.805	
Beta 10,100	-0.982/-0.992	-0.955/-0.967		
Beta 100,100	-0.999/-1.000			

5 Comparison of Algorithms

To be able to judge the speed of Algorithm UNIVCI we compared our C-implementation with Algorithm UNIVD which is at least as fast as the fastest known specialized algorithms for the gamma- and beta distributions (see [8]) and with an inversion algorithm. It is based on the "regula falsi" with a large auxiliary table for good starting values. The results show that the speed of Algorithm UNIVCI is well comparable with fast specialized algorithms. For the case that correlation induction is needed the gain in speed is dramatic. UNIVCI is between 20 and 80 times faster than inversion which makes the

Table 2 Execution times in μ-seconds

	UNIVD($n=33$)	UNIVCI($n=33$)	inversion
Gamma 2	4.1	7.0	160
Gamma 10	4.1	7.0	240
Gamma 100	4.1	7.0	550
Beta 2,2	4.1	7.0	190
Beta 2,100	4.1	7.0	320
Beta 10,100	4.1	7.0	410
Beta 100,100	4.1	7.0	440

savings substantial even for simulations where only a small portion of computing time is used for random variate generation.

We are convinced that for many applications universal algorithms based on transformed density rejection have important advantages compared with standard random variate procedures. They are – coded and debugged only once – applicable to a wide range of distributions. They have good speed for all of these distributions almost independent of the speed of the evaluation of f and they have correlation induction capabilities practically equal to (very slow) inversion algorithms. We hope that this paper will therefore increase the future use of those algorithms in practice.

References

[1] J. H. Ahrens. Sampling from general distributions by suboptimal division of domains. Grazer Mathematische Berichte Nr. 319, Graz, Austria, 1993.

[2] J.H. Ahrens and K.D. Kohrt. Computer methods for sampling from largely arbitrary statistical distributions. *Computing*, 26:19–31, 1981.

[3] P. Bratley, B. L. Fox, and L. E. Schrage. *A Guide to Simulation*. Springer-Verlag, New York, 1983.

[4] L. Devroye. *Non-Uniform Random Variate Generation*. Springer-Verlag, New York, 1986.

[5] W. R. Gilks and P. Wild. Adaptive rejection sampling for gibbs sampling. *Applied Statistics*, 41:337–348, 1992.

[6] W. Hörmann. A rejection technique for sampling from t-concave distributions. *ACM Transactions on Mathematical Software*,(to appear), 1994.

[7] W. Hörmann. A universal generator for discrete log-concave distributions. *Computing*, 51, 1993.

[8] W. Hörmann and G. Derflinger. Universal sampling algorithms for a large class of continuous distributions. Technical report, Institut f. Statistik, Wirtschaftsuniversität Wien, 1994.

[9] V. Kachitvichyanukul, S. J. Cheng, and B. W. Schmeiser. Fast poisson and binomial algorithms for correlation induction. *Journal of Statistical Computation and Simulation*, 29:17–33, 1988.

[10] B. Schmeiser and V. Kachitvichyanukul. Noninverse correlation induction: guidelines for algorithm development. *Journal of Computational and Applied Mathematics*, 31:173–180, 1990.

Local Polynomial Approximation of Rapidly Time-Varying Parameters of Nonlinear Dynamics

Vladimir Y. Katkovnik

Department of Statistics, University of South Africa

ABSTRACT. The kernel approach and the local polynomial approximation (LPA) define a class of nonparametric estimators for unknown time-varying parameters of a dynamic process model.The technique is applicable to the tracking the rapidly time-varying parameters and their derivatives. The a priori information about amplitude values of the parameters and their derivatives can be incorporated in a natural way into the algorithm in order to improve the accuracy of the estimation. Simulation results for tracking a time-varying frequency are presented.

1 Introduction

The paper presents new methods for the estimation of rapidly time-varying parameters. These techniques can handle difficult problems in which there is a combination of a high speed of parameter variations and limited information about models of these variations. A heuristic approach is used to design the estimation algorithms. This approach employs the ideas of local approximation methods of nonparametric regression analysis (e.g Hardle (1990), Katkovnik (1985)). Firstly, a part of a standard series is used to approximate the time-varying parameters and secondly this expansion is exploited to calculate estimates for a single time-instant only. For the next time-instant the calculations should be repeated. This procedure determines the nonparametric character of the point-wise estimation. The localization of the estimation is ensured on the basis of the moving Least Squares Method (LSM) with a window discounting observations outside a neighbourhood of the center of the approximation.

Previous work in this direction was done for time-varying coefficients of linear dynamic models only and is described by a number of authors (Xianya and Evans (1984), Niedzwiecki (1990)). Katkovnik (1991, 1993) used the LPA to estimate both the parameter values and their derivatives and incorporate the a priori amplitude information about parameters and their derivatives.
The contribution of this paper includes an extension of the LPA to models which are nonlinear with respect to

the time-varying parameters.

2 Statement of the problem

Consider the ARX dynamic model with the unknown parameter vector $\theta(t)=(\theta_1(t),\ldots,\theta_n(t))^T \in R^n$. Suppose that input u and output y signals are observed at the sampling instants sT, where T is a sampling interval, and the sampling values are related through a given nonlinear difference equation :

$$y(sT)=F_s(\varphi(sT),\theta(sT)) + v(sT), \quad s=1,2,\ldots N, \qquad (1)$$

where $\varphi(sT)=(y((s-1)T),\ldots y((s-k)T),u((s-1)T),\ldots,$ $u((s-(M-k)T))^T \in R^M$ is a vector of lagged input-output data, and F_s is a given nonlinear function, v(sT) is a noise. Let t and sT be continuous and discrete time-arguments of time-varying functions respectively. The continuous times $\theta_i(t), i=\overline{1,n}$, and their derivatives are to be estimated from discrete measurements $\{y(sT),\varphi(sT)\}$, s=1,...,N. It is supposed that strict mathematical models of $\theta_i(t)$ do not exist and $\theta_i(t)$ are some functions of time, unavailable in analytical form. Let us introduce a vector C consisting of $\theta_i(t)$ and derivatives of $\theta_i(t)$, where the orders of the derivatives are equal to m_i-1:

$$C=(C_{(1)},\ldots,C_{(n)})^T \in R^m, \qquad C_{(i)}=(C_{i1},\ldots,C_{im_i}),$$

$$C_{ij}(t)= \theta_i^{(j-1)}(t), \quad j=\overline{1,m_i}, \quad i=\overline{1,n}, \qquad m=\sum_{i=1}^{n} m_i. \qquad (2)$$

The a priori information is given in the form of the following regularity conditions on $\theta_i(t)$:

(H1). $\theta_i(t) \in C(m_i,B_i)$, where $C(m_i,B_i)$ is a class of processes with bounded m_i- derivatives:

$$|\theta_i^{(m_i)}(t)| \leq B_i, \quad i=\overline{1,n} ;$$

(H2). $\quad C(t)\in Q \subset R^m, \quad m=\sum_{i=1}^{n} m_i,$

where Q is a convex set. For example

$$Q=\left\{C:|\theta_i^{(j)}(t)| \leq B_i^{(j)}, \quad j=\overline{0,m_i-1}, \quad i=\overline{1,n}\right\}, \text{ where the } B_i^{(j)}$$

are given. (H1), the condition of the existence of the presented estimates, is the main assumption about time-

varying parameters. (H2) is an additional assumption
which can be used to improve the estimates.

3 Algorithm

The main idea of the LPA is to apply the truncated
Taylor series in order to approximate the time–varying
parameters and to use this approximation only locally
in time. The localization results in the dependence of
the expansion coefficients on time.
    The following loss function is introduced:

$$J_t(\hat{C}) = \sum_{s=1}^{N} \rho_s \left[y(sT) - \hat{F}_s(\hat{C}) \right]^2, \quad \hat{C} \in R^m, \tag{3}$$

$$\hat{F}_s(\hat{C}) = F_s(\varphi(sT), \Psi(sT-t)\hat{C}), \tag{4}$$

$$\Psi(t) = (\psi_{ij}(t))_{n \times m} = \text{diag}\{\psi_{(1)}(t), \ldots, \psi_{(n)}(t)\}, \tag{5}$$

$$\psi_{(k)}(t) = (f_1(t), \ldots, f_{m_k}(t)) \quad , \quad f_r(t) = t^{r-1}/(r-1)!,$$

$$\rho_s = \rho((t-sT)/\delta).$$

Here $\Psi(sT-t)\hat{C}$ is the Tailor polynomial approximation of

the powers (m_i-1) for $\theta_i(sT)$, $i=1,..,n,$ with the center

of the approximation at the point t. $\rho(t)$ is a window
satisfying the customary properties of a "kernel"
nonparametric estimate. The δ is a width of the window.

We present the high–order LPA estimate of C(t) as a
solution of the constrained optimization problem:

$$\hat{C}_N(t) = \arg \min_{C \in Q} J_N(C). \tag{6}$$

The term "high–order" has a double meaning here.
Firstly, it means that the number of parameters to
estimate is larger than the original number of
parameters in question. Secondly, it can be proved that
the tracking errors are proportional to the higher order
derivatives of the estimated parameters. The aim of the
high-order polynomial approximation in (6) is first
obtaining a high accuracy of the estimation of $\theta(t)$,
with the estimates of the derivatives as a secondary
result, and finally, it enables one to use the a priori
information given in the form (H2). Different numerical
methods, both special and of general purpose, can be used
to solve the problem (3)-(6) and obtain the estimates.

For the unconstrained estimates ($Q \in R^m$) the following
version of the Gauss–Newton algorithm was applied to get

the estimates (6) with a step Δt for $t \in [t_1, \bar{t}]$.

Linearize $\hat{F}_s$ with respect to $\hat{C}$ at the point $\hat{C}^{(i)}$:

$$\hat{F}_s(C) = \hat{F}_s(\hat{C}^{(i)}) + Q(\hat{C}^{(i)})\Delta C, \quad \Delta C = C - \hat{C}^{(i)},$$

$$Q_s(\hat{C}) = \left[\frac{\partial \hat{F}_s(\hat{C})}{\partial \theta} \right]^T \Psi(sT-t)$$

and determine ΔC by minimizing J_t which is quadratic with respect to ΔC, then obtain

$$\Delta C^{(i)} = (\Phi(\hat{C}^{(i)}))^{-1} \sum_{s=1}^{N} \rho_s Q_s^T(\hat{C}^{(i)})[y_s - \hat{F}_s(\hat{C}^{(i)})],$$

$$(7)$$

$$\hat{\Phi}(\hat{C}) = \sum_{s=1}^{N} \rho_s Q_s^T(\hat{C}) Q_s(\hat{C}).$$

The Gauss-Newton algorithm starting from the point t_1 consists of the following steps:

Step 1. Initialization

$\hat{C}_N^{(1)}(t_1) = \hat{C}(t_1)$, where $\hat{C}(t_1)$ is given;

Step k includes: (i). The imbedded recursive procedure

$$\hat{C}_N^{(i+1)}(t_k) = \hat{C}_N^{(i)}(t_k) + \mu \Delta C^{(i)}, \quad i=1,2,\ldots,$$

where $\Delta \hat{C}$ is determined by (7) with the stopping rule

$$i^* = \min_{i}\{i: \|\hat{C}_N^{(i+1)}(t_k) - \hat{C}_N^{(i)}(t_k)\| \leq \varepsilon, \ i \leq \bar{i} \}.$$

$\|.\|$ is the Euclidean norm, $\varepsilon > 0$, $0 < \mu \leq 1$.

(ii). Setting $\hat{C}_N(t_k) = \hat{C}_N^{(i^*)}(t_k)$.

(iii). Initialization of the next time-instant t_{k+1}

$$\hat{C}_N(t_{k+1}) = \hat{C}_N(t_k), \quad t_{k+1} = t_k + \Delta t, \quad \Delta t > 0.$$

The procedure is completed as soon as $t_k \geq \bar{t}$.

Both (3)-(6) and the Gauss-Newton algorithm are off-line computer intensive. There are methods which allow one to get computer less intensive on-line procedures approximating the original off-line ones. However ,for the nonlinear case a switch to an on-line procedure means wasting a piece of information. We study the off-line procedures in order to find out all the potential capabilities of the local approximation approach.

The following asymptotic statement provides insight into the tracking properties of the estimates (3)-(6). Let δ, T and T/δ be small, and TN = I, where I is a length of a fixed observation interval, then N=I/T is large. Consider a particular case when (1) is a regression ($\varphi(sT) = (u((s-1)T),\ldots, u((s-M))) $), and $v(sT)$ is a white noise , Ev=0, $Ev^2 \leq \sigma^2$. Then the order of the asymptotic mean square error (MSE) is of the form

$$E(\hat{\theta}_i(t) - \theta_i(t))^2 = 0(\sigma^2 T/\delta + \delta^{2p} \|B\|^2), \quad 0 < t < I, \qquad (8)$$

$i = 1, .., n, \quad p = \min(m_i), \quad \sum B_i^2 = \|B\|^2.$

The first and second summands in (8) determine variance and bias of the estimates respectively. The choice $\delta = 0(T^{1/2p+1})$ gives an optimal trade-off between tracking properties and noise reduction, then

$$E(\hat{\theta}_i(t) - \theta_i(t))^2 = (T^{p/2p+1}) = 0(N^{-p/2p+1}). \qquad (9)$$

The details, other assumptions, and the proof of the statement are available on request from the author (Katkovnik (1993)).

4 Simulation

Here some results are presented for the frequency tracking problem which can be interpreted as a special case of the general problem of the interest. There is a vast literature on this subject. We do not touch details and use the problem only in order to show some advantages of our algorithms. Recently Huang and Hannan (1993) suggested the new frequency tracking algorithm called REI. In our simulation we make assumptions which are taken for some of the experiments in the paper mentioned above .
Let

$$y(t) = A \cos(t\omega(t)) + v(t), \quad \omega(t) = pi/4 + 0.05 + Gt, \qquad (10)$$

$G = 0.0005$,$v(t)$ is a white Gaussian noise,$N(0,1)$. The time-varying $\omega(t)$ has to be estimated from observations $y(t)$, $t = 1, \ldots, 1000$. The signal-noise relation $\log_{10}(A^2/2)$ are taken 0. The a priori information about the time-varying frequency has not been used in this case and the Gauss-Newton algorithm was applied with:

$m = 2, \quad \rho(x) = 1 - x^2, \quad |x| \le 1, \quad t_1 = \delta \,, \hat{C}_1(t_1) = \omega(t_1), C_2(t_1) = G,$

$\bar{t} = 1000$. All the results in the following table are values of the square root of the mean square error (RMSE) obtained as means over 100 replications, and various sets of t values. The comparison of the REI and LPA algorithms shows that the accuracy of the LPA algorithm is clearly higher.

5 Conclusion

The simulation study of a number of different dynamic and static models has been done. It showed that the LPA has good prospects for solving difficult problems when the tracking accuracy has to be high and the parameters are rapidly varying. Higher order polynomial models can improve the accuracy of estimation when $\theta_i(t)$ are smooth functions of time and the level of noise is not very high.

6 Acknowledgements

I am grateful to the referees and editor for their efforts. This work has been supported by the Foundation for Research Development of South Africa.

Table. RMSE for REI and LPA algorithms.

	Sets of t values		
Algorithm	t=1-100	t=101-500	t=501-1000
REI	.09168	.02137	.01763
LPA,δ=30	.00905	.00183	.00368
LPA,δ=60	.004696	.00129	.003570
LPA,δ=90	.003223	.00106	.0004798

REFERENCES

Hardle,W.,1990, *Applied nonparametric regression.* Cambridge Univ.Press.

Huang,D.,and Hannan,E.J.,1993, *Estimating Time Varying Frequency.* Submitted.

Katkovnik,V.Y.,1985, *Nonparametric Identification and Smoothing of Data (Local Approximation Method)* (Nauka, Moscow, 1985 (in Russian)).

Katkovnik,V.Y.,1991,Nonparametric identification of essentially time-varying dynamics:local approximation methods. *First IFAC Symposium on Design of Control Systems,*Zurich, Switzerland. Preprints.

Katkovnik,V.Y.,1993, High-order local approximation adaptive control of rapidly time-varying dynamics. *12th IFAC Congress,* Sydney ,Australia, 18-23 July, 2, 21-26 . Preprints.

Katkovnik,V.Y.,1993,Estimation of rapidly time-varying physical parameters,Research Report,93/2,June,UNISA.

Niedzwiecki,M.,1990, Recursive functional series modeling estimators for identification of time-varying plants - more bad news than good?'. *IEEE Trans.Automat.Contr.*, AC-35,610-616.

Xianya,X.,and Evans,R.J.,1984,Discrete time stochastic adaptive control for time-varying systems, *IEEE Trans.Aut.Control*, AC-29(7),638-640.

Backfitting and Related Procedures for Nonparametric Smoothing Regression: A Comparative View

Michael G. Schimek, Gerhard P. Neubauer, Haro Stettner

Medical Biometrics Group, University of Graz Medical Schools

Abstract. In this paper we relate backfitting (FRIEDMAN and STUETZLE, 1981) to the standard iterative procedures, i.e. Jacobi and Gauss-Seidel, developed for solving linear equation systems with non-singular system matrices. Aspects of improving the performance of these methods through relaxation are also considered. When fitting additive regression models non-parametrically by linear scatterplot smoothers, we are confronted with a singular system matrix of a certain block structure. Backfitting commonly applied (BUJA, HASTIE and TIBSHIRANI, 1989), although not designed for this situation, is examined. Because of a lack of theory behind this application and unknown convergence behaviour a simulation experiment is carried out applying cubic smoothing splines in a simple additive model. The behaviour of all three iterative procedures is studied by comparing the obtained results with those from a non-iterative Tichonow method (SCHIMEK, STETTNER and HABERL, 1992). The Jacobi iteration turns out to be superior to the other procedures when speed and precision are considered at the same time. Relaxation is of minor importance in Jacobi iteration.

Keywords. Additive model, backfitting, Gauss-Seidel, iterative solution, Jacobi, non-parametric regression, relaxation, scatterplot smoother, simulations, singular system, Tichonow regularization.

1 Iterative methods for solving linear systems

In some non-parametric smoothing regression models linear equation systems of the form $Av = b$ have to be solved in v. Direct methods like Cholesky factorization, Householder transformation or Givens transformation can be applied. Large and sparse systems, typical for regression problems, are solved via iterative methods, especially the Jacobi and the Gauss-Seidel procedure reduce the computational costs. For a discussion of all these methods see GOLUB and VAN LOAN (1989).

1.1 Jacobi and Gauss-Seidel

Equation (1) defines the so-called Jacobi procedure

$$v_i^{(m)} = [b_i - \sum_{\substack{j=1 \\ j \neq i}}^{N} a_{ij} v_j^{(m-1)}]/a_{ii}, \tag{1}$$

with $i = 1, \ldots, N$, an iteration counter m and starting values $v_i^{(0)}$ (frequently with $v_i^{(0)} = 0$). The Gauss-Seidel procedure is defined through equation (2) by

$$v_i^{(m)} = [b_i - \sum_{j=1}^{i-1} a_{ij} v_j^{(m)} - \sum_{j=i+1}^{N} a_{ij} v_j^{(m-1)}]/a_{ii}. \qquad (2)$$

Algorithmically Jacobi is known as the whole step and Gauss-Seidel as the single step procedure. For describing the difference between the two methods we introduce the term *information flow*. We understand it as the way how currently available results (estimates $v^{(m-1)}$) are used to obtain the successive results (estimates $v^{(m)}$). The Jacobi algorithm computes the estimates in iteration m only on the basis of estimates from iteration $m - 1$. All estimates from step m are stored and remain unused within this step. The information update takes place when moving from step m to $m + 1$. Hence the information flow is low (see *Figure 1*). The Gauss-Seidel algorithm takes advantage of all the information currently available, no matter where it comes from (m or $m - 1$). The information update is permanently done. As a result the information flow is high (i.e. maximal; see *Figure 1*).

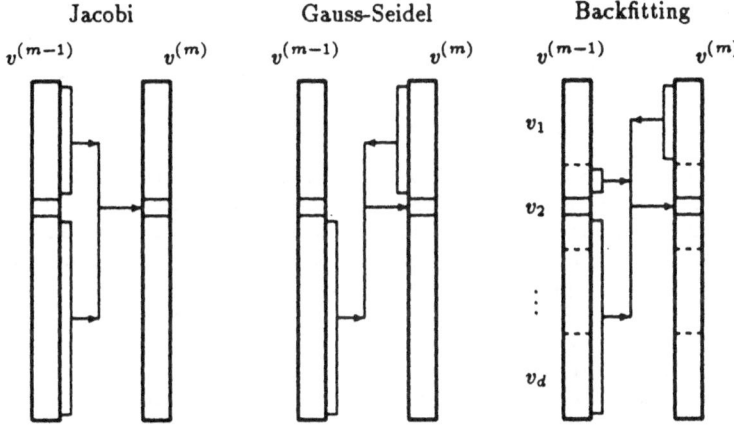

Figure 1: Information flow in iterative methods for linear systems

1.2   Backfitting

Backfitting was developed in the context of multidimensional regression problems (FRIEDMAN and STUETZLE, 1981; HASTIE and TIBSHIRANI, 1986). It is built upon the idea of determining estimates for the covariates successively in a non-parametric manner. It uses the currently available information from all covariates, except the covariate of which estimates are just computed. This leads to a splitting of the system matrix into d blocks B_j of size $N/d \times N$, where each block corresponds to one of the predictor variables X_j, $j = 1, \ldots, d$. Then the Gauss-Seidel procedure is applied to these blocks resulting in d vectors $v_j^{(l)}$ (l denoting the last iteration), each of length N/d. Information update takes place within the iterations, but with a certain lag. Therefore the information flow is higher compared to the Jacobi procedure and lower than in the Gauss-Seidel procedure (*Figure 1*). Hence, given a certain block structure (e.g. in generalized additive models; HASTIE and TIBSHIRANI, 1990), backfitting is a special case of the Gauss-Seidel algorithm.

1.3   Convergence and relaxation

All considerations concerning convergence of the above mentioned iterative procedures are made under the assumption of a non-singular system matrix A. Convergence depends on the eigenvalues λ_i of the iteration matrix (for details see HÄMMERLIN and HOFFMANN, 1992, p.372 and p.376). Acceleration techniques have been developed to improve the speed of convergence. Let us write the iteration schema

$$v^{(m)} = (I - A)v^{(m-1)} + b = v^{(m-1)} - D^{(m-1)}, \qquad (3)$$

with $D^{(m-1)} = Av^{(m-1)} - b$ the deficiency of step $m-1$. Equation (3) reveals that the procedure may be viewed in the light of correcting the m-th estimate for deficiency. Introducing the relaxation parameter ω allows to control the amount of correction

$$v^{(m)} = v^{(m-1)} - \omega D^{(m-1)}. \qquad (4)$$

For relaxed Gauss-Seidel iteration the term *successive overrelaxation (SOR)* is established (GOLUB and VAN LOAN, 1989, p.510). Rewriting (4) leads to the following relationship between the relaxed $v_{rel}^{(m)}$ and unrelaxed $v^{(m)}$ result

$$v_{rel}^{(m)} = (1 - \omega)v^{(m-1)} + \omega v^{(m)}. \qquad (5)$$

For $\omega < 1$ we have underrelaxation, for $\omega > 1$ we have overrelaxation, and for $\omega = 1$ we obtain the unrelaxed algorithm. Theoretical results about the admissible range of ω on the one hand and the optimal choice of ω on the other are available in the numerical literature. Under non-singularity the optimal choice of ω is as follows: (i) Assuming the Jacobi procedure converges, we have convergence for $0 < \omega \leq 1$. The optimal parameter is given by $\omega^* = 2/(2 - \lambda_1 - \lambda_n)$ with $1/2 < \omega^* < \infty$. (ii) Assuming the Gauss-Seidel procedure converges, convergence for *SOR* iteration can only be achieved with $0 < \omega < 2$. There is evidence that this result holds at least for special matrices e.g. symmetric positive definite matrices (see GANDER and GOLUB, 1989, p.530). The optimal choice ω^* is not known.

2   A non-iterative method for solving singular linear equation systems

An alternative non-iterative approach was introduced by SCHIMEK, STETTNER and HABERL (1992) to handle singularity problems concerning the system matrix A. They make use of a specific Tichonow regularization method. Let us again have a singular system $Av = b$ to be solved in v, and a regularization parameter α, then the disturbed system takes the form

$$(A^*A + \alpha I)\tilde{v} = Ab. \qquad (6)$$

A^* denotes the conjugate transpose of A and $\tilde{v}$ the minimum norm solution fulfilling $\min_{Av=b} \|v\|$ for an arbitrary norm $\|.\|$. This disturbed but regular system can now be solved by common techniques.

The deviation $\|v - \tilde{v}\|$ can be estimated in terms of $\sqrt{\alpha}$ and the upper bound of the measurement error $\delta \geq \|b - \tilde{b}\|$ ($\tilde{b}$ denotes the "true" value) of the right-hand side of (6). This type of error occurs in fixed design models (i.e. non-stochastic independent variables). For $\delta \to 0$ the standard result for a non-singular system matrix A is obtained.

3 The simulation study

The algorithms were implemented in GAUSS on a PC 80386/387. For the simulation study we assumed the usual additive regression model

$$Y = \sum_j f(x_j) + \epsilon$$

with a response variable Y, predictor variables x_j and an error term ϵ. The functions f were represented by cubic smoothing splines, evaluated at knots identical with the design points (predictor values). The purpose was to achieve high quality spline fits.

3.1 Outline

Taking into account the specific characteristics of the cubic spline smoother, we produced sample data in the following manner: Independent sets of design points for x_j following a uniform distribution over the interval $[0, 1]$, and $N(0, \sigma_\epsilon^2)$ distributed errors ϵ_i were created (random number generator). The response variable Y was computed based on additive functions $f(x_j)$ of our choice and an additive error term. Specifying a sample size of $n = 100$ we investigated the following two cases, a deterministic additive model M_1 with no error at all, and a stochastic additive model M_2 with $\sigma_\epsilon = 0.25$. The trend functions $f(x_j)$ are given by $f(x_1) = \sin((3\pi/2)x_1^2 + 1)/6$, $f(x_2) = \exp(-5x_2)/3$, and $f(x_3) = x_3/3$. They were selected to describe relationships between variables typical for real data.

The smoothing parameters λ_j for the individual spline fits were determined by the method of robustified generalized cross-validation (ROBINSON and MOY-EED, 1989, p.528). The purpose of this choice was reducing the risk of smoothing parameter misspecification given our functions f.

BUJA, HASTIE and TIBSHIRANI (1989, p.479 and p.493) recommend the relaxation parameter choice of $\omega^* > 1$ for Gauss-Seidel. They claim that Jacobi iteration does not converge and underrelaxation is required. Hence an appropriate range would be $0 < \omega^* < 2/\rho(P)$ with the spectral radius $\rho(P) < d$. Therefore $\omega^* = 2/d$ could serve as a conservative choice. GANDER and GOLUB (1989, p.530f) report that for $d = 2$ an optimal ω^* value can be calculated depending upon the eigenvalues of $S_1 S_2$, where S_1 and S_2 are linear smoother matrices. Concerning SOR and $d > 2$ there is no classical theory for ω^* determination. For our simulations we considered increasing, admissible values of ω. The number of iterations required for convergence was our criterion to specify an adequate region of ω^*.

Technical considerations lead to an optimal regularization parameter value α in the magnitude of δ^2. Knowing the measurement error of the right-hand side of equation (6), we decided for $\alpha = 10^{-3}$.

We used graphical displays, the iteration number and the error sum of squares

$$ESS = \sum_i (Y_i - \epsilon_i - \hat{Y}_i)^2 = \sum_i (\sum_j f(x_{ij}) - \sum_j \hat{f}(x_{ij}))^2$$

to compare the behaviour of the numerical procedures.

3.2 Results

Table 1 gives details from the simulations. The precision of the estimates is expressed through *ESS* and the speed of the iterative procedures is documented by the number of iterations. As a direct method for solving linear systems the Tichonov procedure is of high precision but usually slower than iterative alternatives. Hence, only the iterative methods are compared in speed.

	M₁		M₂		M₁		M₂	
	ESS	Iter	ESS	Iter	ESS	Iter	ESS	Iter
	Tichonov				Tichonov			
	0.062	–	0.676	–	0.062	–	0.676	–
	Jacobi				Gauss-Seidel			
Relaxation	0.062	4	0.672	4	0.202	21	0.792	38
$\omega = 1.5$	no conv.		no conv.		no convergence			
$\omega = 1.4$	no conv.		0.672	6	no convergence			
$\omega = 1.3$	no conv.		0.672	5	no convergence			
$\omega = 1.2$	0.067	4	0.672	5	no convergence			
$\omega = 1.1$	0.067	4	0.672	4	no convergence			
$\omega = 0.9$	0.062	4	0.672	4	0.348	8	0.722	10
$\omega = 0.8$	0.067	4	0.672	4	0.314	6	0.722	6
$\omega = 0.7$	0.067	5	0.672	5	0.314	4	0.722	5
$\omega = 2/3$	0.067	5	0.672	5	0.314	4	0.722	5
$\omega = 0.6$	0.073	5	0.672	5	0.314	5	0.722	5
$\omega = 0.5$	0.084	5	0.672	6	0.314	4	0.722	5
$\omega = 0.3$	0.144	6	0.689	6	0.384	5	0.722	6
$\omega = 0.1$	1.538	2	0.980	2	1.562	2	0.980	2

Table 1: Comparison of Gauss-Seidel and Jacobi iterations to the Tichonov regularization: Results (*ESS* and number of iterations) from the simulations for models M₁ and M₂.

Table 2 summarizes the main results. We see that the convergence behaviour of the methods may deviate from the theoretical results in the literature obtained for regular systems. Some recommendations concerning relaxation (BUJA, HASTIE and TIBSHIRANI, 1989) do not find support in our simulations. Relaxation is less important than expected (almost no effect on Jacobi iteration; Gauss-Seidel iteration speeded up, but loss in precision). Jacobi does not converge for very large values of ω. Very small ω produce fast convergence towards a bad solution for Jacobi as well as Gauss-Seidel. For more simulation results including graphical displays we refer to SCHIMEK, NEUBAUER and STETTNER (1994).

In conclusion, the unrelaxed Jacobi iteration performs as precise as the Tichonow method and as fast as the optimally relaxed Gauss-Seidel iteration. These features are maintained over a wide range of ω values so that relaxation has little impact on Jacobi. On the other hand the unrelaxed Gauss-Seidel iteration requires 5 to 10 times more iterations to converge to less precise results. Relaxation is important for speeding the procedure, but obviously precision is not improved.

Therefore we recommend unrelaxed Jacobi iteration as standard procedure. Relaxation should only be adopted when necessary.

Procedure	Convergence Lit.	Convergence Simul.	Iter. speed	ESS compared to Tichonow
Standard				
Jacobi	no	yes (a)	fast	M_1, M_2: equal
Gauss-Seidel	yes	yes	slow	M_1, M_2: larger
Underrelaxation				
Jacobi	yes	yes	equal	M_1, M_2: equal
Gauss-Seidel	no	yes (a)	faster	M_1: larger, M_2: smaller
Overrelaxation				
Jacobi	no	yes (b)	equal	M_1, M_2: equal
Gauss-Seidel	yes	no (a)	–	–

Table 2: Summary and interpretation of the simulation results.
Legend: Lit.: Literature; Simul.: Simulations; Iter.: Iteration. Results in contradiction to (a) BUJA, HASTIE and TIBSHIRANI (1989), (b) standard results for regular systems.

4 References

BUJA, A., HASTIE, T.J. and TIBSHIRANI, R.J. (1989). Linear smoothers and additive models. *The Annals of Statistics, 17*, pp. 453-510.

FRIEDMAN, J.H. and STUETZLE, W. (1981). Projection pursuit regression. *Journal of the American Statistical Association, 76*, pp. 817-823.

GANDER, W. and GOLUB, G.H. (1989). Discussion of "BUJA, A., HASTIE, T.J. and TIBSHIRANI, R.J. (1989). Linear smoothers and additive models." *The Annals of Statistics, 17*, pp. 529-532.

GOLUB, G.H. and VAN LOAN, C.F. (1989). *Matrix computations.* John Hopkins University Press, Baltimore.

HÄMMERLIN, G. and HOFFMANN, K.-H. (1992). *Numerische Mathematik.* Springer, Berlin.

HASTIE, T.J. and TIBSHIRANI, R.J. (1986). Generalized additive models. *Statistical Science, 1*, pp. 297-318.

HASTIE, T.J. and TIBSHIRANI, R.J. (1990). *Generalized additive models.* Chapman and Hall, London.

ROBINSON, T. and MOYEED, R. (1989). Making robust the cross-validatory choice of smoothing parameter in spline regression. *Communications in Statistics. Theory and Methods, 18*, pp.523-539.

SCHIMEK, M.G., NEUBAUER, G.P. and STETTNER, H. (1994). Backfitting and related procedures for non-parametric smoothing regression in competition. Discussion Paper, Institut für Statistik und Ökonometrie, Humboldt Universität zu Berlin.

SCHIMEK, M.G., STETTNER, H. and HABERL, J. (1992). An operator method for backfitting with smoothing splines in additive models. In DODGE,Y. and WHITTAKER, J. (eds.). *Computational Statistics. Volume 1.* Physica, Heidelberg, pp. 487-491.

Acknowledgement. Research supported by Austrian Science Research Fund grant P8153-PHY.

Part III

Parallel Statistical Computing

Parallelism in Computational Statistics

Herman J. Adèr

Faculty of Medicine, Dept. of Epidemiology and Biostatistics, Vrije Universiteit Amsterdam

Abstract. A parallel implementation of Havránek's search algorithm to find an optimal model in a large model space, is described. Parallel processing is also used to built multi-layered artificial neural networks (ANNs). Both applications are realized using an empty shell that allows for the definition of parallel processes.

We conclude that parallel- and co-processing offer a very natural way to conduct data analytic tasks.

Keywords. Parallelism, Model Search, Neural Network, Empty Shell

1   Introduction

Applications of concurrency and parallelism have barely been studied in relation to Computational Statistics. This, no doubt, is largely due to unavailability of appropriate software for Personal Computers. Furthermore, parallel implementation is often considered only a numerical trick to increase computing performance. Intuitively, however, the idea of interacting tasks, sometimes waiting for each other until necessary information can be shared, sometimes running in parallel, seems a very natural one.

Several kinds of parallelism have been studied (c.f.Wilmarth (1993) for a short overview). In fact there is a variety of models offering possibilities for different applications. Needless to say that implementations are mainly on workstations, mini-computers or main-frames. For example, *Single Instruction, Multiple Data* (SIMD) models are used to built Vectorization Machines. *Multiple Instruction, Multiple Data* (MIMD) models allow for programmable parallelism in which different tasks are assigned to different processors. The (infrequent) global data access is strictly scheduled in MIMD models. In *Massive Parallel Processors* (MPPs) each processor is a complete CPU, and many (typically: more than 100) processors function in parallel.

The emphasis in this paper is on the more general, programmable forms of parallelism.

Since in recent years I have been building and improving an empty shell that allows for the formulation of parallel computation, it seemed attractive to investigate what areas of computational statistics are suitable for the application of parallelism, and then try to built an implementation using this shell.

I will start with a concise description of the structure of PROTOSHELL knowledge bases and their semantics in particular in relation to parallelism. Havránek's proposal to apply a parallel computer architecture to find an optimal model in a large model space will be the subject of section 3. Parallel implementations of Artificial Neural Networks (ANN, for short) will be considered in section 4.

2 Priority Structures

For a general description of PROTOSHELL I refer to Adèr (1992). The rule bases of the shell consist in several *Steps* that may have different functionality. The *Priority Structure* (PS for short) that is described below corresponds to one Step of the shell's rule base.

We start by defining several (finite) sets:

(A) A set Ta of *Tags*. Elements of Ta correspond to names of objects in the rule base;

(B) A set Co of *Conditionals*, or attributes. These are mappings from Ta to the set {true, false};

(C) A set Ac of *Actions*, mappings of Ta to a set of executable operations.

(D) A set Dy of *Dynamic Actions*. These are formulae indicating that changes should be made to the PS during runtime. In section 2.1 an overview over the instructions in Dy is given.

(E) A set Ru of *Condition-Action Rules*. These are mappings $Co \rightarrow Ac \cup Dy$. During runtime an element from $r \in Ru$ of which the Conditional maps on true has the effect that the action part is executed.

(F) A set Im of *Implications*. These are mappings of Co in Co. The effect of such mapping is very close to the usual meaning given to implication.

We define $\mathcal{RI} = Ru \cup Im$. Implications and Condition-Action Rules are indicated as follows:

$$(x|y), \qquad \text{or, if x then y.}$$

A *Prio* p is a structure $p = (\tau, \pi_p, R_p, \gamma)$, in which $\tau \in Ta$ is the name of the Prio, π_p is a whole number called the *priority*, $R_p \subset \mathcal{RI}$ is a row of rules and/or implications and $\gamma \in Co$ is a Conditional indicating whether the elements of R_p have to be processed sequentially or in parallel. Prio's that have $\pi_p \leq 0$ are not executed.

A *Priority Structure*

$$\Pi = (\bigcup_{i=1}^{n} p_i, \ell) \quad \cdot \tag{1}$$

is defined as a row of Prio's and a (real-valued) *priority pointer* ℓ. When the runtime system of the shell is started ℓ has the value $\max_{i=1}^{n} \pi_i$. During runtime Prio's with the higher priority are executed before Prio's with lower priority. Elements with the same priority are executed in parallel. When all Prio's with a priority that matches ℓ have been processed, ℓ is decreased by one. Note that a PS allows for parallel processing at two levels: (a) rules of a Prio with $\gamma = $ `parallel` are processed in parallel, and (b) Prio's that have the same priority are processed in parallel.

2.1 Dynamic Actions

Formulae in Dy can indicate the following operations on the PS:

— Give another value to the Priority Pointer ℓ. This results in a change of flow of control: Prio's that have already been processed, may be processed once more if they have matching priority.

— Change a Priority π_i of an element of Π.

— Give another value to the Condition of a rule in a Prio of Π, for instance change `true` into `false` or vice versa.

— Add a rule to one of the Prio's of Π. The text of the rule is given as a parameter to the formula.

Furthermore, Ac contains an instruction that reads a rule from an external file and processes it. Since this can also be a dynamic formula, changes in Π can be generated in a separate program and then included in the rule-base during runtime. Several Prio's of Π may be initially empty. They only start functioning when they have been filled in during runtime.

3 Havránek's Parallel Model Search Algorithm

In Edwards and Havránek (1987) an ingenious algorithm is described, that searches a large model space to find a subset that is optimal in some pre-defined way. I start with a condensed description, and then highlight some points in the implementation.

Consider a model space $\mathcal{M}$ with a partial weak ordering $\prec$. We look for a partition $\mathcal{M} = A \cup R, A \cap R = \emptyset$, in which A is the set of accepted models, R the set of rejected models. Edwards and Havránek assume *coherence* of $\mathcal{M}$. This means that for models $m_1, m_2 \in \mathcal{M}$:

$$m_1 \prec m_2, m_1 \in A \Rightarrow m_2 \in A, \qquad m_1 \prec m_2, m_2 \in R \Rightarrow m_1 \in R \tag{2}$$

In other words: if a model m is accepted, then models that include m are accepted, and if a model m is rejected, models that are included in m are rejected[1]. For each $S \subset \mathcal{M}$ a *maximum*-set and a *minimal*-set is defined:

$$\max(S) = \{s \in S | s \prec t \Rightarrow t \notin S\}, \qquad \min(S) = \{s \in S | t \prec s \Rightarrow t \notin S\} \quad (3)$$

The algorithm uses the concepts of *a-dual* and *r-dual* of a set $S \in \mathcal{M}$. The a-dual of S(notation: $D_a(S)$) contains the models in $\mathcal{M}$ that are not smaller than any model of S. If S includes the rejected models, $D_a(S)$ holds the simplest models that conceivably may be accepted. A similar definition is given of $D_r(S)$. If S contains the accepted models S, then $D_r(S)$ includes the most complicated models that may conceivably be rejected. Edwards and Havránek prove that during the iterative construction of A and R, the set T of models to be chosen from, obeys:

$$\max(T) = D_r(A) \setminus R, \qquad \min(T) = D_a(R) \setminus A \quad (4)$$

This property can be used to construct a set T of models for the next iteration. The construction of $D_r(A) \setminus R$ and $D_a(R) \setminus A$ is particularly simple, if $\mathcal{M}$ is a (preferably distributive) lattice.

Havránek (1992) considers whether the use of parallel processing is appropriate for the implementation of this algorithm. In his analysis, he considers four levels: (a) *Symbolic planning level*, in which the heuristic translation from application domain to model space takes place. Here the initial set is chosen. Afterwards, evaluation of the meaning of the acceptance of models in terms of the application domain takes place at this level. (b) *Symbolic manipulation*. At this level models are assigned to A or R, using a symbolic representation. (c) *Model evaluation*. Here Goodness-of-Fit tests are conducted on the models in T. (d) *Numerical* and *Data manipulation*. This is the level of the basic numerical operations.

Parallel processing can be used at different levels. Level (a) requires, as Havránek puts it, 'tools from Artificial Intelligence'.

3.1 Implementation

In the implementation, at level (a) an appropriate Goodness-of-Fit test statistic is chosen and the initial set of models is constructed. The level of knowledge formalization of the application domain determines the need for interaction with the user. Dialogues are implemented using Prio's that run alternatingly.

The updating of A and R at level (b) is implemented by two Prio's that are activated dependent on the size of $D_r(A) \setminus R$ and $D_a(R) \setminus A$: the smallest is chosen. At level (c) models in T are tested. First, for each model a copy of

---

[1]Edwards and Havránek in some cases weaken this concept so that violations of the ordering are accepted.

the relevant part of the data-set is taken. For each model a rule is included in which a Goodness-of-Fit test is done in the action part. The result is a Conditional indicating if the model is accepted or rejected. Thus,

$$R_p = \bigcup_{i=1}^{k} (\text{model } m_i | (\texttt{GOF}(m_i) \rightarrow \{\texttt{accepted},\texttt{rejected}\})) \qquad (5)$$

The rules are processed in parallel. Obviously, the procedure GOF must obey coherence. It differs from one technique to another. In our implementation of parallel multiple regression analysis, vectorization techniques are used to compute GOF.

Since the procedure GOF is external to the shell, and only used in one Prio, the rule base for regression analysis can easily be adapted to another technique and the related model space.

4   Neural Networks

Literature on this subject is extensive. The usefulness of Artificial Neural Networks (ANNs) for classification problems is well known. For other applications in the field of statistics see Murtagh (1992).

Parallel processing may be fruitfully used with the implementation of ANNs. Since this a common form of ANNs, we only discuss *multi-layered percep-trons* that are trained using supervised learning. These network is used in two stages: (1) a *training* stage and (2) a *test* stage. In the first stage the input consists in input and output patterns I and O. Weight matrices are built that are optimal in the sense that the network produces with each I the related O with minimal error. In the second stage input is only patterns I.

Both the training stage and the test stage allow for parallel processing, since computation of new weights is conducted per layer, nodewise.

The present implementation uses a separate parallel (back-propagation) algorithm to train the network. This is called in the first Prio of the Priority structure. With the help of the resulting weight matrices the network is generated, using dynamic formulae. The input layer is represented by a (parallel) Prio that activates hidden nodes, represented as Prio's. Likewise, each output node corresponds to a Prio. Activation levels are represented by the priority of the node (thus, the threshold is 0). Again, using dynamic formulae, nodes of a subsequent layers may be activated or inhibited.

Parallel activation is possible for the rules in the Prio that represents the input layer and for the rules in the hidden and output-layer.

5   Conclusions

We considered two applications of different models for parallel processing in Computational Statistics. As a programming tool we used an empty shell which allows to define parallel processes.

Parallelism and co-processing offer a very natural way to describe data analytic tasks, even on a higher conceptual level. Data analysis can correspond to that by using, apart from vectorized processing, more complex forms of parallelism in which each processor has a varied repertoire of operations.

Havránek's proposal to use parallelism in a model search application seems generally applicable. Implementation is relatively simple. Different forms of parallel processing may be used at different levels.

Parallel artificial neural networks may be built using the same shell. In the training stage vectorization is used. During testing, when nodes may contain more complicated operations, MIMD modeling may be indicated.

Hybrid systems in which neural networks and rule based systems are combined can be easily built with PROTOSHELL. Most proposals use the rule base to control the neural net. The shell also allows for the alternative, in which the nodes of the neural network contain a complete rule base.

References

Adèr, H. J. (1992). Protoshell: An empty shell to develop statistical knowledge based systems. In Koening, S. (Ed.), *Computational Statistics. Proceedings of the 10th Symposium on Computational Statistics COMPSTAT. Neuchâtel, Switserland, 24-28 augustus 1992*, Vol. 3, pp. 4–15 Neuchâtel. PAN Presses Académiques.

Edwards, D., & Havránek, T. (1987). A fast model selection procedure for large families of models. *Journal of the American Statstitical Association, 82*(397), 205–213.

Havránek, T. (1992). Parallelization and symbolic computation techniques in model search. In Faulbaum, F. (Ed.), *SoftStat '91. Advances in Statistical Software. The 6th Conference on the Scientific Use of Statistical Software. April 7-12, 1991, Heidelberg*, pp. 219–227. Gustav Fischer Verlag, Stuttgart-Jena-New York.

Murtagh, F. (1992). Neural networks for statistics. In Weichselberger, K., & Klösgen, W. (Eds.), *Preprints of NTTS Conference. Conference on New Techniques and Technologies for Statistics. Feb. 24-26, 1992* Bonn, Germany.

Wilmarth, D. D. (1993). The compute cluster and other parallel programming models. *Computer*, 70–72.

Part IV

Selection Procedures

Construction of Decision Procedures for Selecting Populations

Guido Giani

Abteilung Biometrie und Epidemiologie, Diabetes-Forschungsinstitut an der Universität Düsseldorf

Abstract. Selection procedures are considered in a significance testing point of view. Recent results on the equivalence between selection procedures and multiple tests for a certain class of hypotheses are reviewed and then first applied to construct selection rules for choosing the best population. When under design aspects certain power requirements are incorporated, the classical indifference zone approach of Bechhofer as well as Gupta's subset selection formulation can be shown to fit into this general framework. The equivalence results are moreover demonstrated to provide efficient stepwise procedures for the problem of selecting a subset of populations that contains all good ones. Treating this latter problem by imposing the additional requirement to select none of the bad populations leads to partitioning rules. Selection rules for the one-way layout under a normal distributional set-up are dealt with in detail, and the resulting numerical problems are discussed.

1 Introduction

There are many practical situations in which several populations (or treatments) are being compared with the objective of making a selection among them. In the simplest case the underlying selection goal might be to find the best population. Examples are the search for the most efficient drug dosis, the therapy with the largest success probability, the seed in an agricultural experiment with the largest yield, and so on. In screening experiments the selection of a (random) minimal subset of all populations that contains the best one might be of interest. A more complex goal in this context would be to select a subset that contains all good populations but none of the bad ones, where the properties *good* and *bad* clearly have to be well-defined in advance.

This article concerns two important questions. The first is how to construct selection rules for the mentioned selection goals which meet a correct selection (CS) with preassigned confidence probability. The second question deals with design aspects in planning experiments, when certain additional power requirements are included in the selection formulation that in conditions on the necessary sample sizes. The answer to both questions leads to the numerical task of solving more or less complex integral equations.

In view of the first question a close relationship between selection and multiple testing is given in Section 2. Consideration here is limited to the problem of selecting a subset of all populations that contains all good ones. As shown under mild conditions in Finner & Giani (1994a), there exists a bijective mapping from the class of selection procedures, which guarantee a prespecified level P^* for the probability of correct selection (PCS), onto the subclass of all coherent and consonant multiple tests at multiple level $1 - P^*$ for a family of hypotheses being induced by the selection goal in a very natural way. For the sake of simplicity, in Section 3 as well as in all following sections attention is restricted to continuous distributions with location/scale parameters. The duality between selecting and multiple testing is applied to construct a selection procedure for finding the best population at a confidence level P^*. The resulting procedure is the so-called best-or-all selection rule having previously been studied by Stefansson et al. (1988), among others. When imposing a certain power requirement on this procedure, under the normal distributional set-up one obtains the classical indifference zone approach of Bechhofer (1954) as a special case. It is furthermore demonstrated how Gupta's (1956, 1965) subset selection approach, that is known in the literature as the second main approach (besides that of Bechhofer (1954)) to the formulations of selection problems, also fits into this framework. In Section 4 some results of Giani & Finner (1991) and Finner & Giani (1994a) are reviewed providing stepwise subset selection rules for the goal treated in Section 1. Under the normal distribution these stepwise procedures turn out to select no larger subsets than their classical one-step counterparts. Therefore, the stepwise principle, which was introduced to the selection theory by Naik (1975, 1977) and adapted by Broström (1981) to a special case of the selection problem at hand, leads always to better procedures with smaller expected number of selected populations. In Section 5, besides the aim of subset selection of all good populations, the additional exclusion of all bad ones is required simultaneously at high prespecified confidence level. It is clear that such an additional requirement can only be satisfied via a well-planned experiment with sufficiently large sample sizes. Tong (1969) considered this specific selection problem in the case when good and bad populations are defined by means of comparisons with a specified control. If the reference population is the unknown best, Lam (1986) gave some significant contributions under the normal distributional model with known variance. In this article his procedure is generalized to the unknown variance case. Finally, Section 6 is dedicated to some numerical aspects when determining sample sizes as well as calculating critical values being necessary to carry out the various selection procedures.

2 Selecting and Multiple Testing

Let $I = \{1, \ldots, k\}$ be a finite index set whose elements one can imagine as the labels of $k \geq 2$ populations under consideration, and let $X = (X_1, \ldots, X_k)$

denote a related sampling statistic with values in $I\!\!R^k$. The distribution of X is assumed to be induced by a probability measure from the family $\{P_\vartheta : \vartheta \in \Theta\}$ possessing Θ as parameter space. Good populations are defined in terms of a function $G : \Theta \to \{J : J \subseteq I\}$. For $\Theta \subseteq I\!\!R^k$ important examples are

$$G(\vartheta) = \left\{ i \in I : \max_{j \in I} \vartheta_j - \vartheta_i \leq \epsilon \right\}, \quad \vartheta \in \Theta \tag{2.1}$$

and

$$G(\vartheta) = \left\{ i \in I \setminus \{1\} : \vartheta_i - \vartheta_1 \geq \epsilon \right\}, \quad \vartheta \in \Theta,$$

where in both cases $\epsilon \geq 0$ is a prespecified threshold value. In both these settings the quality of a population is regarded as the higher the larger its associates parameter value is. In the first example good populations do not differ too much with respect to their parameter values, i.e., by at most ϵ, from any of the best populations being associated with the largest parameter value. In the second example the parameter values of all populations are being compared with a control possessing label 1.

The primary objective in this section is to investigate the structure of decision rules which select a subset $S(X) \subseteq I$ of all populations containing all good ones with at least the preassigned confidence probability $P^* \in (0, 1)$, regardless of the true parameter vector. In addition, it is desirable to have the selected subset $S(X)$ as small as possible. Precisely, we are interested in rules $S(X)$ which satisfy the probability requirement

$$\forall \vartheta \in \Theta : \quad P_\vartheta \{S(X) \supseteq G(\vartheta)\} \geq P^* . \tag{2.2}$$

Any such procedure is called to meet a correct selection (CS) at a probability of CS-level P^* or, briefly, at PCS-level P^*.

To get an idea how to construct efficient rules satisfying (2.2), it is useful to bring into plan the connection with a special multiple hypotheses testing problem. For this purpose, set $\mathcal{P} = \{J : \emptyset \neq J \subseteq I\}$, and let us consider the family of hypotheses $\mathcal{H} = \{H_J \in \mathcal{P}\}$ defined by the test problems

$$H_J : J \subseteq G(\vartheta) \quad \text{versus} \quad K_J : J \cap G(\vartheta) \neq J , \quad J \in \mathcal{P} .$$

Furthermore, let A_J be a measurable subset of the domain of X denoting the region of acceptance of a nonrandomized test for H_J. Then, the multiple test for $\mathcal{H}$ with the corresponding set $A_{\mathcal{H}} = \{A_J : J \in \mathcal{P}\}$ of regions of acceptance is said to control the multiple level (of significance) $\alpha \in (0, 1)$ if

$$\forall \vartheta \in \{\vartheta' \in \Theta : G(\vartheta') \neq \emptyset\} : \quad P_\vartheta \left(\bigcap_{J \in \mathcal{P} : \vartheta \in H_J} \{X \in A_J\} \right) \geq 1 - \alpha . \tag{2.3}$$

In other words, the multiple level α is controlled by a multiple test whenever the probability of a rejection of one of the true hypotheses is at most α.

The following results of Finner & Giani (1994a) describe the duality between selecting and multiple testing. Starting out from a multiple level-$(1 - P^*)$ test for $\mathcal{H}$ with $A_{\mathcal{H}}$ as set of regions of acceptance, one obtains a selection procedure $S(X)$ at PCS-level P^* from

$$\forall x \in \mathbb{R}^k : \ S(x) = \bigcup_{J \in \mathcal{P} : x \in A_J} J \ . \tag{2.4}$$

Vice versa, if $S(X)$ has PCS-level P^*, then the multiple test with regions of acceptance given by

$$A_{\mathcal{H}} = \left\{ A_J : A_J = \{ x \in \mathbb{R}^k : J \subseteq S(x) \}, J \in \mathcal{P} \right\}$$

controls the multiple level $1 - P^*$. Noticing that this latter multiple test is coherent and consonant, it moreover turns out that these just described functional relationships provide a bijective mapping from the class of selection procedures at PCS-level P^* onto the class of all coherent and consonant multiple level-$(1 - P^*)$ tests for $\mathcal{H}$. The criteria *coherent* and *consonant* used in this equivalence statement were introduced by Gabriel (1969) to characterize multiple tests whose decision pattern does not possess certain logical inconsistencies which generally are typical of multiple tests. Coherence requires rejection of a hypothesis if all hypotheses implying it are also rejected, whereas consonance means that, whenever a hypothesis is rejected, at least one (if exists) of the hypotheses implied by it is also rejected. From the unsymmetry of interpretation of results from statistical significance testing it is clear that coherence is more desirable than consonance.

To construct a selection procedure subject to (2.2) applying (2.4), it nevertheless remains the task to find a (coherent) multiple test for $\mathcal{H}$ which controls the multiple level $1 - P^*$. The so-called closure principle provides one possible way to reach this. It can be applied whenever the underlying family of hypotheses is closed under intersections what is obvious for the present situation because of

$$\forall J, K \in \mathcal{P} : \ H_J \cap H_K = H_{J \cup K} \ . \tag{2.5}$$

The closure principle states that if, for each $J \in \mathcal{P}$, the set $\tilde{A}_J$ is region of acceptance for testing H_J at level $1 - P^*$, then the multiple test with corresonding class $A_{\mathcal{H}} = \{ A_J : J \in \mathcal{P} \}$ of regions of acceptance defined by

$$A_J = \bigcup_{K \supseteq J} \tilde{A}_K , \quad J \in \mathcal{P} \tag{2.6}$$

hp is coherent and controls the multiple level $1 - P^*$.

3 On Selecting the Best Population

The distributional model in this section refers to independent real-valued random variables $X_1, \ldots, X_k$, where X_i, $i \in I$, has a cumulative distribution function $F(x_i - \vartheta_i)$ with Lebesgue density $f(x_i - \vartheta_i)$, $\vartheta_i \in \mathbb{R}$. Let

$\vartheta_{(1)} \leq \ldots \leq \vartheta_{(k)}$ and $x_{[1]} \leq \ldots \leq x_{[k]}$ denote the ordered components of $\vartheta \in \mathbb{R}^k$ and of the observation $X = x$, respectively. Furthermore, throughout this section let the function G be defined as in (2.1) but for $\epsilon = 0$, i.e., $G(\vartheta) = \{i \in I : \vartheta_i = \vartheta_{(k)}\}$ for $\vartheta \in \mathbb{R}^k$. Following the way of speaking introduced in Section 2, $G(\vartheta)$ consists of just the best population(s).

We now apply the results of Section 2 to construct a so-called one-or-all selection rule $S(X)$ that satisfies (2.2) subject to $|S(X)| \in \{1, k\}$. If for such a procedure $|S(X)| = 1$, then $S(X)$ contains the unique best population with at least probability P^*, whereas the alternative decision $S(X) = k$ provides no usable information and should be interpreted as not enough evidence being available in the data to infer some population as the unique best. For each $i \in I$, let us now consider the ad hoc test for $H_{\{i\}}$ given by the region of acceptance

$$A_{\{i\}}(c) \equiv A_{\{i\}} = \left\{ x \in \mathbb{R}^k : \max_{j \in I \setminus \{i\}} x_j \leq x_{[k-1]} + c \right\},$$

where $c \geq 0$. In view of (2.5), the sets $A_{\{i\}}, i \in I$, induce regions of acceptance for the remaining hypotheses $H_J \in \mathcal{H}$ with $|J| \geq 2$ via

$$A_J(c) \equiv A_J = \bigcap_{j \in J} A_{\{j\}} = \left\{ x \in \mathbb{R}^k : x_{[k]} - x_{[k-1]} \leq c \right\}.$$

Hence, if we determine the critical value c such that

$$\forall J \in \mathcal{P}: \quad \inf_{\vartheta \in H_J} P_\vartheta \{X \in A_J(c)\} \geq 1 - \alpha, \tag{3.1}$$

it follows from the obvious equivalence between (2.3) and (3.1) that $A_\mathcal{H}$ forms the class of regions of acceptance of a multiple level-$(1 - \alpha)$ test for $\mathcal{H}$. The following theorem, which under a more general multiple testing approach to subset selection has also been proved in Finner & Giani (1994b), shows how the critical value c can be calculated.

THEOREM 3.1. *If f is log-concave, then the solution $c \geq 0$ of the equation $P_{(0,0)}\{|X_1 - X_2| \leq c\} = 1 - \alpha$ satisfies (3.1).*

Proof. If $|J| \geq 2$, the assertion (3.1) immediately follows from a lemma of Anderson et al. (1977) (also see Gutman & Maymin (1987) or Kim (1986)) showing that $P_\vartheta \{X_{[k]} - X_{[k-1]} \leq c\}$ is nondecreasing in $\vartheta_{(1)}$. On the other hand, if J contains exactly one element i, say, then consider any $\vartheta \in H_{\{i\}}$ and define a vector $\tilde{\vartheta}$ with components $\tilde{\vartheta}_i = \vartheta_{(k)} - \delta$ and $\tilde{\vartheta}_j = \vartheta_j$ for $j \in I \setminus \{i\}$, where $\delta = \vartheta_{(k)} - \vartheta_{(k-1)}$. From the fact that $x + \tilde{\vartheta} \in A_{\{i\}}(c)$ implies $x + \vartheta \in A_{\{i\}}(c)$ we obtain the first inequality and, from the monotonicity lemma mentioned just before, the third inequality in the expression

$$P_\vartheta\{X \in A_{\{i\}}(c)\} \geq P_{\widetilde{\vartheta}}\{X \in A_{\{i\}}(c)\} \geq P_{\widetilde{\vartheta}}\{X_{[k]} - X_{[k-1]} \leq c\}$$
$$\geq P_{(0,0)}\{|X_1 - X_2| \leq c\} = P^* .$$

Hence the proof is complete. □

With $c = c(P^*)$ determined according to Theorem 3.1 for given $P^* \equiv 1-\alpha$, application of (2.4) now leads to the one-or-all selection rule

$$S(X) = \begin{cases} \{[k]\} & \text{if } X_{[k]} - X_{[k-1]} > c(P^*) \\ I & \text{otherwise ,} \end{cases} \tag{3.2}$$

which meets a CS at confidence level P^*. It is an astonishing result that, whenever the primary interest lies in the search for the best population at a PCS-level P^*, the final decision can be based on only the comparison between the two largest observations. The disadvantage of getting no information in the case of no evidently best population can be partly compensated for by suitable power requirements. One reasonable formulation is to require that, over a prespecified subset Ω_{Pr} of the parameter space where there is strong preference to pick up the unique best population, the one-or-all selection rule $S(X)$ should not decrease the probability $\widetilde{P}^* \in (0,1)$, say, of correctly choosing this unique best population. Precisely, the power requirement should read

$$\forall \vartheta \in \Omega_{\mathrm{Pr}} : \ P_\vartheta\{S(X) \supseteq G(\vartheta) \ \underline{\text{and}} \ |S(X)| = 1\} \geq \widetilde{P}^* . \tag{3.3}$$

In a sampling situation both requirements (2.2) and (3.3) together clearly turn out to be a condition on the sample sizes. This shall be demonstrated in the next section under a normally distributional set-up.

3.1 Designing under the Indifference Zone Approach

Let $\overline{X}_i \sim N(\mu_i, \sigma^2/n)$, $i \in I$, be k independent sample means from normal distributions with common unknown variance $\sigma^2 > 0$. Furthermore, we assume that $\hat{\sigma}/\sigma$ has the cumulative distribution function Q_ν of a $\nu(n)^{-1/2}\chi_{\nu(n)}$ variable, with $\chi^2_{\nu(n)}$ possessing a chi-squared distribution with $\nu(n)$ degrees of freedom depending generally on the sample size n. Under these distributional assumptions, it is natural to base the selection rule (3.2) on the statistic $X \equiv n^{1/2}\overline{X}/\hat{\sigma}$. Because the distribution of this statistic depends only on the standardized mean vector $\vartheta \equiv n^{1/2}(\mu_1/\sigma, \ldots, \mu_k/\sigma) \in \mathbb{R}^k$, it is convenient to assess the quality of each population in terms of σ-units of its mean by using ϑ as parameter vector. Conditioning on $\hat{\sigma}$, from Theorem 3.1 it can easily be seen that the critical value $c(P^*)$ in (3.2) has to be replaced with $2^{1/2}t_{\nu(n),(1+P^*)/2}$ to obtain the one-or-all selection rule at PCS- level

P^* for the unknown variance case, where $t_{\nu(n),\gamma}$ denotes the γ-quantile of the t-distribution with $\nu(n)$ degrees of freedom. Thus, the selection rule reads

$$S(X) = \begin{cases} \{[k]\} & \text{if } \overline{X}_{[k]} - \overline{X}_{[k-1]} > (2/n)^{1/2} t_{\nu(n),(1+P^*)/2} \widehat{\sigma} \\ I & \text{otherwise.} \end{cases} \qquad (3.4)$$

Imposing requirement (3.3) for the special preference zone

$$\Omega_{\text{Pr}} = \{\vartheta \in \mathbb{R}^k : \vartheta_{(k)} - \vartheta_{(k-1)} > n^{1/2}\delta\} \qquad (3.5)$$

leads to a sample size which must not fall below

$$n = \min\{m \in \mathbb{N} : A(m^{1/2}\delta, -2^{1/2}t_{\nu(n);(1+P^*)/2}) \geq \widetilde{P}^*\}, \qquad (3.6)$$

where it is set

$$A(a,b) = \int_0^\infty \int_{-\infty}^\infty \Phi^{k-1}(u + a + bz)d\Phi(u) \, dQ_{\nu(n)}(z), \qquad a,b \in \mathbb{R} \qquad (3.7)$$

with Φ denoting the standardized normal distribution function. This result is easily obtained by noting that any $\vartheta \in \mathbb{R}^k$ with $\vartheta_{(1)} = \ldots = \vartheta_{(k-1)} = \vartheta_{(k)} - \delta n^{1/2}$ is a least favorable configuration (LFC) at which the probability in (3.3) attains its minimum over Ω_{Pr}. Notice that $\vartheta \in \Omega_{\text{Pr}}$ means for the associated actually interesting parameter $\mu = (\mu_1, \ldots, \mu_k)$ that $\mu_{(k)} - \mu_{(k-1)} > \delta\sigma$.

If, for fixed $\widetilde{P}^*$, the PCS-level P^* tends to zero, the quantile of the t-distribution in (3.4) approaches zero. Hence, the resulting selection procedure always chooses the population with the largest sample mean guaranteeing the probability of selecting the best population to be at least $\widetilde{P}^*$ outside the so-called indifference zone $\mathbb{R}^k \setminus \Omega_{\text{Pr}}$. This formulation of a selection problem is a straightforward adaption of the well-known indifference zone approach of Bechhofer (1954) to the unknown variance case.

3.2 Designing under Gupta's Subset Selection Formulation

A natural question, which arises in the context with procedure (3.4), is how to get more usable information in the case of a relatively small difference between the two largest sample means, when this one-or-all selection rule would decide to choose all populations. One possibility to improve the procedure, but mostly at the expense of a slightly larger sample size, is to adapt the classical subset selection formulation of Gupta (1956, 1965). His procedure selects the subset

$$S(X) = \{i \in I : \overline{X}_i \geq \max_{j \in I \setminus \{i\}} \overline{X}_j - d(n, P^*)\widehat{\sigma}n^{-1/2}\} \qquad (3.8)$$

of all populations, where the critical value $d(n, P^*)$ is determined so that the probability requirement

$$\inf_{\vartheta \in \mathbb{R}^k} P_\vartheta\{G(\vartheta) \cap S(X) \neq \emptyset\} = P^* \qquad (3.9)$$

is met, where $P^* \in (1/k, 1)$ is preassigned. Thus, in this approach a CS is defined whenever $S(X)$ contains at least one of the best populations. It is easy to show that (3.9) implies $d = d(n, P^*)$ to fulfill the equation $A(0, d) = P^*$ and that, because of $P^* > 1/k$, it holds $d(n, P^*) > 0$. An incorporation of the power condition (3.3) in Gupta's procedure (3.8) leads for the preference zone (3.5) to the additional requirement

$$\inf_{\vartheta \in \Omega_{\mathrm{Pr}}} P_\vartheta \{ \overline{X}_{(k)} - \overline{X}_{[k-1]} > d(n, P^*) \} \geq \tilde{P}^* . \tag{3.10}$$

Using the same LFC-argument as for showing (3.6), the left-hand side of (3.10) can be seen to be equal to $A(n^{1/2}\delta, -d(n, P^*))$. Thus,

$$n = \min \left\{ m \in I\!N : A(m^{1/2}\delta, -d(m, P^*)) \geq \tilde{P}^* \right\} \tag{3.11}$$

provides the smallest sample size n with corresponding critical value $d(n, P^*)$, for which Gupta's subset selection rule (3.8) satisfies both probability requirements (3.9) and (3.3).

4 Stepwise Selection Procedures for Selecting all Good Populations

Let us return to the distributional set-up of the beginning of Section 3 and to the original definition of good populations as given in (2.1). In this section the aim is to apply the closure principle on a certain test for $\mathcal{H} = \{ H_J : J \in \mathcal{P} \}$ in order to create a stepwise selection rule at PCS- level P^* called closed subset selection procedure. It turns out that this stepwise rule always selects a subset of populations being not larger than that one selected from its one-step counterpart which was proposed by Lam (1986). To construct this stepwise procedure, according to the expositions in Section 2 we first need a suitable level-$(1 - P^*)$ test for each H_J, $J \in \mathcal{P}$. It suggests itself to test H_J by using a region of acceptance of the type $\tilde{A}_J(c) \equiv \tilde{A}_J = \{ x \in I\!R^k : T_J(x) \leq c \}$ that is based on the so-called selection range statistic $T_J(x) = x_{[k]} - \min_{j \in J} x_j$, $x \in I\!R^k$. The critical value c has to be determined as solution of the equation

$$\inf_{\vartheta \in H_J} P_\vartheta \{ X \in \tilde{A}_J(c) \} = P^* . \tag{4.1}$$

Because of the independence of the underlying random variables X_i, $i \in I$, it is obvious that the critical value c defined by (4.1) depends only on the size j, say, of the set J, for which reason we hereafter use the notation c_j instead of c. Once the critical values $c_1, \ldots, c_k$ are calculated in this manner, formula (2.6) provides the regions of acceptance A_J, $J \in \mathcal{P}$, of a coherent multiple level-$(1 - P^*)$ test for $\mathcal{H}$. This multiple test finally yields the desired selection rule via (2.4). However, because Finner & Giani (1994a) showed $c_1 \leq \ldots \leq c_k$, it is simpler to carry out the selection rule in the following stepwise manner:

Step 1.   If $X_{[k]}-X_{[1]}$ $\begin{cases} \le c_k, & \text{set } S(X)=I \text{ and stop;} \\ > c_k, & \text{go to Step 2.} \end{cases}$

Step j.   If $X_{[k]}-X_{[j]}$ $\begin{cases} \le c_{k-j+1} \text{ set } S(X)=\{[k],\dots,[j]\} \text{ and stop;} \\ > c_{k-j+1} \text{ go to Step } j+1. \end{cases}$ (4.2)

Step $k-1$. If $X_{[k]}-X_{[k-1]}$ $\begin{cases} \le c_2, & \text{set } S(X)=\{[k],[k-1]\}; \\ > c_2, & \text{set } S(X)=\{[k]\}. \end{cases}$

Nevertheless, it remains the difficult task to solve (4.1) for c. For this purpose, it is necessary to determine the LFC at which the minimum of $P_\vartheta\{X \in \tilde{A}_J(c)\}$ is attained over H_J. The following theorem provides a partial solution leading to a finite number of possible LFC-candidates.

THEOREM 4.1.   (Finner & Giani 1994, Theorem 3.1). *For $|J| = j \in I$ and $r \in I$, let $c_j(\vartheta(r))$ denote the solution of $P_{\vartheta(r)}\{X \in A_J(c_j(\vartheta(r)))\} = P^*$, where $\vartheta(r) = (0,\dots,0,\epsilon,\dots,\epsilon) \in I\!\!R^k$ is a parameter vector whose first r components are equal to zero. Furthermore, assume that the density f is log-concave and symmetric. Then, the solution $c \equiv c_j$ of (4.1) is given by*

$$c_j = \begin{cases} c_k(\vartheta(m)) & \text{if } j = k, k \in \{2m, 2m+1\} \\ \max\{c_j(\vartheta(r)) : r = 1,\dots,j\} & \text{otherwise.} \end{cases}$$

To calculate c_j for $j < k$, we see from Theorem 4.1 that each of the j parameter points $\vartheta(1),\dots,\vartheta(j)$ jet has to be checked for being least favorable. This can be done numerically by using the formula

$$D_r(\epsilon, c|F) \equiv P_{\vartheta(r)}\{T_J(X) \le c\} =$$

$$r \int_{-\infty}^{\infty} [F(s+c) - F(s)]^{r-1}[F(s-\epsilon+c) - F(s-\epsilon)]^{j-r}$$

$$\times [F(s-\epsilon+c)]^{k-j} dF(s) \qquad (4.3)$$

$$+(j-r) \int_{-\infty}^{\infty} [F(s+\epsilon+c) - F(s+\epsilon)]^r [F(s+c) - F(s)]^{j-r-1}$$

$$\times [F(s+c)]^{k-j} dF(s) .$$

Let us now construct a closed subset selection rule under the normal distributional assumptions with unknown variance, specified in Section 3.1. As in that section we base the decision rule on the statistic $X = \overline{X}n^{1/2}/\hat{\sigma}$ with associated parameter $\vartheta = n^{1/2}(\mu_1/\sigma,\dots,\mu_k/\sigma)$. With these settings the stepwise procedure for selecting all good populations at the PCS-level P^* is performed as described in (4.2), provided the appropriate critical values are available. Unfortunately, it is difficult to determine the critical values as solutions of (4.1) because there exist no LFC-results, except for $J = I$, to make the left-hand side of (4.1) calculable. However, slightly conservative critical values, which are close to the optimum as remarked by Finner & Giani (1994a), can be derived for $|J| = j < k$ with Theorem 4.1 and (4.3) from the inequality

$$\inf_{\vartheta \in H_J} P_\vartheta \{X \in \tilde{A}_J(c)\} \geq \int_0^\infty \min_{1 \leq r < j} D_r(\epsilon, cz|\Phi) dQ_{\nu(n)}(z) \qquad (4.4)$$
$$\equiv E_j(\epsilon, c), \text{ say},$$

whereas for $|J| = k$, i.e., $J = I$, the infimum of $P_\vartheta \{X \in \tilde{A}_I(c)\}$ is equal to

$$E_k(\epsilon, c) \equiv \int_0^\infty D_m(\epsilon, cz|\Phi) dQ_{\nu(n)}(z), \quad k \in \{2m, 2m+1\}. \qquad (4.5)$$

Choosing $c \equiv c_j$ as solution of $E_j(\epsilon, c) = P^*$, $j \in I$, ensures that the stepwise procedure (4.2) meets the PCS-level P^*.

5 Partitioning Procedures

In this section we are concerned with the design problem of determining sample sizes when a certain power requirement is involved in the selection formulation of Section 4. Consideration is restricted to the normal distributional situation with unknown variance as described in Section 3.1. Good populations are defined in terms of (2.1), but replacing the threshold value $\epsilon \geq 0$ there with $n^{1/2}\delta_0 \geq 0$, i.e., $G(\vartheta) = \{i \in I : \vartheta_{(k)} - \vartheta_i \leq n^{1/2}\delta_0\}$. Hence, from $i \in G(\vartheta)$, for the μ-vector corresponding to ϑ it follows the inequality $\mu_{(k)} - \mu_i \leq \delta_0\sigma$ which is independent of the sample size n. Analogously, for each $\vartheta \in \mathbb{R}^k$ let bad populations be defined by $B(\vartheta) = \{i \in I : \vartheta_{(k)} - \vartheta_i > n^{1/2}\delta_1\}$ with a preassigned $\delta_1 > \delta_0$. The task we are faced with is to select with high probability $P^* \in (0,1)$ a subset $S(X)$ containing all good populations but none of the bad ones. In other words, we are interested in partitioning the set I of all populations in two subsets, one of which contains all good populations and the other one all bad ones. To be precise, the probability requirement has the form

$$\forall \vartheta \in \mathbb{R}^k : \quad P_\vartheta \{G(\vartheta) \subseteq S(X) \subseteq B^c(\vartheta)\} \geq P^*. \qquad (5.1)$$

To satisfy (5.1) let us consider the one-step procedure

$$S(X) = \{i \in I : X_{[k]} - X_i \leq cn^{1/2}\},$$

which results from (4.2) by setting all the critical values equal to $cn^{1/2}$, say. It is obvious that (5.1) constitutes a condition on the sample size n

as well as the critical value c. Unfortunately, because $E(\vartheta) \equiv \{x \in \mathbb{R}^k : G(\vartheta) \subseteq S(x) \subseteq B^c(\vartheta)\}$ is not a convex subset of $\mathbb{R}^k$ for each $\vartheta \in \mathbb{R}^k$, none of the minimization results of the convex analysis nor majorization theorems can be applied to obtain the LFC of the probability of a correct partition. Alternatively, one can try to find appropriate lower bounds $L_n(c)$ on $\inf_{\vartheta \in \mathbb{R}^k} P_\vartheta\{X \in E(\vartheta)\}$ and then to determine the sample size n with associated critical value c from

$$n = \min\left\{m \in \mathbb{N} : \max_{c > 0} L_m(c) \geq P^*\right\}, \quad c = \max_{c' > 0} L_n(c'). \quad (5.2)$$

For known common variance σ^2, Lam (1986) gives a lower bound $L_n(c) = \tilde{L}_n(c)$, say, on the probability of a correct partition having the form

$$\tilde{L}_n(c) = P\{U_{[k]} - U_{[1]} \leq n^{1/2} \min\{c - \delta_0, \delta_1 - c\}\}, \quad c \in (\delta_0, \delta_1),$$

where U_i, $i \in I$, denote k independent standardized normal variables. Note that, for each n, the value $c = (\delta_0 + \delta_1)/2$ maximizes $\tilde{L}_n(c)$ and can therefore be regarded as an optimal choise of the critical value within Lam's lower bound approach. Conditioning on $\hat{\sigma}/\sigma$, a generalization to the unknown variance case can be derived from the inequality

$$P_\vartheta\{X \in E(\vartheta)\} \quad = \quad \int_{-\infty}^{\infty} P_\vartheta\{X \in E(\vartheta) \mid \tilde{\sigma}/\sigma = z\} dQ_{\nu(n)}(z)$$

$$\geq \quad \int_0^\infty \tilde{L}_n(cz) dQ_{\nu(n)}(z) \equiv L_n(c).$$

The following formula for the lower bound $L_n(c)$ may be used for purpose of numerical calculation:

$$L_n(c) = k \int_{\delta_0/c}^{(\delta_0+\delta_1)/(2c)} \int_{-\infty}^{\infty} [\Phi(u+n^{1/2}(cz-\delta_0)) - \Phi(u)]^{k-1} d\Phi(u) \, dQ_{\nu(n)}(z)$$

$$\quad (5.3)$$

$$+k \int_{(\delta_0+\delta_1)/(2c)}^{\delta_1/c} \int_{-\infty}^{\infty} [\Phi(u + n^{1/2}(\delta_1 - cz)) - \Phi(u)]^{k-1} d\Phi(u) \, dQ_{\nu(n)}(z)$$

When applying this $L_n(c)$ as lower bound on the probability of correct partition, (5.2) yields the necessary sample size n and critical value c to satisfy (5.1).

6 Numerical Aspects

A computer program has been developed in which all proposed selection rules, among others, are implemented. To determine minimum sample sizes in the design approaches described in Section 3.1, Section 3.2, and Section 5, one has to calculate the double integrals (3.7) and (5.3) numerically, whereas under the normal model the critical values of the stepwise procedure (4.2) require the integrals (4.4) and (4.5) to be computed. These integrals are of a special form being typical of selection problems in the normal case, namely

$$\int_a^b \left[\int_{-\infty}^\infty g_\Phi(u, z) d\Phi(u) \right] \gamma_\nu(z) dz, \quad 0 \leq a < b \leq \infty, \tag{6.1}$$

where γ_ν denotes the density of a chi-squared distribution with ν degrees of freedom, and $g_\Phi(u, z)$ is a linear combination of products formed with factors of the type $\Phi^i(a_1 u + a_2 z + a_3)$, $i < k$, $a_1, a_2, a_3 \in \mathbf{R}$. The computation of the function g_Φ is carried out by applying the algorithm in Hill (1973) for the evaluation of Φ. To calculate the inner integral in the squared brackets of (6.1), 136-point Gauss-Hermite quadrature is used, but the integrand is evaluated in only 58 points, because the remaining part of the quadrature falls below the accuracy of the machine. For $a = 0$, $b = \infty$, $\nu \leq 2000$, the outer integral of (6.1) is calculated by a Gaussian quadrature formula using the zeros of generalized Laguerre polynomials of degree 68 and parameter ν as points (cf. Stroud & Secrest, 1966). Note that these points with their corresponding weights have to be calculated for each degree of freedom ν under consideration. If $\nu > 2000$, the chi-squared distribution in (6.1) is approximated by a normal distribution and then calculated by Gauss-Hermite formula. In case of $0 < a$, $b < \infty$, the outer integral is numerically determined by an adaptive quadrature proposed by Kahaner & Stoer (1983).

To specify the varies selection procedures, one furthermore is faced with the numerical problem of determining $\min\{\zeta > 0 : V(\zeta) \geq P^*\}$, where $V(\zeta)$ is of the form (6.1) and is either a real function with discrete argument ζ as given in (3.6), (3.11), and (5.2), or a continuous function as in Section 4 when calculating critical values. This problem is tackled by solving the equation $V(\zeta) = P^*$, treating ζ as continuous variable in the discrete case. To iterate to the solution, the *Pegasus* algorithm being a modification of the *regula falsi* is employed.

For maximization of $L_n(\cdot)$ over the positive real line (see (5.2)), that has to be performed to determine the minimum sample size for the partitioning problem, an algorithm which combines Golden section search and successive parabolic interpolation is used. The advantage of this method described in Brent (1973, pp. 72-79) is that it does not make use on the knowledge of the derivates of L_n.

References

ANDERSON, P. O., BISHOP, T. A., AND DUDEWICZ, E. J. (1977). Indifference-zone ranking and selection: confidence intervals for true achieved $P(CD)$. *Commun. Statist. - Theor. Meth.* **A6**, 1121-1132.

BECHHOFER, R. E. (1954). A single-sample multiple decision procedure for ranking normal populations with known variances. *Ann. Math. Statist.* **25**, 16-39.

BRENT, R. P. (1973). Algorithms for minimization without derivates. *Prentice Hall, Englewood Cliffs.*

BROSTRÖM, G. (1981). On sequentially rejective subset selection procedures. *Comm. Statist. - Theor. Meth.***A10**, 203-221.

FINNER, H. AND GIANI, G. (1994a). Closed subset selection procedures for selecting good populations. *J. Statist. Planning Infer.* **38**, 179-199.

FINNER, H. AND GIANI, G. (1994b). Duality between multiple testing and selecting. *To appear in a special issue of: J. Statist. Planning Infer.*

GABRIEL, K. R. (1969). Simultaneous test procedures - some theory of multiple comparisons. *Ann. Math. Statist.* **40**, 224-250.

GIANI, G. AND FINNER, H. (1991). Some general results on least favourable parameter configurations with special reference to equivalence testing and the range statistic. *J. Statist. Planning Infer.* **28**, 33-47.

GUPTA, S. S. (1956). On a decision rule for a problem in ranking means. *Ph. D. Thesis (Mimeo Ser. No. 150). Inst. of Statist., Univ. of North Carolina, Chapel Hill.*

GUPTA, S. S. (1965). On some multiple decision (selection and ranking) rules. *Technometrics* **7**, 225-245.

GUTMANN, S. AND MAYMIN, Z. (1987). Is the selected population the best? *Ann. Statist.* **15**, 456-461.

HILL, I. D. (1973). Algorithm AS66: The normal integral. *Appl. Statist.* **22**, 424-427.

KAHANER, D. K. AND STOER, J. (1983). Extrapolated adaptive quadrature. *Siam J. Sci. Statist. Comput.* **Vol. 4**, No. 1.

KIM, W.-C. (1986). A lower confidence bound on the probability of a correct selection. *J. Amer. Statist. Assoc.* **81**, 1012-1017.

LAM, K. (1986). A new procedure for selecting good populations. *Biometrika* **73**, 201-206.

NAIK, U. D. (1975). Some selection rules for comparing p processes with a standard. *Comm. Statist.* **4**, 519-535.

NAIK, U. D. (1977). Some subset selection problems. *Comm. Statist. - Theor. Meth.* **A6**, 955-966.

STEFANSSON, G. KIM, W.-C. AND HSU, J. C. (1988). On confidence sets in multiple comparisons. *In: Statistical Decision Theory and related Topics, Vol. IV. Ed.: S. S. Gupta and J. O. Berger. Academic Press, New York*, 89-104.

STROUD, A. H. AND DON SECREST (1966). Gaussian quadrature formulas. *Prentice-Hall, Englewood Cliffs.*

TONG, Y. L. (1969). On partitioning a set of populations by their locations with respect to a control. *Ann. Math. Statist.* **40**, 1300-1324.

Part V

Computational Aspects in Optimization

A New Robust Design Strategy for Sigmoidal Models Based on Model Nesting

Timothy E. O'Brien

Department of Statistics, Washington State University

1   Introduction

For a given process, researchers often have a specific nonlinear model in mind and perhaps a reasonable initial estimate of the p model parameters. In this situation, optimal design theory produces designs which typically have only p support points even when the final sample size (n) is chosen to exceed the number of parameters. Since p-point designs assume that the model function is known with complete certainty and provide no opportunity to test for the adequacy of the assumed model, they are clearly not "optimal" in most practical settings.

The focus of this paper is to provide an algorithm to obtain efficient designs with "extra" design points, or "robust" designs, by nesting a given model function (the "original" model function) in a larger one (the "super-model") which reduces to the original model for certain parameter choices. This design approach, called the nesting design strategy, has been applied to linear models in [1], [2], [7], [14], and [15], and only in very simple instances to nonlinear models in [3] and [6]. The application to nonlinear models is more difficult since the super-model is often less apparent (and more *ad hoc*) and requires a keen understanding of the nature of the various model functions; for this reason, the discussion here is limited only to sigmoidal growth models (e.g., those given in Chapter 4 of [11] and Chapter 7 of [13]).

2   Sigmoidal Growth Models

Most growth models in current use fall into one of three families: the Weibull, the Log-Logistic, and the Richards. Special cases of these models include the Logistic, Gompertz, Michaelis-Menton, Mischerlich, and Simple Exponential models, and often several model functions from these families can be used to adequately describe a given set of data. For example, the two-parameter Weibull function (W2),

$$\eta_{w2} = \exp\{-[(x/\theta_1)^2]^{\theta_2}\}, \tag{1}$$

and the two-parameter Log-Logistic model function (LL2),

$$\eta_{ll2} = \frac{1}{1 + (x/\theta_3)^{\theta_4}}, \tag{2}$$

behave quite similarly over $\Re_+$ for certain parameter choices. Since optimal designs for either of these model functions typically have only two support points (see [9]), functions which generalize the Weibull, Log-Logistic, and Richards families are required.

One important generalization of (1) is the three-parameter humped Weibull model function (HW3),

$$\eta_{hw3} = \exp\left\{-[(\frac{x - \theta_3}{\theta_1})^2]^{\theta_2}\right\}. \tag{3}$$

In some instances, the HW3 model function fits the data obtained in ozone dose response studies better that the W2 function (see [12]), although the biological interpretation of the corresponding "hump-effect" is not readily apparent. In addition, an important generalization of the Weibull, Log-Logistic and Richards model functions is the six-parameter Eclectic model (E6),

$$\eta_{e6} = \frac{\theta_1}{\left\{1 + \frac{1}{\theta_6} \exp\left[\frac{(\frac{x-\theta_2}{\theta_3})^{\theta_4\theta_5} - 1}{\theta_5}\right]\right\}^{\theta_6}}, \tag{4}$$

studied in [10]. Conditions under which this function reduces to the Weibull, Log-Logistic and Richards family members are given in [10]; to illustrate one of these cases, note that the three-parameter Log-Logistic model function (LL3),

$$\eta_{ll3} = \frac{\theta_1}{1 + (x/\theta_3)^{\theta_4}}, \tag{5}$$

is obtained from (4) by taking $\theta_2 = 0, \theta_5 \rightarrow 0$, and $\theta_6 = 1$.

3 Optimal Design Theory

The design problem for the homoskedastic Gaussian nonlinear model

$$y_i = \eta(\mathbf{x}_i, \phi) + \epsilon_i \qquad \epsilon_i \sim iid\ N(0, \sigma^2) \text{ for } i = 1, ..., n$$

typically involves choosing an n-point design, ξ, to estimate some function of the p-dimensional parameter vector, ϕ, with high efficiency. This design associates the design weights ω_1, ω_2, ..., ω_n with the design points (or vectors) $\mathbf{x}_1$, $\mathbf{x}_2$, ..., $\mathbf{x}_n$, respectively, and the corresponding (Fisher) information matrix is given by

$$\mathbf{M}(\xi, \phi) = \sum_{i=1}^{n} \omega_i \frac{\partial \eta(\mathbf{x}_i)}{\partial \phi} \frac{\partial \eta(\mathbf{x}_i)}{\partial \phi'} = \mathbf{V}' \Omega \mathbf{V},$$

where $\mathbf{V}$ is the n×p Jacobian of η and $\Omega = \text{diag}\{\omega_1, \ldots, \omega_n\}$.

First-order optimal designs typically minimize some convex function of $\mathbf{M}^{-1}$. For example, designs which minimize the determinant $|\mathbf{M}^{-1}(\xi, \phi^o)|$ are called locally D-optimal, where the term "locally" is used to emphasize the fact that an initial estimate of the parameter vector has been used. Other first order design criteria are discussed in [2] and [4], and second-order design criteria–or procedures which provide designs that attempt to reduce curvature in addition to efficiently estimating parameters–are presented in [8], [9], and [10]. Since optimal designs typically have only p support points regardless of the design criteria and final sample size used (see [9] and [16]), we seek a practical design algorithm which efficiently estimates the model parameters and also provides "extra" design points to check for model mis-specification.

4 A First-Order Nesting Design Strategy

Suppose that a researcher feels that the original model function $\eta_{or}(\phi_1)$ adequately describes a given process, but desires a design which, in addition to efficiently estimating the p_1 model parameters of η_{or}, also may be used to check for the adequacy of the assumed model. As a first step, we search for a relevant super-model, $\eta_{sm}(\phi_1, \phi_2)$, such that η_{sm} reduces to η_{or} when the p_2-vector ϕ_2 is equal to some (possibly extended) real vector. It is important that the super-model contains a reasonable generalization of the original model; thus, for example, by nesting the Log-Logistic model function in the Eclectic function, departures from the Log-Logistic model function in the direction of the Weibull and Richards functions may be detected.

A measure of the inefficiency that the design ξ has in estimating ϕ_1 in η_{or} is given by $|\mathbf{M}_{11}{}^{-1}|$, and a measure of it's inefficiency regarding detecting departures from η_{or} in the direction of η_{sm} is given by $\left|\left(\mathbf{M}_{22} - \mathbf{M}_{21} \mathbf{M}_{11}^{-1} \mathbf{M}_{12}\right)^-\right|$ where $\mathbf{M}_{ij} = \mathbf{V}_i \Omega \mathbf{V}_j$ (i,j = 1, 2); see [2]. We combine these measures into the single (first-order) inefficiency measure,

$$\psi_1(\xi, \lambda, \phi) = \frac{\lambda}{p_1} \left|\mathbf{M}_{11}{}^{-1}\right| + \frac{1-\lambda}{p_2} \left|\left(\mathbf{M}_{22} - \mathbf{M}_{21} \mathbf{M}_{11}^{-1} \mathbf{M}_{12}\right)^{-1}\right|, \quad (6)$$

and seek designs to minimize ψ_1 for given choices of λ and $\phi = (\phi_1', \phi_2')'$, designs which are called locally D_λ-optimal here. Note that λ controls the amount of information obtained regarding estimation of ϕ_1 relative to detecting departures from the original model function, so that when we take $\lambda = 1$, we obtain information only regarding ϕ_1, whereas when we choose $\lambda = 0$, we obtain information only about departures from the original function. Our recommendation is to obtain the locally D_λ-optimal design, ξ_λ, such that it's D-efficiency relative to the locally D-optimal design, ξ_D, $\{M_{11}^{-1}(\xi_D) / M_{11}^{-1}(\xi_\lambda)$ is around 0.95.

5 Examples

5.1 Example 1

An environmental scientist believes that the W2 function in (1) with $\theta_1 = \theta_2 = 1$ adequately describes the dose response relationship of a given cultivar exposed to ozone (x), but wishes to allow for the hump-effect of the HW3 function in (3). Our procedure provides the locally D_λ-optimal design using the super-model (3) with $\theta_3 = 0$ and $\lambda = .95$, which associates the weights $\omega = 0.10, 0.43$ and 0.47 with the points x = 0.33, 0.63 and 1.30, respectively. This design is preferred to the locally D-optimal design for estimating the W2 model parameters, ξ_D, since this latter design has only two support points (at x = 0.59 and 1.28 each with equal weight), yet is "close" to ξ_D since it's D-efficiency is 95%.

5.2 Example 2

An animal scientist feels that the LL3 model function in (5) with $\theta_1 = \theta_3 = 1$, and $\theta_4 = 2$ reasonably describes the amount of food left in a cow's rumin x minutes after ingestion, but seeks a design with more than three support points so as to check for model-mis-specification. Using the E6 model function in (4) as the super-model with $\theta_2 = 0, \theta_5 = .001, \theta_6 = 1$ and $\lambda = .95$, the corresponding locally D_λ-optimal design, ξ_λ, places the weights $\omega = 0.18, 0.13\ 0.05, 0.30, 0.14$ and 0.20 at the points x = 0, 0.04, 0.23, 0.60, 1.35 and 1.97, and results in a D-efficiency of nearly 97%. This design is preferred to the locally D-optimal design for estimating the LL3 model parameters since this latter design has only three support points (at x = 0, 0.59 and 1.68 each with equal weight). It is important to note that since the E6 function also has the Weibull and Richards families as special cases, ξ_λ protects against departures from the LL3 function in the direction of practically all other sigmoidal curves.

6 A Second-Order Nesting Design Strategy

The first-order nesting strategy presented above is easily extended to provide efficient robust designs with reduced marginal curvature by using a penalty function approach. Based on Clarke's criterion for the seriousness of curvature (in [5]), we let $k_i = \max\{|\,\mathrm{mc}_i\,| - 0.10, 0\}$, $l_i = \max\{|\,\mathrm{mc}_i\,| - 0.30, 0\}$, and $\pi(\xi, \phi) = \sum_{i=1}^{p_1} \exp\{\alpha_i\, k_i + \beta_i\, l_i\}$; here mc_i is the marginal curvature associated with the ith parameter in η_{or}, $\pi(\xi, \phi)$ is our curvature penalty function, and the α_i's and β_i's are chosen to emphasize certain components of the marginal curvature vector over others. Thus,

$$\psi_2(\xi, \gamma, \lambda, \phi) \;=\; \gamma\, \psi_1(\xi, \lambda, \phi) \;+\; (1 - \gamma)\, \pi(\xi, \phi),$$

for ψ_1 in (6), is a second-order inefficiency measure, and $\gamma \in [0, 1]$ controls the degree of emphasis placed on parameter estimation relative to curvature reduction. Designs which minimize ψ_2 for specific choices of γ, λ, and ϕ, called locally D_γ-optimal here, are usually preferred to locally D_λ-optimal designs when curvature is a concern. For example, the marginal curvatures associated with θ_3 using the D_λ-optimal design given in Example 2 is 0.3066, indicating serious curvature by Clarke's criterion. In contrast, the locally D_γ-optimal design obtained by minimizing ψ_2 (with $\gamma = .5$, $\lambda = .95$, each $\alpha_i = 2$, and each $\beta_i = 0$) is such that no component of the marginal curvature vector exceeds 0.10. Further, this latter design, which associates the weights $\omega = 0.08$, 0.16 0.12, 0.23, 0.21 and 0.20 with the points x = 0, 0.05, 0.13, 0.75, 1.53 and 1.91, is also efficient in estimating the LL3 model parameters since it's D-efficiency is 90%.

7   Discussion

The design strategies presented here provide researchers with a reasonable compromise between so-called "optimal" designs, which typically cannot be used to check whether the assumed model is indeed valid, on the one hand, and designs comprised of arbitrarily chosen design points (for example, those with a geometric spacing of points), on the other. The first nesting design strategy is intended to be used with "close-to-linear" nonlinear models (see [11]); when designs with reduced curvature are desired, the second nesting design strategy should be used. Although our focus here has only been on sigmoidal growth curves, the application of the above procedures to other classes of nonlinear models is obvious provided relevant super-models can be found.

References

[1] Atkinson, A.C. (1972). Planning experiments to detect inadequate regression models. *Biometrika* **59**, 275-293.

[2] Atkinson, A.C. and Donev, A.N. (1992). *Optimum Experimental Designs*. Oxford: Clarendon Press.

[3] Box, G.E.P. and Lucas, H.L. (1959). Design of experiments in non-linear situations. *Biometrika* **46**, 77-90.

[4] Chaloner, K. and Larntz, K. (1989). Optimal Bayesian design applied to logistic regression experiments. *J. Stat Plann. Infer.* **21**, 191-208.

[5] Clarke, G.P.Y. (1987). Marginal curvatures and their usefulness in the analysis of nonlinear regression models. *J. Amer Stat. Assoc.* **82**, 844-850.

[6] Cochran, W.G. (1973). Experiments for nonlinear functions. *J. Amer. Stat. Assoc.* **68**, 771-781.

[7] DeFeo, P. and Myers, R.H. (1992). A new look at experimental design robustness. *Biometrika* **79**, 375-380.

[8] Hamilton, D.C. and Watts, D.G. (1985). A quadratic design criterion for precise estimation in nonlinear regression models. *Technometrics* **27**, 241-250.

[9] O'Brien, T.E. (1992). A note on quadratic designs for nonlinear regression models. *Biometrika* **79**, 847-849.

[10] O'Brien, T.E. (1993). Design strategies for nonlinear regression models. Ph.D. dissertation, North Carolina State University, Raleigh, N.C.

[11] Ratkowsky, D.A. (1983). *Nonlinear Regression Modeling*. New York: Marcel Dekker.

[12] Reinert, R.A., Rufty, R.C. and Eason, G. (1988). Interaction of tobacco etch or tobacco vein mottling virus and ozone on biomass changes in Burley tobacco. *Envir. Pollution.* **53**, 209-218.

[13] Seber, G.A.F. and Wild, C.J. (1989). *Nonlinear Regression*. New York: Wiley.

[14] Stigler, S.M. (1971). Optimal experimental design for polynomial regression. *J. Amer. Stat. Assoc.* **66**, 311-318.

[15] Studden, W. J. (1982). Some robust-type D-optimal designs in polynomial regression. *J. Amer. Stat. Assoc.* **77**, 916-921.

[16] Vila, J.P. (1991). Local optimality of replications from a minimal D-optimal design in regression: a sufficient and quasi-necessary condition. *J. Stat. Plann. Infer.* **29**, 261-277.

Part VI

Computational Aspects in Spatial Statistics and GIS

Some Dynamic Graphics for Spatial Data (with Multiple Attributes) in a GIS

Dianne Cook, Noel Cressie, James Majure, Jürgen Symanzik

Department of Statistics, Iowa State University

Abstract. This paper discusses some multivariate exploratory spatial data analysis tools for detecting spatial dependence. The ideas explored are related to canonical correlation analysis and the graphical tools are related to the dynamic method called the grand tour. The work is implemented with a link between a Geographic Information System, ARC/INFO™, and software for exploring multivariate data, XGobi.

Keywords. canonical correlations, correlation tour, exploratory spatial data analysis, spatial dependence, spatial statistics

1 Introduction

1.1 Terminology and Notation for Spatial Data

Spatial methods involve acknowledgement of the spatial location s of the attribute $Z(s)$; $s \in D$, where D is a subset of d-dimensional space. In what follows we shall mainly be interested in a multivariate attribute:

$$Z(s) = (Z_1(s), \ldots, Z_p(s))'.$$

When we investigate how *average* values of Z vary with s, then we are considering the *trend* or *large-scale variation* in the attribute. One could think of such summaries as estimating

$$\mu(s) = \mathrm{E}[Z(s)] \; ; \;\; s \in D.$$

Define the *small-scale variation* as $\delta(s) \equiv Z(s) - \mu(s)$. Typically, some form of second-moment stationarity of $\delta(s)$ is assumed to allow the *spatial dependence* to be characterized and estimated. Therefore, in the spatial statistical model-building stage, it is important to have available methods for detecting regional variation in the attribute values, "pockets" of non-stationarity, and spurious data points (Cressie, 1984; Getis and Ord, 1992). Prediction based on the squared error loss function relies on the covariance properties of $Z(s)$. In the multivariate setting we need to know $\mathrm{cov}(Z_j(s), Z_{j'}(s'))$, $\forall j, j'$ and $\forall s, s'$; that is, we need to know $\mathrm{cov}(Z(s), Z(s')) = \mathrm{E}[\delta(s)\delta(s')'] \equiv C(s, s')$, where $C(\cdot, \cdot)$ is a $p \times p$ covariance matrix. A stationarity assumption on C leads to

$$E[\delta(s + h)\delta(s)'] = C(h),$$

a matrix function only of h, the spatial lag. Thus, to characterize the *spatial dependence* in the small-scale variation we look at co-dependencies between $\delta(s+h)$ and $\delta(s)$. This will be developed further in Section 2.1 and 2.2. Section 2.3 discusses interactive and dynamic graphical methods to search for spatial dependence. A new component of our work is to combine the strengths of interactive and dynamic graphical methods with the spatial data base management and spatial functionalities of a Geographic Information System (GIS). A GIS is a computer hardware and software system designed to carry out a number of important tasks on spatially referenced data (see, for example, Maguire, 1991; Goodchild, 1991). This work is discussed in Section 3.

1.2 Examples of Interactive and Dynamic Graphics Applied to Spatial Data

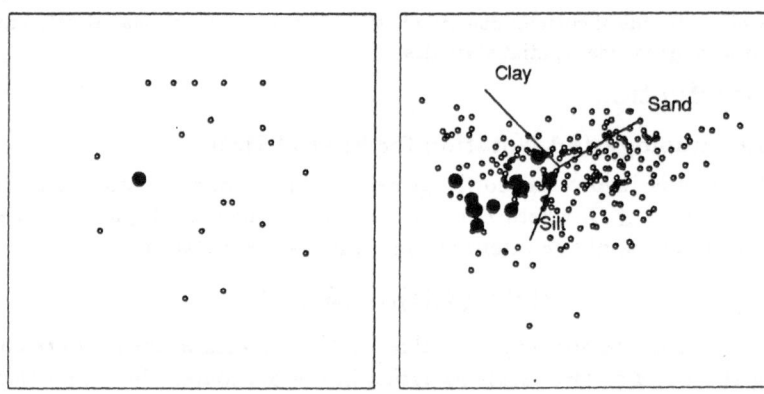

Figure 1: *California soils data - linking the map view to the attribute display.*

A natural approach to visualizing spatial data is to display the two groups of variables $(s, Z(s))$ using appropriate graphics in two separate windows. For example, consider the situation in Figure 1 where three measurements have been collected at many geographic sites; the collection sites are displayed as a 2-dimensional scatterplot of the geographic location (called a "map view" by Haslett et al., 1991), and the measurement data displayed as a 3-dimensional rotating or grand touring (Asimov, 1985; Buja and Asimov, 1986) point cloud. Because it is difficult to illustrate and describe in a static medium like the printed page, an important part of our research involves dynamic motion of some or all of the displays. The rotating and grand touring referred to earlier are two of a number of dynamic graphical methods that are extremely effective in illustrating structure in data, particularly multivariate data.

Each plot in Figure 1 is linked dynamically so that highlighting (or brushing or identifying) in one plot simultaneously causes equivalent highlighting in the other plots. Plot (a) shows the geographic data corresponding to 20 locations in an undisclosed area of California and plot (b) is a 2-dimensional snapshot of a 3-dimensional point cloud of the variables sand, silt, and clay. (The data can be found in Andrews and Herzberg, 1985, set 16. Measurements are also taken at 12 different depths at each site.) At the site highlighted (•) the soil can be (and actually is) classified as silty clay, an almost equal mixture of silt and clay particle sizes with very little sand.[1]

Finding clusters in the image space has been fairly successfully done by using multivariate methods for clustering on the attribute data and linking this to the spatial variable plot. McDonald and Willis (1987) (discussed also in Buja et al., 1991) use the grand tour on 4 Landsat MSS spectral bands to identify forest, grasslands, muddy water of the Rio Solimões, and the dark water of the Rio Negro in an image taken of an area near Manaus, Brazil. They show that a simple linear mixing model for the confluence of the two rivers is sufficient.

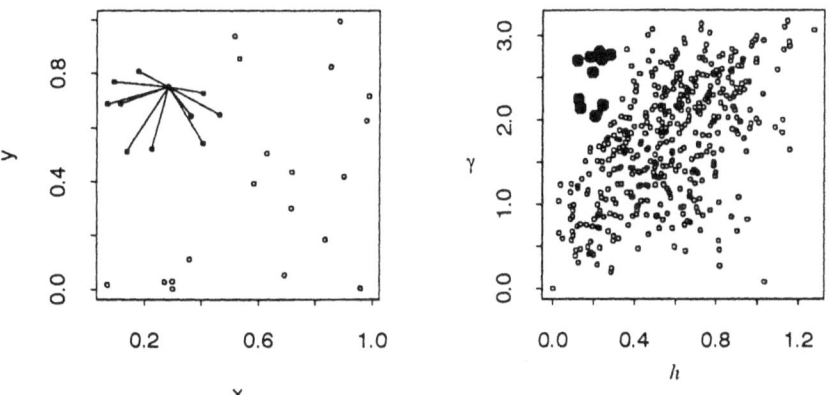

Figure 2: *Using linking between the map view and variogram cloud to detect an outlying point. (The data is simulated purely to illustrate the method.)*

Graphical methods developed for spatial data are enhanced by incorporation into the linked window approach. When there is only one attribute variable, $Z(s)$, an isotropic variogram cloud (Chauvet, 1982; Cressie, 1984), $\gamma_{ij} = |Z(s_i) - Z(s_j)|^{1/2}$ versus $h_{ij} = ||s_i - s_j||^{1/2}$ can be linked to the spatial variable plot. Brushing in the variogram cloud links to *pairs* of points

---

[1] An interesting aside to the data is that there are two reasonably large errors in the recording of the soil particle percentages. Because they are percentages, the points should lie in a 2-dimensional simplex in R^3. During rotation, though, it is clear that 2 points do not lie in this region. Using identification and referring back to the data confirms that the (plot 4, depth 3) percentages sum to 109.1% and the (plot 3, depth 11) percentages sum to 96.3%.

connected by lines in the spatial variable plot. Interesting points (potential outliers or points defining the borders of spatial regions) that are spatially close but differ greatly in the attribute variable value are easily identified by this technique (Haslett et al., 1991; Bradley and Haslett, 1992). In Figure 2, a spurious point is identified. The points (•) in the variogram cloud (plot (b)) that have large values of γ_{ij} despite low spatial distance, h_{ij}, link back to the pairs of points in the map view (plot (a)) connected by lines. All the lines have one common end point indicating a location that has an attribute value differing substantially from its neighbours.

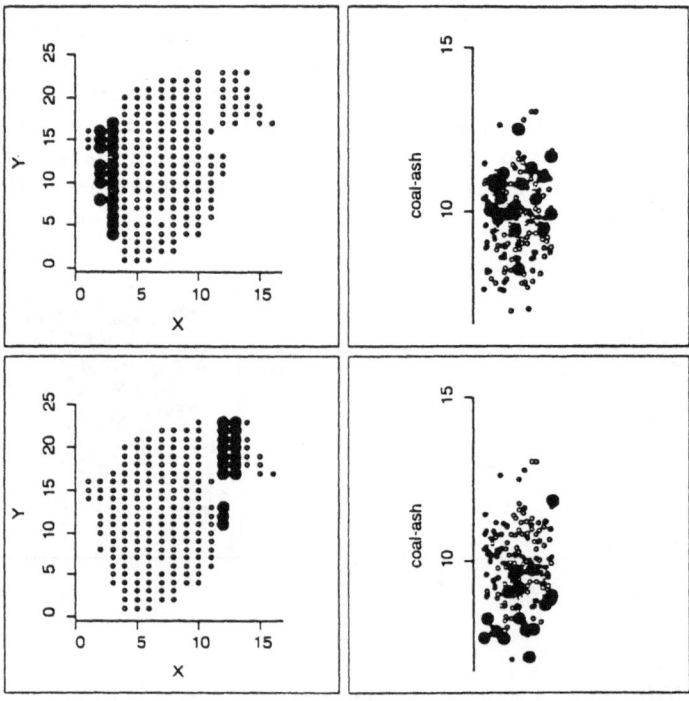

Figure 3: *Detecting east-west trend in the coal-ash data (Cressie, 1993) by linked brushing.*

In the coal ash example in Cressie (1993, Section 2.2), an east-west trend is detected by averaging attribute values, that is, taking means and also medians, over the north-south direction. Graphically (and interactively) this is equivalent to dragging a long narrow brush sideways and watching the change in distribution of highlighted points in the plot of the attribute variable (Figure 3). For higher dimensional (≥ 3) attribute data, a crude approach is to divide the data into several colored groups by brushing regionally in the spatial plot and examining the relative positions of the colored groups during a grand tour.

2 Detecting Spatial Dependence

2.1 Analytical Tools

The essence of most investigations into spatial dependence is to consider relationships between attributes and their lagged (neighbouring) versions. For this section, we shall assume that the trend component, pockets of non-stationarity, and spurious points have been removed from $Z(s)$.

For the moment, let us fix h. We may vary h later if it is appropriate. Our approach is to concatenate $Z(s+h)$ to $Z(s)$ and apply statistical methods to $(Z(s+h)', Z(s)')'$. For example, consider

$$\text{var}\left(\begin{array}{c} Z(s+h) \\ Z(s) \end{array}\right) = \left[\begin{array}{cc} \Sigma(s+h) & C(s+h,s) \\ C(s,s+h) & \Sigma(s) \end{array}\right]$$
$$= \left[\begin{array}{cc} \Sigma & C(h) \\ C(-h) & \Sigma \end{array}\right],$$

under an assumption of stationarity of the covariance matrix. Henceforth, this stationarity is assumed. Notice that $C(h)$ is not necessarily symmetric, although it is square, and $C(-h) = C(h)'$. When $h = 0$, $C(0) = \Sigma$, which is symmetric.

Consider the singular value decomposition (SVD) of $\Sigma^{-1/2}C(h)\Sigma^{-1/2}$, that is,

$$\Sigma^{-1/2}C(h)\Sigma^{-1/2} = UTV',$$

where $T = \text{diag}(\tau_1,\ldots,\tau_r)$, $\tau_i > 0 \ \forall i$, r is the rank of $C(h)$, U is $p \times r$, $U'U = I$, V' is $r \times p$, $V'V = I$. The $\{\tau_i^2\}$ are eigenvalues of $\Sigma^{-1/2}C(h)'\Sigma^{-1}C(h)\Sigma^{-1/2}$, or of $\Sigma^{-1/2}C(h)\Sigma^{-1}C(h)'\Sigma^{-1/2}$. The r columns of U are the eigenvectors of $\Sigma^{-1/2}C(h)\Sigma^{-1}C(h)'\Sigma^{-1/2}$ and the r rows of V' are the eigenvectors of $\Sigma^{-1/2}C(h)'\Sigma^{-1}C(h)\Sigma^{-1/2}$. Then, $U'\Sigma^{-1/2}C(h)\Sigma^{-1/2}V = T = \text{diag}(\tau_1,\ldots,\tau_r)$ Suppose we define $P = \Sigma^{-1/2}U$ and $Q = \Sigma^{-1/2}V$. Then:

$$U'U = I \ \Leftrightarrow \ P'\Sigma P = I$$
$$V'V = I \ \Leftrightarrow \ Q'\Sigma Q = I$$
$$U'\Sigma^{-1/2}C(h)\Sigma^{-1/2}V = T \ \Leftrightarrow \ P'C(h)Q = T = \text{diag}(\tau_1,\ldots,\tau_r).$$

Now consider the problem of maximizing

$$\text{corr}(l'Z(s+h), m'Z(s))$$

with respect to m,l. Notice that the minimum/maximum autocorrelation factors (MAFs) of Switzer and Green (1984) are a special case with $l = m$ (Cressie and Helterbrand, 1994). Hence the problem above is more general and the correlation can achieve a higher maximum. We see that

$$\text{corr}(l'Z(s+h), m'Z(s)) = \frac{l'C(h)m}{\{l'\Sigma l \cdot m'\Sigma m\}^{1/2}}.$$

Under the conditions $l'\Sigma l = m'\Sigma m = 1$, it is immediate that

$$\mathrm{var}(l'Z(s+h) - m'Z(s)) = 2\{1 - \mathrm{corr}(l'Z(s+h), m'Z(s))\}.$$

Hence maximising the correlation is equivalent to minimising the variability of (a linear combination of) $Z(s)$ contrasted with (a different linear combination) of the neighbouring vector $Z(s+h)$.

From standard results in in canonical correlation analysis (e.g., Dillon and Goldstein, pp. 340-2), the problem is equivalent to solving the following canonical equations:

$$(\Sigma^{-1}C(h)\Sigma^{-1}C(h)' - \lambda I)l = 0 \text{ and } (\Sigma^{-1}C(h)'\Sigma^{-1}C(h) - \lambda I)m = 0.$$

The largest eigenvalue, λ, of the product matrix $\Sigma^{-1}C(h)\Sigma^{-1}C(h)'$ or of $\Sigma^{-1}C(h)'\Sigma^{-1}C(h)$ is the squared canonical correlation coefficient. It can be shown that the associated eigenvectors satisfy

$$l = \Sigma^{-1}C(h)m/\sqrt{\lambda} \quad \text{and} \quad m = \Sigma^{-1}C(h)'l/\sqrt{\lambda}.$$

We normalise l and m so that $l'\Sigma l = m'\Sigma m = 1$. Thus, the SVD analysis and the canonical correlation analysis are the same, where the canonical correlation coefficient $\sqrt{\lambda} = \tau_1$, $l = p_1$ (the first column of P from the SVD), and $m = q_1$ (the first column of Q from the SVD).

2.2 Multivariate Spatial Models

The use of canonical correlations and principal components to determine the presence of common factors is well known in multivariate analysis (Anderson, 1958, for example). There is also a considerable literature on canonical correlations for multivariate time series, where the presence of temporal dependence gives a new perspective to the approach (for example, Akaike, 1976; Box and Tiao, 1977; Cooper and Wood, 1982; Pena and Box, 1987; Aoki, 1987; Tiao and Tsay, 1989). There are two related features of this time series literature that make their generalization to spatial data unnatural. First, the notion of "past, present, and future" is important for the methods that use canonical correlations of past behaviour with present and future behaviour. Second, the models used are of the autoregressive-moving-average type, which has no natural analogue in space (Cressie, 1993, Section 6.3). Wackernagel (1988) has proposed spatial common-factor models that do not require autoregressions or moving averages for their definitions. Both result in symmetric $C(h)$ and so are considered to be too specialized. We now consider a class of models that includes Wackernagel's, but, in general, allows a non-symmetric covariance matrix function $C(h)$.

The class of models considered are the *shifted-lag* models, defined as

$$Z_j(s) = \sum_{i=1}^{k} a_{ij} W_i(s + \Delta_{ij}) + \epsilon_j(s), \quad j = 1,\ldots,p,$$

where $\{W_1(\cdot), \ldots, W_k(\cdot), \varepsilon_1(\cdot), \ldots, \varepsilon_p(\cdot)\}$ are pairwise independent (or at least uncorrelated). There are $p \times k$ lag vectors $\{\Delta_{ij} : i = 1, \ldots, p; j = 1, \ldots, k\}$ that are in general unknown parameters of the model. The unknown parameters $\{a_{ij} : i = 1, \ldots, p; j = 1, \ldots, k\}$ are made identifiable by assuming, for example, that $\mathrm{var}(W_i(s)) \equiv 1$. Finally, $\varepsilon_j(\cdot)$ is a white-noise process with $v_j^2 \equiv \mathrm{var}(\varepsilon_j(s))$; $j = 1, \ldots, p$. For example, if $k = 1$ and $p = 2$, $Z_1(s)$ and $Z_2(s)$ derive their co-dependence from a common spatial process W_1. Without loss of generality, assume that $\Delta_{11} = 0$ and $a_{11} = 1$; then $Z_1(s) = W_1(s) + \varepsilon_1(s)$ and $Z_2(s) = a_{12} W_1(s + \Delta_{12}) + \varepsilon_2(s)$.

Assume further that each W_i is second-order stationary. Then, for $j \neq j'$, the (j, j')-th element of $C(h)$ is

$$\mathrm{cov}(Z_j(s + h), Z_{j'}(s)) = \sum_i a_{ij} a_{ij'} C_i(h + \Delta_{ij} - \Delta_{ij'}),$$

where $C_i(h) \equiv \mathrm{cov}(W_i(s + h), W_i(s))$; $i = 1, \cdots, k$. Also, the j-th diagonal element of $C(h)$ is

$$\mathrm{var}(Z_j(s)) = \sum_i a_{ij}^2 + v_j^2; \quad j = 1, \ldots, p.$$

Notice that $C(h) \neq C(-h)$ unless Δ_{ij} does not depend on j. When Δ_{ij} is actually Δ_i, $C(h)$ is symmetric and the shifted-lag models reduce to those of Wackernagel (1988). Further, $\mathrm{cov}(Z_j(s), W_i(s + h)) = a_{ij} C_i(h - \Delta_{ij})$. This function is symmetric about $h = \Delta_{ij}$. Thus, an indication of the value of Δ_{ij} is obtained by shifting $C_i(\cdot)$ until it is symmetric.

However, $W_i(\cdot)$ is unknown. For the purpose of exploratory analysis, we propose to use a part of the spatial process that "looks like" $W_i(\cdot)$ in a number of important ways. We shall now establish that $p_i' Z(s + h) - q_i' Z(s)$ is a substitute for $W_i(s)$, where $p_i = \Sigma^{-1/2} u_i$, $q_i = \Sigma^{-1/2} v_i$, $\{u_i\}$ are the r eigenvectors of $\Sigma^{-1/2} C(h) \Sigma^{-1} C(h)' \Sigma^{-1/2}$, and $\{v_i\}$ are the r eigenvectors of $\Sigma^{-1/2} C(h)' \Sigma^{-1} C(h) \Sigma^{-1/2}$. We require that $\mathrm{cov}(W_i(s), W_{i'}(s')) = 0 \; \forall i \neq i', \forall s, s'$. This implies that $\mathrm{cov}(p_i' Z(s+h) - q_i' Z(s), p_{i'}' Z(s'+h) - q_{i'}' Z(s')) = 0$ for $s = s'$ and $\forall i \neq i'$. That is, we require

$$p_i' \Sigma p_{i'} + q_i' \Sigma q_{i'} - p_i' C(h) q_{i'} - q_i' C(-h) p_{i'} = 0, \forall i \neq i'.$$

But this is true because we know that $P \Sigma P = I = Q' \Sigma Q$, and $P' C(h) Q = T = \mathrm{diag}(\tau_1, \ldots, \tau_r)$.

For h fixed, the SVD (equivalently, canonical correlation) analysis yields a set of processes $\{p_i' Z(s + h) - q_i' Z(s)\}$ that are substitutes for the spatial factors $\{W_i(s)\}$. These depend on a choice of h in much the same way that Switzer and Green's (1984) MAFs do. We investigate their stability as h is varied and their power as an exploratory tool for building shifted-lag models.

2.3 Interactive and Dynamic Graphical Tools

Two aspects of the methods discussed in the previous sections are ideally approached with interactive and dynamic graphical tools. The first is studying linear combinations of $Z(s+h)$ versus $Z(s)$ in a more flexible manner than that allowed by canonical correlation analysis alone, and the second is studying the effect of varying h by providing interactive controls. The work that we describe is new but there is some history to the ideas which we now relate.

In canonical correlation analysis, as discussed in the previous sections, the problem is to predict $y = (y_1, \ldots, y_q)'$ from $x = (x_1, \ldots, x_p)'$. When the solution is restricted to linear predictors, similar reasoning to that given in Section 2 shows that the optimal predictor is given by the linear combinations $u = Ax$ and $v = By$ which minimize $E[||v - u||^2]$, when assuming $E[v - u] = 0$, and where A and B each have r linearly independent rows satisfying $A \cdot \text{var}(x) \cdot A' = I$ and $B \cdot \text{var}(y) \cdot B' = I$. This can be solved algebraically (for example, via SVD) but a more general and flexible approach is to visualize the minimization process and to incorporate user interaction.

Consider arbitrary linear combinations of x and y: $\alpha'x$ and $\beta'y$ where $\alpha \in S^{p-1}$ (S^{p-1} is a $(p-1)$-sphere in $\mathbb{R}^p$) and $\beta \in S^{q-1}$. The sequence of projections

$$T = \{(\alpha(t)'x, \beta(t)'y) : t \text{ is time-continuous on } \mathbb{R}\}$$

shown dynamically is defined as a *correlation tour* (Buja et al., 1988). The term "correlation" derives directly from canonical correlation analysis. The word "tour" makes the analogy to the automatic motion algorithm generating movement over T developed for the grand tour (Asimov, 1985). The application of a correlation tour has broad scope. For example, in the regression setting ($q = 1$) let the vertical axis display be the residuals after fitting the linear model $y = x'\theta + \varepsilon$. The correlation tour plots the residuals against arbitrary linear combinations of x allowing more general checking of the model fit than that given by plotting the residuals versus $x_1, \ldots, x_p$, and the fitted values. For our purpose, set

$$x = Z(s) \quad \text{and} \quad y = Z(s+h),$$

and consider linear combinations of the form

$$\alpha_\Sigma = \Sigma^{-1/2}\alpha \ (\in \mathcal{E}^{p-1}) \quad \text{and} \quad \beta_\Sigma = \Sigma^{-1/2}\beta \ (\in \mathcal{E}^{q-1}),$$

where $\mathcal{E}^{p-1}$ is an ellipse in $\mathbb{R}^p$ of the form $\{w \in \mathbb{R}^p : w'\Sigma w = 1\}$ (and similarly for $\mathcal{E}^{q-1}$). We are interested in finding the linear combinations of $Z(s)$ which best predict $Z(s+h)$. The sequence of projections

$$T_\Sigma = \{(\alpha_\Sigma(t)'x, \beta_\Sigma(t)'y) : t \text{ is time-continuous on } \mathbb{R}\}$$

is shown dynamically. In the implementation, Σ will be estimated by the sample variance-covariance matrix of $\{Z(s_1), \ldots, Z(s_n)\}$.

In addition to allowing the directions of movement, α, β, to vary randomly (like that provided in the "Data Viewer" (Buja et al., 1988)), we use the quantity $-E[|| \beta'_\Sigma Z(s+h) - \alpha'_\Sigma Z(s)||^2]$ (or, equivalently, $\text{corr}(\beta'_\Sigma Z(s+h), \alpha'_\Sigma Z(s)))$ as an index I_{SC} (Index of Spatial Continuity) to direct the movement of the correlation tour in the direction that maximises I_{SC}. In practice, $-E[|| \beta'_\Sigma Z(s+h) - \alpha'_\Sigma Z(s)||^2]$ is estimated by a sample average. The maximum of I_{SC} is the largest eigenvalue, τ_1, and $(\alpha_\Sigma, \beta_\Sigma)$ are the associated eigenvectors (called p_1, q_1 in Section 2.1). To be identical to the procedure discussed in Section 2.1 we would proceed iteratively, to tour in the orthogonal subspaces to find the remaining $r - 1$ eigenvalues and corresponding eigenvectors. There are several advantages of this graphical approach. We have finessed the need to estimate the covariance matrix $C(h)$. We have a visual summary of the lack of symmetry in $C(h)$ by comparing results for h and $-h$, as described below. Other indices, for example, $\text{med}|\beta'_\Sigma Z(s+h) - \alpha'_\Sigma Z(s)|^{1/2}$, may also be considered. Finally, features of the data, such as outliers, affecting the numerical solutions, can be detected visually and addressed as an integral part of the process. If $\alpha_\Sigma, \beta_\Sigma$ are synchronized so that $\alpha_\Sigma \equiv \beta_\Sigma$ then the process is analogous to finding the MAFs suggested by Switzer and Green (1984).

The nature of fixing h is effectively subsetting on pairwise spatial displacement and augmenting the attribute variables with a spatially lagged set of variables, effectively doubling the number of variables. Allowing h to vary enables one to study and compare different subsets. To see this, consider

$$N(h) = \{(s_i, s_j) : s_i - s_j = h; s_i, s_j \in D\}.$$

For example, let $\dim(D) = 1$; then varying h can be achieved by interactively brushing in a plot of h. Figure 4 shows an example on a data set with observations taken at regularly spaced points. All pairwise locations, $\{(s_i, s_j) : s_i, s_j \in D\}$, are plotted and these form a grid in $\mathbb{R}^2$; points of zero distance lie along the diagonal $s_i = s_j$, points with $s_i < s_j (h > 0)$ are in the upper triangle, and points with $s_i > s_j (h < 0)$ are on the lower triangle. This form is awkward to brush using the usual rectangular brush, so we have rotated the points $45°$ and split them at $h = 0$ so that the triangle of points on the left represent those with $h > 0$ and points on the right represent those with $h < 0$. This format also allows brushing on $|h|$ to check symmetry of $C(h)$. Figure 4 shows two window dumps, one displays the spatial lags in the manner described above with lag $h = 1$ brushed as $\times$ and the second plot shows a snapshot, $(Z_1(s), Z_1(s+h))$, of a correlation tour on $(Z(s)', Z(s+h)')'$ with corresponding points brushed. (The data is simulated from a shifted-lag model containing no covariance structure.) Erase brushing can be used instead of glyph brushing to concentrate attention on specific lags only. The method is also general enough to be used when the observations are made on irregular intervals. When $\dim(D) = 2$, it may be of more interest to consider h as an angle, a magnitude, and allow brushing on angle or magnitude or both.

The usefulness of varying h can be appreciated by considering the methods discussed in Cressie and Majure (1993). There, plots of 9 pairwise com-

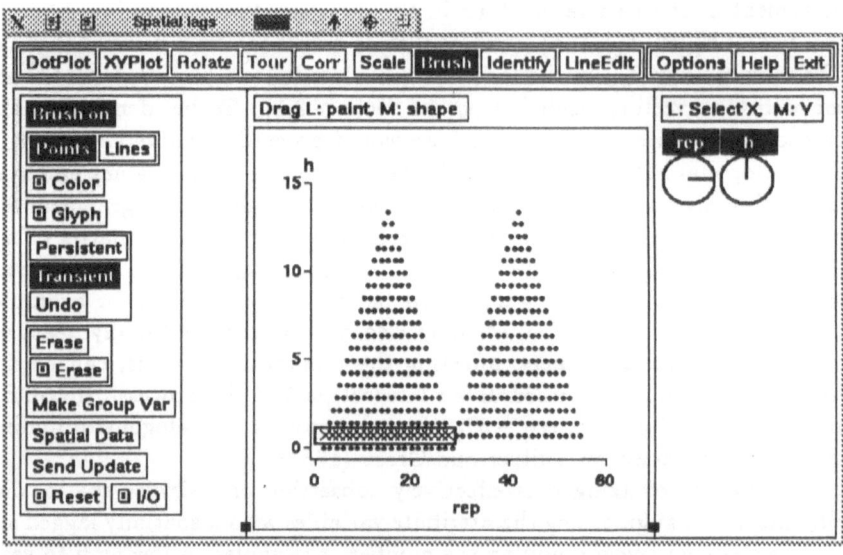

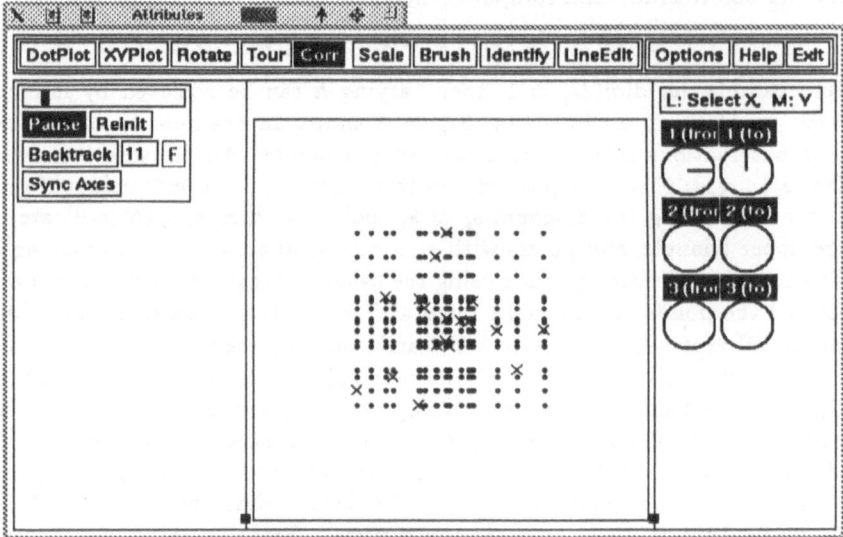

Figure 4: *Brushing on spatial lags linked to a correlation tour in XGobi. The top plot shows XGobi in* Brush *mode with all spatial lags plotted, and lag $h = 1$ are brushed ($\times$). The bottom plot shows XGobi in* Corr *mode, at the starting projection $(Z_1(s), Z_1(s + h))$ of a correlation tour on $Z(s)$ versus $Z(s + h)$.*

binations of 3 attributes are examined at different spatial lags. To reduce the visual overload, summary measures are computed for each pair and lag. These numerical measures are plotted by lag for each pair of variables and these plots are examined to gain insight into spatial dependence over lags. This is a process beautifully suited to an interactive environment, where the user can change pairwise plots at will and brush (subset) on spatial lags interactively.

3 Integration of Interactive and Dynamic Graphics Tools into a GIS

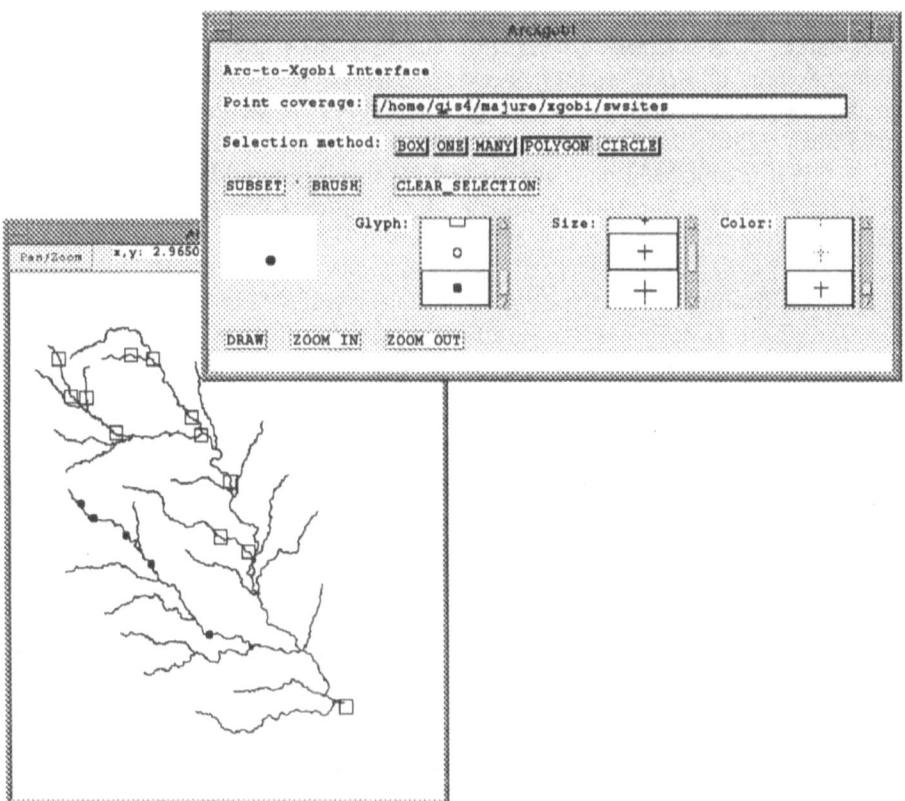

Figure 5: *ARC/INFO control panel and example map view used in the link with XGobi.*

Emphasis in GIS development has been on the input of data, its management (storage, retrieval) and display of maps, graphs, and tables. GISs have some capability to allow statistical analyses but it is generally limited. A number of recent suggestions have been made (Openshaw, 1991; Anselin and Getis,

1992; Ding and Fotheringham, 1992; Fotheringham and Rogerson, 1993, for example) to redress this imbalance. Our research addresses the extremely important problem of multivariate exploratory spatial data analysis in a GIS. GIS data structures allow the representation of areal features for the storage of information reported at an aggregated spatial level, such as counties or census tracts, and linear features for the storage of information collected from a stream or a transportation network, for example. The topological data structure of a GIS makes it possible to determine spatial relationships between sampling locations, such as stream sites, that would be difficult to determine otherwise. The display capabilities of a GIS allow the spatial variables to be overlaid on a background of hydrography, transportation, population, land use, or other information relevant to the attributes being considered. For example, in Figure 5 the map view shows sampling sites along streams in Erath county, Texas. Information about the topography or land use near a sample site can give valuable insights into the values of attributes collected at the site.

A GIS is intrinsically multivariate and yet this is ignored by the univariate statistical analyses currently available. By building an interface between a GIS and software for dynamic graphics, we will also provide a good platform for developing new graphical methods (for example, those discussed in Section 2.3) by facilitating use and testing on data sets available in the GIS.

Our efforts have focused on interfacing the GIS software, ARC/INFO™ with XGobi (Swayne et al., 1991). XGobi provides interactive and dynamic graphical tools in the X Window System™environment for exploring multivariate data through the manipulation of scatterplots. ARC/INFO is used to maintain the data base and to display the geography, while XGobi primarily is used to explore the relationships amongst the attributes. Figure 6 shows how the communication between these two programs is established.

Before any connection can be established, ARC/INFO and an ARC/XGobi interface process must be activated on the same host where the ARC/INFO data base is maintained. An XGobi process, that is, a client either residing on the same host or anywhere else, that wants to use some ARC/INFO data sets has to connect to the ARC/XGobi interface process, that is, the server. The XGobi client can select whether it wants the data set that is currently selected within ARC/INFO and possible updates of this selection or only static data that is available in the ARC/INFO data base. In the latter case it is not even required that ARC/INFO is running. The interprocess communication is based on Stevens' (1990) concurrent server example, and uses a TCP (Internet stream) socket. This setup implies that upon receiving a connection request from an XGobi client, the ARC/XGobi server forks an identical child process, and thus establishes a one–to–one connection between an ARC/XGobi server and an XGobi client. A more detailed description of

---

™ *ARC/INFO* is a trademark of Environmental Systems Research Institute, Inc.

™ *X Window System* is a trademark of MIT.

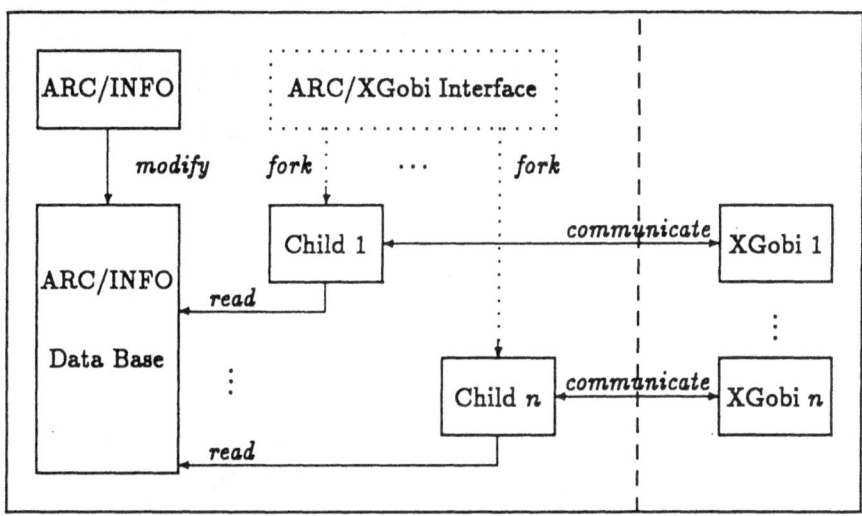

Figure 6: *Interface linking ARC/INFO with XGobi.*

this link can be found in Symanzik et al. (1994).

The main purpose of the ARC/XGobi servers (children) is to continuously check whether the ARC/INFO data base has been changed. Changes to the ARC/INFO data base reflect the results of brushing and subsetting operations conducted through the ARC/INFO control panel shown in Figure 5. When changes are detected the newly brushed or subsetted points are immediately passed to each XGobi client. The user can brush the sampling locations in the map view using a variety of brush types and choice of glyphs, with different size and color combinations. The choice has been coordinated to reflect those available in XGobi. The attribute values currently visible in XGobi that are linked to the marked coordinates will instantaneously obtain the same glyph. So, this interface allows the user to link views interactively such that modifications of one view automatically change the other views in the different XGobi clients.

4 Summary

In the analysis of spatial data, it is desirable to keep the spatial variables firmly rooted in the geography of the area of interest and also provide a high level of interaction with the data using dynamic graphics tools. With this in mind, we have built a link between the GIS software ARC/INFO and software for interactively exploring multivariate data with dynamic graphics, XGobi. The combination provides tools for multivariate exploratory spatial data analysis for examining both large-scale variation (trend) and small-scale variation (spatial dependence) in the attribute variables, $Z(s)$, and a platform for exploring new methods. We have focused on developing tools for detecting spatial dependence, in particular, as applied to shifted-lag models. The graphical tools are dynamic involving displays of linear combinations of $Z(s+h)$ versus $Z(s)$, in a correlation tour. The effectiveness and usefulness of the methods shall be assessed on simulated and actual data.

Acknowledgements

Cressie's research was supported by the National Science Foundation, the National Security Agency, and the Office of Naval Research. Symanzik was partially supported by a German "DAAD–Doktorandenstipendium aus Mitteln des zweiten Hochschulsonderprogramms".

References

Akaike, H. (1976). Canonical Correlation Analysis and the Use of an Information Criterion. In Mehra, R. K. and Lainiotis, D. J., editors, *Systems Identification: Advances and Case Studies*. Academic Press, New York, NY, pages 27–96.

Anderson, T. W. (1958). *An Introduction to Multivariate Statistical Analysis.* Wiley, New York, NY.

Andrews, D. F. and Herzberg, A. M. (1985). *Data - A Collection of Problems from Many Fields for the Student and Research Worker.* Springer-Verlag, New York, NY.

Anselin, L. and Getis, A. (1992). Spatial Statistical Analysis and Geographic Information Systems. *Annals of Regional Science*, 26:19–33.

Aoki, M. (1987). *State Space Modelling of Time Series.* Springer, New York, NY.

Asimov, D. (1985). The Grand Tour: A Tool for Viewing Multidimensional Data. *SIAM Journal Scientific and Statistical Computing*, 6(1):128–143.

Box, G. E. P. and Tiao, G. C. (1977). A Canonical Analysis of Multiple Time Series. *Biometrika*, 64:355–365.

Bradley, R. and Haslett, J. (1992). High-interaction Diagnostics for Geostatistical Models of Spatially Referenced Data. *The Statistician*, 41:371–380.

Buja, A. and Asimov, D. (1986). Grand Tour Methods: An Outline. *Computing Science and Statistics*, 17:63–67.

Buja, A., Asimov, D., Hurley, C., and McDonald, J. A. (1988). Elements of a Viewing Pipeline for Data Analysis. In Cleveland, W. S. and McGill, M. E., editors, *Dynamic Graphics for Statistics*, pages 277–308. Wadsworth, Monterey, CA.

Buja, A., McDonald, J. A., Michalak, J., and Stuetzle, W. (1991). Interactive Data Visualization using Focusing and Linking. In Nielson, G. M. and Rosenblum, L., editors, *Proceedings of Visualization '91*. IEEE Computer Society Press, Los Alamitos, CA, pages 156–162.

Chauvet, P. (1982). The Variogram Cloud. In *Proceedings of the 17th APCOM Symposium*. Colorado School of Mines, Golden, CO, pages 757–764.

Cooper, D. M. and Wood, E. F. (1982). Identifying Multivariate Time Series Models. *Journal of Time Series Analysis*, 3:153–164.

Cressie, N. (1984). Towards Resistant Geostatistics. In Verly, G., David, M., Journel, A., and Marechal, A., editors, *Geostatistics for Natural Resources Characterization, Part 1*. Reidel, Dordrecht, pages 21–44.

Cressie, N. and Helterbrand, J. D. (1994). Multivariate Spatial Statistical Models. *Geographical Systems*, 1. Forthcoming.

Cressie, N. and Majure, J. (1993). Visualizing Spatial Dependence in Multivariate Data. Technical Report 93-4, ISU Statistical Laboratory Preprint Series.

Cressie, N. A. C. (1993). *Statistics for Spatial Data (revised edition)*. Wiley, New York, NY.

Dillon, W. R. and Goldstein, M. (1984). *Multivariate Analysis: Methods and Applications*. Wiley, New York, NY.

Ding, Y. and Fotheringham, A. S. (1992). The Integration of Spatial Analysis and GIS. *Computers, Environment and Urban Systems*, 16:3–19.

Fotheringham, A. S. and Rogerson, P. (1993). GIS and Spatial Analytical Problems. *International Journal of Geographical Information Systems*, 7:3–19.

Getis, A. and Ord, J. K. (1992). The Analysis of Spatial Association by Use of Distance Statisics. *Geographical Analysis*, 24:189–206.

Goodchild, M. F. (1991). The Technological Setting of GIS. In Maguire, D. J., Goodchild, M. F., and Hind, D. W., editors, *Geographical Information Systems: Principles and Applications, vol. 1*. Longman, London, pages 9–20.

Goodchild, M. F., Haining, R. P., and Wise, S. (1992). Integrating GIS and Spatial Analysis - problems and possibilities. *International Journal of Geographical Information Systems*, 6:407–423.

Haslett, J., Bradley, R., Craig, P., Unwin, A., and Wills, G. (1991). Dynamic Graphics for Exploring Spatial Data with Application to Locating Global and Local Anomalies. *The American Statistician*, 45(3):234–242.

Maguire, D. J. (1991). An Overview and Definition of GIS. In Maguire, D. J., Goodchild, M. F., and Hind, D. W., editors, *Geographical Information Systems: Principles and Applications, vol. 1*. Longman, London, pages 9–20.

McDonald, J. A. and Willis, S. (1987). Use of the Grand Tour in Remote Sensing. ASA Video Collection, American Statistical Association, Alexandria, VA.

Openshaw, S. (1991). Developing Appropriate Spatial Analysis Methods for GIS. In Maguire, D. J., Goodchild, M. F., and Hind, D. W., editors, *Geographical Information Systems: Principles and Applications, vol. 1*. Longman, London, pages 389–402.

Pena, D. and Box, G. E. P. (1987). Identifying a Simplifying Structure in Time Series. *Journal of the American Statistical Association*, 82:836–843.

Stevens, W. R. (1990). *UNIX Network Programming*. Prentice–Hall, Englewood Cliffs, NJ.

Swayne, D. F., Cook, D., and Buja, A. (1991). XGobi: Interactive Dynamic Graphics in the X Window System with a Link to S. In *ASA Proceedings of the Section on Statistical Graphics*. American Statistical Association, Alexandria, VA, pages 1–8.

Switzer, P. and Green, A. A. (1984). Min/max Autocorrelation Factors for Multivariate Spatial Imagery. Technical Report 6, Department of Statistics, Stanford University.

Symanzik, J., Majure, J., Cook, D., and Cressie, N. (1994). Dynamic Graphics in a GIS: A Link between ARC/INFO and XGobi. *Computing Science and Statistics*, 26. To appear.

Tiao, G. C. and Tsay, R. S. (1989). Model Specification in Multivariate Time Series. *Journal of the Royal Statistical Society, Series B*, 51:157–195.

Wackernagel, H. (1988). Geostatistical Techniques for Interpreting Multivariate Spatial Information. In Chung, C. F., Fabbri, A. G., and Sinding-Larsen, R., editors, *Quantative Analysis of Mineral and Energy Resources*. Reidel, Dordrecht, pages 393–409.

Stochastic Modelling of Spatial and Dynamic Patterns. Applications in Ecology

Hans-Peter Bäumer, Heidrun Ortleb, Dietmar Pfeifer, Ulrike Schleier-Langer

HRZ–Angewandte Statistik, Carl von Ossietzky Universität Oldenburg

Abstract. Time dependent Poisson point processes are considered to describe spatial point patterns in time. Asymptotic equilibrium, extinction and explosion as long-time behaviour of the system may (eventually) occur. To lay the appropriate foundation for programming adaption of such processes to discrete and finite structures will be investigated. A further type of data from ecological case studies are spatially aggregated counts. An open stochastic network model will be proposed to describe the time dependent abundance of different species in a finite number of observation windows considering simple kinds of interaction.

Keywords. Stochastic point processes, stochastic networks, spatial and dynamic point patterns, spatially aggregated counts

1 Time Dependent Poisson Point Processes on R^d

Time dependent Poisson point processes $\{\xi_t\}_{t \geq 0}$ on R^d, $d \in N$, of the following form

$$\xi_t = \sum_{k=1}^{C(t)} 1_{\{T_k > t\}} \varepsilon_{X_k(t)}, \ t \geq 0$$

are investigated in the paper by Pfeifer, Bäumer, and Albrecht (1993).

$\{C(t)\}_{t \geq 0}$ denotes a Poisson counting process on $R^+ \cup \{0\}$ with finite cumulative intensity measure $E[C(t)] = \Lambda(t)$. $\Lambda(t)$ is some weakly increasing absolutely continuous function with $\Lambda(0) \geq 0$, $t \geq 0$. This Poisson process governs birth of particles. $I_k(t) = 1_{\{T_k > t\}}$ is indicator variable of the event $\{T_k > t\}$ where $\{T_k\}_{k \in N}$ is a family of iid random variables with absolutely continuous cdf F, $F(0)=0$, and density $f=F'$. T_k controls life length of each particle. $\{X_k(t) \mid t \geq 0\}_{k \in N}$ denotes a family of iid random variables with $X_k(t) \in R^d$ which controls location of each particle in R^d. For each Borel set B ε_x denotes the Dirac measure with mass 1 at $x \in B$.

In short, given a Borel set B $\xi_t(B)$ describes the number of particles with $X_k(t)$ in B which have an expiring date later than t.

Weak convergence of Poisson point processes ξ_t to some Poisson point process ξ with an appropriate intensity measure μ is proven (Pfeifer, Bäumer, and Albrecht (1993), p. 251 ff.).

Long-time behaviour of such Poisson point processes $\{\xi_t\}_{t \geq 0}$ is studied with the following specifications. The cumulative intensity $\Lambda(t)$ of the Poisson birth process $\{C(t)\}_{t \geq 0}$ is $t^{d/2}\exp(c\tau t)+\Lambda(0)$, $t \geq 0$, $c \geq 0$, $\tau > 0$. Lifetime density is $f(t)=\tau\exp(-\tau t)$. Initial location is distributed $N(0,\sigma_0^2 I)$, $\sigma_0^2 > 0$, I identity matrix. Movement is governed by componentwise independent Brownian motion B_t distributed $N(0,\sigma^2 t I)$, $\sigma^2 > 0$. Intensity measure μ of ξ is proportional to m_d, the d-dimensional Lebesgue measure. In the case of c=1, asymptotic equilibrium of the system will be achieved. For c < 1 extinction (eventually) occurs, and for c > 1 the system (eventually) explodes (s. Pfeifer, Bäumer, and Albrecht (1993), p. 252 ff.).

As an introduction to random point processes on R^d the monograph of König and Schmidt (1992) is recommended.

For computational purposes the model has to be adapted to discrete and finite structures.

2   Adaption to Discrete and Finite Structures

The Poisson point processes $\{\xi_t\}_{t \geq 0}$ will be considered in discrete time steps of length h, h > 0. As spatially finite structure in R^2 a circle of finite radius R around the origin is proposed. To determine an appropriate R movement and lifetime of particles as well as hardware restrictions have to be taken into account. To visualize realizations of the Poisson point processes ξ_{kh}, k=0,1,..,n, on a screen a suitable square submatrix of the pixel matrix available is assumed a square of side length SL with the origin as midpoint. Particles trying to cross the boundary of the circle of radius R, $R > 2^{1/2}SL$, become caught at the boundary. Then, the Poisson point processes $\{\xi_t\}_{t \geq 0}$ are splitted up into three independent Poisson point processes. For the Poisson point processes in the open circle weak convergence to some Poisson point process ξ_C with an intensity measure proportional to m_2 is preserved.

The probability β, $0 < \beta < 1$, to survive m successive time steps h is $\exp(-\tau mh)$. Truncating $\ln\beta/(-\tau h)$ to integer is a conservative choice of m.

Location of particles at time t+h is controlled by iid random variables $X_k(t+h)=X_k(t)+Z_k(h)$ which are distributed $N(0,\sigma_0^2+\sigma^2(t+h)I)$ where $Z_k(h)$ is $N(0,\sigma^2 hI)$. Let L_{max} denote the Euclidean distance between $X_k(t)$ and $X_k(t+mh)$ which will maximally be covered by a particle during m successive time steps h with probability 1-α. The random variable $Y_m=\|Z_k(mh)\|^2/(m\sigma^2 h)$ is distributed $\chi^2_{df=2}$ implying $L_{max}=(m\sigma^2 h\chi^2_{df=2;q=1-\alpha})^{1/2}$. Then, each particle with an

Euclidean distance to the origin equal to or greater than $R=L_{max}+2^{1/2}SL$ will be captured at the boundary of the circle.

The net increase of particles in the open circle of radius R after a discrete time step h will be determined next. The Poisson birth processes $C(t)$, $C(t+h)$ are independent implying $C(t+h)-C(t)$ to be a Poisson process ψ with intensity measure $E\psi=E[C(t+h)]-E[C(t)]=\Lambda(t+h)-\Lambda(t)$. In the case of $d=2$ intensity measure $E\psi$ is $exp(c\tau t)[t(exp(c\tau h)-1)+hexp(c\tau h)]$. With lifetime density $f(t)=\tau exp(-\tau t)$ cdf is $F(t)=1-exp(-\tau t)=P(T_k \leq t)$. Regarding p-thinning of the Poisson process ψ with $p=P(T_k>t+h)=1-F(t+h)=exp(-\tau(t+h))$ a Poisson process ζ results with intensity measure $E\zeta=pE\psi=exp(-\tau(t+h))E\psi$. Considering spatial restriction to the open circle around the origin with radius R implies probability $p^*=P(\|X_k(t+h)\|<R)=1-exp(-R^2/(2(\sigma_0^2+\sigma^2(t+h))))$. Then, a Poisson process ζ^* with intensity measure $E\zeta^*=p^*E\zeta=(1-exp(-R^2/(2(\sigma_0^2+\sigma^2(t+h)))))exp(-\tau(t+h))E\psi$ results by p^*-thinning of the Poisson process ζ. The Poisson process ζ^* describes the net increase of particles in the open circle of radius R in the time interval $[t,t+h]$.

Location of particles belonging to the net increase is governed by iid random variables $R_k(t+h)$ with $R_k(t+h) \in R$ and iid φ_k which is uniformly distributed over the interval $[0,2\pi]$. The conditional cdf of $R_k(t+h)$ given $\|X_k(t+h)\|<R$ is obtained as $F_{t+h}(r)=(1-exp(-r^2/(2(\sigma_0^2+\sigma^2(t+h)))))/(1-exp(-R^2/(2(\sigma_0^2+\sigma^2(t+h)))))$ with $r \in [0,R]$ and $F_{t+h}=0$ elsewhere.

Let M_t be the number of particles with $T_k > t$, then the number of particles with $T_k \leq t+h$ is distributed $B(M_t,p_b)$ with $p_b=1-exp(-\tau h)$. By inversion of the binomial distribution $B(M_t,p_b)$ the number of particles with $t < T_k \leq t+h$ is obtained.

3   Applications to Ecological Case Studies

On the basis of the theoretical results mentioned above a programme was implemented in Borland Pascal, version 7.0, using the object oriented library Vision. Input parameters are c, τ, σ_0, σ, $\Lambda(0)$, and h with default 1. User specified scale of time and scale of length are optional input. As a facultative feature of the programme time dependent planar Poisson Voronoi diagrams are generated.

For spatial tesselations and especially for applications of Poisson Voronoi diagrams the monograph of Okabe, Boots, and Sugihara (1992) may be consulted.

In the monograph of Richter and Söndgerath (1990) as well as in some recent papers (e.g. Pfeifer, Schleier-Langer, and Bäumer (1994)) applications of point process theory to ecological case studies are discussed. Simulation of features of the ecological process of repopulation of artificially depopulated experimental fields in the Lower Saxon Wadden Sea is a further example (cf. Bäumer, H.-P. (1994)). Substantive interpretation of the time dependent variation of the abundance of the gastropod *Hydrobia ulvae* repopulating the experimental fields implies the hypothesis of equilibrium as long-time behaviour. Adults of this species are smaller than 1 mm. Setting c to 1, time dependent planar Poisson point processes are applied starting with no individuals in the experimental field. Their number is gradually increased approaching an equilibrium with $E\xi_C(B) \approx m_2(B)$, $B \in \mathfrak{R}^2$, the Borel σ-field over R^2.

But often in ecological case studies spatially aggregated data are gathered. As an example the abundance of two among several benthic species in each square of 100 cm^2 of a multicorer sample is presented in the following scheme.

245	142	326	52	293
222	368	84	18	67
239	25	477	213	204
18	570	183	494	47
238	119	126	591	20

0	1	2	0	2
4	4	1	3	2
1	0	5	1	1
0	1	1	2	0
0	0	1	2	0

Then, for a number of disjoint observation windows W_i, $i=1,...,J$, the abundance of different species $s_1,...,s_N$ in the window at time t is known, but the location of each individual is not. The observation windows are neither necessarily adjacent nor of equal area as in the scheme above.

This configuration is basic to multiclass queueing networks consisting of a set $J=\{1,...,J\}$ of queues and a set $C=\{c_1,...,c_N\}$ of customer classes. A network is said to be open if customers are allowed to leave or enter it.

A relatively simple open stochastic network model will be proposed to describe the dynamics of an assemblage of several species.

4 Rethinking an Open Stochastic Network Model

Let $W=\{W_1,..,W_J\}$ denote a set of observation windows and $S=\{s_1,...,s_N\}$ a set of faunal species. At a given time t, $t \geq 0$, each organism in the open system belongs to one of the N species and is in one of the J observation windows. Then, $n_{s,i}(t)$, $s=1,...,N$, denotes the abundance of species s in the window W_i at time t. The sum of the $n_{s,i}(t)$, $s=1,...,N$, is n_i, the total count of organisms in window W_i. Immigration, movement of individuals between observation windows, emigration, distribution and consumption of resources with regard to simple kinds of interaction as aggregation and repulsion are considered in the model.

For each species s the number of individuals immigrating into window W_i at time t is described as an independent Poisson count process $C_{s,i}(t)$, $t \geq 0$, with intensity $\mu_{s,i}=E[C_{s,i}(t)] \geq 0$. Then, the transposed vector $A_s=(\mu_{s,i})_{i=1,...,J}$ is called vector of immigration rates. The conditional probability for an immigrating individual to arrive in window W_i is $p_{s,oi}=\mu_{s,i}/\mu_s$ where μ_s denotes the sum of the $\mu_{s,i}$, $i=1,...,J$. An individual emigrating from the system will never return.

Movement of organisms of species s in the system is described by the routing matrix $P_s=(p_{s,ik})_{i,k=1,...,J}$. The probabilities $p_{s,ik}$ govern the internal transition of an individual of species s from W_i to W_k. Let $p_{s,i}$ denote the sum of the $p_{s,ik}$, $k=1,...,J$. Then, $p_{s,io}=1-p_{s,i}$ is the probability for an individual of species s to emigrate directly from window W_i.

Assuming the probability for an organism of species s to leave the system directly or indirectly to be larger than 0 ensures that matrix $I-P_s$ has an inverse which is a convergent series. Therefore, a unique transposed vector

$$Q_s = A_s(I-P_s)^{-1} = A_s \sum_{k=0}^{\infty}(P_s)^k$$

exists. The nonnegative components $q_{s,i}$ of vector Q_s may be considered the equilibrium immigration rate for individuals of species s in observation window W_i.

For each species s let $\{R_{s,i}|i=1,...,J\}$ denote a family of independent random variables with $R_{s,i}\in R^+$, cdf $F_{s,i}$, and finite expectation $E[R_{s,i}]=r_{s,i}$. $R_{s,i}$ controls the resources available for species s in window W_i. The independent random variable $X_{s,i}=R_{s,i,j}$ distributed according to $F_{s,i}$ governs the portion of these resources available to the j-th organism of species s in window W_i. At the arrival of the j-th individual of species s in window W_i at time t_0 its resources $\rho_{s,i}(t_0)=r_{s,i,j}(t_0)$ are a realization of random variable $X_{s,i}$. The individual resources $\rho_{s,i}(t_0)$ are spent with a rate which is proportional to some given function $v_i(n_i)$. In the case of $\rho_{s,i}(t)=0$ the j-th individual immediately moves to another window W_k according to $p_{s,ik}$ of the routing matrix P_s.

To investigate the long-time behaviour of the system the processes $\eta(t)=(\eta_1(t),...,\eta_J(t))$ with $\eta_i(t)=(n_i,\varphi_{i,1},...,\varphi_{i,ni})$ are introduced. The organisms are indexed with regard to their arrival time. With $\varphi_{i,j}=\varphi_{i,j}(t)=(s_j,r_{i,j},z_{i,j}(t))$, s_j species of j-th organism, $r_{i,j}$ individual resources, and $z_{i,j}(t)$ resources already spent by the j-th organism at time t, the processes $\eta(t)$ are Markovian. Then, several important equilibrium properties of the model exist (cf. Pollet (1986), p. 395 ff.).

To assert the most relevant property set $b_{s,i}=q_{s,i}r_{s,i}$ which may be considered the average amount of resources carried by organisms of species s in the window W_i and let b_i denote the sum of the $b_{s,i}$, $i=1,...,J$. Let further $d_i(n)$ denote the product of $jv_i(j)$, $j=1,...,n$, with $d_i(0)=1$. Then, if

$$e_i^{-1} = \sum_{n=0}^{\infty}b_i^n/d_i(n)$$

is finite for each window W_i an equilibrium distribution for $\eta(t)$ exists. In this case the n_i are independent under equilibrium distribution and the probability that n organisms are in window W_i is $P(n_i=n)=e_ib_i^n/d_i(n)$.

Under the simplifying conditions that the windows W_i are adjacent squares of equal area and all parameters are independent of the choice of index i the equilibrium distribution is Poissonian if no interaction between organisms is assumed. In the case of aggregation the equilibrium distribution is geometric. Assuming repulsion the equilibrium distribution is binomial. Combining no interaction and aggregation in the specification of the function $v=v_i$ allows to model the well known empirical relation of sample mean and variance observed for spatially aggregated counts.

Acknowledgement

The authors thank Dr. K. Borovkov for scientific cooperation. We are grateful to P. Sporea for programming and to all the colleagues in the Ökosystemforschung Niedersächsisches Wattenmeer who offered and discussed their data from ecological case studies.

References

Bäumer, H.-P. (1994): Stochastic Simulation of Dynamic Point Patterns. In Faulbaum, F. (ed.): SoftStat '93. Advances in Statistical Software 4. The 7th Conference on the Scientific Use of Statistical Software. Heidelberg, March 14-18, 1993. Stuttgart etc.: Fischer

König, D. and Schmidt, V. (1992): Zufällige Punktprozesse. Eine Einführung mit Anwendungsbeispielen. Stuttgart: Teubner

Okabe, A., Boots, B. and Sugihara, K. (1992): Spatial Tesselations. Concepts and Applications of Voronoi Diagrams. Chichester etc.: Wiley

Pfeifer, D., Bäumer, H.-P., and Albrecht, M. (1992): Spatial Point Processes and their Applications to Biology and Ecology. Modeling of Geo-Biosphere Processes, 1, 145-161

Pfeifer, D., Bäumer, H.-P., and Albrecht, M. (1993): Moving Point Patterns: the Poisson Case. In Opitz, O., Lausen, R., and Klar, R. (eds.): Information and Classification - Concepts, Methods, and Applications - (pp. 248-257). Proceedings of the 16th Annual Conference of the "Gesellschaft für Klassifikation e.V.". Dortmund, April 1-3, 1992. Studies in Classification, Data Analysis, and Knowledge Organization. Berlin etc.: Springer

Pfeifer, D., Schleier-Langer, U., and Bäumer, H.-P. (1994): The Analysis of Spatial Data from Marine Ecosystems. In Bock, H. H., Lenski, W., and Richter, M. M. (eds.): Information Systems and Data Analysis. Proceedings of the 17th Annual Conference of the "Gesellschaft für Klassifikation e.V.". Kaiserslautern, March 3-5, 1993. Studies in Classification, Data Analysis, and Knowledge Organization. Vol. 4. Heidelberg etc.: Springer

Pollett, P. K. (1986): Some Poisson Approximations for Departure Processes in General Queueing Networks. statistics, 17, 393-405

Richter, O. and Söndgerath, D. (1990): Parameter Estimation in Ecology. The Link between Data and Models. Weinheim: VHC

Part VII

Computational Aspects in Discrimination and Classification

Part VII

Computational Aspects in
Discrimination and
Classification

Classification and Discrimination: the REC-PAM Approach

Antonio Ciampi

Department of Epidemiology & Biostatistics, McGill University, Montreal and McGill University/Montreal Children's Hospital Research Institute

Abstract. RECPAM is a method for constructing regression trees from data which permits explicit treatment of a great variety of response variables. It is shown that it is possible to use the RECPAM methodology to find solutions to two basic problems of data analysis: discovering and identifying classes in data (classification), and finding simple and economical rules to assign individuals to classes (discrimination). The RECPAM based approach to classification finds classes which are economically described in terms of some of the variables and such that the joint distribution of all variables is, at least approximately, homogeneous within classes and distinct across classes. RECPAM based discrimination simultaneously treats categorical and continuous variables, while respecting their distinct nature: it builds discrimination models which can be seen as a generalization of tree-based models and multi-category logistic regression.

Key words: tree-growing, conceptual clustering, data-driven modelling, AI and Statistics

1 Introduction

Class formation and the development of rules to assign individuals to classes are two basic activities performed by the human mind when relating to the exterior world. It should therefore come as little surprise to find that these two activities are among the most popular both in statistical data analysis and Artificial Intelligence (AI): they are known to statisticians under the name of *classification* (or *clustering*) and *discrimination*, while in the AI community they are called *unsupervised* and *supervised learning*. It is somewhat more surprising to notice that statisticians and AI researchers seem to know little of each other's successful efforts in developing automated versions of these two basic activities.

The emphasis is different and complementary. Statisticians develop classification and discrimination algorithms having as primary goal that of properly taking into account the variation within the population and, possibly, measurement error; comparatively less attention is given to the correspondence between statistical models and human cognitive structures. As a result, *interpretation* is needed after the results of the algorithm are obtained, an effort that can be quite laborious and, occasionally, fruitless. This difficulty is especially apparent in the application of clustering algorithms: clusters are often obtained on the basis of geometries that are natural to the mathematician but 'look odd' to the scientist on whose behalf the clustering is performed.

On the other hand, the AI researcher working in the area of machine learning is mainly concerned with 'making sense', partially or entirely neglecting both natural variations and errors of measurement. For instance, in several machine learning algorithms, where individuals are called *examples*, the goal is to find a *generalization* which is perfectly consistent (hence without error) with *all* examples. Also, learning algorithms are developed with the aim of imitating some aspects of the learning human mind. Thus, the treatment of discrete variables is privileged, in keeping with the fact that continuous variables play very little role in natural concept formation (an ordinary child learns to distinguish between large and small dogs without resorting to weighing or measuring the size of neighborhood dogs!).

Meaningful communication, however, is now beginning to take place in the area known as tree-growing. The work of Breiman *et al.* (1984) from the statistical community, and that of Quinlan (1986), from the AI community, have jointly inspired much recent research (Buntine, 1992, Crawford, 1989, Chou, 1991, Chou *et al.* 1989, to quote but a few examples). Though omnipresent in AI, trees are recognized by statisticians mostly as tools for discrimination and prediction (Although the acronym CART, introduced by Breiman *et al.* (1984), stands for classification and regression trees, the word 'classification' is used here in the sense of 'assigning to known classes', hence discrimination in the language used throughout this paper. This is the same usage as in Gordon, 1981). It is one of the two purposes of this paper to demonstrate that trees offer a natural approach to both classification and discrimination: indeed, automated tree-growing mimics a basic strategy of the human mind, that underlies both class formation and the development of rules to assign individuals to classes. The other purpose is to show that a natural generalization of tree-growing can considerably improve the range of successful applications of this technique in discrimination.

The approach to tree-growing known as RECPAM is particularly suited to both purposes. It was originally developed to generalize AID (Sonquist and Morgan, 1964), CART's ancestor, to survival data (Ciampi *et al.*, 1981). Eventually, Breiman *et al.* (1984) work stimulated a more ambitious development, aimed at generalizing CART to other situations occurring in biostatistics (Ciampi *et al.*, 1987, Ciampi, 1991, 1992, Ciampi *et al.*, 1991, 1993). The generalization is approached from the regression angle (Ciampi, 1991). The aim of early RECPAM work was to construct a predictive model for a one-dimensional parameter of a distribution, the prediction being piece-wise constant on a partition of the predictors space. As a further generalization, vector parameters were considered. RECPAM is an ongoing project and its development proceeds with that of a software package designed according to concepts of *reusable software engineering:* this enables the programmer to add new models to an existing basic structure with minimal work other than that of programming task-specific functions, see Lou & Ciampi (1992).

From the point of view of machine learning, RECPAM does *supervised* learning; unlike most supervised learning algorithms, however, the object of the prediction is not a class variable, but a parameter, which is usually continuous and possibly multi-dimensional; even in the case of a class variable, the aim is to predict the class probability distribution rather than the class itself. As we shall see, it is because of this and of some other typical features, that RECPAM can serve as a bridge between supervised and unsupervised learning, hence between discrimination and clustering.

The next section of this paper reviews the basic concepts of the RECPAM methodology. Section 3 is devoted to the RECPAM approach to classification (unsupervised learning). In section 4 we discuss the RECPAM approach to discrimination (supervised learning). We conclude with a discussion in which we return to the general issues underlined above.

2　The RECPAM approach to tree-growing

A complete exposition of RECPAM can be found in Ciampi (1992). Here we will limit ourselves to emphasize the ideas that are needed in the development of our approach to discrimination and classification.

RECPAM adaptively builds a statistical model from a data matrix of the form $D = [U \mid Z]$, obtained from measurements of the vector variables (u, z) on N individuals. The variables of u are known as *criterion variables*, and those of z as *predictors*.

It is assumed that the criterion variables, given the predictors, have a homogeneous joint distribution, the form of which is defined below, after introducing some notation. If P is any subset of the predictor space, we will denote by $I_P(z)$ the indicator function of the set P. Let $\mathcal{P}_1$ and $\mathcal{P}_2$ denote partitions of the predictor space, with $\mathcal{P}_1$ not finer than $\mathcal{P}_2$, and let $P \in \mathcal{P}_1$, and $P' \in \mathcal{P}_2$ denote generic sets of the partitions; also, $P \supseteq P' \in \mathcal{P}_2$ will denote that P' is a subset of P.

Then, with this notation, the distribution of u is:

$$f(u|z; \mathcal{P}_1, \mathcal{P}_2) = \sum_{P \in \mathcal{P}_1} \left\{ \sum_{P \supseteq P' \in \mathcal{P}_2} f(u|\gamma_P, \alpha_{P'}, \beta)) I_{P \cap P'}(z) \right\} \quad (2.1)$$

where the γ's, the β's, and α's are parameters to be determined from the data. (Notice that the inner summation is over P' for P fixed, and the outer summation is over P). The tree structure is contained in the term $I_{P' \cap P}(z)$. Thus, the distribution function is piece-wise constant with respect to z. To determine it, we need to know a number of pieces equal to the cardinality of the finer partition $\mathcal{P}_2$. Another direct implication of the model (2.1) is that the distribution is known, except for the nuisance parameter, once we know the two partitions and a number of (vector) parameters equal to card($\mathcal{P}_1$) + card($\mathcal{P}_2$). RECPAM adaptively determines the partitions and estimates these parameters from the data.

The β's, referred to as *global parameters*, represent those characteristics of the distribution which are known *a priori* not to be dependent on z, while the α's, referred to as *local parameters*, represent those characteristics which are known to depend very finely on z. On the other hand, the γ's, or *criterion parameters*, represent characteristics of the distribution which are suspected to vary somewhat as a function of z, but to an extent which is not known. Indeed, the purpose of the RECPAM algorithm is to determine just how finely these characteristics vary across the prediction space. In this sense, the γ's may be considered of primary interest, while the other parameters play the role of *adjustments*, as it will appear more clearly in section 4.

Notice that if $\mathcal{P}_1 = \mathcal{P}_2 = \mathcal{P}$, then there are no local parameters and (2.1) reduces to:

$$f(u|z;\mathcal{P}) = \sum_{P \in \mathcal{P}} f(u|\gamma_P,\beta)I_P(z) \tag{2.2}$$

All steps of the RECPAM construction are based on the *(observed generalized) information increment of a finer partition with respect to a coarser partition*, $\Delta I(\mathcal{P}_2 : \mathcal{P}_1|D)$. This is defined as follows. Let $LF(\mathcal{P}_1,\mathcal{P}_2;U)$ denote a measure of lack-of-fit of $f(u|z;\mathcal{P}_1,\mathcal{P}_2)$, to the data U. Then:

$$\begin{aligned}\Delta I(\mathcal{P}_2 : \mathcal{P}_1|D) = &\inf\{LF(\mathcal{P}_1,\mathcal{P}_2;U);\gamma_P, P \in \mathcal{P}_1, \alpha_{P'}, P' \in \mathcal{P}_2, \beta\} - \\ &\inf\{LF(\mathcal{P}_2,\mathcal{P}_2;\beta);\gamma_P, \alpha_P, P \in \mathcal{P}_2, \beta\}\end{aligned} \tag{2.3}$$

Clearly, the two terms of this equation correspond to the two hypotheses on the variation of $f(u|z)$ across the predictor space, expressed, respectively, by (2.1) and (2.2). In both terms, β and α preserve their character, which is, respectively, global (no variation) and local (variation on the finest partition). In contrast, the γ's vary, in the first term, across the coarser partition only, and in the second term, across the finer one. Thus ΔI represents the improvement in fit due to allowing the criterion to vary across the finer partition. In this paper we assume that f is completely determined once the parameters are known and take as lack-of-fit measure the negative log-likelihood of the data, which corresponds to the usual definition of (sample) information (Kullback, 1959).

All the quantities needed in RECPAM are defined as special cases of ΔI. Thus the *information content of a partition* $\mathcal{P}$ is the information increment of $\mathcal{P}$ with respect to the trivial partition consisting of one set, and the information loss of $\mathcal{P}_1$ with respect to $\mathcal{P}_2$ is just the information increment of $\mathcal{P}_2$ with respect to $\mathcal{P}_1$. By a slight abuse of language, but with no risk of confusion, the *information content of a tree (question)* will mean the information content of the associated partitions.

The RECPAM algorithm is an approximate search for the best predictive model of the family (2.1). The first step of the search proceeds by recursively partitioning the predictor space into binary splits, corresponding to binary questions involving the predictors. At each node, the question with highest information content is chosen. The search is restricted to questions belonging to an *a priori* specified class or SDQ (Split Defining Questions) family, see Ciampi (1991), and satisfying an admissibility conditions depending on the node and the data set. In this paper we will consider as SDQ family that of the *simple questions*, *i.e.* questions involving an individual predictor z_i, of the form 'is z_i in A?', where A is any set of possible values of z_i, if z_i is nominal, and of the form $[a, \infty]$, or $[\infty, a)$, if z_i is ordinal. The admissibility condition will be usually a restriction on the number of individuals in the data set that belong to a node. This creates a tree with the root node representing the whole population, with (internal) nodes representing subpopulations, and with branches (splits) associated to questions. The nodes at which no question of the SDQ family is admissible are the terminal nodes of the tree, or leaves. The result of this first step is referred to as the *large tree*. The partition associated to it is the finer partition of equation (2.1). If there are local parameters, these will vary in any case across the leaves of the large trees. If there are global parameters, these do not vary at all, but are independent of the leaves and the nodes of the large tree.

The best fitting model is not necessarily the best predictive model. The tendency to produce best fits with inferior predictive accuracy is called *overfitting*. The next step, known as *pruning*, aims to reduce the size of the large tree in order to reduce overfitting. It results in a smaller tree known as the *honest tree*. The latter has the advantage, when compared with the large tree, of a greater *generalizability* to future data, and hence of a better predictive accuracy, an advantage that offsets the loss in goodness-of-fit to the data at hand. Pruning means a) constructing a sequence of subtrees: this is done by successively eliminating branches that are 'parents' of leaves, in order of increasing information content of the associated question; b) choosing out of this sequence the tree with best predictive accuracy, calculated by appropriately correcting the information content of the subtrees.

The third and final step, not found in other tree-growing algorithms, is known as *amalgamation*: further simplification of the model associated to the honest tree is attempted, by successively amalgamating leaves (terminal nodes) issuing from different 'parents', in order of increasing information loss. This yields a sequence of nested partitions of the predictor space, out of which the one with best predictive accuracy is selected: this is called the *RECPAM classification*. Thus, as a result of this final step, RECPAM identifies a number of *classes* which are *homogeneous* and *distinct* with respect to the criterion, *i.e.* there is one unique γ associated to a class, and γ's corresponding to different classes are different. It is this special character of the RECPAM classes which allows the adaptation of the general methodology to *classification*.

It should be noted here that if there are local parameters, these have been allowed to vary across the leaves of the large tree at any step of the construction. Thus the final model is of the form (2.1), where $\mathcal{P}_1$ is the partition associated to the RECPAM classes, and $\mathcal{P}_2$, as remarked above, is the subpartition of $\mathcal{P}_1$, the sets of which are the RECPAM classes. In the particular case in which there are no local parameters, the two partitions coincide and equation (2.2) gives the form of the statistical model associated to the RECPAM classification.

The correction for overfitting of the information contents, needed to choose the honest tree and the RECPAM classifications out of the appropriate sequences, may be performed in RECPAM, just as in CART, by v-fold cross-validation, or, as in other tree-growing approaches (Crawford, 1989), by the bootstrap. However, RECPAM allows the use of some specific large-sample options for choosing the honest tree and the amalgamation classes which avoid the intensive calculations of resampling approaches. One of these, the one to be used in this paper, is known as the AIC *elbow rule*, which consists in attaching an AIC (Akaike Information Criterion), see Akaike (1974), to each tree of the pruning sequence (each partition of the amalgamation algorithm), looking at the graph of the AIC as function of the step of the algorithm, and choosing the tree (partition) corresponding to the (usually present) elbow of the curve. This is not a strictly automatic choice and some degree of subjectivity is allowed, in keeping with the exploratory nature of the REC-PAM algorithm. We refer again to Ciampi (1991, 1992) for details, since the emphasis of the paper is on models underlying the proposed approach to classification and discrimination rather than on the details of model choice.

3   RECPAM models for classification

The search for classification is driven by the need to simplify what we observe. This is done by grouping individuals into classes which are homogeneous and distinct, so that we can consider individuals in the same class as being 'the same' and individuals from any two different classes to be 'different'. The quotation marks underline the fact that we need to specify *in what sense* individuals of the same class are the same and individuals from different classes are different.

In the typical formulation of the classification problem (clustering), one seeks to identify classes from a data matrix of the form $[Z]$, with columns representing measurements of a vector variable z on N individuals. Usually the problem is overdetermined and further specification is needed. One appealing approach consists in seeking a *partition* of the variable space which optimizes a specified measure of goodness-of-clustering. This approach is fully developed in the seminal book by Diday (1980). The problem reduces to an appropriate choice of the measure and on the development of a successful algorithm of optimization. Both steps are difficult, although the nature of the difficulty is quite different. The choice of the measure involves making explicit the 'in what sense' above. The development of the optimization algorithm requires side-stepping the combinatorial computational effort: in many cases this can only be done by limiting the search by appropriate heuristics, and/or by introducing a random component in the search, with the result that the optimization is only *approximate* and/or *local*.

The RECPAM approach proposed here shares this limited goal. It requires one preliminary step, which coincides with what is done in practice in many applications of traditional clustering algorithms: extraction of factors from the matrix $[Z]$. The result of this preliminary step is a matrix of the form $D = [U|Z]$, where U is the matrix of the factors extracted from Z. After this is done, the space of the predictors is partitioned by RECPAM so that in each class the data may be represented by a homogeneous conditional distribution $f(u|z)$. In particular, RECPAM finds a representation of the distribution in the form given by equation (2.2) (only one partition, or $\mathcal{P}_1 = \mathcal{P}_2$). The process is shown schematically in Figure 1 below, which contains, at the left, a hypothetical RECPAM tree, and, at the right, a graph showing the original data points as stars, and their projections on the factorial plane $\mathcal{F}$ as dots. Clearly, RECPAM identifies a partition of the z-space, and hence of the rows of the original data matrix, in such a way as to 'best' cluster the representations of the points on the factorial plane. The term 'best' is implicitly defined by the information measure. As we have seen, RECPAM attempts to approximate a 'best fit' within a model family to the data: in this case, the information measure is a dissimilarity measure between a model and the empirical distribution of the data.

What remains to be discussed is how to determine the factors and how to model the distribution of the factors. As for the first step, we can refer the reader to the rich literature on principal components (Joliffe, 1986), if all the z's are continuous, and on correspondence analysis (Greenacre, 1984), if they are all discrete. Furthermore, recent generalizations of correspondence analysis permit the treatment of a mixture of discrete and continuous variables (Escofier & Pagès, 1988, chapter 7, section 6). One simple approach, however, used in the example below, consists in discretizing the continuous variables and applying multiple correspondence analysis to the resulting data matrix.

The RECPAM model used here is based on assuming that the distribution of the factors can be approximated by a mixture of multivariate normal distributions. We can then use a very simple particular case of the theory developed in Ciampi *et al.* (1993), see also Ciampi *et al.* (1992). We assume that there are no local and no global parameters, and that the criterion is $\gamma = (\mu, \Sigma)$, *i.e.* the mean vector and the elements of the variance-covariance matrix.

Given the form of the multinormal distribution, we can see that the proposed approach essentially uses RECPAM as a method for 'bump hunting' in the factor plane, but with reference to the original variables. There are other possibilities for modeling $f(u|z)$, but they would require further development of the RECPAM software and they will not be discussed here.

Figure 1:

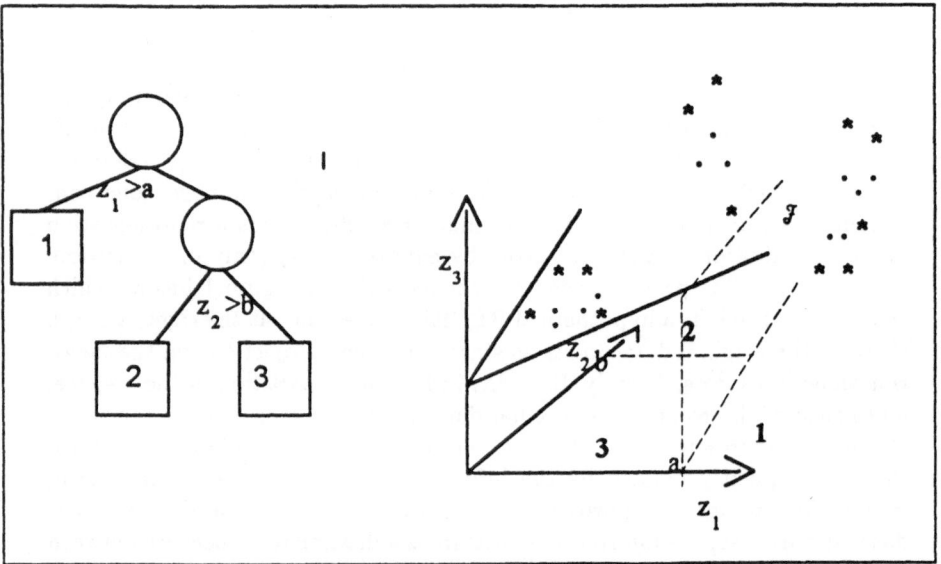

The following example is given strictly for purpose of illustration. The data are obtained from measurements of a number of clinical variables on a population of children with Insuline Dependent Diabetes Mellitus (IDDM), taken at diagnosis, except for one, which is the result of a test at 6 months from diagnosis. A summary of these variables is given in Table 1 below. An important clinical question is whether it is possible to identify clinical subgroups of the disease. A full discussion of the problem is found in (Ciampi et al., 1990), where a classical clustering analysis is performed in order to propose an IDDM classification. The RECPAM approach proposed here yields a different type of answer, summarized in Figure 2, and quite easy to interpret.

The figure is self-explanatory. Nodes are represented by circles, leaves by squares and RECPAM classes by rounded squares. The numbers denote the number of individuals of the data set which falls in each node, leaf or class. The tree suggests that there are two distinct subgroups of IDDM patients. We observe, on the one hand, a group of patients (79 in the data

Table 1

Variable	Type
Demographics	
• Age group	discrete (3 levels)
• Sex	binary
Genetic Marker	
• DR	discrete (5 levels)
β-cell function	
• production of C-peptides	continuous
• Insuline does	continuous
• Evidence of function at 6 months	binary
Autoimmune activity: presence/absence of	
• Islet cells antibodies	binary
• Insuline antibodies	binary
• Thyroid antibodies	binary
Endocrine system function: blood level of	
• T4	continuous
• TSH	continuous
• Cortisol	continuous
Metabolic control	
• Hbal	continuous
• Cholesterol	continuous
• Triglycerides	continuous
• Glucose	continuous
Clinical presentation	
• Coma	binary
• Blood pH	continuous
• Symptoms duration	continuous

set) represented by the left-most rounded square, which is the union of the two leaves: these are patients with Coma at presentation or a very low level of C-peptides, evidence of rapidly disappearing function of the insuline-producing pancreatic cells (β-cells). The other group (58 in the data set) consists of patients with a disease with slow and/or less severe onset. It should be noted that the earlier analysis also produced two classes. These were, however, much harder to interpret, whereas the classes obtained here are obtained *together with their description*. It should be noted that a detailed study of the means and the variance-covariance matrices of the factors for the two groups would add clinical interest to this description. For example one could see that the less severe group appears to be constituted of older children, with more girls than expected, and with a higher degree of metabolic imbalance

Figure 2:

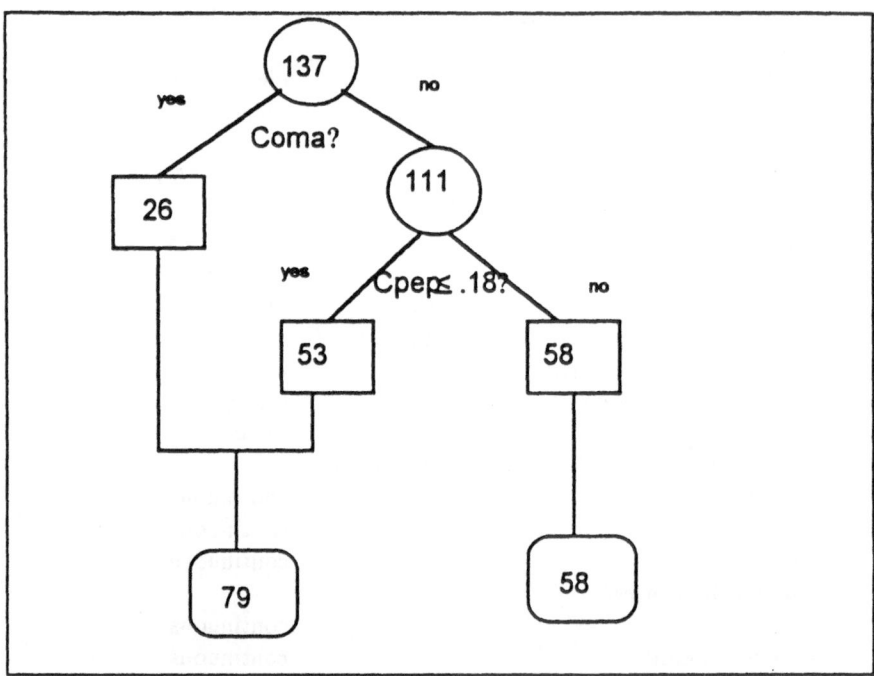

than the rapid-onset group.

The multinormal model may seem in some cases inadequate, and it may be felt that the underlying assumptions impose serious distortions on the data. It is not too difficult to avoid this criticism, at least partially. Instead of using the log-likelihood of the multinormal model, we may use, as the lack-of-fit measure appearing in the definition of information increment (2.3), the *inertia* of the data (Greenacre, 1984). As a further refinement, the data reduction, or factor extraction, could be repeated at each node during tree construction. On the other hand, with any one or both of these refinements, pruning and amalgamation could not be done by approximate measures such as the AIC; instead, one would have the choice between cross-validation/bootstrap and new approximate indices that would necessitate *ad hoc* development.

Before concluding this section, it is interesting to introduce another approach to the problem of classification which is closely related to the REC-PAM methodology. The density of z, $f(z)$, can always be decomposed as a product: $f(z) = f(y|x)f(x)$, for any decomposition $z = (y, x)$. It can be argued that, at least in certain cases, the goal of the classification exercise is to find the decomposition of z which maximizes the information of x about y. In fact, when faced with a large number of variables, the analyst naturally wants to single out those variables which best help predict as much as can be predicted of the remaining ones. Gower's *predictive classification*

(Gower, 1974) is motivated exactly by this type of reasoning. Latent variables methodology (Everitt, 1984) pursues a very similar goal, but with the decomposition: $f(z) = f(z|\xi)f(\xi)$, where ξ is a vector of latent variables and $f(z|\xi) = f(z_1|\xi)f(z_2|\xi)\cdots f(z_m|\xi)$ (conditional independence); this is possible, of course, only if the $f(z)$ has a special form. Gower's model and the latent variable model suggest a third decomposition: $f(z) = f(z|\xi)f(\xi)$, where ξ is sought in a specific class of functions of z so as to maximize the information about z. If the class of functions is chosen of the form:

$$\xi(z) = \sum_{P \in \mathcal{P}} \gamma_P I_P(z) \tag{3.1}$$

and furthermore the model:

$$f(z|\xi; \mathcal{P}) = \sum_{P \in \mathcal{P}} f(z|\gamma_P) I_P(z) \tag{3.2}$$

is assumed, then a RECPAM-type strategy can easily be developed to construct such a model in a tree-shaped form.

4   RECPAM models for discrimination

After defining or discovering classes, it is natural to become concerned with the problem of finding simple and economical rules to decide to which class a newly observed individual belongs. This gives rise to the problem of *discrimination* or *supervised learning*. The typical formulation of the problem starts from a data matrix of the form $D = [C|Y]$, obtained by measuring the vector variable (c, y), where c is a discrete variable taking K values: c_1, $c_2, \cdots, c_K$ (which denote to mutually exclusive class labels), and y a vector, often a mixture of discrete and continuous variables.

The decision theoretic approach to discrimination is well known (Hand, 1981). It can be shown that optimal decisions for any specified loss-function can be made if one knows either: i) $p(c|y)$, the conditional distribution of c given y and the marginal distribution of y; or, equivalently, ii) the conditional distributions of y given the classes and the marginal distribution of the class variable. Dawid (1976) refers to the former situation as the *diagnostic paradigm* and to the latter as the *sampling paradigm*. The RECPAM approach to discrimination based on the sampling paradigm is outlined in Ciampi et al. (1993). Here we will concentrate on discussing the use of RECPAM within the diagnostic paradigm. In particular, we will limit ourselves to a discussion of several RECPAM models for $p(c|y)$ and on model construction from a data matrix of the form $[C|Y]$.

The simplest and best known tree-structured model for $p(c|y)$ is, as one based on (2.2):

$$p(c|y; \mathcal{P}) = \sum_{P \in \mathcal{P}} p(c)_P I_P(y)$$

$$pp(c) \geq 0, \quad pp(K) = 1 - \sum_{k=1}^{K-1} p(c_k) \qquad (4.1)$$

This is essentially the model built by CART (Breiman *et al.*, 1984), ID3 (Quinlan, 1986) and most other tree-growing algorithms. The link with equation (2.2) will appear more clearly below.

In spite of its simplicity and popularity, the above model has two major shortcomings. First, it treats the continuous variables as though they had on p just a threshold effect. Intuitively, for example, it is odd to assume that exactly on my 50th birthday my probability of dying within a week makes a sudden jump, yet that may well be a prediction obtained from model (4.1), if the Age variable appears in the tree, dichotomized at 50. Indeed at least some continuous variables are best treated as continuous, producing a linear effect, perhaps after a monotonic transformation. Secondly, one great advantage of the tree-model, that of emphasizing interactions among variables, may be its undoing when we consider variables which just do *not* interact in any significant way with the others. For instance, returning to age, it may well be that it steadily increases the probability of dying regardless of the presence of other risk factors.

Now, the traditional linear discriminant analysis approach does not have these shortcomings. In the logistic regression version (diagnostic paradigm), continuous variables are treated as continuous and discrete variables as discrete; moreover, although a few interactions may be added selectively, the most natural assumption in that framework is the absence of interactions. Of course, the weak point of the (generalized) linear approach is that interactions are often important and they appear in complex patterns ... as in a tree structured model!

Let us return now to equation (2.1) and show that, after careful analysis, one can use it to at least partially remedy the shortcomings of model (4.1). First, we need to consider our vector y and partition it as $y = (l, g, x, z)$, where:

1. z denotes the vector of the *main predictors*, those that will define the RECPAM classes explicitly;

2. l is a vector of *local covariates*, i.e. auxiliary predictors, *not interacting within themselves*, but known to affect class probabilities in an extremely variable way, according to main predictors patterns;

3. g is a vector of *global covariates*, i.e. auxiliary predictors, *not interacting within themselves*, and that, like Age in the above discussion, may be assumed to affect class probabilities in a way that does not depend on the other factors;

4. x is a vector of *criterion covariates*, i.e. auxiliary predictors, *not interacting within themselves*, and of uncertain degree of interaction with the main predictors.

We will now re-write: $(c, y) = (u, z)$, with $u = (c, x, l, g)$, and, correspondingly for the data matrix: $D = [U|Z]$, with $U = [C|X|L|G]$. We seek a model for $p(c|x, l, g, z)$ which reflects the above analysis. The model of equation (2.1) seems a reasonable candidate. We shall write: $p(c|x, l, g, z) = f(u|z; \mathcal{P}_1, \mathcal{P}_2)$, with: $f(u|\gamma_P, \alpha_{P'}, \beta) = f(c, x, l, g|\gamma_P, \alpha_{P'}, \beta)$ defined as follows. Let: $\gamma_P = \{\gamma_P^{(c)}, c \neq c_{K-1}\}, \alpha_{P'} = \{\alpha_{P'}^{(c)}, c \neq c_{K-1}\}, \beta = \{\beta^{(c)}, c \neq c_{K-1}\}$, where $\dim(\gamma_P^{(c)}) = \dim(x)$, $\dim(\alpha_{P'}^{(c)}) = \dim(l)$ and $\dim(\beta^{(c)}) = \dim(g)$. (Notice that each class, except the one we have arbitrarily labeled c_K, a sort of "reference class", contributes parameters: this is because the constraint that the probability sum to 1 causes the probability of one class to be entirely determined by those of the other classes). Let:

$$\eta(c, x, g, l; \gamma_P, \alpha_{P'}, \beta) = x \cdot \gamma_P^{(c)} + l \cdot \alpha_{P'}^{(c)} + g \cdot \beta^{(c)} \qquad (4.2)$$

Then, for $c \neq c_K$:

$$f(c, x, l, g|\gamma_P, \alpha_{P'}, \beta) = \frac{\exp(\eta(c, x, g, l; \gamma_P, \alpha_{P'}, \beta))}{1 + \sum_{k=1}^{K-1} \exp(\eta(c_k, x, g, l; \gamma_P, \alpha_{P'}, \beta))} \qquad (4.3)$$

while for $c = c_K$, owing to the constraint that the probabilities sum to 1, the numerator of the above equation becomes 1.

Usually in discriminant analysis, x will contain the constant term. Notice that the 'simple' CART-like model (4.1) is one in which there are neither local nor global variables and x reduces to the constant term. In this case, the left-hand side of (4.2) and (4.3) reduce, respectively, to the scalar parameter $\gamma_P^{(c)}$ and to:

$$f(c|\gamma_P) = \frac{exp(\gamma_P^{(c)})}{1 + \sum_{k=1}^{K-1} exp(\gamma_P^{(c_k)})}$$

which, if we set $f(c|\gamma_P) = p_{P(c)}$, and compare with equation (4.1), is essentially the classical CART model for discrimination.

The minimization underlying the RECPAM construction, contained in equation (2.3), is performed without major problems. In fact, it can be seen that the model (4.3) can be re-written in the more familiar form:

$$\log(\frac{p(c|x, g, l, z)}{p(c_K|x, g, l, z)}) = \sum_{P \in \mathcal{P}_1} I_P(z) x \cdot \gamma_P^{(c)} + \sum_{P \in \mathcal{P}_2} I_{P'}(z) l \cdot \alpha_{P'}^{(c)} + g \cdot \beta^{(c)} \qquad (4.4)$$

(for $c \neq c_K$), which is immediately recognized as a K-class logistic regression. Thus any software for doing such regression can be used to interactively build the RECPAM model for discriminant analysis given by equation (4.3). The case $K = 2$ was treated in (Ciampi et al., 1991), and the appropriate software has been developed within the RECPAM project.

An example will now be discussed, again for purpose of illustration only. The data are the low birth-weight data, given in Hosmer & Lemenshow (1990)

142

Figure 3:

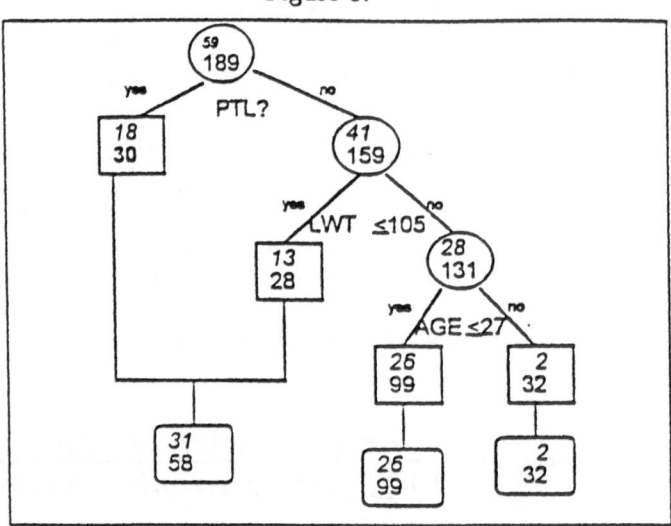

to illustrate use of logistic regression, and already analyzed in Ciampi *et al.*, (1993) with a different approach (sampling paradigm). The individuals are women who have had a baby, and on which several variables were observed in the course of their pregnancy; the class variable is binary and distinguishes mothers whose babies had abnormally low birth-weight from those whose baby had normal birth-weight: $c = 1$ if birthweight is less than 2500 g, 0 otherwise. It is desired to find rules to discriminate between the two classes based on the observed variables. The data contain two continuous variables: AGE and LWT, the weight (in lbs) at the last menstrual period before pregnancy. They also contain six categorical variables: RACE (1 = white, 2 = black, 3 = other), SMOKE (smoking during pregnancy, 1 = smoker, 0 = non-smoker), PTL (history of premature labor, 1 = yes, 0 = no), HT (history of hypertension, 1 = yes, 0 = no), UI (uterine irritability, 1 = yes, 0 = no), and PLTV (physician visits during last trimester, 1 = one or more, 0 = none).

The tree corresponding to model (4.1) is given in Figure 3. As discussed, it ignores the difference between continuous and categorical variables.

In this figure, as in the following one, the numbers in ordinary characters at each node, leaf or RECPAM class, denote the number of individuals of the data set that belong to the corresponding node or leaf or RECPAM class. The numbers in italics denote the number of babies with low birthweight in the corresponding node, leaf or RECPAM class.

Figure 4 shows the tree which is obtained when treating LWT and AGE as global variables.

Exactly the same tree is obtained when treating LWT and AGE as criterion. The two associated models can be written, respectively, with obvious

Figure 4:

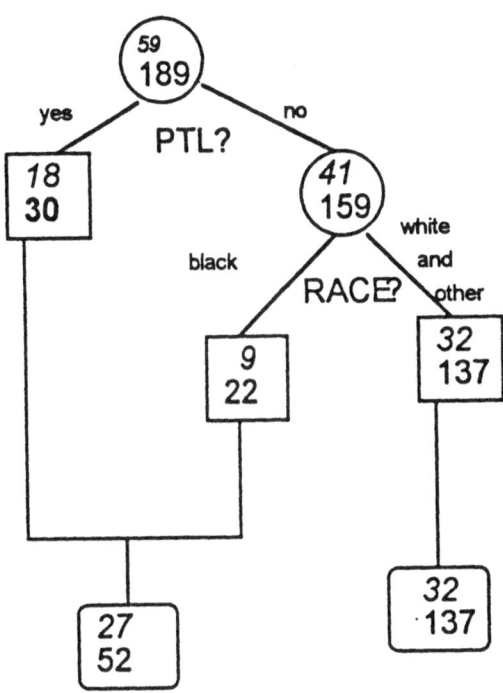

meaning of symbols:

$$\log(\tfrac{p}{1-p}) = (1.66)I[\text{no PTL \& not black}](z) + (3.18)I[\text{no PTL \& black or PTL}](z) - (0.06)\text{AGE} - (0.01)\text{LWT}$$

and:

$$\begin{aligned} log(\tfrac{p}{1-p}) &= (.79 + .04\text{AGE} - .02\text{LWT})I[\text{no PTL \& not black}](z) + \\ & \quad (2.98 - .02\text{AGE} + .002\text{LWT})I[\text{no PTL \& black or PTL}](z) \\ &\cong (.79 - .02\text{LWT})I[\text{no PTL \& not black}](z) + \\ & \quad (2.98 - .20\text{AGE})I[\text{no PTL \& black or PTL}](z) \end{aligned}$$

We will not enter the details of these calculations, nor will we discuss the problem of determining the 'best' model among these three and others that could be estimated easily. Suffice it to point out that the models seem to fit equally well, but that on grounds of interpretability, the one that considers LWT and AGE as criterion seems to be the most satisfactory, because it avoids thresholding the continuous variables, and also because it shows the predictive role of RACE, which is masked in the tree of Figure 3. It indicates that in the first RECPAM class (mothers who are not black and have no history of premature labor) the baseline risk for low birth-weight is rather low, is not affected by the mother's age, and is greater the lower the weight of the mother before pregnancy. On the other hand, the second RECPAM class (complement to the first) has a substantially higher baseline risk, is not affected by the mother's weight before pregnancy, and is greater the younger the mother is.

In general, as it has also been remarked by other authors, continuous variables tend to occur more than they should in tree construction, since they contribute many more splits than discrete variables. One remedy that has been proposed (Sauerbrei, Schumacher) is to choose the split not on the basis of the information content, but rather on the basis of an associated significance level corrected for multiple comparisons. The solution we propose here is, in a sense, simpler, and respects the special nature of continuous variables.

5 Discussion

We have shown that two fundamental problems of data analysis, classification and discrimination, can find general approximate solutions within the tree-growing paradigm, and, in particular. in its particular version called RECPAM. That tree-growing is useful in discrimination is well known. Less well known is the potential of the methodology as a method of classification: indeed, RECPAM produces a method that belongs to the family of *conceptual clustering* algorithms, see the appropriate chapter in Gale (1986), since it produces classes together with their 'conceptual' descriptions. As for discriminant analysis, the unique features of RECPAM tree-growing allow the construction from data of models which are a combination of multi-category logistic regression and tree-structured predictors of class probabilities.

More generally, tree-growing, and in particular RECPAM, combines features of machine learning methods, such as those developed within the AI community, and traditional statistical methods. It tends therefore towards the integration of two distinct but complementary goals: i) discovering models from data that fall within basic cognitive structures; and ii) identifying models, within flexible classes which, as experience or theory suggest, seem to handle uncertainty reasonably well. Perhaps the integration of these two goals may be described to be central to the area of research called 'AI and Statistics' (Gale, 1986).

In the introductory remarks of this paper we have contrasted the point of view prevalent in supervised and unsupervised learning with the one prevalent in (statistical) discrimination and classification. These remarks also apply to a more general comparison between the basic goals of Statistics, on the one hand, and of AI on the other. At the risk of oversimplifying, the basic philosophy of statistical model building may be summarized as: 'Find models which best account for the uncertainties inherent in the problem under study, and then interpret them, simplifying when necessary, to best communicate to the client'. From this point of view, interpretation is a matter of 'art' rather than 'science'. Likewise, the inclusion into the modeling effort of preliminary knowledge, other than that which pertains to uncertainty, is left to the intuition and the imagination of the analyst. In contrast, knowledge representation and mimicry of human cognition are central preoccupation and object of systematic enquiry in AI, while uncertainty is dealt with as a nuisance, often in an *ad hoc* manner.

The history of tree-based algorithms offers a fascinating example of both convergences and divergences between statistical and AI points of view. Trees first attracted statisticians because they offer a natural way to discover inter-actions among variables (AID = Automatic Interaction Detection). Within the 'purely' statistical community trees have opened new important hori-zons in adaptive model building. MARS (Friedman, 1991) and PIMPLE (Breiman, 1991) are recent efforts of two of the authors of CART to model data by flexible approaches that go well beyond trees. They result in extraor-dinarily effective model fitting strategies with excellent statistical properties. Unfortunately, however, the results of these modeling efforts are not nearly as interpretable as trees. This is an inevitable consequence of moving away from the imitation of human cognitive strategies.

On the other hand, work on discrimination and classification carried out by the AI community under the name of supervised and unsupervised learn-ing, and often based on trees, is dominated by the 'combinatorial explosion'. This is especially so in unsupervised learning, where conceptual clustering offers a very ambitious approach to discovering classes but at crippling com-putational costs. Arguably, this *impasse* is reached because uncertainty is not properly treated.

The RECPAM approach is statistical in that it is model based. The model, however, remains close to both human cognition and empirically sound ways of handling uncertainty; moreover, it encourages the introduction of preliminary knowledge (see the case of local, global and criterion predictors in discriminant analysis). The approach falls somewhere between supervised and unsupervised learning. The supervision is provided by the model it-self. The latter can be chosen as 'directive' as is desired, to reflect existing knowledge. The models for discrimination discussed in section 4 are strongly 'directive' supervisors, since it is implied that we start from data sets of *classified* individuals. The supervision provided by classical regression- tree models (AID and the regression part of and CART) is weak: it consists only in stating which variables may be predicted from the others. The RECPAM models of section 3 provide supervisors which are rather 'non-directive', since no variable is *a priori* special; however, the approach selects those variables which best situate the 'cloud of points' formed by the data in the appropriate multidimensional space.

RECPAM is a very open project that can benefit from the input of both AI (improving search strategies) and Statistics (improving statistical properties of the algorithms, further extending treatment to other data structures). It is hoped that in turn, both Statistics and AI will benefit from such development.

References

Akaike, H. (1974). A New Look at the Statistical Model Identification. *IEEE Transactions on Automatic Control*, **AC-19**, 716–723.

Breiman, L. (1991). The II-method for Estimating Multivariate Functions from Noisy Data. *Technometrics*, **33**, 125–160.

Breiman, L., Friedman, J.H., Olshen, R.A. and Stone, C.J. (1984). *Classification and Regression Trees*. Waldsworth International Group, Belmont, California.

Buntine, W. (1992). Learning Classification Trees. *Statistics and Computing*, **2**, 63–73.

Chou, P. (1991). Optimal Partitioning for Classification and Regression Trees. *IEEE Transactions of Pattern Analysis*, bf 13, 340–354.

Chou, P., Lookabough, T., and Gray, R.M. (1989). Optimal Pruning with Applications to Tree-structured Source Coding and Modeling. *IEEE Transactions of Information Theory*, **35**, 299–315.

Ciampi, A. (1991). Generalized Regression Trees. *Computational Statistics and Data Analysis*, **12**, 57–78.

Ciampi, A. (1992). Constructing Prediction Trees from Data: the RECPAM Approach. *Proceedings from the Prague 1991 Summer School on Computational Aspects of Model Choice*, 105–152. Physica-Verlag, Heidelberg.

Ciampi, A., Bush, R.S., Gospodarowicz, M. and Till, J.E. (1981). An Approach to Classifying Prognostic Factors Related to Survival Experience for Non-Hodgkins Lymphoma Patients. *Cancer*, **47**, 621–627.

Ciampi, A., Chang, C.-H., Hogg, S.A., McKinney, S. (1987). Recursive Partition: a Versatile Method for Exploratory Data Analysis in Biostatistics. In I. MacNeil, G.J. Umphrey (eds.), *Festschrifts in Honor of Prof. Joshi*, **5**, *Biostatistics*, 23–50. D. Reidel, Dordrecht.

Ciampi, A., Hendricks, L. and Lou, Z. (1992). Tree-growing for the Multivariate Model: the RECPAM Approach. In Y. Dodge, J. Whittaker (eds.), *Computational Statistics*, **1**, 131–136. Physica-Verlag, Berlin.

Ciampi, A., Hendricks, L. and Lou, Z. (1993). Discriminant Analysis for Mixed Variables: Integrating Trees and Regression Models. In C.M. Cuadras, C.R. Rao (eds.), *Multivariate Analysis: Future Directions*, **2**, 3–22. North-Holland, Amsterdam.

Ciampi, A., Lou, Z., Lin, Q. and Negassa, A. (1991). Recursive Partition and Amalgamation with the Exponential Family: Theory and Applications. *Applied Stochastic Models and Data Analysis*, **7**, 121–137.

Ciampi, A., Schiffrin, A., Thiffault, J., Quintal, H., Weitzner, G., Poussier, P. and Lalla, D. (1990). Cluster Analysis of an Insuline-dependent Diabetic Cohort: Towards the Definition of Clinical Subtypes. *Journal of Clinical Epidemiology*,

43, 701–715.

Crawford, S. (1989). Extensions to the CART Algorithm. *International Journal of Machine-Man Studies*, **31**, 197–217.

Dawid, A.P. (1976). Properties of Diagnostic Distributions. *Biometrics*, **32**, 647–658.

Diday, E. (1980). Optimisation en Classification Automatique. *INRIA*, Le Chesnay.

Escofier, B. and Pagès, J. (1988). Analyses Factorielles Simples et Multiples. Dunod, Paris.

Everitt, B.S. (1984). *An Introduction to Latent Variable Models*. Chapman & Hall, London.

Friedman, J.H. (1991). Multivariate Adaptive Regression Splines (with discussion). *Annals of Statistics*, **19**, 1–141.

Gale, W.A. (ed.) (1986). *Artificial Intelligence and Statistics*. Addison-Wesley, Reading, Mass.

Gordon, A.D. (1981). Classification: Methods for the Exploratory Analysis of Multivariate Data. Chapman & Hall, London.

Gower, J.C. (1974). Maximal Predicitive Classification. Biometrics, **30**, 643–654.

Greenacre, M.J. (1984). *Theory and Application of Correspondence Analysis*. Academic Press, London.

Hand, D.J. (1981). *Discrimination and Classification*. J. Wiley & Sons, New York.

Hosmer, D.W. and Lemenshow, S. (1990). *Applied Logistic Regression*. J. Wiley, New York.

Joliffe, I.T. (1986). *Principal Component Analysis*. New York, Springer-Verlag.

Kullback, S. (1959). *Information Theory and Statistics*. J. Wiley, New York.

Lou, Z. and Ciampi A. (1992). Reuse Oriented Approach in Developing Statistical Software. In H.J. Newton (ed.), *Proceedings of the 24th Symposium on the Interface, Computing Science and Statistics*, **24**, 40–44.

Quinlan, J. (1986). Induction of Decision Trees. *Machine Learning*, 1, 81–106.

Sauerbrei, W., Schumacher, M. Private communication.

Sonquist, J.A. and Morgan, J.N. (1964). The Detection of Interaction Effects. Ann Arbor: Institute for Social Research, University of Michigan.

Classification of Tumour ^{1}H NMR Spectra using Nonlinear Approaches

Axel Benner

Biostatistik, Deutsches Krebsforschungszentrum, Heidelberg

Abstract. Different linear and nonlinear techniques for the discrimination and classification of multivariate data sets were applied to a set of data from 231 urine samples of healthy volunteers and patients with various forms of cancer (four disease classes), inflammatory and infectious diseases.

Keywords. NMR spectroscopy, nonlinear discriminant analysis

1 Introduction

Discrimination and classification of a sample into one of a number of classes based upon a series of measurements (features) is an important problem in many scientific fields. "Classical" statistical methods for these tasks have been developed under the constraint of being computable with a relatively small amount of calculations and were therefore usually restricted to linear models. The enormous improvements in the development of hard- and software in the last 15 years has resulted in many new statistical methods for the analysis of multivariate data, which are computationally more expensive than their classical predecessors, but which are also less dependent on distributional assumptions and allow a much more flexible treatment of complicated situations. The use of neural networks for pattern recognition and classification nowadays is very popular (especially with non-statisticians), whereas new statistical methods are less widely known to the public.

The purpose of this paper is to compare the application of different statistical classification techniques (including the neural network approach) to a complex medical data set, which consists of 231 observation vectors for 22 variables. Because the main medical interest is in prognosis the main interest in comparison is in the *generalization abilities* of the different classifiers.

The data set was kindly provided by Dr. Reininghaus from the German Cancer Research Center (DKFZ), Heidelberg.

2 Data Description

High resolution ^{1}H NMR spectra of urine are displaying resonances from a wide variety of low molecular weight molecules which feature in important

biochemical processes. ^{1}H NMR spectra were obtained from 24-hour urine samples. The spectra are quantified by measuring a series of peak heights or areas for 22 specific known endogenous metabolites:
`3-hydroxybutytrate, lactate, alanine, acetate, acetone, acetoacetate, unknown#1 (at 2.35 ppm), citrate, sarcosine, dimethylglycine, creatine, creatinine, acetylcarnitine, carnitine, betaine, choline, glycine, unknown#2 (at 3.66 ppm), unknown#3 (at 3.82 ppm), hippurate, formate, histidine`

The concentrations of the metabolites were evaluated by use of the creatinine peak at about 3.07 ppm as an internal standard (Reininghaus, 1992). Prior to analysis all metabolite data were expressed as base 10 logarithms and standardized to have unit variance.

Seven disease groups were built according to prior clinical diagnoses (cp. Table 1).

Table 1. Definition of the disease groups

Class number	Clinical Diagnoses	# Urine samples
1	Healthy Volunteers	18
2	Inflammatory Diseases	33
3	Infectious Diseases	21
4	Lymphoma	26
5	Tumours of the Urinary Tract	38
6	Tumours of the Gastrointestinal Tract	63
7	Tumours of the Respiratory Tract	32
	Total	231

The statistical analysis of the urine spectra started with a univariate inspection of the data. After outlier correction and missing values imputation four metabolites (`acetone, unknown#2, unknown#3` and `histidine`) seemed to add only noise to the data and should be excluded from further analysis.

From an earlier analysis of these data (Reininghaus, 1992) six metabolites seemed to be the most useful for the classification task (`acetoacetate, the unknown compound at 2.35 ppm, acetylcarnitine, carnitine, citrate and sarcosine`).

Classification was done using (a) all, (b) the remaining 18, and (c) the six selected metabolites.

3 Statistical Methods for Classification

In general discriminant analysis can be used to find a rule to classify an observation into one of J unordered but known classes on the basis of p measures of variables (features). The methods used here are

Linear discriminant analysis (LDA)
It is the standard tool for discrimination and classification

Generalized discriminant analysis (GDA)
This is a nonlinear approach which uses the Alternating Conditional Expectation (ACE) methodology (Breiman & Ihaka, 1985)

Classification Trees (CART)
This technique uses binary recursive partitioning for classification (Breiman et al., 1984)

BRUTO
An automatic backfitting algorithm for fitting smooth additive models (Hastie & Tibshirani, 1990)

Multivariate adaptive regression splines (MARS)
An adaptive regression technique which includes interactions in a hierarchical manner (Friedman, 1991)

Feed-forward neural networks
A generalization of linear regression. A feed-forward network with two functional layers (one hidden layer and the output layer), which also allows skip-layer connections is used here (Ripley, 1993)

A new statistical approach to classification which was used in the present analysis is called "Flexible Discriminant Analysis" (FDA; Hastie et al., 1993), which provides a class of nonparametric versions of discriminant analysis by using multi-response regression with optimal scoring, replacing linear regression by nonparametric regression methods. BRUTO, MARS and the neural network approach are used here as special options in the FDA framework.

All computations were done using S-Plus, Version 3.1 (Statistical Sciences Inc., 1992) on a Sun SPARCstation. The code for the generalized discriminant analysis was from the S library gdiscr, the code for BRUTO, MARS and FDA was from the S library fda, and the code for the feed-forward neural network was from the S library nnet, all available from StatLib (statlib@lib.stat.cmu.edu).

4 Criteria for evaluation

The purpose of the present analysis is to examine the application of classification procedures to a complex medical data set and to evaluate the performance of these procedures by their prediction error (misclassification rate). To get a realistic estimate of the misclassification rate cross-validation is used (Stone, 1974). Cross-validation uses part of the available data to train the classifier, and a different part for testing the performance of the classifier. It provides a good assessment of how effective a classification rule will be in classifying a new observation. In K-fold cross-validation the data set is divided in K roughly equal-sized groups, and each group is used as the test set for a classifier trained on the other $K - 1$ groups. The observed error rates on the K groups are averaged to estimate the total misclassification rate. Two special versions of K-fold cross-validation are used here: 10-fold cross-validation and leave-one-out cross-validation, with K equal to the number of observations in the data set.

5 A simulation study

To compare the six selected classification techniques they were first applied to simulated data. Here the waveform data set simulation (Breiman et al., 1984) was used. A training sample of size 90 and a test sample of size 500 were constructed using equal prior probabilities for the three classes. The classification methods were applied to these samples and error rates were computed. This procedure was repeated 250 times. The mean error rates with their standard errors resulting from the 250 simulations are given in Table 2.

Table 2 Error rates from 250 simulations of the waveform data

Method	Mean error rates (standard error)	
	Training	Test
LDA	0.070 (0.002)	0.237 (0.002)
GDA	0.005 (0.001)	0.278 (0.002)
CART	0.102 (0.002)	0.347 (0.002)
FDA/ BRUTO	0.099 (0.002)	0.227 (0.001)
FDA/ MARS (degree=1)	0.095 (0.003)	0.266 (0.002)
FDA/ NNET (2 hidden units + skip-layer)	0.018 (0.001)	0.335 (0.003)

The neural network approach gave the best results for the training set, but appeared to be no more superior when applied to the test samples. The methods showing the smallest error rates for the test samples are FDA/BRUTO and LDA.

6 Results and discussion

Discrimination of the NMR data set worked well with GDA, CART or NNET (see e.g. the result of the generalized discriminant analysis as displayed in Fig. 1), but failed if LDA or MARS was used.

Classification on the basis of the remaining 18 metabolites (see section 2) gave best results, using only the six selected metabolites was worst. The

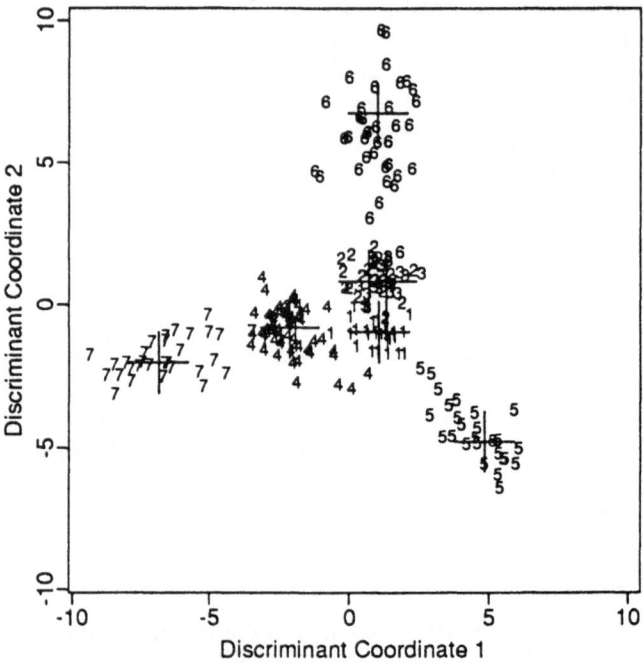

Fig. 1 Generalized discriminant analysis applied to the whole NMR data set. The plotting characters used are the class numbers as defined in Table 1. The "+" signs describe the centers of the classes as generated by generalized discriminant analysis

cross-validated misclassification rates for the classification using the remaining 18 metabolites are given in Table 3. The results for MARS and NNET were comparable to "random" classification, whereas LDA, unable to grasp the nonlinearity in the data, classified new observations mainly into class 6 (see Table 1).

Table 3 Cross-validated error rates: NMR data set

Method	Error rates	
	Leave-one-out	10-fold
LDA	0.749	0.823
GDA	0.199	0.208
CART	0.169	0.221
FDA/ BRUTO	0.368	0.472
FDA/ MARS (degree=3)	0.653	0.827
FDA/ NNET	0.861	0.883

The method which worked best for discrimination (the neural network approach with 15 hidden units) lost its superiority if generalization was the goal. This result may be expected because flexible fitting procedures like the feed-forward neural network tend to overfit the training data and generalization of the results maybe impaired.

High error rates in classification of new NMR spectra resulted probably for several reasons:

1. The groups were built on the basis of clinical diagnosis with the groups of inflammatory and infectious diseases being very heterogeneous

2. The comparison of the classification results cannot be treated symmetrically for the NMR data set, because wrong classification between the control groups cannot be compared with wrong classifications of tumour patients. This leads to the need to compare error rates according to false positive or false negative classifications of tumour patients.

3. The NMR spectra were taken "as they are" before selection of the individual peaks.

References

Breiman, L., Friedman, J.H., Olshen, R.A. and Stone, C.J. (1984). *Classification and Regression Trees.* Monterey: Wadsworth & Brooks/Cole.

Breiman, L. and Ihaka, R. (1985). *Nonlinear Discriminant Analysis via ACE and Scaling.* Technical Report #40, U.C. Berkeley Statistics Department.

Friedman, J. (1991). Multivariate adaptive regression splines (with discussion). *Annals of Statistics*, Vol. 19, 1-141.

Hastie, T.J. and Tibshirani, R.J. (1990). *Generalized Additive Models.* London: Chapman & Hall.

Hastie, T.J., Tibshirani, R.J. and Buja, A. (1993). Flexible Discriminant Analysis by Optimal Scoring. Preprint.

Reininghaus, F. (1992). NMR-spektroskopische Untersuchungen von Körperflüssigkeiten: Studie zur Unterscheidung verschiedener Erkrankungen und Krebs mittels ^{1}H-NMR-Spektroskopie menschlichen Urins. Dissertation, Universität Witten/Herdecke.

Ripley, B. D. (1993). Statistical aspects of neural networks. In *Chaos and Networks – Statistical and Probabilistic Aspects* (eds O. E. Barndorff-Nielsen, D. R. Cox, J. L. Jensen and W. S. Kendall), 40-123. London: Chapman & Hall.

Stone, M. (1974). Cross-validation choice and assessment of statistical predictions. *Journal of the Royal Statistical Society*, B, Vol. 36, 111-147.

Fuzzy Clustering and Mixture Models

Gilles Celeux, Gérard Govaert

Centre Hospitalier Universitaire de Grenoble, Service S.I.I.M.

Abstract. We analyze finite mixture models as fuzzy clustering models. This point of view allows us to present fuzzy clustering as a natural extension of hard clustering. Moreover, it provides a great variety of possible fuzzy clustering criteria.

Keywords. Gaussian mixture, fuzzy clustering, maximum likelihood

1 Two Approaches for Mixture Analysis

Basing cluster analysis on Gaussian mixture models has become a classical and powerful approach. Data $x_1, \ldots, x_n$ in $\mathbf{R}^d$ are assumed to arise from a random vector with density

$$f(\mathbf{x}) = \sum_{k=1}^{K} p_k \Phi(\mathbf{x}|\mu_k, \Sigma_k) \tag{1}$$

where the p_k's are the mixing proportions ($0 < p_k < 1$ for all $k = 1, \ldots, K$ and $\sum_k p_k = 1$) and $\Phi(\mathbf{x}|\mu, \Sigma)$ denotes the density of a Gaussian distribution with mean vector μ and variance matrix Σ.

In this context, two commonly used maximum likelihood (m.l.) approaches have been proposed: the mixture approach and the classification approach. Loosely speaking, the mixture approach is aimed to maximize the likelihood over the mixture parameters, whereas the classification approach is aimed to maximize the likelihood over the mixture parameters and over the identifying labels of the mixture component origin for each point.

1.1   The Mixture Approach

In the mixture approach, the parameter $\theta = p_1, \ldots, p_{K-1}, \mu_1, \ldots, \mu_K, \Sigma_1, \ldots, \Sigma_K$ is chosen to maximize the loglikelihood

$$L(\theta) = \sum_{i=1}^{n} \ln \left[\sum_{k=1}^{K} p_k \Phi(\mathbf{x}_i | \mu_k, \Sigma_k) \right], \qquad (2)$$

using generally the EM algorithm (Dempster et al. 1977). Starting from an initial parameter θ^0, an iteration of the EM algorithm consists in computing the current conditional probabilities $t_k(\mathbf{x}_i)(1 \le i \le n, 1 \le k \le K)$ that $\mathbf{x}_i$ arises from the kth mixture component for the current value of θ, according to the equation (3) (E step); then the m.l. estimates $\hat{p}_k, \hat{\mu}_k, \hat{\Sigma}_k$ are computed using the conditional probabilities $t_k(\mathbf{x}_i)$ as conditional mixing weights (M step)

$$t_k(\mathbf{x}_i) = \frac{\hat{p}_k \Phi(\mathbf{x}_i, \hat{\mu}_k, \hat{\Sigma}_k)}{\sum_{\ell=1}^{K} \hat{p}_\ell \Phi(\mathbf{x}_i, \hat{\mu}_\ell, \hat{\Sigma}_\ell)}. \qquad (3)$$

1.2   The Classification Approach

In the classification approach, the indicator vectors $\mathbf{z}_i = (z_{ik}, k = 1, \ldots, K)$ with $z_{ik} = 1$ or 0 according as $\mathbf{x}_i(1 \le i \le n)$ has been drawn from the kth component or from another one, identifying the mixture component origin, are treated as unknown parameters. Then the parameters θ and $\mathbf{z}_1, \ldots, \mathbf{z}_n$ are chosen to maximize the Classification Maximum Likelihood (CML) criterion (Symons 1981)

$$CL(\theta, \mathbf{z}_1, \ldots, \mathbf{z}_n) = \sum_{k=1}^{K} \sum_{\mathbf{x}_i \in P_k} \ln \left[p_k \Phi(\mathbf{x}_i | \mu_k, \Sigma_k) \right]. \qquad (4)$$

where $P = (P_1, \ldots, P_K)$ is the partition of $\mathbf{x}_1, \ldots, \mathbf{x}_n$ associated to the indicator vectors $\mathbf{z}_1, \ldots, \mathbf{z}_n : P_k = \{\mathbf{x}_i / z_{ik} = 1\}$.

This criteria can be optimized by making use of a classification version of the EM algorithm, the so-called CEM algorithm (Celeux and Govaert 1992), that we described now.

Starting from an initial partition P^0, an iteration of the CEM algorithm consists in computing the current conditional probabilities $t_k(\mathbf{x}_i)(1 \le i \le n, 1 \le k \le K)$ according to the equation (3) (E step); then a updated partition is calculated by assigning each $\mathbf{x}_i$ to the cluster which provides the maximum current conditional probability $t_k(\mathbf{x}_i)(1 \le k \le K)$ (C step); and the m.l. estimates $(\hat{p}_k, \hat{\mu}_k, \hat{\Sigma}_k)$ are computed using the cluster P_k as sub-sample $(1 \le k \le K)$ (M step).

2 Mixture Approach and Fuzzy Clustering

First, it is worth noting that the *classification* matrix $c = (c_{ik} = t_k(x_i), i = 1,\ldots,n; k = 1,\ldots,K)$ defines a fuzzy clustering since the conditions $0 \leq c_{ik} \leq 1$ and $\sum_{k=1}^{K} c_{ik} = 1$ are verified.

Then, it can be shown that each time the M step can be achieved through closed form equations, the estimate $\hat{\theta}$ can be written as a function $\phi(c)$. Thus, the non decreasing sequence $(L(\theta^m), m > 0)$ of the EM algorithm can be regarded as a non decreasing sequence $(L \circ \phi(c^m), m > 0)$. It means that the EM algorithm is maximizing a fuzzy clustering criterion.

2.1 EM and CEM: Fuzzy and Hard Clustering

The following discussion on the fuzzy clustering criteria derived from the mixture approach is based on the interpretation of EM proposed by Hathaway (1986). This author showed that the EM algorithm for the mixture problem can be interpreted as a method of coordinate descent on a particular objective function.

First, we extend the CML criterion to fuzzy clustering

$$CL(\theta, c) = \sum_{i=1}^{n} \sum_{k=1}^{K} c_{ik} \ln [p_k \Phi(x_i|\mu_k, \Sigma_k)]. \tag{5}$$

and we define

$$LC(\theta, c) = CL(\theta, c) + E(c) \tag{6}$$

where

$$E(c) = - \sum_{i=1}^{n} \sum_{k=1}^{K} c_{ik} \ln c_{ik}$$

is the entropy of the classification matrix c.

The EM algorithm can be regarded as an alternating optimization of the function $LC(\theta, c)$

E step : c^{r+1} maximizes $L(\theta^r, c)$ over c : $c^{r+1} = \Psi(\theta^r)$,

M step : θ^{r+1} maximizes $L(\theta, c^{r+1})$ over θ : if this step can be achieved through closed form equations, we have $\theta^{r+1} = \Phi(c^{r+1})$.

It can be shown that $L(\theta) = LC(\theta, \Psi(\theta))$. If the classification matrix c defines a partition (hard clustering), we have $LC(\theta, c) = CL(\theta, c)$. Thus, the CEM algorithm can be regarded as a particular version of EM under the constraint that c is associated to a partition.

As a consequence, both the mixture approach and the classification approach in mixture analysis can be interpreted as optimizing the fuzzy clustering and the hard clustering versions of the same criterion.

2.2 Fuzzy Clustering Criteria

Figure 1 shows the EM algorithm as described in the previous section. As it is apparent from this figure, when the M step is closed form, EM can be considered as optimizing two different fuzzy clustering criteria: $L \circ \phi(c)$ or $LC(c, \Phi(c))$. The second criterion can be preferred since, contrary to the first, it is closely related to classical hard clustering criteria in many situations.

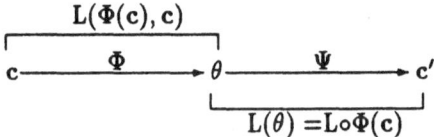

Fig. 1. Diagram of an EM iteration

For instance, using Gaussian mixtures, it is possible to derive fuzzy versions of inertia-type criteria $\mathrm{tr}(W)$, $|W|$ where W is the within cluster scattering matrix. For simplicity, the mixing proportions are assumed to be equal. For Gaussian mixture, LC takes the form

$$LC(\theta, c) = \sum_{k=1}^{K} \sum_{i=1}^{n} c_{ik} \ln \left[p_k \frac{1}{(2\pi)^{\frac{d}{2}}} \frac{1}{|\Sigma_k|^{\frac{1}{2}}} \exp(-\frac{1}{2}(x_i - \mu_k)' \Sigma_k^{-1}(x_i - \mu_k)) \right] + E(c),$$

which can be written as

$$LC(\theta, c) = \frac{1}{2} \sum_{k=1}^{K} \sum_{i=1}^{n} \mathrm{tr}[(c_{ik}(x_i - \mu_k)(x_i - \mu_k)' \Sigma_k^{-1}] - \frac{1}{2} \sum_{k=1}^{K} n_k \ln |\Sigma_k| + \mathrm{Cst} + E(c),$$

where

$$n_k = \sum_{i=1}^{n} c_{ik}.$$

In the particular case where the component variance matrices are equal and proportional to the identity matrix, which leads to the minimization of $\mathrm{tr}(W)$ in a hard clustering context, this criterion becomes

$$LC(\Phi(c), c) = -\frac{nd}{2} \ln \mathrm{tr}(W) + E(c) + \mathrm{Cst},$$

where W is now a fuzzy version of the within cluster scattering matrix,

$$W = \sum_{k=1}^{K} \sum_{i1}^{n} c_{ik}(x_i - \bar{x}_k)(x_i - \bar{x}_k)'.$$

with

$$\bar{\mathbf{x}}_k = \frac{\sum_{i=1}^n c_{ik} \mathbf{x}_i}{n_k}..$$

Remark. In this situation,

$$L \circ \Phi(\mathbf{c}) = \sum_{i=1}^n \ln \sum_{k=1}^K \exp(-\frac{1}{2}(\mathbf{x}_i - \mu_k)'(\mathbf{x}_i - \mu_k)) + \text{Cst}$$

does not lead to an easy interpretation.

When the component variance matrices are equal but unknown, which leads to the minimization of $|W|$ in a hard clustering context, the criterion becomes

$$LC(\Phi(\mathbf{c}), \mathbf{c}) = -\frac{d}{2} \ln |W| + E(\mathbf{c}) + \text{Cst}.$$

3 Discussion

Another approach of fuzzy clustering consists in optimizing fuzzy versions of inertia-type criteria. For instance, the fuzzy version of $\text{tr}(W)$ is

$$\sum_{k=1}^K \sum_{i=1}^n (c_{ik})^m \|\mathbf{x}_i - \mathbf{g}_k\|^2, \tag{7}$$

where the $\mathbf{g}'_k$s are vectors in $\mathbf{R}^d$ and m is real number greater than one which has to be defined by the user. Usually $m = 2$ is recommended (Bezdek 1981). In fact, this fuzzy clustering criterion can be thought of as unnatural and artificial. The choice of m is very sensible and difficult to manage. Moreover, it can induce numerical difficulties when $\|\mathbf{x}_i - \mathbf{g}_k\|^2 = 0$.

By contrast, fuzzy clustering through mixture model is a fruitful line of approach. Many mixture models are of interest from a clustering point of view (for instance, the latent class model for discrete data, see Everitt 1984) and most of those mixture models lead to a closed form M step allowing our construction. Moreover, Gaussian mixture models have been proved to be efficient for handling many clustering situations in a simple and elegant manner since they allow easy specification of the volumes, the orientations, the shapes and the sizes of clusters (see Banfield and Raftery 1993 and Celeux and Govaert 1993). These Gaussian mixture models give rise to quite as much interesting fuzzy clustering criteria.

References

Banfield, J. D. and Raftery, A. E. (1993). Model-based Gaussian and non Gaussian clustering. *Biometrics*, **48**.

Bezdek J. C. (1981). *Pattern recognition with fuzzy objective function algorithms.* New York: Plenum.

Celeux, G. and Govaert, G. (1992). A classification EM algorithm for clustering and two stochastic versions. *Computational Statistics and Data Analysis*, **14**, 315-332.

Celeux, G. and Govaert, G. (1993). Gaussian parsimonious clustering models. *Rapport INRIA* 2028.

Dempster, A. P., Laird, N. M. and Rubin, D. B. (1977). Maximum likelihood from incomplete data via the EM algorithm (with discussion). *J. R. Statis. Soc. B*, **39**, 1-38.

Everitt, B. (1984). *An introduction to latent variable models.* London: Chapman and Hall.

Hathaway, R. J. (1986). Another interpretation of the EM algorithm for mixture distributions. *J. Statistics & Probability Letters*, **4**, 53-56.

Symons, M. J. (1981). Clustering criteria and multivariate normal mixtures. *Biometrics*, 37, 35-43.

Clustering in an Interactive Way

Hans-Joachim Mucha

IAAS, Berlin

Abstract. Some of the most important cluster analysis techniques are available in the interactive statistical computing environment *XploRe*. Furthermore, new adaptive clustering methods can be carried out. They seem to be a little bit intelligent because of their ability for learning the appropriate distance measures. Moreover, adaptive distances should also be used in order to obtain multivariate plots (Mucha 1992). In that way, both the interpretation of clustering results and highly interactive work becomes much easier.

Keywords. Cluster analysis, K-means method, Statistical software

1 Introduction

XploRe aims mainly at e*Xplo*ratory *Re*gression and nonparametric data analysis (Härdle 1990). Often, cluster analysis is also used rather in an exploratory manner than in order to fit a given general model which has to be assumed in advance. Cluster analysis techniques attempt to detect unknown structures in the data. They aim at the partition of a large set of unarranged objects into smaller, homogeneous classes, groups or clusters. Herein, usually, many variables (features) are considered simultaneously. At least, the cluster analysis provides a practical useful reduction or description of data. Some of the most widely used clustering techniques (four partitioning methods and eight hierarchical clustering algorithms, respectively) are available in *XploRe* (Mucha 1994). For instance, the partitioning K-means method as well as Ward's hierarchical technique will be considered in a little more detail here.

2 Variance Criteria

Concerning a fixed number of clusters K, some practical useful clustering methods attempt to minimize the sum of within-cluster variances

$$V_K = \sum_{k=1}^{K} \sum_{i=1}^{I} \delta_{ik} m_i d_Q^2(\mathbf{x}_i, \overline{\mathbf{x}}_k). \tag{1}$$

Here, we consider I observations (row points) containing measurements in J metric-scaled variables each, i. e. the structure of the I row points of a data table $\mathbf{X}=(x_{ij})$, $i = 1, 2, ..., I$, $j = 1, 2, ..., J$ has to be investigated. The indicator function δ_{ik} is equals 1 if the observation $\mathbf{x}_i$ (we write shortly: i) comes from cluster k, or 0 otherwise. Generally, as a result of cluster analysis we get either a partition $P(I, K)$ of I row points into K clusters ($K \ll I$) or a sequence of partitions, that is a so-called hierarchy. A partition is simply a categorical variable $\mathbf{p}$ which allocates a positive integer value (state) $p_i \in \{1, 2, ..., K\}$ to every row point $\mathbf{x}_i$. It is the basis of the indicator function

δ_{ik} mentioned above. Furthermore, the element $\bar{x}_{kj}$ of the vector $\bar{x}_k$ is the mean value of the variable j in the cluster k

$$\bar{x}_{kj} = \frac{1}{n_k} \sum_{i=1}^{I} \delta_{ik} m_i x_{ij}, \qquad (2)$$

where n_k denotes the mass of the cluster k, which is equal to the sum of the masses of all observations i belonging to the cluster k. In formula (1) above, the squared weighted Euclidean distance

$$d_Q^2(x_i, x_l) = (x_i - x_l)'Q(x_i - x_l) = \|x_i - x_l\|_Q^2 \qquad (3)$$

between two observations x_i and x_l (and as well as between an observation x_i and a cluster centroid $\bar{x}_k$) is used. This is a well-known dissimilarity measure in cluster analysis and principal components analysis (PCA), respectively. Here Q is diagonal. Usually the weights $q_{jj} = 1/s_j^2$ (called standard weights; hereby quite different scales of the variables become comparable one with another), or $q_{jj} = 1$ (the scales of the variables should be already comparable one with each other) are used, where s_j^2 is the total variance of the variable j:

$$s_j^2 = \frac{1}{M} \sum_{i=1}^{I} m_i (x_{ij} - \bar{x}_j)^2. \qquad (4)$$

Herein M is the total mass of all I observations. With the aim of interpretation of the clustering results a graphical representation in a low-dimensional space is a helpful tool. In the case of the standard weights $q_{jj} = 1/s_j^2$ in (3) the PCA of the correlation matrix (instead of the covariance matrix) is computed. As a result each variable becomes an equal importance in cluster analysis as well as in PCA. Moreover the use of the inverse total variances in (3) as weights in a cluster analysis contradicts to the fundamental assumption that there are several populations (clusters) with different parameters each we look for. As a final consequence of using these standardizations the detection of an underlying cluster structure in X becomes difficult or even impossible.

In practice, (1) is minimized either by exchange an observation between clusters at a time (K-means method: Mucha 1992; exchange algorithm: Späth 1985) or by a step by step hierarchical agglomeration with minimal increase of within-cluster variances $V_{L-1} - V_L$, $L = I, I-1, ..., K+1$, (Ward 1963).

Example. There are 1000 random generated observations in total (with 9 variables each): 300 in class 1, 150 in 2, 350 in 3, and 200 in 4. The classes differ in their location and variance. The variables give different contributions to the distinction of the classes. These *XploRe* commands are used:

```
Z =normal(350 9)      ; 350*9 standard normal distributed random numbers
X1=Z[1:300,].*1.2   -*(0,0,0.5,3.5,(-1),(-1), 0 , 0 ,0)   ; class 1
X2=Z[201:350,]      +*(0,0,0.5, 2 , 0.5, 0.5,0.3,0.5,0)   ; class 2
X3=Z[1:350,].*1.5   +*(0,0, 0 , 1 , 5.5, 2 ,0.5,0.5,0)   ; class 3
X4=Z[101:300,].*0.9 -*(0,0,1.5,1.5, 2 , 0 , 0 ,0.5,0)   ; class 4
X =X1 | X2 | X3 | X4
q =1./var(X)
```

162

```
r =ceil(uniform(rows(X)).*4)
( p C V n ) = kmeans( X r  0 q)
```

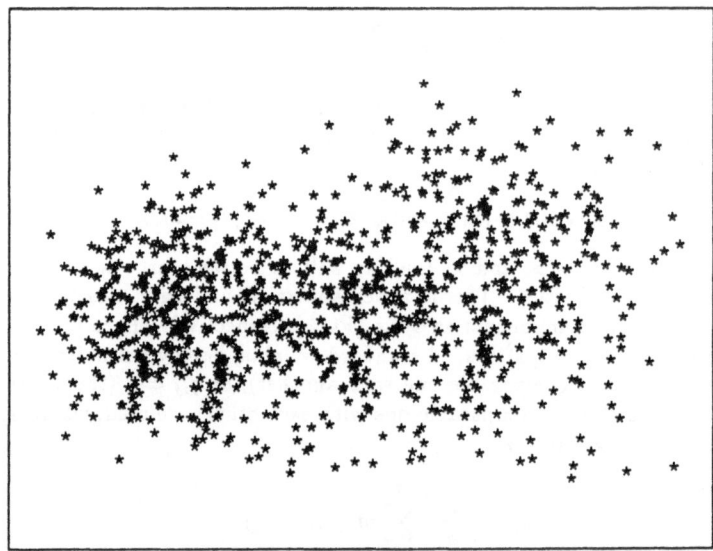

Figure 1: PCA plot of the example; 36.5% of the total variance is explained.

The output parameters of *kmeans* are the final partition **p** $(= P(I, K))$ which minimizes (1), the $(K * J)$-matrix **C** of cluster centroids (2), the $(K * J)$-matrix **V** of within-cluster variances divided by the weight of the corresponding cluster

$$v_{kj} = \frac{1}{n_k} \sum_{i=1}^{I} \delta_{ik} m_i q_{jj} (x_{ij} - \bar{x}_{kj})^2 \ , \tag{5}$$

and the vector **n** of weights of the K clusters. If $m_i = 1, (i = 1, 2, ..., I)$, then n_k counts the number of observations in the cluster k. These parameters can be used by the matrix language in an easy and convenient way. For instance, **p** is useful for multivariate and dynamic *XploRe* graphics. One can either transform the states of **p** into colors or symbols.

Figure 1 gives you a first idea about the data of our example; it shows projections onto the first PCA-plane. The PCA is based on the correlation matrix which is an usual option of statistical software. The class labels are ignored. But even in the case of adding the given class labels into the plot, it doesn't becomes much clearer that there is some structure in the data.

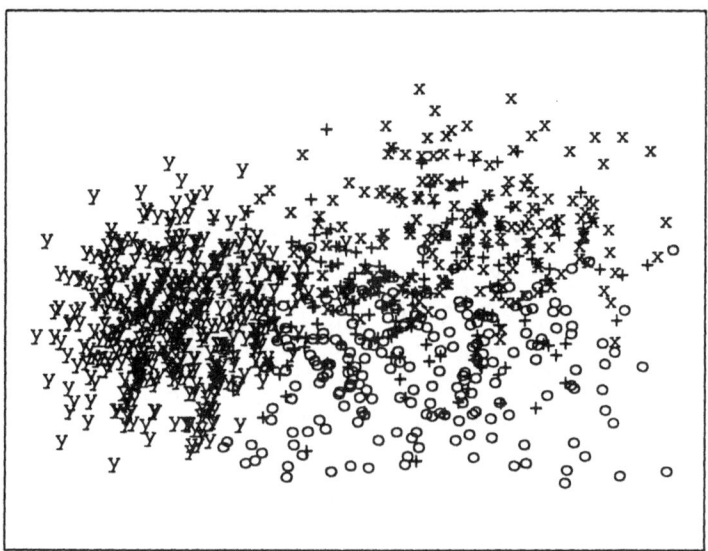

Figure 2: PCA plot of the cluster labels of the observations after an usual K-means cluster analysis. The error rate is a quite terrible one: 55.3%.

Cluster analysis and PCA make no use of the class labels, but on the other side it will be of course interesting to compare the a priori given groups with any structure found in the data by these techniques. Here we are interested in a K-means cluster analysis of the 1000 points into 4 clusters. Because of the known true class membership we are able to assess the performance of this clustering technique by the error rate simply. Figure 2 shows the cluster membership marked by different symbols. The K-means method based on the standard weights $q_{jj} = 1/s_j^2$ in (3) fails obviously.

3 Adaptive Cluster Analysis

For increasing the stability in cluster analysis specific or adaptive weights were recommended (Mucha and Klinke 1993). For example, the simple adaptive weights

$$q_{jj} = 1/\bar{s}_j^2 \qquad (6)$$

can be used in (3), where $\bar{s}_j$ denotes the pooled standard deviation of the variable j

$$\bar{s}_j^2 = 1/M \sum_{k=1}^{K} \sum_{i=1}^{I} \delta_{ik} m_i (x_{ij} - \bar{x}_{kj})^2. \qquad (7)$$

Because of simplicity one can use M (instead of $M - K$) equals the sum of all weights m_i of the observations x_i, $i = 1, 2, ..., I$, i. e. the adaptive weights become independent from the number of clusters K.

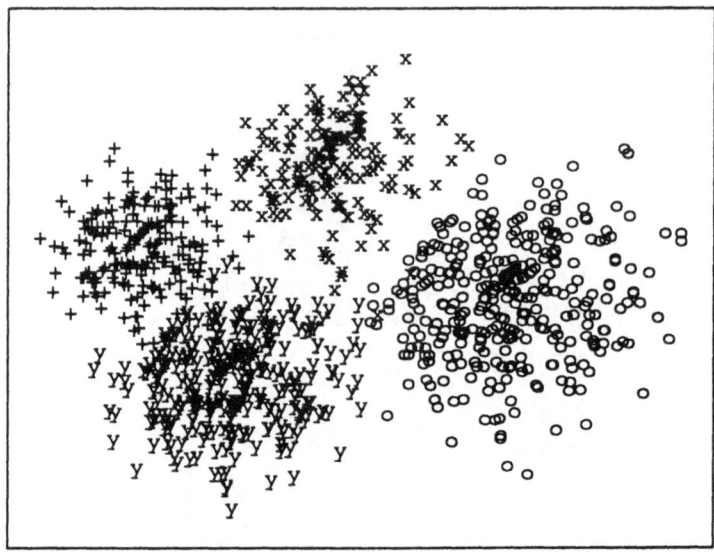

Figure 3: PCA plot of the result of the adaptive K-means method. The PCA is based on the corresponding adaptive distances; the first PCA plane accounts for 62.4% (46.5% + 15.9%) of the total variance. The error rate is 3.8%.

Unfortunately, the pooled variances can't be computed because of the unknown true cluster structure. It is well known in the case of $K \ll I$ that the pooled standard deviations concerning a random partition are nearly equal the total standard deviations. Therefore, starting with the weights $q_{jj}^{(0)} = 1/s_j^2$ and a random initial partition $P^{(0)}(I, K)$ the K-means method computes a (local) optimum partition $P^{(1)}(I, K)$. In a repeated K-means clustering step (with a new random initial partition) the weights $q_{jj}^{(1)} = 1/\bar{s}_j^2$ are used, where $\bar{s}_j$ is the pooled standard deviation of the variable j over K clusters regarding to the new partition $P^{(1)}(I, K)$. After carrying out the second K-means run we get again a partition denoted by $P^{(2)}(I, K)$. The new adaptive weights $q_{jj}^{(2)} = 1/\bar{s}_j^2$, which correspond to the partition $P^{(2)}(I, K)$, are used in a next K-means clustering step, and so on. We repeat this procedure as long as a chosen stop rule is fulfilled (for example, if no changes in the optimum partition occur, or concerning the adaptive weights, if no or very small variations are pointed out.

In figure 3, the final partition of the adaptive K-means method is shown in the first PCA plane. Six iterations are necessary for the convergence. Obviously, the clusters become visible and the error rate decreases rapidly. Moreover, the importance of each variable for clustering can be assessed (Mucha 1992). Here, the final weights are: 1, 1, 1.28, 4.12, 7.11, 1.35, 1.04, 1.10, 1.

4 Interactive Use

XClust is a so-called *XploRe* library containing a collection of macros for cluster analysis and classification. The macros are based mainly on functions like *kmeans*, *distance* (computation of a matrix of distances such as (3), χ^2, Jaccard,...), *agglom* (hierar-

165

chical cluster analysis), *discrim* (linear discriminant analysis), and *cart* (classification
and regression trees). So-called teaching macros are an easy get in for unexperienced
users. Otherwise there are several interactive possibilities for experienced users. For
example, one can cut the dendrogram obtained by a hierarchical cluster analysis or a
recursive partitioning regression at an "interesting" cluster distance level which corresponds to a partition of all clustering objects into, say for instance, K clusters.
Another highly interactive action is clustering as well as regrouping the observations
by visual inspection a multivariate graphic. Generally, the brushing facilities enable
to select, identify (show labels or values), color, mark,...

However, the exploratory interactive data analysis gives you a first insight only.
Additionally, simulation software can be highly recommended in order to validate the
the number of clusters or the contributions of the variables.

References

Härdle, W. (1990): Applied Nonparametric Regression. *Cambridge University Press*, Cambridge

Mucha, H.-J. (1992): Clusteranalyse mit Mikrocomputern. Akademie Verlag, Berlin

Mucha, H.-J., Klinke, S. (1993): Clustering Techniques in the Interactive Statistical Computing Environment *XploRe*. Discussion Paper 9318. Institute de Statistique, Universite
Catholique de Louvain, Louvain-la-Neuve

Mucha, H.-J. (1994): Clustering Techniques in the Computing Environment *XploRe*. Proc.
17th. Annual Conference of the GfKl, Univ. of Kaiserslautern, 1993, (Eds.: Bock, H. H.,
Lenski, W., and Richter, M. M.), 259–268, Springer-Verlag, Heidelberg

Späth, H. (1985): Cluster Dissection and Analysis. Theory, FORTRAN Programs, Examples.
Ellis Horwood Limited, Chichester

Ward, J. H. (1963): Hierarchical grouping methods to optimize an objective function. Journal
of the American Statistical Association 58, 236-244

Stability of Regression Trees

Karl-Peter Pfeiffer, Bernhard Pesec, Robert Mischak

Ludwig Boltzmann-Institut für Epidemiologie & Gesundheitssystem-forschung, Graz

Abstract
Growing regression trees on randomly selected subsamples of a data set may result in very different trees. This fact is used to perform an analysis of the stability of regression trees by an evaluation of the frequency of paths. Furthermore the frequency of splits is considered as an alternative procedure for tree growing.

Keywords: Regression tree, stability, frequency of paths

1 Introduction

Tree-based methods for regression and classification generally require fewer assumptions than classical methods and a wide variety of data structures can be handled. For the growing of regression or classification trees a multistage or sequential hierarchical decision scheme is used. The tree is constructed by recursive partitioning of the feature space into two groups. Compared to standard regression or classification problems, where only a single "optimal" feature set is used, the advantage of the tree structure is, that different subsets of features can be used at different decision levels.

The binary tree structure also provides an easy way of understanding the predictive structure of the data.

The aim of this study is to find a parsimoneous tree by a growing algorithm based on the analysis of the frequency of paths of random subsamples and to develop criteria for the stability of an regression tree. The procedures described here have been applied for the identification of homogenous groups of patients for a prospective hospital financing system in Austria.

2 Regression Trees

The aim of a regression tree is to predict a continuous response variable from a set of independent variables. The nodes of the tree represent a hierarchy of subpopulations. To grow a regression tree, three rules are necessary:

- a splitting criterion for model evaluation
- a stopping rule, which determines which nodes are terminal
- a function, which determines a characteristic value for each node.

Usually a regression tree is formed by recursively splitting nodes according to this variable which optimizes a certain model evaluation criterion. This growing procedure is followed by a pruning process to identify the right-sized tree [Breim84, Ciamp91]. To perform the growing and pruning process usually the data set has to be divided into two subsets, one for growing and one for pruning. This two data sets can also be considered as a learning and a test sample. Gelfand [Gelfa91] critizes the algorithm proposed by Breiman [Breim84] and suggests an iterative procedure where the subsamples for growing and pruning are successively interchanged. It can be shown that this algorithm converges and requires less computational time. Beside this point the selection of the evaluation criterion and split criterion is one of the main problems. In place of the classical least square criterion other similarity or dissimilarity criteria, like the median absolute distance or also test statistics for the difference of the two groups after a split can be used. A search algorithm is used to find the best splitting variable and the best cut-off point for continuous or ordinal variables. Regression trees are considered to be very robust [LeBla92]. But they are only robust in the way the model evaluation criteria and split procedures are robust.

As a stopping rule a certain relative increase or decrease of the model evaluation criterion, the significance level of the two-group test and/or the minimal number of observations in a node can be used.

Finally characteristic values for each node, like the mean or median can be computed. Usually this characteristic value is choosen in agreement with the optimization criterion and vice versa.

3 Evaluation of Regression Trees

One of the main problems for the identification of regression trees is to identify the "best" tree, that means the optimal subsets of variables. Similar to stepwise variable selection in multiple regression analysis a model evaluation criterion and a stopping rule have to be defined to identify the most important variables. "Best" tree means a right sized tree which neither underfits nor overfits the data.

One possibility is the application of tree growing and pruning algorithm based on subsamples of the data set [Gelfa91]. But also this iterative procedure must not lead to the best trees in that sense that the most likely paths have been selected. Therefore another definition of "best" tree based on the frequency of the paths is proposed.

3.1 Analysis of the frequency of paths

The reason for the following analysis is that an application of the tree-growing procedure on different randomly selected subsamples of a data set may result in different regression trees (Tab.1). The analysis of many data sets has shown that in the case of a skew distribution of the dependent variable or if outliers are present the splits in a node may be biased. Therefore the evaluation of the representativeness of the paths of a regression tree is proposed as stability criteria. A solution is considered stable, if the paths of this tree have a high frequency beyond a serie of trees from random subsamples. Furthermore the frequency, how often a node is terminal, can be used as a stopping rule.

In this example the regression trees of a series of B=100 randomly selected subsamples - for each of them L%=33% of the data set have been selected randomly - have been computed and an analysis of the frequency of paths of different length L(i), i=1,2,.... has been performed. For the analysis of the frequency of paths the commutative law can be applied, because the order of the split criteria is not important. The frequency of the paths can now be used to evaluate a regression tree grown according to some optimization criterion and stopping rule (Fig.1). This means, the tree with the paths with the highest probability is considered as the "most stable" tree.

Tab.1: Example of frequency of paths of length L(i), i=1,2,3,4 for a series of B=100 randomly selected subsamples of the data set ojm156 (N=1686, L%=33%). nX ... not X

Length of paths	Absolute frequency of node	Frequency, if this node is terminal	Sequence of split variables
1	89	0	A
1	89	16	nA
1	11	0	B
1	11	0	nB
2	99	28	A & B
2	92	43	A & nB
2	52	52	nA & C
2	52	52	nA & nC
2	18	18	nA & D
2	18	18	nA & nD
3	40	38	A & nB & E
3	40	40	A & nB & nE
3	55	50	A & B & F
3	55	38	A & B & nF
4	10	8	A & B & E & nF
4	9	9	A & B & E & F

3.2 Tree growing by frequency of split-criteria

An algorithm was developed which uses the frequency of paths and the frequency of terminal nodes for tree growing. This is a foreward selection process and therefore the commutative law cannot be applied. For each node the split with the highest absolute frequency is selected. A node is considered to be terminal if the relative frequency of terminal nodes in this node is greater than p(term). In Tab.2 the results of the tree growing process using the frequency of splits is shown.

Tab.2: Example of tree growing by frequency of paths (see Tab.1)

Length of paths	Absolute frequency of node	Frequency, if this node is terminal	Sequence of split variables
1	89	0	A
2	88	25	A & B
3	48	45	A & B & F
3	48	32	A & B & nF
4	8	8	A & B & nF & E
4	8	6	A & B & nF & nE
2	88	39	A & nB
3	40	38	A & nB & E
3	40	40	A & nB & nE
1	89	16	nA
2	52	52	nA & D
2	52	52	nA & nD

3.3 Program package

The aim of this study was to develop a new prospective hospital financing system in Austria. A large data set (N=400.000 patients) consisting of medical diagnosis, medical procedures and some other personal data is available. For each of this patients the cost, which is the dependent variable, was also available. The whole data set has been divided into about 400 subgroups according to medical aspects.

To handle large data sets (N>15000) and to use medical diagnosis and medical procedures as split variables a special program package was developed. Different split and model evaluation criteria and search procedures have been implemented and it is also possible to use a hierarchy of the variables to consider medical information. There is an option for B-times random selection of a subsample of L% (with replacement). The results of this analysis are represented in a table of the frequency of paths with and without the application of the commutative law.

4 Application

The procedures described above have been applied to the data sets of groups of patients to identify homogenous groups for a new prospective hospital payment system in Austria. The aim was to predict the cost of a patient from information about diagnosis, medical procedures and some personal data. Also if groups which are homogenous from a medical viewpoint are analyzed, very frequently the distribution of the dependent variable is asymmetric and outliers can be present. Therefore nonparametric tests or the median absolute distance have been used for splitting. Looking for more robust growing procedures random subsamples of a data set have been analyzed and sometimes very different sequences of variables in the pathes of the trees have been found (Tab.1). This leads to the idea to examine the stability of a regression tree by the analysis of the frequency of paths.

Fig.1: Regression tree for the data set *ojm156*. Sample sizes (N) and mean values of the nodes and frequency of the paths (k) for B=100 random subsamples.

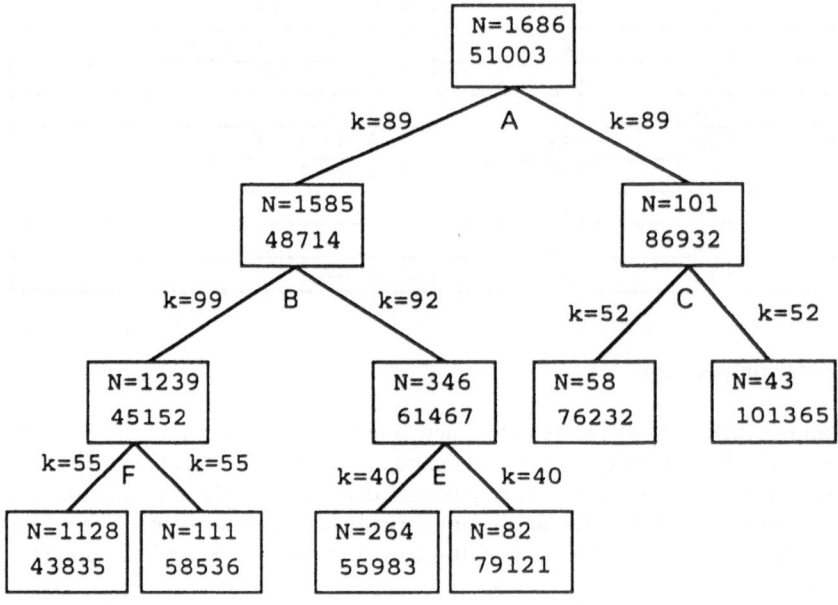

Fig.1 shows the regression tree which was identified by a growing process applied to the whole data set (N=1686). As a splitting criterion the Wilcoxon-U-test statistic was used and as a stopping rule the minimal number of cases in a subgroup n(min)=10 or a value of the Wilcoxon-U-test-statistic not significant for $p<0.05$ was defined. An evaluation of the stability by B=100 regression trees from random subsamples shows that already the first split criterion was found in 89 random subsamples, for a second criterion the frequency is 99 respectively 92 and that for longer paths the frequency decrease substantially until mainly terminal nodes are reached. If the frequency of paths is used for tree-growing, the commutative law for the split criterias can not be applied. In this example the frequency based algorithm leads to a very similar tree (Tab.2). But in many other examples the analysis of the frequency of paths has lead to different trees compared to classical tree-growing procedures, especially for longer paths.

5 Discussion

Regression and classification trees are a powerful tool for data analysis and the tree structure is an attractive way to present the dependency of a predictor from a set of predictor variables. Different growing and pruning algorithms have been described in the literature but none of them examines the stability of the solutions. An analysis of the frequency distribution of the paths of randomly selected subsamples of a data set shows, that the resulting regression trees can be very different. The calculation of the frequency of paths is one attempt to evaluate the stability of a regression tree and a possibility to find the most likely solution. Instead of a pruning process the frequency, how often a node is terminal, can be used as a stopping rule.

References

[Breim84] Breiman L., Friedman J.H., Olshen R.A., Stone C.J.: Classification and Regression Trees. Waldsworth International Group, Belmont, Ca., 1984
[Ciamp91] Ciampi A.: Generalized regression trees. Computational Statistics & Data Analysis, 12, 1991, 57-78
[Gelfa91] Gelfand S.B., Ravishankar C.S., Delp E.J.: An Iterative Growing and Pruning Algorithm for Classification Tree Design. IEEE Trans.on Pattern Analysis and Machine Intelligence, 13/2, 1991, 163-174
[LeBla92] LeBlanc M., Crowley J.: Relative Risk Trees for Censored Survival Data. Biometrics, 48, 1992, 411-425

Acknowledgement: This work was supported by the Austrian Ministery of Health, Sports and Consumer Protection.

Modelling for Recursive Partitioning and Variable Selection

Roberta Siciliano, Francesco Mola

Dipartimento di Matematica e Statistica, Università di Napoli Federico II

Abstract. We present a binary segmentation methodology in which it is possible to select simultaneously sub-groups of variables as well as sub-groups of cases in each node of the binary tree. To a recursive partition procedure which defines either a classification tree or a regression tree we add a hierarchy of models. The main advantages of this approach, especially for large samples, are: to abandon immediately unsignificant variables; to reduce rapidly the number of possible splits in each node; the amalgamation procedure becomes faster and is of higher interpretative value.

Keywords. Classification tree, regression tree, test of total homogeneity

1 Introduction

Binary segmentation is a powerful tool for both classification and regression purposes in case of large samples. Breiman, Friedman, Olshen and Stone (1984) have obtained important results in this field with the well-known CART (Classification and Regression Trees). Nevertheless, recently new methodological and computational results have been worked out: see for example the RECPAM (Recursive Partition and Amalgamation) of Ciampi et al. (1987); Ciampi (1992). Mola and Siciliano (1992; 1993) have proposed a two-stage predictive splitting procedure for classification trees. In each node, the first stage consists in choosing the best predictor and some competitors, whereas the second stage consists in choosing the best splitting variable among the possible split of the selected predictor(s). The approach can be fruitfully used in CART as well as in RECPAM.

In this paper, we introduce a recursive partition procedure, based on the two–stage predictive splitting algorithm, to select in each node simultaneously sub-groups of predictors as well as sub-groups of cases. To this end, statistical models are tested to select the most significant predictors as well as to define the splitting criterion. The choice of the statistical model (regression, logistic regression etc.) depends on the type (numerical and/or categorical) of the variables. We present the general methodology in section 2. In section 3 we discuss the case of ordinary regression trees providing a new splitting rule based on the test for total homogeneity in regression. In section 4 we remark the advantages of our method and we show some prospectives.

2 General methodology

2.1 Definitions and rules

Consider a learning sample $L = \{y|X\}$ where the N–vector y includes the observations of the dependent variable Y and the matrix X includes N row vectors $x_n = (x_{1n}, \ldots, x_{Mn})$ of measurements of M explanatory variables $(X_1, \ldots, X_M)$ of a numerical or categorical type, with N the number of observed cases. The N values y_n of the dependent variable are independent realizations of a random variable with a known distribution function.

In regression situations the expected value of the dependent variable is a function of a set of explanatory variables. The same will be true here, since we assume that the random variable y_n has expectation $E(y_n) = f(x_n; \theta)$, where the vector θ include the unknown parameters to be estimated on the basis of the observed sample. A statistical model is specified upon definition of the random variable y_n and of the functional relationship $f(\cdot)$.

Clearly, depending on the type (numerical and/or categorical) of the variables the well–known models can be specified: analysis of variance, ordinary regression, covariance analysis and logistic regression.

For a given learning sample L we define: one of the above–mentioned models $\mathcal{F} : E(y_n) = f(x_n; \theta)$, the statistical test Λ to validate the model to the data, the statistic λ to test the prediction value of a predictor, the α significance level. Under the null hypothesis that a predictor does not provide any contribution in prediction, we consider the p–value that corresponds to the smallest significance level that can be considered to reject the null hypothesis: we say that a predictor is significant for values of $p \leq \alpha$. Furthermore, we choose an unbiased and consistent test based on the statistic λ in such way that by increasing the sample size its power function tends to unity in the alternative set. In the following, the index "t" is used as index of a variable (or a set of variables), of a number of observations, of a model, that concern to the node t ($t = 1, \ldots, H$).

We start with a general model in which all the predictors are significant and their number depends on the sample size N. At the top node t_0, we consider the accepted model $\mathcal{F}_0 : E(y_n) = f(x_n; \theta)$ with M_0 significant predictors. We integrate the two–stage binary segmentation procedure with a sequential fitting of sub–models of $\mathcal{F}_0$ to sub–groups of cases, where a sub–model includes a sub-group of the predictors belonging to the model $\mathcal{F}_0$. The idea is to fit sub–models of the general model to sub–groups of cases such that sub-groups of cases are internally homogeneous and externally heterogeneous with respect to the selected sub–models.

Denote by t_l and t_r respectively the left node and the right node of the node t. In the procedure, the models $\mathcal{F}_{t_l}$ and $\mathcal{F}_{t_r}$ are sub–models of the model $\mathcal{F}_t$. In this way, a hierarchy of models will be associated to the nodes of the binary tree according to the following rules: (a) if a model at node t_l for a sample of N_{t_l} cases is accepted, then the model at node t for a sample of N_t cases has been accepted, and (b) if a model at node t for a sample of N_t cases is rejected, then the models at t_l and t_r are also rejected. While rule (a) ensures that the hierarchy is respected, rule (b) ensures to stop the procedure when a model is rejected.

The above mentioned rules resemble the rules used by Edwards and Havranek (1987) for a fast model selection procedure. The difference with their approach is that we do not compare a set of models fitted to a fixed sample but we select among models fitted to ever changing sub–samples. This procedure we call a model selection tree.

2.2 Two–stage splitting procedure

We consider a splitting procedure based on two stages: the first stage consists in selecting some best predictors, whereas the second stage consists in choosing the best splitting variable.

At node t, a subset of the learning sample $\mathbf{L}$ is observed: $\mathbf{L}_t = \{\mathbf{y}_t | \mathbf{X}_t\}$, where the set $\mathbf{X}_t$ includes the M_t available predictors $(X_1, \ldots, X_{M_t})$ for predicting $\mathbf{y}_t$ using the N_t cases, where $M_t \leq M$. We have the accepted model $\mathcal{F}_t^* : E(y_{n_t}) = f(\mathbf{x}_{n_t}; \theta_t)$, where $\mathbf{x}_{n_t}$ is the n_t–th row vector of measurements of the M_t available predictors in the set $\mathbf{X}_t$ and θ_t is the vector of the unknown parameters to be estimated on the basis of the observed node sample of N_t cases. Let $\mathbf{Z}_t$ be the matrix that include N_t row vectors $\mathbf{z}_{n_t} = (z_{1n_t}, \ldots, z_{K_t n_t})$ of measurements of the K_t significant predictors $(Z_{t1}, \ldots, Z_{tK_t})$ for $1 \leq K_t \leq M_t$. The best predictor (the most significant variable) at node t is the predictor Z_t^* with the smallest p–value. When there are many predictors, we retain a subset of best predictors formed by the most significant predictors with p–value below a fixed p^*–value $(p^* < \alpha)$.

In the following, we consider the more general case of selecting a subset of J_t best predictors $(Z_{t1}^*, \ldots, Z_{tJ_t}^*)$ for $1 \leq J_t \leq K_t$. For each best predictor $Z_{tj}^*(j = 1, \ldots, J_t)$ we define the sub-set $\mathbf{S}_{tj}$ of splitting variables $(S_{t1}, \ldots, S_{tq_j}, \ldots, S_{tQ_j})$ which includes the Q_j possible "dichotomizations" of the best predictor Z_{tj}^*. The measurement $s_{n_t q_j}$ of the splitting variable S_{tq_j} on the n_t–th case is a dummy variable with values 0 or 1. Let the set $\mathbf{S}_t$ be the union of the J_t sub–sets: $\mathbf{S}_{t1} \cup \ldots \mathbf{S}_{tj} \cup \ldots \cup \mathbf{S}_{tJ}$ with $Q = \sum_j Q_j$ splitting variables. In spite of the usual approaches, the two–stage approach considers at each node a reduced number of splitting variables since these are derived from a reduced set of predictors, namely the best predictors.

Let $_{(j)}\mathbf{z}_{n_t}$ be the row vector $\mathbf{z}_{n_t}$ in which the measurement of the best predictor Z_{tj}^* is omitted. For each splitting variable $S_{tq_j}(j = 1, \ldots, J_t; q_j = 1, \ldots, Q_j)$ we consider the model $\mathcal{F}_{tS_{q_j}} : E(y_{n_t}) = f(s_{n_t q_j}, _{(j)}\mathbf{z}_{n_t}; \theta_t)$, in which the measurement of the splitting variable S_{tq_j} replaces the measurement of the best predictor Z_{tj}^*. The best splitting variable S_t^* at node t is the most significant splitting variable that provides the smallest p–value. Thus, among all possible splits the best split provides the highest increment in prediction according to the statistical test λ. The best splitting variable S_t^* sends a proportion $p_{t_l} = N_{t_l}/N_t$ of cases in t to the sub--sequent left node t_l and a proportion $p_{t_r} = N_{t_r}/N_t$ of cases to the subsequent right node t_r. We temporary associate to the left node the model $\mathcal{F}_{t_l} : E(y_{n_{t_l}}) = f(s_{n_t q_j},_{(j)} z_{n_{t_l}}; \theta_t)|s_{n_{t_l} q_j} = 0)$ and to the right node the model $\mathcal{F}_{t_r} : E(y_{n_{t_r}}) = f(s_{n_t q_j},_{(j)} z_{n_{t_r}}; \theta_t)|s_{n_{t_r} q_j} = 1)$.

We consider a stopping criterion which ensures that in all terminal nodes the associated models are accepted. In the left node we test the model $\mathcal{F}_{t_l}$: if we reject it then we declare the node t to be a terminal one, because according to rule (b), all following sub–models will be rejected; otherwise, we define the accepted model $\mathcal{F}_{t_l}$ with the set of available predictors as the model to be considered in the next iteration. In the right node we test the model $\mathcal{F}_{t_r}$ and we proceed similarly. In addition, the procedure also takes also into account other "natural" stopping criteria: in the terminal node the sample size should be greater than a minimum value and the degrees of freedom in the model should be greater than zero. When it is impossible to make further splits, we have grown the maximal binary tree as well as a hierarchy of models.

2.3 Amalgamation procedure

We consider a kind of amalgamation procedure that allows to select an optimal binary tree from the maximal binary tree. We use as a criterion the classical test to measure the difference in averages of the dependent variable into the two sub–nodes which are terminal nodes with the assumption of homogeneity of the variances. The null hypothesis says that the two subgroups belong to the same population: in rejecting this hypothesis we do not eliminate the split; otherwise we eliminate the split. This criterion is applied starting from the lowest terminal nodes until no splits can be eliminated.

3 Example for numerical variables

3.1 Variable selection and splitting rule in ordinary regression

We specialize now the above methodology to the regression problem. Consider the usual regression model $\mathcal{F} : E(y_n) = \beta_1 + \beta_2 x_{2n} + \ldots + \beta_M x_{Mn}$ with the normal assumptions. The statistical test Λ to validate the regression model is given by the usual F–test, while the statistical test λ to validate a predictor is given by the partial F–test (see for example Johnston, 1972). We describe in detail the two stages of the splitting procedure at node t.

To node t is associated the accepted model $\mathcal{F}_t^* : E(y_{n_t}) = \beta_1 + \beta_2 x_{2n_t} + \ldots + \beta_{M_t} x_{M_t n_t}$. We have the fit of this model to the sample of N_t cases. From the partial F_p–tests we take the K_t significant predictors $(Z_1, \ldots, Z_{K_t})$ with $p \leq \alpha$. Let without loss of generality Z_1 result to be the best predictor.

For the best predictor Z_1^* we define the set $\mathbf{S}_t$ of Q splitting variables. For each splitting variable $S_{tq}(q = 1, \ldots, Q)$ we consider the model $\mathcal{F}_{tS_q}$:

$E(y_{n_t}) = \beta_1 s_{n_t} + \beta_2 z_{2n_t} + \ldots + \beta_{Mt} z_{K_t n_t}$, in which the measurement of the splitting variable S_{t_q} replaces the measurement of the best predictor Z_1^*. The model $\mathcal{F}_{tS_q}$ is a model for covariance analysis with a dummy variable that discriminates two sub–groups of N_{t_l} and N_{t_r} cases and with (K_{t-1}) numerical covariates. The increment in explained total variation of the dependent variable due to the splitting variable can be validated by the F_h–test for total homogeneity in regression (Johnston, 1972). The statistical test at node t can be defined in our context as:

$$ F_{h_t} = \frac{(RSS_t - ESS_t)/(K_t - 1)}{RSS_t/(N_t - 2K_t - 2)} $$

where RSS_t is the residual sum of squares of the regression model $E(y_{n_t}) = \beta_1 + \beta_2 z_{2n_t} + \ldots + \beta_{Mt} z_{K_t n_t}$ $(n_t = 1, ..., N_t)$, while ESS_t is the residual sum of squares of the set of two independent regressions: $E(y_{n_{t_l}}|s_{n_t} = 0) = \beta_1^0 + \beta_2^0 z_{2n_{t_l}} + \ldots + \beta_{Mt}^0 z_{K_t n_{t_l}} (n_{t_l} = 1, ..., N_{t_l})$ and $E(y_{n_{t_r}}|s_{n_t} = 1) = \beta_1^1 + \beta_2^1 z_{2n_{t_r}} + \ldots + \beta_{Mt}^1 z_{K_t n_{t_r}} (n_{t_r} = 1, ..., N_{t_r})$. Low values of the F_{h_t}–statistic mean that there is no difference in regression between the two sub–groups, while high values mean that the two sub–groups are heterogeneous in regression. The best splitting variable S_t^* is the splitting variable that provides the highest value of the F_{h_t}–statistic or the smallest p–value.

3.2 Application

To illustrate our methodology we show an example with, only for sake of brevity, a small number of cases and variables. However, we like to emphasize that our procedure is especially efficient and useful for very large samples.

We consider a sample of 62 italian communes on which are measured the following 6 numerical variables: *average income* y, *% of graduated* X_2, *% of analphabetism* X_3, *% of autonomous labor force* X_4, *% employees and workers* X_5, *% directors* X_6. We define the regression model with the *average income* as dependent variable and the remaining variables as predictors. The results are in table 1.

Tab. 1: splits sequence and terminal nodes information

				split sequence				terminal nodes	
t	N_t	N_{t_l}	N_{t_r}	available predictors	significant predictors	best pred.	split value	$\bar{y}$	$\bar{s}$
1	62	38	24	X_2, X_3, X_4 X_5, X_6	X_2, X_3, X_4 X_5, X_6	X_3	3.1		
2	38	21	17	X_2, X_3, X_4 X_5, X_6	X_2, X_3 X_4, X_5	X_2	2.8		
3	24	17	7	X_4, X_5	X_4, X_5	X_5	43		
4	21	10	11	X_3, X_4, X_5	X_4, X_5	X_5	45.2		
5	17							8.27	.79
6	17							5.85	.41
7	7							5.27	.26
8	10							8.92	.85
9	11							7.96	.86

4 Concluding remarks and prospectives

The two–stage recursive partition procedure has allowed to build up a hierarchy of models associated to the binary tree. In this way, this binary segmentation methodology can resolve simultaneously two problems: it selects recursively the most typological predictors and stabilizes a partition of the sample into sub–groups which are internally homogeneous and externally heterogeneous. For the interpretation of the results we grow two binary trees: the maximal binary tree which represents the maximal splitting of only the significant variables, neglecting the others, and the optimal binary tree which discriminates a minimal set of sub–groups of cases determined by modelling and not by the usual impurity measures. Each node in the binary tree represents a special situation with its own population, with its own model, with its own variables, which allows independent statistical investigation.

Modelling for recursive partitioning and variable selection provides important computational advantages: (a) to abandon immediately and definitively unsignificant variables; (b) to reduce rapidly the number of possible splits in each node using only significant variables; (c) the amalgamation procedure becomes faster being based on a statistical test and is of higher interpretative value since it guarantees a small set of sub–groups of cases.

In this paper the methodology has been applicated to the case of numerical variables with the use of the ordinary regression model providing a new splitting rule. However the same methodology can be used with other statistical models for any type of variables.

Acknowledgements: The present paper has been supported by a grant of MURST 60%, 1993. The authors are indebted to Prof. Carlo Lauro for helpful comments.

References:

Breiman, L., Friedman, J.H., Olshen, R.A. and Stone, C.J. (1984). Classification and Regression Trees. Wadsworth International Group, Belmont, California.

Ciampi, A., Chang, C.H., Hogg, S.A. & McKinney, S. (1987). Recursive Partition: A Versatile Method for Exploratory Data Analysis in Biostatistics. *Joshi Feistschrift, vol. 5, Biostatistics, I.B.Mac Neil and G.J. Umphrey (eds.), 23-50.*

Ciampi, A. (1992). Tree-Growing for the Multivariate Model: The RECPAM Approach. In: Y. Dodge and J. Whittaker (eds.): *Computational Statistics.* Vol. 1. (Compstat '92 Proceedings). Physica Verlag.

Edwards, D. & Havranek, T. (1987). A Fast Model Selection Procedure for Large Families of Models. *Journal of the American Statistical Association*, 82, 205-213.

Johnston, J. (1972). Econometrics Methods. New York: McGrow Hill.

Mola, F. & Siciliano, R. (1992): A Two-stage Predictive Splitting Algorithm in Binary Segmentation. In: Y. Dodge and J. Whittaker (eds.): *Computational Statistics.* Vol. 1. (Compstat '92 Proceedings). Physica Verlag.

Mola, F. & Siciliano, R. (1993). Alternative Strategies and CATANOVA Testing in Two-stage Binary Segmentation. *Proceedings of the 4th Conference of IFCS.*

Part VIII

Computational Aspects in Bayesian Statistics

Approximate Bayesian Inferences for the Reliability in the Weibull Case: Some Aspects of Reparametrization

Jorge Alberto Achcar

Universidade de São Paulo, ICMSC

Keywords. Reliability, Weibull distribution, reparametrization

1 Introduction

Usually, inferences on the reliability function $R(t_0)$ at time t_0 assuming different parametrical models and censored lifetime observations are obtained by using asymptotical methods. One of these asymptotical results is given by the asymptotical normality of the maximum likelihood estimators. Under the Bayesian approach, we get marginal posterior densities or posterior moments for $R(t_0)$ based on numerical or approximation methods. These results, usually depend on an appropriate transformation of $R(t_0)$, to get accurate results. One way to find an appropriate reparametrization, is to search for a one-to-one transformation of $R(t_0)$ that gives close "normality" for the likelihood function (see for example, Anscombe, 1964; Sprott, 1973, 1980; Kass and Slate, 1992; or Hills and Smith, 1993). Assuming a Weibull distribution for the lifetimes in a reliability experiment, we explore the use of some popular transformations for proportions (see for example, Guerrero and Johnson, 1982; or Aranda-Ordaz, 1981) and a measure to nonnormality of likelihood functions or posterior densities given by the standardized form of the third derivative of the logarithm of the likelihood or posterior density (see Sprott, 1973; or Kass and Slate, 1992). We also check the adequability of the proposed reparametrization, by using the t-plot proposed by Hills and Smith (1993).

2 A Reparametrization for $R(t_0)$

One way to improve the "normality" of the likelihood function for the reliability function at time t_0, is to consider different transformations for $R(t_0)$. For example, we could consider the logit reparametrization $\phi_L = \ln[R/(1-R)]$, where $R = R(t_0)$. Some parametric families of transformations for proportions (see for example, Atkinson, 1985) also could be used to improve the "normality" of the likelihood for $R(t_0)$.

To obtain an invertible family of transformations which includes the logit, Guerrero and Johnson (1982) suggested the transformation

$$\phi_{GJ}^*(\lambda) = \left\{ \left(\frac{R}{1-R} \right)^\lambda - 1 \right\} \bigg/ \lambda \, . \tag{1}$$

For a given value of λ, we can consider a modified form of Guerrero and Johnson transformation given by

$$\phi_{GJ}(\lambda) = \left(\frac{R}{1-R} \right)^\lambda - 1 \tag{2}$$

which should not produce different results as considering (1).

The great advantage of transformation (2), is that it is readily inverted. With $\phi_{GJ} = \phi_{GJ}(\lambda)$, we obtain,

$$R = \frac{(\phi_{GJ}+1)^{1/\lambda}}{1+(\phi_{GJ}+1)^{1/\lambda}} \, . \tag{3}$$

To find an appropriate value of λ that gives good "normality" for the likelihood function of $\phi_{GJ}(\lambda)$, we choose λ in (2) that gives third derivative of the log-likelihood function $\ell(\phi_{GJ})$ at the maximum likelihood estimator $\hat{\phi}_{GJ}$ in a standardized form,

$$\mathrm{STD}(\hat{\phi}_{GJ}) = \left| \ell'''(\hat{\phi}_{GJ}) \left(-\ell''(\hat{\phi}_{GJ}) \right)^{-3/2} \right| \tag{4}$$

close to zero (see Sprott, 1973; or Kass and Slate, 1992).

We also could explore another invertible family of transformations for proportions proposed by Aranda-Ordaz (1981), to get similar results.

3 Reliability Function at Time t_0 Considering a Weibull Distribution and Censored Data

Suppose there is a random sample of n units with lifetimes $T_1, T_2, \ldots, T_n$, but that associated to each unit is also a fixed censoring time $L_i > 0$ (type I censored data). We observe T_i only if $T_i \leq L_i$ and the data consists of pairs (t_i, δ_i), $i = 1, \ldots, n$ where $t_i = \min(T_i, L_i)$ and $\delta_i = 1$ if $t_i = T_i$ or $\delta_i = 0$ if $t_i = L_i$.

Considering the Weibull distribution with density,

$$f(t; \alpha, \beta) = \frac{\beta}{\alpha} \left(\frac{t}{\alpha} \right)^{\beta-1} \exp \left\{ - \left(\frac{t}{\alpha} \right)^\beta \right\} \tag{5}$$

where $t > 0$; α, $\beta > 0$ and type I censored data, the log-likelihood function for the reliability function at time t_0, $R = R(t_0) = \exp\left\{-\left(\frac{t_0}{\alpha}\right)^\beta\right\}$ and β, is given by

$$\ell(R,\beta) = d\ln\beta - d\beta\ln t_0 + d\ln\{-\ln R\} + (\beta-1)\sum_{i\in D}\ln t_i + \frac{T(\beta)}{t_0^\beta}\ln R, \quad (6)$$

where $T(\beta) = \sum_{i=1}^n t_i^\beta$, $d = \sum_{i=1}^n \delta_i$ is the observed number of lifetimes and D denotes the set of units for whom lifetimes are uncensored.

The maximum likelihood estimator for $R(t_0)$ is given by

$$\widehat{R}(t_0) = \exp\left\{-\frac{d\,t_0^{\widehat{\beta}}}{\sum_{i=1}^n t_i^{\widehat{\beta}}}\right\} \quad (7)$$

where $\widehat{\beta}$ satisfies,

$$\frac{d}{\widehat{\beta}} - d\ln t_0 + \sum_{i\in D}\ln t_i - \frac{d\sum_{i=1}^n t_i^{\widehat{\beta}}(\ln t_i - \ln t_0)}{\sum_{i=1}^n t_i^{\widehat{\beta}}} = 0. \quad (8)$$

Usually, inferences on $R(t_0)$ are based on the asymptotic normality of $\widehat{R}(t_0)$ and $\widehat{\beta}$.

In practical work, this normal approximation can be very poor, especially for small sample sizes (see for example, Lawless, 1982).

With type II censored data, the form of the likelihood function (6) is the same, but d is fixed and $T(\beta) = \sum_{i=1}^d t_{(i)}^\beta + (n-d)t_{(d)}^\beta$, where $t_{(1)},\ldots,t_{(d)}$ are the first d ordered observations of a random sample of size n from the Weibull density (5).

4 The Guerrero-Johnson Transformation for $R(t_0)$ with β Known

Assuming β known, the log-likelihood function for $R(t_0)$ is given by,

$$\ell(R)\alpha\,d\ln\{-\ln R\} + \frac{T}{t_0^\beta}\ln R \quad (9)$$

where $T = \sum_{i=1}^n t_i^\beta$.

In the original parametrization $R(t_0)$, the standardized third derivative (4) of the log-likelihood function at the maximum likelihood estimator $\widehat{R}(t_0) = \exp\left\{-dt_0^\beta/T\right\}$ is given by

$$\mathrm{STD}(\widehat{R}) = \left|d^{-1/2}\left(3\ln\widehat{R} + 2\right)\right|. \quad (10)$$

Observe that if t_0 is large, that is, $R(t_0)$ is small, we could have large values for $\text{STD}(\widehat{R})$, which indicates bad "normality" for the likelihood function.

Considering the Guerrero-Johnson transformation, the log-likelihood function for $\phi_{GJ}(\lambda)$ is given by

$$\ell(\phi_{GJ}) \alpha \, d \ln B(\phi_{GJ}) - \frac{T}{t_0^\beta} B(\phi_{GJ}) , \tag{11}$$

where $T = \sum_{i=1}^n t_i^\beta$ and $B(\phi_{GJ}) = \ln\left[1 + (\phi_{GJ} + 1)^{-1/\lambda}\right]$.

The standardized third derivative (4) of $\ell(\phi_{GJ})$ locally at the maximum likelihood estimator $\widehat{\phi}_{GJ} = \left[e^{\,dt_0^\beta/T} - 1\right]^{-\lambda} - 1$, is given by

$$\text{STD}(\widehat{\phi}_{GJ}) = \left| d^{-1/2} \left(2 - \frac{3B(\widehat{\phi}_{GJ})B''(\widehat{\phi}_{GJ})}{\left(B'(\widehat{\phi}_{GJ})\right)^2} \right) \right| \tag{12}$$

The value for λ such that $\text{STD}(\widehat{\phi}_{GJ}) = 0$, is given by

$$\lambda = \left(\frac{2T}{3dt_0^\beta} + 1\right)\left(1 - e^{-dt_0^\beta/T}\right) - 1 . \tag{13}$$

With this value of λ, we can consider the asymptotic normality of $\widehat{\phi}_{GJ}$, to get better inferences on $R(t_0)$.

When β is unknown, we should search for a joint transformation of β and $R(t_0)$ that gives joint "normality" for the likelihood function. Since this transformation cannot easily be obtained (see for example, Kass and Slate, 1992) and our parameter of interest is $R(t_0)$, an alternative way is to search for a reparametrization of $R(t_0)$ that gives close "normality" for the profile likelihood.

5 An Example

Consider a type II censoring data set consisting of $n = 12$ units where the experiment terminated when it was observed $d = 8$ failures (data set introduced by Lawless, 1982, p.103). The observed lifetimes (in hours) are given by 31,58,157,185,300,470,497 and 673. Assuming an exponential distribution with density (5) and $\beta = 1$, the maximum likelihood estimator for the reliability function at time $t_0 = 5$ is given by $\widehat{R}(5) = 0.9921$. From the normal limiting distribution for $\widehat{R}(5)$, we find a 95% confidence interval for $R(5)$ given by $(0.9867;09976)$.

It is interesting to observe that $2T/\alpha$ has an exact chi-square distribution with $2d$ degrees of freedom. An exact 95% confidence interval for α is given by

(351.6;1465.4) which corresponds to a 95% confidence interval for $R(5)$ given by (0.9859;0.9966).

Considering the Guerrero and Johnson transformation (2), we could improve the "normality" of the likelihood function considering an appropriate value for λ in $\phi_{GJ}(\lambda)$. With $t_0 = 5$, we find from (13), $\lambda = -0.3281$.

From the normal limiting distribution for the maximum likelihood estimator $\widehat{\phi}_{GJ}(-0.3281) = -0.7955$, we find an approximate 95% confidence interval for $\phi_{GJ}(-0.3281)$ given by (-0.8422; -0.7487), which corresponds to a better 95% confidence interval for $R(5)$ given by (0.9854;0.9964).

We also could check the "normality" of the likelihood function in the parametrization $\phi_{GJ}(-0.3281)$ considering the t-plot (see Hills and Smith, 1993) $T(\phi_{GJ})$ against some values of ϕ_{GJ}, where

$$T(\phi_{GJ}) = \text{sgn}(\phi_{GJ} - \widehat{\phi}_{GJ}) \left\{ -2\ell(\phi_{GJ}) + 2\ell(\widehat{\phi}_{GJ}) \right\}^{1/2}$$

and $\widehat{\phi}_{GJ}$ is the maximum likelihood estimator of ϕ_{GJ}.

Since we observe a straight line (see figure 1), we conclude by the "normality" of the likelihood function for $\phi_{GJ}(-0.3281)$. In the original parametrization $R(5)$, the plot of $T(R(5))$ against $R(5)$ is markedly curved (see figure 2), which indicates the nonnormality of the likelihood function for $R(5)$.

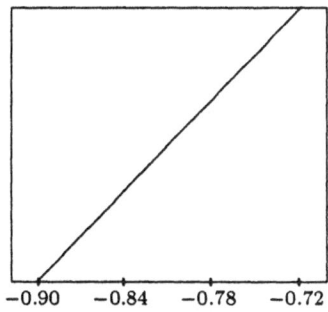

−0.90   −0.84   −0.78   −0.72

Fig.1 - t-plot for $\phi_{GJ}(-0.3281)$

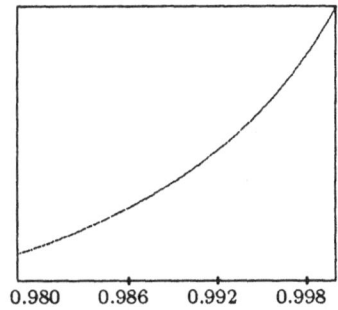

0.980   0.986   0.992   0.998

Fig.2 - t-plot for $R(5)$

In table 1, we have exact and approximate 95% confidence intervals for $R(t_0)$ with $t_0 = 30, 500$ and 2000, considering the parametrizations $R(t_0)$ and $\phi_{GJ}(\lambda)$, respectively. We observe good inference results considering the parametrization $\phi_{GJ}(\lambda)$ with the appropriate values for λ.

Table 1.   95% confidence intervals for $R(t_0)$

t_0	Using Exact Distribution for $2T/\alpha$	Asymptotical Normality for $\widehat{R}(t_0)$	λ Given by (13)	Asymptotical Normality for $\widehat{\phi}_{GJ}(\lambda)$
30	(0.9182;0.9797)	(0.9224;0.9850)	-0.3026	(0.9155;0.9786)
500	(0.2412;0.7109)	(0.2054;0.7023)	0.0071	(0.2332;0.6930)
2000	(0.0034;0.2554)	(-0.5050;0.1353)	0.1596	(0.0026;0.2373)

In table 2, we have Laplace's approximations (see for example, Tierney and Kadane, 1986) for the posterior mean $E(R(t_0)/\text{data})$ considering a Jeffreys noninformative prior density for $R(t_0)$, $\pi_0(r)\alpha\, 1\,/[(-\ln r)r]$ (see for example, Martz and Waller, 1982). Since in this case, we have an exact expression for $E(R(t_0)/\text{data})$, we observe very accurate approximate Bayesian inferences for $R(t_0)$ considering the parametrization $\phi_{GJ}(\lambda)$, and with λ given by (13).

Table 2. Posterior means for $R(t_0)$

		LAPLACE'S APPROXIMATIONS			
t_0	Exact	Parametrization $R(t_0)$	Percentage Errors	Parametrization $\phi_{GJ}(\lambda)$	Percentage Errors
5	0.9921	0.9931	0.101	0.9921	0.000
30	0.9538	0.9595	0.598	0.9538	0.000
500	0.4708	0.5210	10.663	0.4710	0.042
2000	0.0697	0.1804	158.824	0.0687	1.435

References

ANSCOMBE, F.J. (1964). Normal likelihood functions, Ann. Inst. Stat. Math., 16, 1-19.

ARANDA-ORDAZ, F.J. (1981). On two families of transformations to additivity for binary response data, Biometrika, 68, 357-363.

ATKINSON, A.C. (1985). Plots, transformations and regression. Oxford: Clarendon press.

GUERRERO, V.M.; JOHNSON, R.A. (1982). Use of the Box-Cox transformation with binary response models, Biometrika, 69, 309-314.

HILLS, S.E.; SMITH, A.F.M. (1993). Diagnostic plots for improved parametrization in Bayesian inference, Biometrika, 80, 1, 61-74.

KASS, R.E., SLATE, E.H. (1992). Reparametrization and diagnostics of posterior nonnormality. In Bayesian Statistics 4, Ed. J.M. Bernardo, J.O. Berger, A.P. Dawid and A.F.M. Smith, pp. 289-306. Oxford University Press.

LAWLESS, J.F. (1982). Statistical models and methods for lifetime data. New York: John Wiley & Sons.

MARTZ, H.F.; WALLER, R.A. (1982). Bayesian Reliability Analysis. New York: John Wiley & Sons.

SPROTT, D.A. (1973). Normal likelihoods and their relation to large sample theory of estimation, Biometrika, 60, 457-465.

SPROTT, D.A. (1980). Maximum likelihood in small samples: estimation in the presence of nuisance parameters, Biometrika, 67, 515-523.

TIERNEY, L.; KADANE, J.B. (1986). Accurate approximations for posterior moments and marginal densities, Jour. of American Stat. Association, 81, 82-86.

Simulation of a Bayesian Interval Estimate for a Heterogeneity Measure

Derrick N. Joanes, Christine A. Gill, Andrew J. Baczkowski

Department of Statistics, University of Leeds

Keywords. Bayesian estimation, heterogeneity, diversity

1 Introduction

A well-known measure of the heterogeneity of a population is provided by the statistic

$$H = -\sum_{i=1}^{s} p_i \log p_i$$

where it is assumed that s possible categories have respective probabilities $p_1, ..., p_s$.

This summary statistic has been used in statistical ecology as a measure of the diversity of a population consisting of s different species (e.g. Pielou, 1975). It has also been used in a medical context to describe the degree of uncertainty concerning the probabilities of possible diseases in a diagnosis (Patil and Taillie, 1982).

Given a random sample of n observations from data falling into the s categories, we consider the problem of determining interval estimates for H.

Gill and Joanes (1979) adopted a Bayesian approach, in which it is assumed that the joint prior density of the p_i's may be represented by a conjugate Dirichlet distribution of the form

$$\pi(p_1, ..., p_s) = \frac{\Gamma(ks)}{[\Gamma(k)]^s} \Pi_{i=1}^{s} p_i^{k-1}$$

where k is a suitably chosen constant. This prior is widely used in this context (e.g. Boender and Rinnooy Kan, 1987).

Although the joint posterior density of $p_1, ..., p_s$ is easily derived, the posterior distribution of H cannot be obtained in closed form. However, moments of this distribution are available, and turn out to involve polygamma functions.

2 Moments of the distribution of H

The mean of the posterior distribution of H may be obtained directly from the marginal posterior distribution of p_i.

Using the above Dirichlet prior, the joint posterior distribution of $p_1, ..., p_s$ is given by

$$f(p_1, ..., p_s) \propto \Pi_{i=1}^s p_i^{n_i+k-1}$$

where n_i is the number of observations in the sample falling into category i.

It follows that the marginal posterior distribution of p_i has a beta distribution with parameters $n_i + k, n - n_i + ks - k$. To simplify the notation we will in future write $k_i = n_i + k$, and $\Sigma k_i = K$, so that the parameters of the beta distribution become $k_i, K - k_i$.

Thus,

$$
\begin{aligned}
E(H) &= E\left(-\sum_{i=1}^s p_i \log p_i\right) \\
&= -\sum_{i=1}^s \frac{1}{B(k_i, K - k_i)} \int_0^1 p_i^{k_i}(1 - p_i)^{K-k_i-1} \log p_i dp_i \\
&= -\sum_{i-1}^s \left[\frac{\frac{\partial}{\partial k_i} B(k_i + 1, K - k_i)}{B(k_i, K - k_i)}\right].
\end{aligned}
$$

This expression can, in turn, be evaluated in terms of digamma functions as

$$E(H) = -\sum_{i=1}^s \left[\frac{k_i}{K}\{\psi(k_i + 1) - \psi(K + 1)\}\right].$$

For the higher moments it is convenient to define

$$B(k_1, ..., k_s) = \frac{\Pi_{i=1}^s \Gamma(k_i)}{\Gamma(K)}$$

so that for $s \geq 3$ the joint posterior density of p_i, p_j may be written as

$$f(p_i, p_j) = \frac{p_i^{k_i-1} p_j^{k_j-1}(1 - p_i - p_j)^{K-k_i-k_j-1}}{B(k_i, k_j, K - k_i - k_j)}.$$

The second moment of H, given by

$$E(H^2) = E\left\{\sum_i p_i^2(\log p_i)^2 + \sum_i \sum_j p_i p_j \log p_i \log p_j\right\}, \quad i \neq j$$

then becomes

$$E(H^2) = \sum_i \left[\frac{\frac{\partial^2}{\partial k_i^2} B(k_i + 2, K - k_i)}{B(k_i, K - k_i)} \right] + \sum_i \sum_j \left[\frac{\frac{\partial^2}{\partial k_i \partial k_j} B(k_i + 1, k_j + 1, K - k_i - k_j)}{B(k_i, k_j, K - k_i - k_j)} \right]$$

$$, i \neq j, \ (s \geq 3).$$

Joanes and Gill (1992) proposed that approximate highest posterior density *(HPD)* intervals for H can be derived by fitting an appropriately scaled beta distribution by the method of moments.

The precision with which the fitted beta distribution approximates to the theoretical distribution of H may be judged by comparing higher moments of the theoretical and fitted distributions.

3   Simulation of the distribution of H

Following an approach adopted by Engen (1978) we consider a transformation involving beta-type variables which facilitates simulation of the theoretical distribution of H.

We have previously found that the marginal posterior distribution of p_i has a beta distribution with parameters $k_i, K - k_i$, i.e. dependent upon the particular partition of the n observations into each category. Further, the joint posterior distribution of $p_1, ..., p_s$ is of the non-symmetric Dirichlet form

$$f(p_1, .., p_s) \propto \Pi_{i=1}^s p_i^{k_i - 1}, \qquad (k_i > 0).$$

The transformation we consider is:

$$
\begin{array}{llll}
p_1 & = & V_1 & \text{where} \\
p_2 & = & V_2(1 - V_1) \\
p_3 & = & V_3(1 - V_1)(1 - V_2) \\
& & \cdot \\
& & \cdot \\
& & \cdot \\
p_{s-1} & = & V_{s-1}(1 - V_1)....(1 - V_{s-2})
\end{array}
\qquad
\begin{array}{ll}
V_1 \sim \text{Beta} & (k_1, K - k_1) \\
V_2 \sim \text{Beta} & (k_2, K - k_1 - k_2) \\
V_3 \sim \text{Beta} & (k_3, K - k_1 - k_2 - k_3) \\
\\
\\
\\
V_{s-1} \sim \text{Beta} & (k_{s-1}, k_s) \quad ,
\end{array}
$$

the V_i's $(i = 1, ..., s - 1)$ being mutually independent.

Note also that $p_s = 1 - p_1 - ... - p_{s-1} = (1 - V_1)...(1 - V_{s-1})$.

Thus a set of independent beta-type variables may be used to generate the non-independent p_i's, and hence the distribution of H.

Clearly the simulated distribution of H will be dependent, through k_i, upon the particular sample partition of the n observations into the s categories.

The parameter k can be described as a flattening constant in the prior distribution for the p_i values (Good, 1966). Of particular interest are the

190

Table 1: Comparison of 95% HPD limits for H when n = 10, s = 3 based on (1) simulation, (2) method of moments fitting to a beta distribution (bracketed figures)

Partition	k=0		k=1/2		k=1	
	Lower	Upper	Lower	Upper	Lower	Upper
10 0 0			0.00 (0.00)	0.66 (0.67)	0.09 (0.10)	0.86 (0.85)
9 1 0			0.10 (0.12)	0.87 (0.87)	0.25 (0.25)	0.99 (0.97)
8 2 0			0.26 (0.27)	0.98 (0.95)	0.39 (0.40)	1.05 (1.03)
8 1 1	0.14 (0.15)	0.97 (0.96)	0.30 (0.30)	1.05 (1.03)	0.42 (0.42)	1.08 (1.06)
7 3 0			0.40 (0.41)	1.03 (0.98)	0.52 (0.51)	1.08 (1.05)
7 2 1	0.35 (0.35)	1.08 (1.06)	0.48 (0.47)	1.10 (1.08)	0.57 (0.56)	1.10 (1.09)
6 4 0			0.52 (0.50)	1.05 (1.00)*	0.62 (0.58)	1.09 (1.05)*
6 3 1	0.52 (0.50)	1.10 (1.08)	0.61 (0.60)	1.10 (1.09)	0.68 (0.67)	1.10 (1.10)
6 2 2	0.55 (0.55)	1.10 (1.10)	0.64 (0.63)	1.10 (1.10)	0.70 (0.70)	1.10 (1.10)
5 5 0			0.57 (0.53)	1.05 (0.99)*	0.65 (0.61)	1.09 (1.05)*
5 4 1	0.63 (0.59)	1.10 (1.08)	0.69 (0.67)	1.10 (1.09)	0.74 (0.73)	1.10 (1.10)
5 3 2	0.70 (0.69)	1.10 (1.10)	0.75 (0.75)	1.10 (1.10)	0.80 (0.79)	1.10 (1.10)
4 4 2	0.75 (0.75)	1.10 (1.10)	0.80 (0.79)	1.10 (1.10)	0.83 (0.83)	1.10 (1.10)
4 3 3	0.81 (0.81)	1.10 (1.10)	0.85 (0.85)	1.10 (1.10)	0.88 (0.87)	1.10 (1.10)

values $k = 1$, giving the uniform joint prior density; $k = \frac{1}{2}$, giving the invariance theory prior; and $k = 0$, where the posterior mean of p_i equals the usual maximum likelihood estimator.

Simulation of the distribution of H was based on 1,000,000 $(p_1, ..., p_s)$ s-tuples with HPD limits being estimated from a frequency distribution of the resulting values of H based on 200 class intervals. The estimation procedure first involved a sorting method to find the classes with most similar frequencies corresponding to approximate 95% HPD limits. These limits were further refined by simultaneous solution of quadratic equations involving adjacent classes. Special cases such as one-sided HPD intervals were dealt with separately.

Three independent simulations were carried out, one involving 2,000,000 values of H over 400 classes, and these gave consistent results accurate to two decimal places.

The simulation method provides an alternative way of estimating approximate HPD intervals for H. Such estimates may be compared with those obtained using the method outlined in Section 2.

4 Results

Table 1 compares the approximate 95% HPD limits based on simulation of the distribution of H with those produced by Joanes and Gill (1992) based on fitting a beta distribution by the method of moments (bracketed figures). All possible partitions when $n = 10$ and $s = 3$ are presented. There are no entries for partitions including a zero when $k = 0$, because the Dirichlet

Table 2: Comparison of skewness and kurtosis for selected partitions when $n = 10, s = 3$

k	Partition	skewness			kurtosis		
		exact	simulated	fitted beta	exact	simulated	fitted beta
0	6 2 2	-0.83	-0.83	-0.86	0.43	0.42	0.40
	6 4 0	0.14	0.14	-0.38	0.31	0.31	-0.14
1/2	5 5 0	0.38	0.38	-0.39	0.14	0.13	-0.09
	6 2 2	-0.97	-0.97	-1.02	0.86	0.85	0.87
	6 4 0	-0.14	-0.14	-0.59	-0.19	-0.19	0.11
1	5 5 0	-0.01	-0.01	-0.61	-0.45	-0.45	0.17
	6 2 2	-1.08	-1.08	-1.13	1.23	1.24	1.29

distribution does not exist for $k_i = 0$.

Generally the two sets of results agree very well. For the most uneven partitions, the intervals based on simulation are slightly wider than those based on the fitted beta distribution, but are narrower and shifted to the right slightly for the more even partitions. As k increases, any differences between comparable intervals are reduced.

Two partitions in particular give differing results, (6,4,0) and (5,5,0). Table 2 presents examples of the values of skewness and kurtosis calculated from the exact moments of $H/\log_e s$ (exact) and estimated from the simulated distribution of $H/\log_e s$ (simulated) and the fitted beta distribution of $H/\log_e s$ (fitted beta).

The means and variances for $H/\log_e s$ estimated by the two methods agree with the exact values to three decimal places and are not presented here. The skewness and kurtosis based on simulation are much closer to the exact values than those from the fitted beta distribution for all partitions. Skewness tends to be slightly more extreme in the fitted model and partitions (6,4,0) and (5,5,0) provide the most extreme examples. Fitted values of kurtosis also differ from the exact values in many cases but these do not appear to affect the size or location of the HPD intervals, whereas skewness does. For partition (6,2,2) all three sets of values agree well as do the HPD intervals produced by the two methods.

Results suggest that the simulated distribution of H is a better approximation to the exact distribution than the fitted beta distribution and therefore the HPD intervals based on the simulation are more realistic. Similar results and conclusions arose for other values of n and s.

References

[1] BOENDER, C.G.E. and RINNOOY, KAN, A.H.G. (1987) *A multinomial Bayesian approach to the estimation of population and vocabulary size*, Biometrika, 74, pp.849-856.

[2] ENGEN, S. (1978) *Stochastic abundance models*. Chapman and Hall, London.

[3] GILL, C.A. and JOANES, D.N. (1979) *Bayesian estimation of Shannon's index of diversity*, Biometrika, 66, 1, pp.81-85.

[4] GOOD, I.J. (1966) *How to estimate probabilities*, Journal of the Institute of Mathematics and its Applications, 2, pp.364-383.

[5] JOANES, D.N. and GILL, C.A. (1992) *Interval estimation of a measure of diversity*, Journal of Applied Statistics, 19, 3, pp.323-327.

[6] PATIL, G.P. and TAILLIE, C. (1982) *Diversity as a concept and its measurement*, Journal of the American Statistical Association, 77, pp.548-567.

[7] PIELOU, E.C. (1975) *Ecological Diversity*. Wiley, New York.

Part IX

Computational Aspects in Sequential Analysis Including Quality Control

Multivariate Process Control Through the Means of Influence Functions

Luan Jaupi, Gilbert Saporta

Département de Mathématiques; Conservatoire National des Arts et Métiers, Paris

Abstract. A diversity of multivariate control charts for the process mean and dispersion have been proposed recently to distinguish between random and assignable causes of process variability. This paper deals with the same problem and the use of influence function is proposed to distinguish between chance and special causes of process variability. When the process has reached a state of statistical control, the process mean and the structure of dispersion matrix should be stable over time. The effect of observations or subgroups on these parameters may be evaluated, among others, by the means of influence functions. Hence, special causes of variation could be identified by an unusual influence of observations or subgroups on the process mean and/or dispersion parameters.

1 Introduction

There are many practical situations in which the simultaneous control of several quality characteristics is necessary for an adequate description of each item quality. The procedure most generally employed in the past has been to built up one control chart for each variable. Unfortunately, when the variables are highly correlated, this may be misleading. The problem therefore requires a mutivariate approach to take into account the relationships among the variables. In this aim recently, Jackson [1991], Alt & Smith[1988], Pignatiello & Runger [1990], among others, have suggested a diversity of multivariate control charts to monitor process mean and describe process variability.

Consider a manufacturing process where each item is characterised by a p quality characteristic vector $X = (X_1, X_2, .., X_p)$. Because of variation causes X is a random vector. When the process has reached a state of statistical control the underlying distribution function of p quality characteristics is assumed to be F, with mean μ and covariance matrix Σ, ideally multivariate normal. Unfortunately when disturbances occurs in the process, the distribution function of X is no longer F. Therefore, when special causes exist in a process the distribution of X is assumed to be an arbitrary distribution noted G. A convenient model which describes quite well the two sources of variation is the contaminated one, Hampel et al. [1986]. That is, we consider that the distribution function of quality vector X is

$$F_{\epsilon H} = (1 - \epsilon) F + \epsilon G$$

with $\qquad 0 \leq \epsilon \leq 1$

We consider also that the distribution of output subgroups taken under a stable process is F and the distribution of output segments taken when special causes exist in the process is G. Therefore, if one considers k output subgroups, for large k roughly proportion ϵ of subgroups will be contaminants. When process is under control we have $\epsilon=0$. A control chart intents to suggest the time at which such contaminants occurs in a process. Since different statistics can be used to monitor process mean and describe process variability, several different control charts can be proposed.

The present paper deals with multivariate control problems and the use of the influence function is proposed to distinguish between random and special causes. In section 2 we derive the influence function of a mutivariate location-scale parameter and those of each eigenvalue and eigenvector of a covariance matrix. Multivariate control charts for process mean and process variability are proposed in section 3. In section 4 a numerical example is given to illustrate our method and compare it with existing ones.

2 Influence Function of Mean, Covariance Matrix, Eigenvalues and Eigenvectors

Let X be a p-dimensional random variable with distribution function F. We consider the following class of covariance estimators. Let Σ be a pxp covariance matrix and μ a p-dimensional location vector defined by

$$\int (x - \mu)dF = 0$$

(1)

$$\int \left[(x - \mu)(x - \mu)^T - \Sigma \right] dF = 0$$

Let δ_z denote the distribution giving unit mass to the point $z \in R^p$. The perturbation of F by δ_z is denoted as

$$F_{\epsilon z} = (1 - \epsilon) F + \epsilon \delta_z$$

In this manner, the parameter (μ, Σ) can be regarded as a pair of statistical functional. In order to calculate the influence function of μ and Σ we substitute F by $F_{\epsilon z}$ in Eq.1 and take the derivative with respect to ϵ at $\epsilon=0$. The differentiation of the mean equation gives

$$IF(z, \mu) = z - \mu$$

(2)

and the differentiation of the covariance matrix equation gives

$$IF(z; \Sigma, F) = (z - \mu)(z - \mu)^T - \Sigma$$

(3)

(cf. Huber [1981], Hampel et al. [1986], Jaupi & Saporta [1993])
Under regularity conditions Eq.3 implies the following expansion for the covariance matrix

$$\Sigma(F_{\epsilon z}) = \Sigma(F) + \epsilon \, IF(z ; \Sigma, F) + o(\epsilon)$$

(4)

Let $\lambda_1(F)$, $\lambda_2(F)$,.., $\lambda_p(F)$ be the eigenvalues of the covariance matrix $\Sigma(F)$. We denote by $\alpha_1(F)$, $\alpha_2(F)$,..., $\alpha_p(F)$ the associated eigenvectors. From Eq.4, with the assumption that all eigenvalues of Σ are simple, we have the following expressions for the influence function of the jth eigenvalue and the jth eigenvector, $(j=1,...,p)$, of the covariance matrix Σ

$$\text{IF} (z ; \lambda_j, F) = \alpha_j^T \text{ IF} (z ; \Sigma, F) \alpha_j \qquad (5)$$

$$\text{IF} (z ; \alpha_j, F) = \sum_{\substack{k=1 \\ k \neq j}}^{p} (\lambda_j - \lambda_k)^{-1} \alpha_k \alpha_k^T \text{ IF} (z ; \Sigma, F) \alpha_j \qquad (6)$$

(cf. Sibson [1979], Critchley [1985], Tanaka [1988], Jaupi [1992])

The importance about the influence function lies in its heuristic interpretation: it describes the effect of an infinitesimal contamination at point z on the estimate. One could say it measures the rate of change of the estimate as F is shifted infinitesimally in the direction of δ_z.

Usually, during the initial stages of process control, k subgroups of size n are taken from the process output. For reasons of simplicity we contemplate here only the case when individuals are used to monitor the process mean and describe variability. That is we consider that n=1. In applications of the influence function the unknown distribution function F has to be estimated by $\widehat{F}$ the empirical distribution function based on a random sample $X_1, X_2,..,X_k$ from F. Replacing F by $\widehat{F}$ and taking $z=X_i$ in Eq. (2) and (5), we have for the empirical influence function of the mean and the eigenvalues of a covariance matrix the following

$$\text{IF}(x_i; \mu, \widehat{F}) = x_i - \bar{x} \qquad (7)$$

$$\text{IF} (x_i, \lambda_j, \widehat{F}) = \widehat{\alpha}_j^T \text{ IF}(x_i; \Sigma, \widehat{F}) \widehat{\alpha}_j = c_{ij}^2 - \widehat{\lambda}_j \qquad (8)$$

where c_{ij} is the jth principal component, $(j=1,...,p)$, of the ith observation.

3   Multivariate Control Charts
3.1   Chart for Process Mean

When one is monitoring the process mean, what is calculated and plotted on a chart is the value of a quadratic form. In general, a quadratic form can be written as $X^T M X$, where M is a definite positive matrix and $X \in R^p$. Taking

$$M = \widehat{\Sigma}^{-1}$$

with $\widehat{\Sigma}^{-1}$ the inverse of sample covariance matrix we find a first sample version of a dignified statistic, Hotelling's T^2. Thus, for the ith subgroup, $(i=1,...,k)$, one would plot

$$T_i^2 = \text{IF} (x_i; \mu, \widehat{F})^T \widehat{\Sigma}^{-1} \text{ IF} (x_i; \mu, \widehat{F}) = (x_i - \bar{x})^T \widehat{\Sigma}^{-1} (x_i - \bar{x}) \qquad (9)$$

Each of T_i^2 values, $(i=1,...,k)$, of Eq. (9) would be compared with

$$UCL = \frac{(k-1)p}{k-p} F_{\alpha, p, k-p} \tag{10}$$

(cf. Jackson [1991]).

Points lying outside the control limit indicate the time at which the potential assignable causes occur in the process. Knowing the time of such highly influential occurrence is the best evidence for searching out and eliminating a process disturbance.

3.2 Influential Shewhart Charts for Process Variability

The total dispersion of quality characteristic vector X is equal with the sum of the eigenvalues of the covariance matrix Σ. It can be decomposed into components along each eigenvector of Σ. Generally assignable causes that affect the variability of the output do not increase significantly each component of total variance of X. Instead, they may have a large influence in the variability of some components and small effect in the remaining directions. For this reason, it seems promising to detect any significant departure from the stable level of variability of each component. One approach in developing a control chart that describes the variability on each direction consists in the use of the influence functions of the eigenvalues of the covariance matrix Σ. That is, if one is monitoring the process variability according to the jth, $(j=1,...,p)$, direction, what would be calculated and plotted on an individual control chart is the value of the influence function of the jth, $(j=1,...,p)$, eigenvalue of the covariance matrix Σ. Thus, for the ith subgroup, $(i=1,...,k)$, one would plot

$$IF(x_i, \lambda_j, \widehat{F}) = c_{ij}^2 - \widehat{\lambda}_j \tag{11}$$

The limits of these influential charts are three sigma limits. That is, they lie at a distance from the expected value of $IF(z, \lambda_i, \widehat{F})$, equal three times the standard error of this statistic.

4 Numerical Example

To illustrate our influential charts we use the chemical data example published in Jackson (1991, Tables 1.1 and 1.2). For each of the two variables, there are 15 original observations. To exibit some out-of-control condition four additional points have been included, observations n° 16 - 19. Figure 1 shows control charts for the original variables, x_1 and x_2 and the influential charts for the first and the second eigenvalue. All these charts have three sigma limits based on the variability of the 15 original observations. For these four Shewhart control charts, these limits are

$$x_1: \qquad 10 \pm 3(.95) = 7.13; \quad 12.86$$
$$x_2: \qquad 10 \pm 3(.84) = 7.47; \quad 12.52$$
$$IF(x_i, \lambda_1, \widehat{F}): \quad 0 \pm 3(1.68) = -5.04; \quad 5.04$$
$$IF(x_i, \lambda_1, \widehat{F}): \quad 0 \pm 3(.069) = -0.282; \quad 0.282$$

Table 1 shows values of the empirical influence function for the two eigenvalues and values of T^2 statistic evaluated at the 19 observations. The upper control limit of T^2 chart is

$$UCL = \frac{(k-1)p}{k-p}F_{\alpha,p,k-p} = \frac{(15-1)2}{15-2}F_{.0054,2,13} = 17.26$$

The graphical display of T^2 chart is shown in Figure 2.

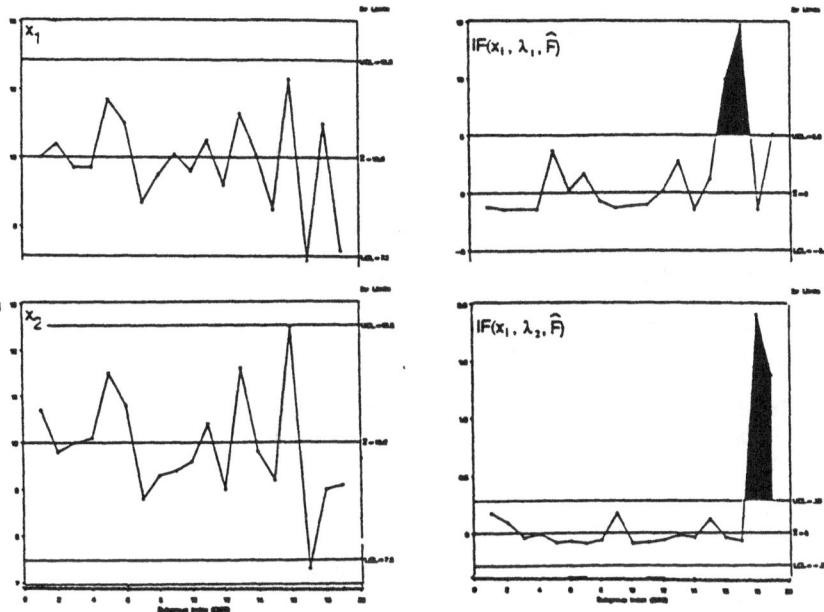

Figure 1: Control charts for original variables and influence charts for eigenvalues.

oas	$IF(x_i, \lambda_1, \widehat{F})$	$IF(x_i, \lambda_2, \widehat{F})$	T^2 statistic
1	-1.210000	0.1700000	3.120000
2	-1.420000	0.0900000	2.060000
3	-1.400000	-0.0400000	0.520000
4	-1.420000	-0.0100000	0.920000
5	3.690000	-0.0800000	3.620000
6	0.180000	-0.0700000	1.270000
7	1.680000	-0.0900000	2.170000
8	-0.730000	-0.0600000	0.790000
9	-1.330000	0.1700000	3.000000
10	-1.130000	-0.0900000	0.220000
11	-1.040000	-0.0800000	0.320000
12	0.160000	-0.0600000	1.460000
13	2.740000	-0.0200000	3.680000
14	-1.440000	-0.0400000	0.540000
15	1.240000	0.1200000	4.250000
16	10.040000	-0.0400000	8.510000
17	14.830000	-0.0700000	11.380000
18	-1.450000	1.9100000	23.140000
19	5.180000	1.3800000	21.550000

Table 1: Values of the empirical influence function for the two eigenvalues and T^2 statistic evaluated at the nineteen observations.

Figure 2: Multivariate T^2 chart.

200

An inspection of these charts shows that: (a) the same message is received from the two control charts for original variables. The values of observation n° 17 exceed the UCL of these charts, but no out-of-control signal is received at observations n° 16, 18 and 19; (b) there are two very highly influential observations n° 16 and 17 for the first influential chart and also the influence of observations n° 18 and 19 far exceeds the UCL of the second influential chart. Thus, there would be reason to suspect that something is wrong in the process when the data in the observations n° 16, 17, 18 and 19 were obtained; (c) the T^2 chart shows that there are two points that exceed the UCL n° 18 and 19. Conversely, no out-of-control message is received at observations n° 16 and 17.

Bibliography

ALT, F. B. & SMITH, N. D. (1988) *Multivariate Process Control*
Handbook of Statistics Vol. 7, 333-351; North-Holland

CRITCHLEY F. (1985) *Influence in Principal Components Analysis*
Biometrika 72, 627-636

HAMPEL F.R., RONCHETTI E.M., ROUSSEEUW P.J., STAHEL W.A. (1986) *Robust Statistics - The Approach Based on Influence Functions*; Wiley, New-York

HUBER P. (1981) *Robust Statistics*; Wiley, New-York

JACKSON, J. E. (1991) *A User's Guide in Principal Components*
Wiley, New York

JAUPI, L. & SAPORTA, G. (1993) *Using the Influence Function in Robust Principal Components Analysis*; In S. Morgentaler, E. Ronchetti and W.A. Stahel, eds., New Directions in Statistical Data Analysis and Robustness, 147-156; Birkhäuser Verlag Basel

JAUPI, L. (1992) *Méthodes robustes en analyse en composantes principales*; Thèse , Conservatoire National des Arts et Métiers, Paris

PIGNATIELLO, J. J. & RUNGER, G. C. (1990) *Comparaison of Multivariate CUSUM Charts* ; Journal of Quality Technology 22, 173-186

SIBSON R. (1979) *Studies in the Robustness of Multidimensional Scaling Perturbation Analysis of Classical Scaling*; J.R. Statist. Soc., **B, 41**, 2, 217-229

TANAKA Y. (1988) *Sensivity Analysis in PCA: Influence on the Subspace Spanned by Principal Components*; Comm. Statist. Theory-Methods 17, n° 9, 3157-3175

TRIQ – A PC-Program for Design and Analysis of Triangular Sequential Trials

J. Schmidtke, B. Schneider

Biorat GmbH, Rostock

1. The Triangular Sequential Trial
The Triangular Sequential Trial is used to compare a new medical treatment with a standard.
For the purpose of designing and evaluating a trial a patients response must be selected (or else seveal responses must be combined to form a single composite response) as the effect variable of the trial. The magnitude of this response will vary from patient to patient according to some probability distribution. It is important to have some knowledge of the form of this distribution. It will come from the predictions of clinicians and from the results and experience of previous studies and trials. Within the probability model of patient response one parameter, denoted here by Θ, will serve as a measure of treatment difference. It can always be arranged that the value zero of Θ corresponds to equivalence of the treatments, that positive values correspond to an advantage for the new treatment and that negative values correspond to and advantage for the standard.
For inference about Θ two sample statistics play an important role: V and Z. Z is a measure for the evidence of treatment differences accumulated, and V indicates the amount of information about Θ contained in Z. Z is the "efficient score" for Θ and V is "Fisher´s information". Both can be calculated at any stage of the trial, V will increase during the trial. On the basis of the test statistics a decision about the continuation or termination of the trial will be met.

1.1. One-sided test
A statistical model for the response of the patients, where the parameter of interest Θ represents the difference between the responses of the two treatments, is used. The null hypothesis (H0), that the two treatments are equal ($\Theta=0$), is to be tested against the one-sided alternative hypothesis (H1), that the new treatment is superior ($\Theta>0$). The minimum difference of interest is denoted by Θ_1, the significance level by α, the risk of the second kind by ß.
At each inspection point $x(i)$ the statistics $Z(i)$ and $V(i)$ are calculated from the observation available. Using a (Z,V)-coordinate system the information accumulated up the i-th can be represented as the point $(Z(i), V(i))$ in the Z-V-plane. The sample path of the $(Z(j),V(j))$, j=1,...,i, represents the development of the study so far.

For the test mentioned above a triangle with its baseline on the Z-axis symmetrically about 0 is defined in the (Z,V)-plane. If at step i the point $(Z(i),V(i))$ reaches or crosses one of the boundaries of the triangle (i.e. the path of the $(Z(j),V(j))$ leaves the triangular region), the trial is terminated and either the null or the alternative hypothesis is accepted - depending on which boundary (the lower or the upper) is crossed. This means that the triangular boundaries are the **stopping boundaries** for the trial.

202

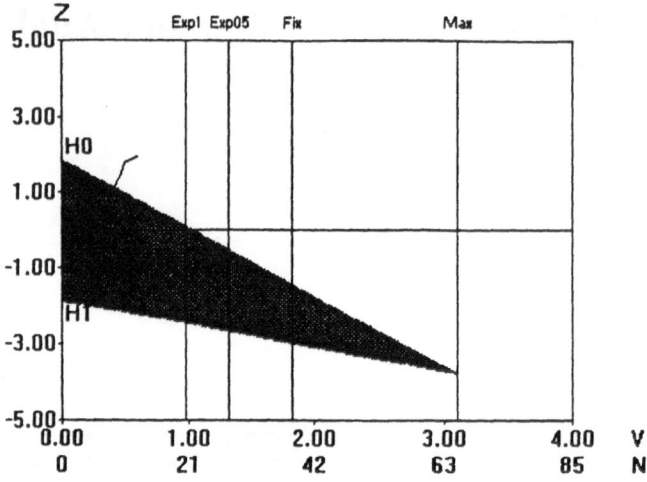

Fig. 1 Decision triangle for the one-sided test

1.2. Two-sided test

In the two-sided case the null hypothesis (H0), that the two treatments are equal ($\Theta=0$), has to be tested against the two-sided alternative hypotheses (H1+), that the new treatment is superior ($\Theta>0$) and (H1-), that the new treatment is inferior. The minimum difference of interest is denoted by Θ_1, the significance level by α, the risk of the second kind by β.

A two-sided test can be constructed as a combination of two appropriate one-sided tests. Then in the (Z,V)-plane the triangular region for H1+ will be defined like that for the one-sided test, the triangular region for H1- results from reflection on the V-axis. Again the path of the points (Z(i),V(i)) is drawn, as soon as this path leaves the region defined by the union of both triangles, the trial is stopped and either the null hypotheses is accepted or rejected.

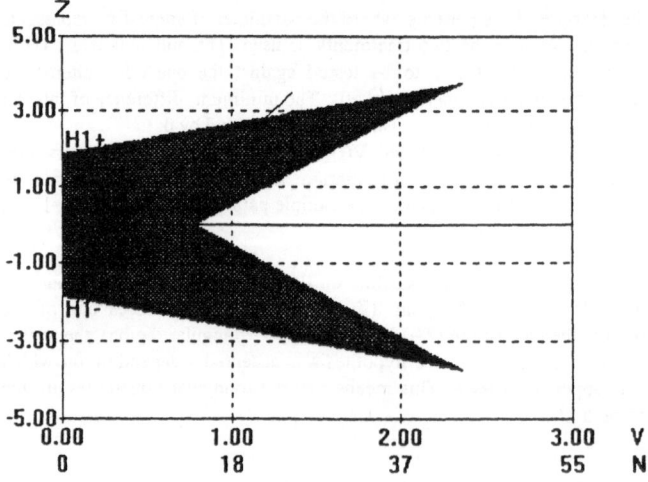

Fig. 2 Decision triangle for the two-sided test

2. TRIQ-A PC-Program for Triangular Sequential Trials

2.1. System configuration for TRIQ
Hardware
For running TRIQ you need:
- a computer with 286-, 386- or 486-processor, compatible to Microsoft® Windows version 3.x
  (recommendation: at least a 386SX).
- an EGA-, VGA- or Super-VGA-graphics card, compatible to Microsoft® Windows version
  3.x.
- a mouse compatible to Microsoft® Windows is recommended, but not essential.
- a 1,2MB-,5.25''disk drive or a 1.44MB-,3.5''disk drive.

System software
To use **TRIQ** Microsoft® Windows version 3.x or higher has to be installed on the computer.

Memory
TRIQ requires at least 1 MB RAM (recommendation: at least 2 MB).

Capacity of the hard disk
TRIQ requires at least 600 KB of free memory on the hard disk.

Program
TRIQ is a modul of the program **Cademo for Windows**. It was written object-oriented in
Turbo-Pascal for Windows .

2.2. Types of response
The data taken from the patients to characterize the most appropriate response can have different
distributions. The program **TRIQ** offers solutions for four types of response:

Binary response variables
Given: Two independent samples form identically and independently Bernoulli distributed
        random variables.
        $x_{1i} \sim$ Bernoulli(p_1) ; $x_{2i} \sim$ Bernoulli(p_2)

Test Parameter: $\Theta = \log \dfrac{p_1(1-p_2)}{p_2(1-p_1)}$

Quantitative response variables
Given: Two independent samples of identically and independently distributed normal
        random variables with common variance σ^2.
        $x_{1i} \sim N(\mu_1,\sigma^2)$; $x_{2i} \sim N(\mu_2,\sigma^2)$

Test Parameter: $\Theta = \mu_1 - \mu_2$

Ordinal response variables
Given: Two independent samples of distributed random variables X_1 and X_2
        distributed over k ordinal response categories
        $C_1, C_2, ... C_k$ (k>2) with $C_1 < C_2 < ... < C_k$ and probabilities

        $P(X_1 = C_j) = p_{1j}$, $P(X_2 = C_j) = p_{2j}$

The cumulative probabilitiy to fall in a catagory less then or equal C_j is:

$$Q_{1j} = \sum_{l=1}^{j} p_{1j} \quad , \quad Q_{2j} = \sum_{l=1}^{j} p_{2j}$$

Test Parameter: $\Theta = \log \dfrac{Q_{1j}(1-Q_{2j})}{Q_{2j}(1-Q_{1j})}$

It is assumed that Θ is independent of j (j=1,2,...,k-.1).

Censored response variables (survival data)

A sample of m patients is treated with T_1 and an independent sample of n patients with T_2. Response variables are the survival times t_i which are assumed as relations of identically and independently distributed random variables T_1 and T_2 with function $F_1(t)$ for T_1 and $F_2(t)$ for T_2. The proportional hazard rate model is assumed to hold; i.e.:

$$F_1(t) = (F_2(t))^{\exp(-\Theta)}$$

If Q_1 is the probability to die during the total observation period, if treated with T_1 and Q_2 the corresponding probability, if treated with T_2, then Θ is defined as:

$$\Theta = \log \frac{\log(1-Q_2)}{\log(1-Q_1)}$$

2.3. Program handling

With TRIQ you can design and evaluate Triangular Sequential trials.

2.3.1. Designing

If you want to design your trial you have to input the parameter (α, β, ...) to construct the design. The design parameters are listed in form of tables. Here you can find the random sample sizes to be expected for your test problem and the fixed sample size for non-sequential trial.

2.3.2. Analysis

You can create, edit and save data files. You can also work with data created by other (Windows)-software and transfer the data to TRIQ about the clipboard .

For the analysis you have to input the parameter (α, β, ...) to calculate the design and analyze data. The program calculates both cumulative frequencies and the test-statistics Z and V for each monitoring step and lists these results in form of tables in an output-window. You can also open a graphic window to represent these results. With the help of the tables or the graphic you can check your test. You can edit, save or print all the outputs .

Example:

1. Table

Decision:
Triangular Sequential Design
Response type: Binary
Alternative: one-sided test

Risks

| alpha | 0.1000 |
| beta | 0.2000 |

Response Probabilities

| Group 1 | 0.9000 |
| Group 2 | 0.6000 |

Design

Parameter	V	N
Exp(1)	0.7709	16
Exp(0.5)	0.9498	20
Fix	1.3607	29
Max	2.6669	56

File: BINARY.DAT

H[0] : reject
H[1] :

Calculate data

No	group	response	SUM1	SUM2	N1	N2	Z	V	ZH0	ZH1
1	1	1	1	0	1	0	0.0000	0.0000	-1.4649	1.4649
2	2	0	1	0	1	1	0.5000	0.1250	-1.2589	1.5336
3	1	0	1	0	2	1	0.3333	0.1481	-1.2208	1.5463
4	1	1	2	0	3	1	0.5000	0.1875	-1.1559	1.5679
5	2	0	2	0	3	2	0.8000	0.2880	-0.9903	1.6231
6	1	0	2	0	4	2	0.6666	0.2962	-0.9767	1.6277
7	2	0	2	0	4	3	0.8571	0.3498	-0.8884	1.6571
8	2	0	2	0	4	4	1.0000	0.3750	-0.8470	1.6709
9	1	0	2	0	5	4	0.8888	0.3840	-0.8320	1.6759
10	1	1	3	0	6	4	1.2000	0.5040	-0.6344	1.7418
11	2	0	3	0	6	5	1.3636	0.5409	-0.5735	1.7621
12	1	0	3	0	7	5	1.2500	0.5468	-0.5637	1.7653
13	1	1	4	0	8	5	1.5384	0.6554	-0.3848	1.8250
14	2	0	4	0	8	6	1.7142	0.6997	-0.3119	1.8493
15	1	0	4	0	9	6	1.6000	0.7040	-0.3048	1.8516
16	2	0	4	0	9	7	1.7500	0.7382	-0.2483	1.8705
17	2	0	4	0	9	8	1.8823	0.7620	-0.2091	1.8835
18	1	0	4	0	10	8	1.7777	0.7681	-0.1990	1.8869
19	2	0	4	0	10	9	1.8947	0.7872	-0.1675	1.8974
20	2	0	4	0	10	10	2.0000	0.8000	reject	***

2.Graphic:

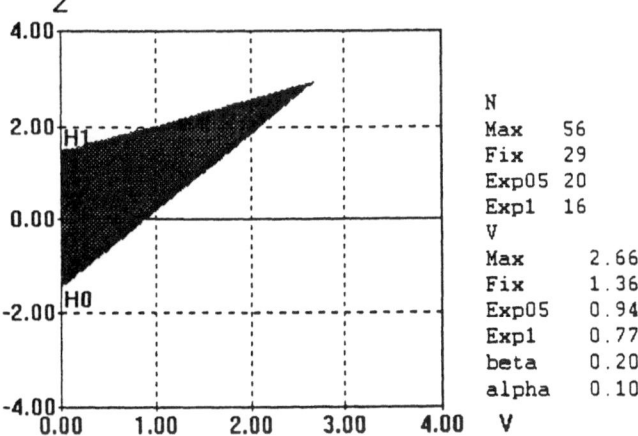

Fig. 3 Design One-sided test -binary Data

References

Borland GmbH (1991) Turbo Pascal für Windows - Programmierhandbuch (2nd edition). München

Rasch, D. (1990) Einführung in die mathematische Statistik, I+II (3rd edition).
VEB Deutscher Verlag der Wissenschaften, Berlin

Schneider, B. (1992) An Interactive Computer Program for Design and Monitoring of Sequential Clinical Trials In: Proc. of the 1992 (XV/th) International Biometrie Conference, Hamilton, New Zealand, vol. 1. Invited papers.

Wald, A. (1947) Sequential Analysis. New York: Wiley

Whitehead, J. (1983) The Design and Analysis of Sequential Clinical Trials (1st edition). Chichester: Ellis Horwood Ltd.

Whitehead, J. (1992) The Design and Analysis of Sequential Clinical Trials (2nd edition). Chichester: Ellis Horwood Ltd.

Whitehead, J. and Brunier, H. (1989) Pest2.0 Operating Manual. Reading University

Part X

Computational Aspects in Reliability and Survival

Estimation of Parameters of the Inf. of Weibull Distributed Failure Time Distributions

Mostafa Bacha

INRIA-Rocquencourt, Domaine de Voluceau

Abstract. In this paper, we present an EM-type algorithm to estimate the parameter vector θ of the *inf.* of k independent non identical *Weibull* distributions.

Keywords. *Weibull* distribution, non-postmortem data, censored data, EM algorithm, Competing risks, series system, Maximum likelihood estimate

1 Introduction and notations

We consider a series system S composed of k independent and different components $S_1, \ldots, S_k$. We associate with each component S_j, a positive random variable T_j with *Weibull* pdf

$$f_j(t \mid \theta_j = (\beta_j, \eta_j)) = R_j(t \mid \theta_j) \, h_j(t \mid \theta_j), \quad \beta_j > 0, \, \eta_j > 0$$

where

$$R_j(t \mid \theta_j) = \exp\left(-\left(\frac{t}{\eta_j}\right)^{\beta_j}\right)$$

and

$$h_j(t \mid \theta_j) = \frac{\beta_j}{\eta_j}\left(\frac{t}{\eta_j}\right)^{\beta_j - 1}$$

are respectively the survivor and failure rate functions of S_j. Our task is to estimate $\theta = (\theta_1, \ldots, \theta_k)$ the parameter vector of $T = \inf(T_1, \ldots, T_k, C)$ where C is a right censoring time. For this goal, n independent items are put on life test. In the case of postmortem life test, observations are obtained in the form $(t; \Delta) = (t_1, \ldots, t_n; \delta(t_1), \ldots, \delta(t_n))$, where $\delta(t_i) = (\delta_1(t_i), \ldots, \delta_k(t_i))$ is the indicator vector whose jth element is unity if t_i is a failure instant due to the jth component and 0 otherwise. The likelihood function is

$$L(\theta; t, \Delta) \propto \prod_{i=1}^{n} \prod_{j=1}^{k} \left[R_j(t_i \mid \theta_j) \, h_j(t_i \mid \theta_j)^{\delta_j(t_i)} \right] = \prod_{j=1}^{k} L_j(\theta_j; t, \Delta)$$

where

$$L_j(\theta_j; t, \Delta) = \prod_{i=1}^{n} \left[R_j(t_i \mid \theta_j) \, h_j(t_i \mid \theta_j)^{\delta_j(t_i)} \right] \ .$$

Therefore, the maximum likelihood estimate (MLE) of θ is $\hat{\theta} = (\hat{\theta}_1, \ldots, \hat{\theta}_k)$ where $\hat{\theta}_j$ is the MLE of θ_j corresponding to $L_j(\theta_j; t, \Delta)$ (see [4] for more details).

For the non postmortem data, only t and $s = (s_1, \ldots, s_n)$ are observed, where $s_i \equiv s(t_i) = 1$ if failure, and $s_i = 0$ if censored. Using the relation $\underline{f}(t \mid \theta) = \underline{R}(t \mid \theta) \, \underline{h}(t \mid \theta)$ where

$$\underline{R}(t \mid \theta) = \prod_{j=1}^{k} R_j(t \mid \theta_j)$$

and

$$\underline{h}(t \mid \theta) = \sum_{j=1}^{k} h_j(t \mid \theta_j)$$

are respectively the survivor and failure rate functions of S, the joint pdf of t_i and s_i is

$$\underline{f}(t_i \mid \theta)^{s_i} \underline{R}(t_i \mid \theta)^{1-s_i} = \underline{R}(t_i \mid \theta) \, \underline{h}(t_i \mid \theta)^{s_i} \ .$$

In the following, we try to deal with this model in practical manner.

2   The estimation problem

The identifiability of the model stated above was proved by Basu & Ghosh [2] for $k = 2$ provided $\beta_1 \neq \beta_2$. Their result can be extended for $k > 2$ under the supposition, which we will assume, that all β_j are different.

To find the MLE $\hat{\theta}$, one has to maximize the log-likelihood

$$l(\theta; t, s) = \sum_{i=1}^{n} s_i \log\left(\underline{h}(t_i \mid \theta) \right) + \sum_{i=1}^{n} \log\left(\underline{R}(t_i \mid \theta) \right)$$

under the constraints $\beta_j > 0$ and $\eta_j > 0$. This is a hard work and its difficulty increase with the number of components. Furthermore, it has been our experience that $l(\theta; t, s)$ is flat over large parameter range, so any optimization method involving derivation may give rise to numerical problems. We propose an algorithm to estimate θ through the EM algorithm (see [3]).

3 The algorithm

Let $Q(\theta \mid \tilde{\theta}) = E[l(\theta; t, \Delta) \mid \tilde{\theta}, t, s]$ denote the conditional expectation of $l(\theta; t, \Delta)$ knowing $\tilde{\theta}$, t and s. Given the current estimate $\theta^{(m)}$ of θ, the EM algorithm consists of finding $\theta^{(m+1)}$ which maximizes $Q(\theta \mid \theta^{(m)})$. The important characteristic of the EM algorithm is that $l(\theta; t, s)$ is not decreasing at each iteration. In our case and after calculations, $Q(\theta \mid \tilde{\theta})$ is

$$Q(\theta \mid \tilde{\theta}) = \sum_{j=1}^{k} Q_j(\theta_j \mid \tilde{\theta})$$

where

$$Q_j(\theta_j \mid \tilde{\theta}) = \sum_{i:unc} p_j(t_i \mid \tilde{\theta}) \log(h_j(t_i \mid \theta_j)) + \sum_{i=1}^{n} \log(R_j(t_i \mid \theta_j))$$

where $\sum_{i:unc}$ indicates sum over uncensored observations and

$$p_j(t_i \mid \tilde{\theta}) = \frac{h_j(t_i \mid \tilde{\theta}_j)}{\underline{h}(t_i \mid \tilde{\theta})} \tag{1}$$

is the conditional probability that the failure come from the jth component knowing that it failed at time t_i. Taking the derivatives of $Q_j(\theta_j \mid \tilde{\theta})$ with respect to η_j and β_j and rearrange

$$\frac{\partial Q_j(\theta_j \mid \tilde{\theta})}{\partial \eta_j} = \frac{\beta_j}{\eta_j} \left[\sum_{i=1}^{n} \frac{t_i^{\beta_j}}{\eta_j^{\beta_j}} - \sum_{i:unc} p_j(t_i \mid \tilde{\theta}) \right] \tag{2}$$

and

$$\frac{\partial Q_j(\theta_j \mid \tilde{\theta})}{\partial \beta_j} = \frac{1}{\beta_j} \sum_{i:unc} p_j(t_i \mid \tilde{\theta}) + \sum_{i:unc} p_j(t_i \mid \tilde{\theta}) \log\left(\frac{t_i}{\eta_j}\right)$$
$$- \sum_{i=1}^{n} \left(\frac{t_i}{\eta_j}\right)^{\beta_j} \log\left(\frac{t_i}{\eta_j}\right) \tag{3}$$

setting (2) equal to 0 leads to

$$\eta_j^{\beta_j} = \frac{\sum_{i=1}^{n} t_i^{\beta_j}}{\sum_{i:unc} p_j(t_i \mid \tilde{\theta})}. \tag{4}$$

Substituting (4) into (3) and setting the latter to 0 gives

$$\frac{1}{\beta_j} - \frac{\sum_{i=1}^{n} t_i^{\beta_j} \log(t_i)}{\sum_{i=1}^{n} t_i^{\beta_j}} + \frac{\sum_{i:unc} p_j(t_i \mid \tilde{\theta}) \log(t_i)}{\sum_{i:unc} p_j(t_i \mid \tilde{\theta})} = 0, \tag{5}$$

an equation that we can solve iteratively.

Thus, the algorithm we propose is

step 1. start with $\theta^{(0)}$

In iteration $(m+1)$,

step 2. for uncensored observations t_i, calculate the $p_j(t_i \mid \theta^{(m)})$, $j = 1, \ldots, k$ using (1)

step 3. calculate $\theta^{(m+1)}$ by solving (5) and (4) given $p_j(t_i \mid \theta^{(m)})$

step 4. repeat steps 2 and 3 until $\left| l(\theta^{(m+1)}; t, s) - l(\theta^{(m)}; t, s) \right| < \epsilon$

This algorithm is simple to implement and does not require any computations on the Hessian matrix like Newtonian methods or other optimization algorithms.

4   Simulations results

To save time, a controlled Monte Carlo experiment has been done. $Ns = 50$ simulation runs are performed with $k = 2$ and $n = 100$. We calculate the empirical mean (E) and standard deviation (Std) over the Ns samples for each parameter β_j and η_j. We increase Ns by 50 and update the empirical statistics until the calculated quantities does not vary considerably. We have found that 150 samples $(Ns = 150)$ are sufficient for all cases in our simulation study. The first table shows the results for the case where the two components are well represented, the second table for the case where one component dominates in the sample (see unc_j, the average number of uncensored observations coming from the jth component). At the sight of these results, we can draw the following remark.

• The estimates are good if the two components are well represented in the sample.

• The estimation of scale parameter η_j is very good if the corresponding shape parameter β_j is greater than one. The reason is that η_j, considered as function of β_j, varies much more slowly with $\beta_j > 1$ than with $\beta_j < 1$.

• If the number of observations coming from the jth component is very small, the algorithm over-estimates β_j. A possible explanation to this phenomenon is that $\sum_{i:unc} p_j(t_i \mid \tilde{\theta})$ tends to be small so that the β_j, solution of equation (5), becomes large. For example, in the second table with censoring time $c = 1.5$, $\hat{\beta}_2 = 18.145$ and $\sum_{i:unc} p_2(t_i \mid \hat{\theta}) = 4.851$ were observed.

• The average number of iterations needed to reach the absolute error of 10^{-4} was about 22 for uncensored samples and about 26 if censored. Also, we have remarked, for 6 samples among 150 in the second table with $c = 1.5$, that this number can increase considerably exceeding 200 iterations if one component is "jammed" by the other.

We can conclude that the proposed algorithm provides accurate estimation with regard to our severe situation. It is possible to improve this estimation by using the algorithm in the *Bayesian* context as presented in the next section.

n & c	unc_j	$E(\beta_j)$	$Std(\beta_j)$	$E(\eta_j)$	$Std(\eta_j)$
$n = 100$ uncensored	$unc_1 = 49$ $unc_2 = 51$	0.517 4.287	0.084 0.848	1.245 0.502	1.062 0.028
$n = 100$ $c = 0.5$	$unc_1 = 48$ $unc_2 = 34$	0.515 4.311	0.101 1.208	1.377 0.553	1.524 0.029

Table 1. True parameter $\theta^* = ((0.5, 1.0), (4.0, 0.5))$, initial guess $\theta^{(0)} = ((0.8, 1.5), (3.5, 1.0))$ and $\epsilon = 10^{-4}$

n & c	unc_j	$E(\beta_j)$	$Std(\beta_j)$	$E(\eta_j)$	$Std(\eta_j)$
$n = 100$ uncensored	$unc_1 = 73$ $unc_2 = 27$	0.510 4.583	0.062 1.241	1.149 1.995	0.326 0.137
$n = 100$ $c = 1.5$	$unc_1 = 69$ $unc_2 = 9$	0.497 5.399	0.066 4.044	1.401 2.103	2.998 0.846

Table 2. True parameter $\theta^* = ((0.5, 1.0), (4.0, 2.0))$, initial guess $\theta^{(0)} = ((0.7, 1.5), (3.7, 2.3))$ and $\epsilon = 10^{-4}$

5   About *Bayesian* framework

Generally, in reliability analysis, a priori information is available as expert opinion and/or life test results on similar systems. To take into consideration this information, we choose an a priori distribution, $\pi(\theta)$ which reflects our knowledge about θ. This prior is transformed, by the data $t = (t_1, \ldots, t_n)$, to the posterior $\pi(\theta \mid t) \propto L(\theta; t) \pi(\theta)$. Now, one must be able to infer about θ from $\pi(\theta \mid t)$. Frequently, MCMC methods like Gibbs sampler are needed. These iteratives methods are time consuming and not simple to implement in this context. Recently, Newton and Raftery (see [5]) have proposed an algorithm called WLB-SIR allowing to generate samples from the posterior $\pi(\theta \mid t)$. The algorithm works as follows

step 1. sample weight vector w of length n from some probability distribution, Dirichlet distribution for example.

step 2. calculate $\tilde{\theta}$ maximizing $L_w(\theta; t) = \prod_{i=1}^{n} (f(t_i \mid \theta))^{w_i}$.

Repeat steps 1 and 2, M times, for M large. After calculating a kernel density estimate $\hat{g}(\theta)$ at each $\tilde{\theta}$, resample the $\tilde{\theta}$ using the weights $\pi(\tilde{\theta})L(\tilde{\theta}; t)/\hat{g}(\tilde{\theta})$. We have applied this algorithm for *Weibull* models (see [1]) and the result was very satisfactory. Using our algorithm in step 2 of WLB-SIR algorithm, we can made *Bayesian* inference for the model of the *inf.* of *Weibull* distributions without great difficulty. Formally, the step 2 above can be done considering equations

$$\frac{1}{\beta_j} - \frac{\sum_{i=1}^{n} w_i t_i^{\beta_j} \log(t_i)}{\sum_{i=1}^{n} w_i t_i^{\beta_j}} + \frac{\sum_{i:unc} w_i p_j(t_i \mid \tilde{\theta}) \log(t_i)}{\sum_{i:unc} w_i p_j(t_i \mid \tilde{\theta})} = 0$$

and

$$\eta_j^{\beta_j} = \frac{\sum_{i=1}^{n} w_i t_i^{\beta_j}}{\sum_{i:unc} w_i p_j(t_i \mid \tilde{\theta})}$$

instead of equations (5) and (4) in step 3 of our algorithm.

Acknowledgements :   I am specially indebted to Dr. Celeux, G. and Prof. Robert, C.. Sincere thanks are given to Dr. Diebolt, J. and Dr. Idee, E..

References

[1] BACHA, M. and CELEUX, G. (1994). Contribution to discussion of paper by NEWTON, M. A. and RAFTERY, A. E. *JRSS*, B (to appear).

[2] BASU, A. P. and GHOSH, J. K. (1980). Identifiability of distributions under competing risks and complementary risks model. *Communications in Statistics, Theory and Methods*, A 9(14), 1515-1525.

[3] DEMPSTER, A. P., LAIRD, N. M. and RUBIN, D. B. (1977). Maximum likelihood from incomplete data via the EM algorithm. *JRSS*, B 39, 1-38.

[4] HERMAN, R. J. and PATELL, K. N. (1971). Maximum likelihood estimation for multi-risk model. *Technometrics*, 13(2), 385-396.

[5] NEWTON, M. A. and RAFTERY, A. E. (1994). Approximate Bayesian inference with the weighted likelihood bootstrap. *JRSS*, B (to appear).

Survival Graphics Software: Current Status, Future Needs and Criteria of Assessment

Constanza Quintero, Axel Benner, **Lutz Edler**, Mihaela Blaga

Biostatistik, Deutsches Krebsforschungszentrum, Heidelberg, Abt. 0820

1. Introduction

Survival analysis might never have become such a famous statistical method in medicine had there not been the Kaplan-Meier curve (KMC) visualizing the survival time of patients also in the presence of incomplete (censored) data. The Kaplan-Meier estimator satisfies good statistical properties, simplicity, intuitive interpretability and capability to stimulate imagination. Hence, each renowned survival software package allows plotting the KMC. Beyond KMC, survival software is rather heterogeneous in analytical and graphical methods. A comprehensive comparison of survival software although important and welcome to many users seems out of reach, and in face of the many commercial and public domain software systems it might not be manageable at all. Therefore we restricted ourselves to well disseminated and important software packages. Among those, we chose SAS as a widely distributed package with the pretention of general applicability also to less statistically trained persons, and we chose S-Plus as a more open and more rapidly developing system designed to be used predominantly by statisticians. The questions were: *Which graphics are available for the analysis of survival data? Which graphics are needed and should be realized? What support is given for obtaining easily good graphical outputs?*

2. Methods

The contents of SAS (SAS, 1992) and S-Plus (Statsci, 1992) for survival analysis were retrieved from the manuals and examined by applications. For the determination of future needs of survival software we used as comparative standard functions implemented in our own developed software package SURVIVAL (Weber, 1986) and proposals in biometrical literature. We considered a graphical tool as available if that could be obtained either directly by calling a survival procedure or if it could be obtained by applying a plotting routine on a well defined output data set/object of a survival program like *gplot* in SAS or *plot* in S-Plus. Graphical displays which are only possible after programming the output or by elaborating complex macros were not considered because of their limited value for the common user.

The Example: *Survival after Acute Lymphoblastic Leukemia (ALL).* A total of 104 children newly diagnosed between 1981 and 1983 in the Childrens' University Hospital of Jena, Germany, with non-B acute lymphoblastic leukemia (NB-ALL) entered a study for prognostic significance of protein kinase C (PKC) on the duration of relapse free survival time and on total survival time (Volm et al.(1994)).

Table 1: Statistical Procedures and Results Available in Survival Software

STATISTICAL PROCEDURE	ANALYTICAL		GRAPHICAL	
	SAS	S-Plus	SAS	S-Plus
DESCRIPTION				
Survival Curve				
Kaplan-Meier Estimate $\hat{S}_{KM}$	yes	yes	yes	yes
Actuarial Survival Curve	yes	0	yes	0
Harrington-Fleming	0	yes	0	yes
Parametric models: fit	yes	0	yes	0
Affiliation to a Parametric Model:				
plot : log $\hat{S}$	yes	yes	yes	yes
plot : normal $\hat{S}$	yes	yes	yes	yes
plot : log-normal $\hat{S}$	yes	yes	yes	yes
plot : log -log $\hat{S}$	yes	yes	yes	yes
Censoring Pattern	yes	yes	yes	yes
Standard Deviation				
Greenwood	yes	yes	0	0
Tsiatis	no	yes	0	0
Confidence Bands	yes	yes	yes	yes
Quantiles				
Median	yes	0	0	0
others	yes	0	0	0
(Cumulative) Hazard Function				
From KM-Estimate	yes	0	yes	0
Nelson Estimate	0	yes	0	0
CONFIRMATORY / COMPARISONS				
Logrank Test	yes	yes	-	-
Gehan-Wilcoxon	0	yes	-	-
Harrington-Fleming family	0	yes	0	yes
EXPLANATORY				
Cox Regression	yes	yes	yes	yes
with strata	yes	yes	yes	yes
with model selection	yes	0	yes	0
time dependent covariates	yes	yes	yes	yes
Residuals				
deviance residuals	yes	yes	yes	yes
martingale residuals	yes	yes	yes	yes
score residuals	0	yes	0	yes
Schoenfeld residuals	0	yes	0	yes

Elevated PKC levels -present in 48 (46%) of our cases- have been associated with multidrug resistance and therefore PKC activity was investigated. Besides that peripheral blast cell count (PBC), a well known clinical prognostic factor, was examined as a quantitative continuous covariate as well as a variable dichotomized at a cut off value of 50000 blast cells/mm^3. Patients with more than 50000 cells/mm^3 were thought to do worse than those with lower values (n= 43 (41%)). The data were kindly provided by Prof. Volm, DKFZ.

3. Importance of Graphics and Their Assessment

Graphics are important for the application of survival analysis in biomedicine and they raise the acceptance of statistical analyses. Primarily this is due to the basic mechanisms of human perception, where most information gets mediated by the visual system. Another reason is that a graphic is able to stimulate imagination, especially if used as an exploratory tool for data analysis. Graphics are often required to complement an analysis and to increase the chance of acceptance of findings, and they can transmit results to the public much better than any other medium. Graphical representations have to obey laws of human visual perception and systems processing. In presenting survival curves one should be aware that the human system is more prone to horizontal than to vertical comparison, see Cleveland & McGill (1984).

4. Methods in Survival Analysis and Implementation in Statistical Software

SURVIVAL DATA: Survival analysis is based on observations $(X_i,\delta_i,Z_i)_{(i=1,...,n)}$ of an individual i from a study population of size n. X_i is the observed failure time, δ_i the censoring status and $Z_i=(Z_{i1},...,Z_{ip})$ a vector of covariates. If $\delta_i=1$, X_i is equal to an observed failure time $T_i=t_i$, if $\delta_i=0$ X_i is equal to the censoring time C_i competing with the failure time T_i: $X_i=\min\{T_i,C_i\}$, $\delta_i=1$ iff $T_i\leq C_i$ and $\delta_i=0$ else. $S(t)=P(T>t)$ denotes the survival distribution, $\lambda(t)$ the hazard function, $\Lambda(t)$ the cumulative hazard function, and $e(t)=E[T|T>t]$ the mean residual survival time. We write $S(t,z)$, $\lambda(t,z)$ etc. to denote the dependency from the covariate $Z=z$.

Table 1 shows the procedures and estimates implemented in SAS (procedures LIFEREG, LIFETEST and PHREG) and S-Plus (surv.fit, surv.diff, coxreg, agreg) as well as availability of graphics. Basic visualization is plotting the Kaplan-Meier estimate curve $\hat{S}_{KM}(t)$ on $t:<t,\hat{S}_{KM}(t)>$. As an example of the use of graphical tools in survival analysis we show here only two residual plots in Fig 1.

5. More Graphical Tools Are Needed

Regression models have been used in survival analysis with great success for the adjustment of covariate effects if treatments are compared, for the determination of prognostic factors, and for predicting prognosis. Graphical displays might help in examining the relation between covariates and survival time. Desirable graphical tools have been listed in Table 2. As an example we would like to mention the censored data boxplots (Gentleman&Crowley, 1991) which are comprehensive means for comparing several survival strata by a boxplot derived from the Kaplan-Meier estimate. That provides the median and the upper and lower quartile of the survival time distribution and the survival probability $S_{KM}(t)$ at that minimum/maximum observed survival time.

Fig. 1. Martingale residuals, with a Loess smoother, from the model using only PKC are plotted against the PBC (S-Plus on the left and SAS on the right).

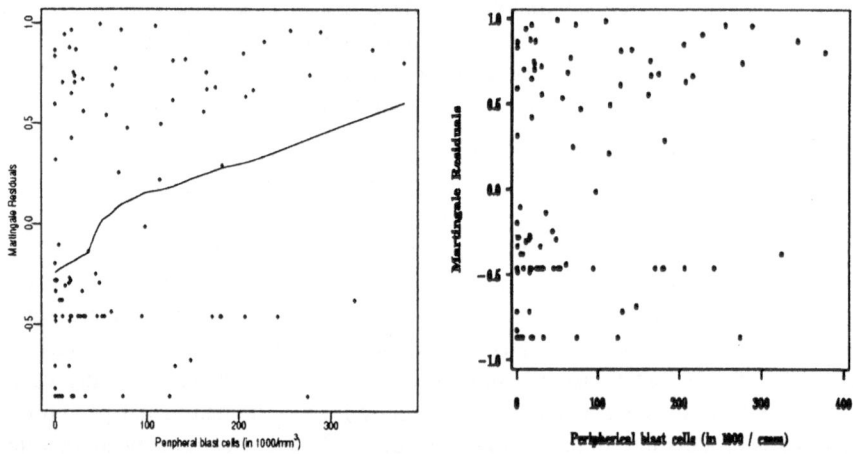

Of urgent need are graphics in the explanatory analysis of the proportional hazards model. This ranges from diagnostics of the maximum likelihood iteration to adequate residual plots and model diagnostics.

Table 2: Desirable Graphics in Survival Software

DESCRIPTIVE
Equivalents of the survivor function
   Hazard function $\lambda(t)$
   Cumulative hazard function $\Lambda(t)$
   Mean residual survival $e(t)$
Global Confidence bands
Pointwise Confidence bands
COMPARISON
   Censored data Box plot for z-strata
EXPLANATORY
Iteration diagnostic: Plot of log-likelihood function(z fixed)
Covariate information: survival-covariate-scatter plot
Results: baseline survivor function (z=0)
            relative risk scatter plot
            Survival functionals
Survival functionals : $<t, F(S(t,z))>$ and $<z, F(S(t,z))>$
Model diagnostics:
            proportional hazards assumption
            Case influence plots

In the SURVIVAL-COVARIATE-SCATTER PLOT a continuous covariate Z_j is illustrated by the scatter plot $<z_{ij}, x_i>, i=1\ldots n$. Censoring information can be expressed graphically by using different symbols for censored and uncensored times X_i. The survival-covariate-scatter plot smoother gives the mean survival time depending on the covariate. The RELATIVE-RISK-SCATTER PLOT is a special case of a plot of a functional F of the survivor function S(t,z) on the covariate value of Z: $F(S(t,Z))=\beta Z$. Functionals of the survival function S depending on a covariate z generalize this type of plotting. The SURVIVAL-QUANTILE PLOT (SQP) and COVARIATE-QUANTILE PLOT (CQP) are members of that family (Abel et al., 1986). A survival-median plot as obtained by our SURVIVAL package is shown in Fig.2 for the ALL example.

Graphical representation of S(t,z) should take into account the basically three-dimensional aspect concerning the objects t, z, S(t,z). Contour plots of S(t,z) on (t,z) would be more adequate than two-dimensional plots $<t,z>$, $<t,S(t,z)>$, and $<z,S(t,z)>$ but have not been implemented. Generalizations are given when the components t, z, and S(t,z) are replaced by functionals H(t), G(z) and F(S(t,z)). This opens another variety of plot options which has not been explored. Finally, there is the need for **model diagnostics**. The log-log-plot has been used for assessing the proportionality in the Cox Regression although it can be assessed by eye only. Other model diagnostics based on generalized residuals have been criticized recently by Elandt-Johnson&Smith (1989) who got a conservative behavior of graphical tests based on them.

Fig. 2. Survival-Median Plot. Estimated median survival of ALL patients dependent on PBC using the Cox Model

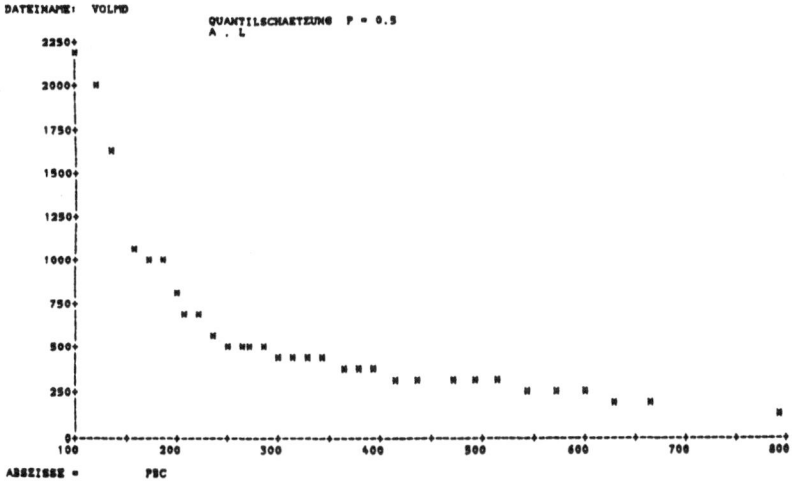

6. Discussion

We reviewed survival software in two major program packages emphasizing graphical tools. This was guided by our own experience as well as new developed methods in statistical literature. A direct comparison of the examined survival packages with our own product was considered as unfair for two reasons:

(i) SURVIVAL was developed in APL language which is not so widely used,

(ii) such a comparison would naturally be biased by selection of features and of criteria chosen in the developement of our own package.

As expected there were no major differences between SAS and S-Plus. Similarities may be partly due to the fact that both systems use software developed from the same school (F.Harrell and T.M.Therneau). There is already a small advantage of S-Plus in model diagnostics and explanatory plots efficiently supported by its interactivity. This could grow to superiority if new developments as e.g. those of Therneau's most recent package of survival functions for S would be included in a user-friendly way also for non-statisticians.

Bibliography

Abel, U., Berger, J.,and Edler, L. (1986): A method for analysing the dependence of failure-time statistics on quantitative covariates. EDV in Medizin und Biologie 17, 90-92.

Cleveland, W.S., and McGill, R. (1984): Graphical perception: Theory, experimentation, and application to the development of graphical methods. J.Amer.Statist. Assoc. 79, 531-554.

Elandt-Johnson,R.C., and Smith,F.B.(1989): Graphical generalized residuals in fitting distributions: Applications to epidemiological follow up data. Statistics in Medicine, 8, 703-723.

Gentleman, R., and Crowley, J. (1991): Graphical methods for censored data. J.Amer. Statist. Assoc. 86, 678-683.

Kaplan, E.L.,and Meier,P. (1958): Non-parametric estimation from incomplete observations. J. Amer. Statist. Assoc. 53, 457-481.

SAS Institute Inc. (1992): SAS Procedures Guide, Version 6.07, 4th Edition, Cary.

StatSci (1992): S-Plus User's Manual Vol.2, version 3.1. Statistical Sciences, Inc., Seattle.

Therneau,T.M.,Grambsch,P.M.,and Fleming,T.R.(1990).Martingale-based residuals for survival models. Biometrika 77, 147-160.

Volm,M.,Sauerbrey,A.,and Zintl,F.(1994):Prognostic significance of protein kinase C in newly diagnosed childhood acute lymphoblastic leukemia. International Journal of Oncology 4, 363-368.

Weber,E. (1986): Statistische Auswertung Biomedizinischer Daten. Teil II.
SURVIVAL: Analyse zensierter Beobachtungen bei Überlebens- oder Ausfallzeiten. Abteilung Biostatistik, Deutsches Krebsforschungszentrum, Heidelberg.

Part XI

Computational Aspects in Statistics in Neural Networks

Neural Networks: More Than 'Statistics for Amateurs'?

Kurt Hornik

Institut für Statistik und Wahrscheinlichkeitstheorie,
Technische Universität Wien

Abstract. Statisticians often feel that neural networks are just an amateurish attempt to perform statistical inference, quite often using standard tools but always a non-standard terminology. We shed some light on this issue by looking at two leading cases of considerable overlap between statistics and neural networks, namely multilayer perceptrons for nonparametric regression and neural network-type on-line learning algorithms for extracting principal components.

1   Introduction

The last few years have brought a huge number of applications of "neural networks" (NNs) in various fields. As these applications are typically data-based, one might wonder up to what extent neural networks just perform statistical tasks, perhaps in a new guise. In fact, one quite often comes across statements like the following (Anderson, Pellionisz & Rosenfeld (1990), page 541):

> Neural networks are statistics for amateurs. A properly designed network, when learning and responding, performs good statistical inference, based on what it saw when it learned and what it sees when it responds. Most networks conceal the statistics from the user.

Clearly, "amateur" has a pejorative sense. But even though the above has its nice aspects, too (as it implies that users can easily and successfully do statistical inference if they know some basic neural network design principles), it is by far too general to be true.

To start with, it is not clear at all what *exactly* "neural networks" are. Typically, the concept is used as a synonym for 'multilayer perceptrons (MLPs) trained by Back-Propagation (BP)', but many other NN paradigms exist, see e.g. Simpson (1990) and Hertz, Krogh & Palmer (1991). To make matters worse, NN methods are often intertwined with elements from fuzzy logic, genetic algorithms and their likes. Recently, Bezdek (1992) has introduced the term 'Computational Intelligence' (CI) as a common label for these techniques, distinguishing them from classical (symbolic) 'Artificial Intelligence' (AI) and from 'Biological Intelligence' (BI). IEEE standardizations for these concepts are currently being worked out, see also Marks (1993).

In the sequel, we shall only deal with *computational* NNs. Even without giving precise definitions, we can exhibit at least two of their basic characteristics.

Locality NNs are built from simple processing elements (PEs) called nodes, units or neurons. The intelligence of these units consists in their ability to perform certain computations (e.g., to compute a linear threshold function).

Adaptivity NNs contain adjustable parameters (e.g., the weights associated with the connections between the units). The intelligence of the whole system also consists in its ability to learn by adjusting (training) these parameters.

As far as we are concerned, another key feature is that these systems should (eventually) be *realizable in special-purpose hardware* rather than be simulated on conventional digital computers. I.e., in some sense NNs are the revival of analog computing with the new ingredient of intelligence by adaptivity (on-line learning).

Obviously, none of these characteristics is of interest in (mainstream) statistics. Nevertheless, there is considerable overlap—if NNs are used e.g. for automatic classification, they clearly perform a statistical task. Unfortunately, the fact that the NN community has created its own terminology which heavily relies on biological notions often impedes communication with statisticians (in fact, there currently are even more fields like e.g. machine learning which deal with statistical issues from a different point of view using incompatible terminologies).

Hence in the past few years it has become quite attractive and popular to review NN efforts from an "outsider's" (e.g. statistician's) point of view. In the sequel, we shall add to this by looking at two NN paradigms of leading interest and their interactions with statistics in our own (and hence biased) way.

2   Multilayer Perceptrons

We already mentioned that typically, the term "neural network" is taken synonimously for "multilayer perceptrons trained by BP". These NNs are built of neurons which compute maps of the form $x \mapsto \psi(a'x + \theta)$ (in what follows, ' denotes transpose). In NN terminology, a and θ are the weights and the bias associated with the neuron, respectively, and ψ is its activation function. (In some sense, this is related to the way biological neurons work; a gives the synaptic coupling strengths and $-\theta$ the threshold above which the neuron which receives the weighted signal $a'x$ fires.) If these neurons are combined in a way that information is processed in one direction from each layer to the next one, one obtains a multilayer perceptron. (Typically, the term "multilayer feedforward network" is used as well. But clearly, radial basis function networks which use the different activation mechanism $x \mapsto \psi(\gamma|x - m|)$ are also multilayer feedforward networks.)

Perceptrons were introduced in Rosenblatt (1962). In the simplest case, there is only one neuron of the above type, and the network computes the map $x \mapsto \psi(a'x + \theta)$. If ψ is taken as the sign or Heaviside function, this can be used for implementing linear decision functions, and there is a simple learning algorithm for the adjustable parameters a and θ which can be shown to converge in finite time if the problem is linearly separable. The usual story is that perceptrons were given a fatal blow by Minsky & Papert (1968) who showed that many important problems, in particular XOR, are not linearly separable, and thus extinguished NN funding and hence also most research for almost twenty years. In fact, Rosenblatt did consider more complicated architectures, but failed to obtain a usable rule for adjusting the parameters (this is sometimes referred to as the 'Credit Assignment Problem').

Such a rule, the 'generalized delta rule' or 'back-propagation' algorithm was apparently discovered by Bryson & Ho (1969) and Werbos (1974), but rediscovered and popularized by Rumelhart, Hinton & Williams (1986). BP simply performs on-line gradient descent on the error function

$$E(w) = \frac{1}{2} \sum_{t=1}^{T} |y_t - f(x_t, w)|^2,$$

where $((x_t, y_t), t = 1, \ldots, T)$ is the set of training patterns which consist of inputs x and targets (desired outputs) y, w is a vector containing containing the adjustable parameters, and $f(x, w)$ is the output of the net with parameters w from an input x. (Note that f also depends on the network's topology— the numbers of neurons in each layer and their connectivity pattern—and the choices for the activation functions.) I.e., upon presentation of a pattern, w is updated according to

$$\Delta w \propto -\nabla_w |y - f(x, w)|^2.$$

Due to the special (layered) structure of the MLP, the gradient can be computed recursively by propagating the error at the outputs back through the network, whence the name Back-Propagation.

From this description we see that "training a neural network by BP" can (usually) be translated into more familiar technology as fitting a particular, typically nonlinear, parametric model by nonlinear least squares regression; more precisely, by an algorithm which attempts to do so. It is clear that there might be other classes of model functions which achieve superior fit, and that BP, as a gradient descent algorithm, compares rather unfavorably to more sophisticated numerical minimization algorithms and can at best find local minima of the error function. In fact, the NN community has put considerable effort into improving BP, but typically without too much background in numerical analysis. We also see that research on MLPs has to deal with at least the following three questions.

- How flexible is the class of MLPs?

- Which specific MLP should be chosen to fit given data?

- How should a specific MLP be fit to the data?

It seems clear that these questions naturally ask for interaction with approximation theory, statistics (model selection), and numerical analysis (minimization).

However, NN research has been rather self-contained for a long time. The first simple MLP applications basically consisted in generating look-up tables, e.g. implementing boolean functions (in particular, XOR) based on an *exhaustive* set of training patterns. Clearly, if the data is only a sample from an underlying population, an exact fit on the data set (zero bias) will result in huge variance (or, in NN terminology, in bad **generalization** performance). This phenomenon, which has received huge interest in the NN literature, is easily understood within a statistical framework as e.g. given in White (1989), see in particular Levin, Tishby & Solla (1990) and Geman, Bienenstock & Goursat (1992). It ought to be clear that "optimal" fitting of MLPs to data is a problem of nonlinear model selection, but apparently this point has been found hard to adopt by most NN researchers and practitioners (in this sense, they did remain amateurs).

Let us look at the representational capabilities of MLPs first. The first general results were obtained independently and almost simultaneously by Cybenko (1989), Funahashi (1989) and Hornik, Stinchcombe & White (1989) and have meanwhile substantially been generalized, cf. in particular Hornik (1991, 1993a). It was shown that MLPs can approximate any reasonable function arbitrarily well, provided only that enough hidden units are available for internal computations and that their activation function be nonpolynomial. To give a more precise statement, let n be the input dimension and, for $A \subseteq \mathbb{R}^n$ and $\Theta \subseteq \mathbb{R}$, let $\mathcal{N}(\psi; A, \Theta)$ be the set of all functions on $\mathbb{R}^n$ of the form

$$x \mapsto \sum_{i=1}^{k} \beta_i \psi(a_i' x + \theta_i),$$

i.e., the span of $\{\psi(a'x + \theta), a \in A, \theta \in \Theta\}$. This is the set of all mappings from $\mathbb{R}^n$ to $\mathbb{R}$ which can be implemented by a single hidden layer perceptron with input-to-hidden weights in A and hidden unit biases in Θ, respectively; for sake of simplicity, all hidden units have the same activation function ψ and the output unit is linear. The following result is theorem 2 in Hornik (1993a).

> Let ψ be *essentially bounded and nonpolynomial* on some nonde-generate compact interval Θ, and let A contain a neighborhood of the origin. Then for all compactly supported finite measures μ on $\mathbb{R}^d$ and $1 \leq p < \infty$, $\mathcal{N}(\psi; A, \Theta)$ contains a subset which is dense in $L^p(\mu)$.

Similar results can be given for uniform approximation of continuous functions on compacta. Hence, MLPs can basically implement any reasonable function, and for this reason are feasible competitors in nonparametric statistical inference.

Unfortunately, it is much harder to obtain results describing how well classes of functions can be approximated by MLPs of certain complexity. Barron (1993) shows that if the function g to be approximated has $g(0) = 0$ and a Fourier transform with finite first moment and ψ is sigmoidal (i.e., $\psi(-\infty) = 0$ and $\psi(\infty) = 1$), then there is always a function g_k implemented by a single hidden layer perceptron with k hidden units which have ψ as their activation function such that

$$\|g - g_k\|_{L^2(B_r, \mu)} \leq \frac{2\sqrt{\mu(B_r)}}{\sqrt{k}} \int |a\hat{g}(a)| \, da,$$

where B_r denotes the ball of radius r centered at the origin; similar results can be obtained for uniform approximation (Barron, 1992). I.e., the MSE scales like $1/k$. As this is independent from n, it has been argued that this result shows that MLPs can avoid the "curse of dimensionality"; however, it is clear that n enters via the integrability condition on $\hat{g}$. In fact, we should emphasize that as the classes of MLP model functions are nonlinear in some of their parameters, the number of free parameters is not necessarily the correct complexity measure. Murata, Yoshizawa & Amari (1993) argue that, if more generally goodness-of-fit is measured by a loss function $\ell(x, y; w)$ (thus far, we had $\ell(x, y; w) = |y - f(x, w)|^2$, the correct corresponding complexity measure is given by $\text{trace}(SQ^{-1})$, where S is the expected outer product of the gradient of ℓ (with respect to w) and Q is the expected Hessian.

How can an optimal MLP architecture be determined from the data? Using cross-validation would be one possibility; but due to the nonlinear structure of the model classes (and the fact that the parameter estimation may be hard), this may computationally be too expensive. Regularization methods, like the Bayesian "Occam's Razor" of MacKay (1992) or the Network Information Criterion of Murata, Yoshizawa & Amari (1993), may turn out to be more convenient to use.

We already pointed out that MLPs can be used as one method for nonparametric regression. In particular, they are closely related to the model functions

$$x \mapsto \sum_{i=1}^{k} \beta_i \psi_i(u_i'x)$$

used in Projection Pursuit regression (Friedmann & Stützle, 1981); in some sense, MLPs transfer the bandwidth selection problem for the estimation of the ψ_i by scatterplot smoothing into selecting the right number of hidden units. An extensive comparison of various NN techniques and traditional statistical methods has recently been given in Ripley (1993). On several benchmark data sets for classification tasks it is typically found that MLPs are always among the winners as far as their classification power is concerned, but suffer from enormous demands on CPU time. So whereas on the one hand their nonlinear structure provides them with possibly superior representation capabilities, it on the other hand makes actually fitting them to data quite cumbersome. Of course, we already pointed out that several MLPs with independent random initial configurations could be trained in parallel by BP on special purpose hardware—but without doing so, the computational demands definitely put some clear bounds on their usefulness in real-world problems. A remedy would be to find better algorithms for training the parameters which enter nonlinearly; one such approach, based on the relation between MLPs and wavelet representations, was given in Zhang & Benveniste (1992) and produced quite encouraging results.

Which conclusions can be drawn from the above discussion? We have seen that the most prominent use of "neural networks" amounts to employing MLPs in nonparametric statistical inference. However, in order to make them perform good inference when used by amateurs, a whole lot of quite challenging research is still necessary. In the long run, we should see that "neural network modelling will eventually become one of the standard techniques of applied statistics" (Amari, 1993, page 4).

We should point out that MLPs with Heaviside activation functions are of substantial interest to computer scientists and engineers from a non-statistical perspective. In this context, the units which compute the map $x \mapsto H(a'x+\theta)$ are usually referred to as linear threshold gates; the corresponding MLPs are called (linear) threshold circuits. These can be used to define interesting nontrivial complexity classes of boolean functions. Let C_n be a circuit (network) with n boolean inputs and a single boolean output. Consider the following classes of sequences $(C_n, n \geq 1)$:

TC_d^0 C_n has depth $\leq d$ and $\leq$ poly(n) linear threshold gates of arbitrary fan-in

AC^0 C_n has depth $O(1)$ and $\leq$ poly(n) AND, OR and NOT-gates.

Then one can show that e.g. PARITY, MAJORITY, MULTIPLICATION, and SORTING are in $\bigcup_d TC_d^0$, but not in AC^0. In fact, MULTIPLICATION is in $TC_3^0 \setminus TC_2^0$ and ADDITION is in $TC_2^0 \setminus TC_1^0$, cf. e.g. Bruck (1990) and Bruck & Smolensky (1992) and the references therein. These results do not only shed some light on the computational power of MLPs, but also show that standard problems can be realized in simple special-purpose NN hardware.

3   PCA Networks

One of the basic statistical problems is data compression or feature extraction in following sense. Given n-dimensional (random) data vectors x and some $p < n$ (and maybe even $p \ll n$), find a function $f : \mathbb{R}^n \rightarrow \mathbb{R}^p$ such that $y = f(x)$ contains "as much information about x as possible". If mean square error of the optimal linear reconstruction of x from y (i.e., the linear regression of x on y) is employed as optimality criterion, i.e., if it is desired to minimize

$$\mathbb{E} \, |x - Bf(x)|^2$$

over all functions f from $\mathbb{R}^n$ to $\mathbb{R}^p$ and all $n \times p$ matrices B, the optima satisfy

$$(Bf)_{opt}(x) = U_p U_p' x,$$

where $U_p = [u_1, \ldots, u_p]'$ and u_i is a unit length eigenvector associated with the i-th largest eigenvalue λ_i of $\Sigma = \mathbb{E}(xx')$, see e.g. Bourlard & Kamp (1988). I.e., it suffices to consider $linear$ maps $y = Ax$ only. If x is gaussian, the optimal transformations also maximize the mutual information between x and $y = f(x)$.

Assume for simplicity that the x is centered such that Σ is the covariance matrix of x and that all eigenvalues of Σ are simple. Then the above says that up to an invertible linear transformation, the optimal features to be extracted are

$$u_1' x, \ldots, u_p' x,$$

the first p $principal$ $components$ of x. This is well known to statisticians. Given a sample, it seems straightforward to first estimate Σ by the data covariance matrix $\hat{\Sigma}$ and then u_i by a unit length eigenvector $\hat{u}_i$ associated with the i-th largest eigenvalue of $\hat{\Sigma}$, using some suitable numerical routine for eigenvector computations. However, this has at least the following disadvantages:

- The direct computation of the eigenvectors is not necessarily "simple". If new data arrive (or perhaps even keep arriving at some rate), it is desirable to directly update the estimates of the u_i rather than updating the estimate of Σ and recompute the eigenvectors. In other words, a "simple" adaptive algorithm is needed in many applications.

- Storing estimates of the covariance matrix can consume huge amounts of memory. In image processing application, the data could e.g. be 1024×1024 pixel gray values.

Hence, there are applications where "simple" adaptive algorithms for PCA extraction which do not require explicit computation of the full data covariance matrix are needed.

Such algorithms obviously exist. By the above characterization, the optimum can be obtained by minimizing

$$E(A, B) = \mathbb{E}\,|x - BAx|^2$$

over all $p \times n$ matrices A and $n \times p$ matrices B, which could e.g. be attempted by Back-Propagation. In fact, BAx is the output of a *linear* "neural network" with a single bottleneck hidden layer. Baldi & Hornik (1989) have shown that E is without local minima; all critical points which do not give the global minimum of E are saddle points. Hence, we might expect BP to work reasonably well. Another possibility is to follow Xu (1993) and perform gradient descent on $E(A, A')$. This algorithm is more complicated, but avoids the computation of B which is typically unnecessary and can be made "local" at the expense of introducing additional computations at the input and output nodes of $y = Ax$. In both cases, only the space spanned by the first p eigenvectors of Σ is extracted (rather than the eigenvectors themselves), which is insufficient if it is desired to have the outputs (extracted features) uncorrelated.

Within the last few years, a different class of simple PCA algorithms has received broad attention in the neural networks community. These are based on *hebbian learning*. One version of this rule motivated by Hebb (1949) can be stated as follows: "Modify the weights between two units proportional to the product of their activations". In a simple linear network $y = a'x$ with a single output unit, this translates into

$$\Delta a \propto xy.$$

Using stochastic approximation theory, such on-line algorithms can most conveniently be analyzed in terms of their associated ODE (Ordinary Differential Equation) which is obtained by "averaging over all patterns". In our case, this gives

$$\dot{a} = \mathbb{E}(xy) = \mathbb{E}(xx'a) = \Sigma a$$

which performs (continuous-time) gradient ascent on $a'\Sigma a/2$ and tends to infinity unless started at the origin. Now recall the standard variational description of the eigenvectors of Σ: each u_i maximizes $u'\Sigma u$, the variance of $u'x$, over all unit length vectors u perpendicular to $u_1, \ldots, u_{i-1}$. Hence in particular, u_1 maximizes $u'\Sigma u$ over $u'u = 1$. Thus, the hebbian rule should be modified in a way which keeps a close to the unit sphere. Proceeding along these lines, Oja (1982) obtained the rule

$$\Delta a \propto xy - ay^2,$$

typically referred to as the Oja one-unit algorithm. Under suitable assumptions, this rule does extract the first principal component.

Several generalizations of this idea to the case where $p > 1$ are possible. One approach is to provide suitable additive constraint terms for the hebbian rule $\Delta A \propto yx'$ similar to the above. This gives e.g. the symmetric algorithm obtained independently by Williams (1985), Baldi (1988) and Oja (1989) or the asymmetric Generalized Hebbian Algorithm of Sanger (1989). Another idea is to use the Oja one-unit rule for each of the rows of A (i.e., for each output unit individually) and provide an additional output decorrelation mechanism (otherwise, all units would extract the first component only). This can e.g. be accomplished by introducing inhibitory weights between the output units and training them by *anti-hebbian* learning. As examples, we refer to Rubner & Tavan (1989), Kung & Diamantaras (1990) and Leen (1991). These algorithms always extract the first p eigenvectors directly (rather than just the space spanned by them); some are strictly local and hence easily implementable in special-purpose hardware. A reasonably satisfactory but not entirely complete stochastic approximation-type convergence analysis for the whole class of constrained hebbian-learning based PCA algorithms was carried out in Hornik & Kuan (1992); further results are contained in Baldi & Hornik (1994) who provide a comprehensive review of learning in linear networks.

To sum up: there now exists a variety of neural-network type on-line learning algorithms for extracting principal components. These rules avoid direct computation of the data covariance matrix and its eigenvectors and thus are extremely well suited to situations where adaptivity and updating speed are of concern. In these cases, they conveniently replace the "standard" statistical method.

Now suppose that the extracted features are actually given by $y = Ax + e$, where e is "noise" which may e.g. be due to unreliability of the processing elements. Such situations arise naturally if information is transmitted over noisy channels. For simplicity, assume that the noise has mean zero, co-variance matrix R, and is uncorrelated with the signal x. Using the same optimality criterion as before, we would like to minimize

$$E(A, B) = \mathbb{E} |x - B(Ax + e)|^2 = \mathbb{E} |x - BAx|^2 + \text{trace}(BRB').$$

Clearly, this problem has no finite solution; by letting $A = \mu U_p'$ and $B = U_p/\mu$ and $\mu \to \infty$, we see that $\inf_{A,B} E(A, B) = \mathbb{E} |(I - U_p U_p')x|^2$ equals the minimum of the noise-free case, but the infimum clearly is not attained. Intuitively, A should be as large and B as small as possible in order to have the signal part suppress the noise. Hence, to obtain finite solutions to our "noisy PCA" problem, we either have to penalize for large A (the case of soft constraints), e.g. by adding $\mathbb{E} |Ax|^2$ to E, or to explicitly constrain A to some suitable compactum (the case of hard constraints), e.g. by considering only (sub)orthogonal A. The latter case was examined in some detail in Hornik (1993) and Diamantaras & Hornik (1993). It turns out that in the orthogonal case, extraction of the first p principal components (of the data covariance matrix) is always optimal; however, if the restriction is weakened to $\|A\|_F \leq p$ ($\|A\|_F = \sqrt{\text{trace}(AA')}$ is the Frobenius norm of A, then some number q of principal components which depends on the eigenvalues of Σ and R and typically is smaller than p is extracted. In the high noise limit, i.e. if the minimal eigenvalue of R is large enough, only the first principal component should be extracted. Note that both constraints give full PCA in the noise-free case.

These results can nicely be interpreted in a way that "if the network is not too constrained, the units respond to high noise by high coopera-tion/redundancy". But is also has some clear implications for the neural PCA algorithms discussed above: the ones based on output decorrelation are not "robust" to noise in the above sense (as these algorithms are explicitly based on output decorrelation, which is not optimal if the noise is "large"). In fact, it turns out that Oja's one-unit rule itself has its problems. In the noisy case where $y = a'x + e$, the associated ODE is

$$\dot{a} = \mathbb{E}(xy - y^2 a) = \Sigma a - (a'\Sigma a + \rho)a,$$

where ρ is the variance of e. Hence, if ρ is larger than the largest eigenvalue λ_1 of Σ, $a = 0$ is the only equilibrium of the ODE and in fact, as

$$\frac{1}{2} \frac{d|a|^2}{dt} = a'\dot{a} = a'\Sigma a - (a'\Sigma a + \rho)a'a \leq -a'\Sigma aa'a$$

which is strictly negative unless $a = 0$, it is also globally attractive! How to deal with the noise effects algorithmically is currently being investigated.

4   Conclusions

We have looked at two leading cases where statistics and neural networks look at the same problems from a different perspective. Where such considerable overlap occurs, interaction is desirable, although not straightforward due to incompatible terminologies. In the long run, neural network modeling and special-purpose hardware realizations should become a standard tool in applied statistics.

References

Anderson, J. A., Pellionisz, A. & Rosenfeld, E. (1990). *Neurocomputing 2: Directions for Research*. Cambridge, MA: MIT Press.

Amari, S.-I. (1993). Mathematical methods of neurocomputing. In Barndorff-Nielsen, O. E., Jensen, J. L., & Kendall, W. S. (eds.), *Networks and Chaos—Statistical and Probabilistic Aspects* (pp. 1–39). Monographs on Statistics and Applied Probability 50. London: Chapman & Hall.

Baldi, P. & Hornik, K. (1989). Neural networks and principal component analysis: learning from examples without local minima. *Neural Networks*, 2, 53–58.

Baldi, P. (1989). Linear learning: landscapes and algorithms. In D. S. Touretzky (ed.), *Advances in Neural Information Processing Systems 1*. San Mateo, CA: Morgan Kaufmann.

Baldi, P., & Hornik, K. (1994). Learning in linear neural networks: a survey. *IEEE Transactions on Neural Networks*, to appear.

Barron, A. R. (1992). Neural net approximation. In Narendra, K. (ed.), *Proceedings of the 6th Yale Workshop on Adaptive Learning Systems*.

Barron, A. R. (1993). Universal approximation bounds for superpositions of a sigmoidal function. *IEEE Transactions on Information Theory*, **IT-39**, 930–945.

Bezdek, J. C. (1992). On the relationship between neural networks, pattern recognition and intelligence. *International Journal of Approximate Reasoning*, 6, 85–107.

Bourlard, H. & Kamp, Y. (1988). Auto-association by the multilayer perceptrons and singular value decomposition. *Biological Cybernetics*, 59, 291–294.

Bruck, J. (1990). Harmonic analysis of polynomial threshold functions. *SIAM Journal on Discrete Mathematics*, 3, 168–177.

Bruck, J. & Smolensky, P. (1992). Polynomial threshold functions, AC^0 functions and spectral norms. *SIAM Journal on Computing*, 21, 33–42.

Bryson, A. E. & Ho, Y.-C. (1969). *Applied Optimal Control.* New York: Blaisdell.

Cybenko, G. (1989). Approximation by superposition of a sigmoidal function. *Mathematics of Control, Signals and Systems,* **2**, 303–314.

Diamantaras, K. & Hornik, K. (1993). Noisy principal component analysis. In *Proceedings of MEASUREMENT 93* (Smolenice, SK).

Friedman, J. H. & Stützle, W. (1991). Projection pursuit regression. *Journal of the American Statistical Society,* **76**, 817–823.

Funahashi, K. (1989). On the approximate realization of continuous mappings by neural networks. *Neural Networks,* **2**, 183–192.

Geman, S., Bienenstock, E. & Goursat, R. (1992). Neural networks and the bias/variance dilemma. *Neural Computation,* **4**, 1–58.

Hebb, D. O. (1949). *The Organization of Behavior.* New York: Wiley.

Hertz, J., Krogh, A. & Palmer, R. G. (1991). *Introduction to the Theory of Neural Computation.* Redwood City, CA: Addison-Wesley.

Hornik, K., Stinchcombe, M., & White, H. (1989). Multilayer feedforward networks are universal approximators. *Neural Networks,* **2**, 359–366.

Hornik, K. (1991). Approximation capabilities of multilayer feedforward networks. *Neural Networks,* **4**, 251–257.

Hornik, K. & Kuan, C.-M. (1992). Convergence analysis of local feature extraction algorithms. *Neural Networks,* **5**, 229–240.

Hornik, K. (1993a). Some new results on neural network approximation. *Neural Networks,* **6**, 1069–1072.

Hornik, K. (1993b). Noisy linear networks. In R. Mammone (ed.), *Artificial Neural Networks for Speech and Vision* (pp. 37–44). London: Chapman & Hall.

Kung, S. Y. & Diamantaras, K. (1990). A neural network for adaptive principal component extraction (APEX). In *Proceedings of the IEEE International Conference on Acoustics, Speech and Signal Processing* (Albuquerque, New Mexico, April 1990) (pp. 861–864).

Leen, T. (1991). Dynamics of learning in linear feature-discovery networks. *Network,* **2**, 85–105.

Levin, E., Tishby, N. & Solla, S. A. (1990). A statistical approach to learning and generalization in layered neural networks. *Proceedings of the IEEE,* **78**, 1568–1574.

Marks, R. J., II (1993). Intelligence: computational versus artificial. *IEEE Trans-*

actions on *Neural Networks*, **4**, 737–739.

Minsky, M. L. & Papert, S. A. (1968). *Perceptrons*. Cambridge, MA: MIT Press.

Murata, N., Yoshizawa, S. & Amari, S.-I. (1993). Network information criterion for determining the number of hidden units for artificial neural network models. *IEEE Transactions on Neural Networks*, to appear.

Oja, E. (1982). A simplified neuron model as a principal component analyzer. *Journal of Mathematical Biology*, **15**, 267–273.

Oja, E. (1989). Neural networks, principal components and subspaces. *International Journal of Neural Systems*, **1**, 61–68.

Ripley, B. (1993). Statistical aspects of neural networks. In Barndorff-Nielsen, O. E., Jensen, J. L., & Kendall, W. S. (eds.), *Networks and Chaos—Statistical and Probabilistic Aspects* (pp. 40–123). Monographs on Statistics and Applied Probability 50. London: Chapman & Hall.

Rosenblatt, F. (1962). *The Principles of Neurodynamics*. New York: Spartan Books.

Rubner, J. & Tavan P. (1989). A self-organizing network for principal component analysis. *Europhysics Letters*, **10**, 693–698.

Rumelhart, D. E., Hinton, G. E. & Williams, R. J. (1986). Learning representations by back-propagating errors. *Nature*, **323**, 533–536.

Sanger, T. D. (1989). Optimal unsupervised learning in a single-layer linear feed-forward neural network. *Neural Networks*, **2**, 459–473.

Simpson, P. K. (1990). *Artificial neural systems: foundations, paradigms, research and applications*. Elmsford, NY: Pergamon Press.

Werbos, P. (1974). *Beyond Regression: New Tools for Prediction and Analysis in the Behavioral Sciences*. Ph.D. thesis, Harvard University.

White, H. (1989). Learning in artificial neural networks: A statistical perspective. *Neural Computation*, **1**, 425–464.

Williams, R. J. (1985). *Feature discovery through error-correction learning*. Technical Report 8501, Institute for Cognitive Science, University of California, San Diego.

Xu, L. (1993). Least mean square error reconstruction principle for self-organizing neural nets. *Neural Networks*, **6**, 627–648.

Zhang, Q., & Benveniste, A. (1992). Wavelet Networks. *IEEE Transactions on Neural Networks*, **NN-3**, 889–898.

Part XII

Computational Aspects in Robust Statistics

On Robust Nonlinear Regression Methods Estimating Dose Response Relationships

Hans-Peter Altenburg

University of Heidelberg, Fakultät für Klinische Medizin Mannheim, Med. Statistik, Biomathematik und Informationsverarbeitung

Abstract. The paper discusses some modified versions of computational high breakdown algorithms that are useful for the case of nonlinear dose response relationships. They can be used not only in the case of adapting simple calibration curves or expectation formulas such as two or four parameter probit or logit regression models but also models representing special experimental situations such as general assays and parallel line assays, respectively, or to robustify general nonlinear estimation procedures such as the generalized least squares algorithm in case of overdispersion leading to robust effective concentration values (e.g. EC50) characterizing the dose response relationship of the experiment.

1 Introduction

Many problems of inference in biological or medical research concern the relation between a response and an appertained stimulus. Usually two types of response are distinguished: the quantal or all-or-nothing and the quantitative response. An example for quantal response might be the impact of the intensity of light in a solarium on the appearence of different skin lesions such as follicular, teleangiectatic or epidermatic skin injuries. Quantitative response relationships we have for instance in the case of the effect of substance concentrations on the concentration of some metabolite or in radiometric assays that are today a common method for measuring hormones or other material at very small concentrations.

One can model the relationship between the administered doses and the responses by a sigmoid curve that may decrease or increase depending on the response, and that has finite limits as the dose approaches 0 or infinity (i.e. the dose becomes very large). Plotting the response curve against log-dose one usually arrives at a symmetric sigmoid curve in place of the asymmetric. Frequently the curve is characterized by four parameters: the response limits at zero and at infinite dose, the dose corresponding to the median response rate and the rate at which the response changes with respect to dose.

Since experimental dose response data often show outliers one has to use robust estimation procedures to minimize the influence of the outliers on the parameter estimates. Despite many dose response relationships can be analyzed by linearization of the data and therefore also by robust linear regression methods as the least median of squared residuals procedure there is a need for nonlinear modelling of the relationship because a mathematical model representing the statistical features of the experimental data should concentrate attention on the original observations rather than on derived functions such as linear transformations. Although the definitions of the least median of squared residuals or other robust estimators are easily extended to nonlinear regression, the analysis and computation of these estimators in the nonlinear case is far from trivial. The aim of the following paper is to present some robust statistical techniques to estimate nonlinear dose response relationsships.

2 Quantal Dose Response Assays

If the response is quantal, the organism is classified as having responded or not. In order to obtain repeatable and scientifically interpretable results a quantal response needs to be precisely defined as possible. Any object exposed has under controlled conditions a certain intensity with no effect at this level. Such a level is called a threshold or tolerance value. This tolerance value varies in a given population individually. At very low dose levels we will have no or only few individuals that react with the drug and at high levels the portion of effects tends to be near unity. Thus a statistical assessment of dose-response relations requires model assumptions on a underlying frequency or probability distribution of the portion $G(x)$ of individuals with a tolerance value of at most x. In many cases, $G(x)$ is given in terms of a probability density function $g(x)$ as follows:

$$G(x) = \Pr(\text{tolerance value} \leq x) = \int_{-\infty}^{x} g(y)\,dy. \tag{1}$$

The interpretation of the dose-response curve as a cumulative distribution of tolerances cannot be regarded as wholly satisfactory, since the evidence so far given for tolerances is indirect. There are, however, certain special circumstances in which tolerances, or something very like them, can be observed directly. We allow for the tolerance distribution cumulative distribution functions $G(x)$ with $G(-x) = 1 - G(x)$, which includes the commonly used probit or logit models. E.g. the probit model assumes that $G(x)$ depends on the logarithm of the dose by the standard normal distribution $\Phi(z)$:

$$G(x) = \Pr(\text{tolerance value} \leq x) = \begin{cases} \Phi(\beta + \gamma \lg x) & \text{for } x > 0 \\ 0 & \text{for } x = 0 \end{cases} \tag{2}$$

The parameter β determines that dose, on which 50% of the individuals react ($ED50 = 10^\beta$, "50% effective dose"). The parameter γ is a measure for the steepness of the curve at the point $\lg x = \beta$.

In the following we refer to a very general family of dose response models. Let d be the total number of administered doses and n_i the numbers of units treated with dose $x_i, i = 1, \ldots, d$. The number of responders r_i is then binomially distributed with some success probability p_i. We assume that the p_i depend on some unknown m-dimensional parameter vector $\theta = \theta(\beta, \gamma)$ according to

$$p_i = p_i^{\theta}(X) = \Pr(\text{Response at dose level } i | X, \theta) = G_i(X\theta) \qquad (3)$$

with the so-called link function G_i and a $d \times m$ "design matrix" X which consists of nonrandom quantities and describes the underlying experimental situation. The link function G_i is defined in terms of the distribution function G as follows:

$$G_i(x) = G(x_i) \qquad \text{for every } x = (x_1, \ldots, x_d). \qquad (4)$$

The concrete form of the design matrix depends on the problem under consideration. To illustrate this we assume that two treatment groups are under study, where treatment one has been administered at d_1 dose levels, treatment two at d_2 dose levels and $d = d_1 + d_2$. In a "general assay" $m = 4$ curve parameters have to be estimated (two for each group). The matrix X for such a general assay looks like

$$X = \begin{pmatrix} 1 & x_1 & 0 & 0 \\ \vdots & \vdots & \vdots & \vdots \\ 1 & x_{d_1} & 0 & 0 \\ 0 & 0 & 1 & x_{d+1} \\ \vdots & \vdots & \vdots & \vdots \\ 0 & 0 & 1 & x_d \end{pmatrix}, \qquad (5)$$

and the i-th element of $X\theta$ has the form

$$\begin{aligned} \beta_1 + \gamma_1 x_i, & \qquad \text{if } i \leq d_1, \\ \beta_2 + \gamma_2 x_i, & \qquad \text{if } i > d_1. \end{aligned} \qquad (6)$$

Generalization to more than two groups or to a parallel line assay, test of parallelism and so on is apparent. The advantage of such an "design matrix" approach is that many practical questions like the estimation of parameters and relative potencies or the testing of hypothesises can be answered within the same mathematical framework.

In standard situations the parameter estimation can be performed by using an iteratively reweighted least squares technique minimizing

$$\sum_{i=1}^{d} w_i \left(r_i / n_i - p_i^{\theta}(X) \right)^2, \qquad (7)$$

with weights $w_i = n_i/(p_i^\theta(X)(1 - p_i^\theta(X)))$. In the case of outliers in Y-direction robust methods like Huber, Tukey or a winsorized procedure similar to that described in [1] work in most cases.

However, these procedures fail if we have outliers in both, Y-direction and X-direction. Despite we mostly have not so many outliers that there is a need for high breakdown procedures, the probabilistic high breakdown methods known from linear regression work in a similar way. The situation in nonlinear quantal response estimation is rather similar to linear regression and different to general nonlinear regression situations. The following probabilistic method is similar to the algorithms one uses in linear regression to get least median of squares estimates.

(i) Perform a weighted ordinary least squares estimation to get estimates $\hat{\theta}^*$ as initial estimate and let med$\{res^*\}$ the median of the squared residuals med$_i res_i^2$, where $res_i = \hat{p}_i - r_i/n_i$.

Repeat (ii) – (v) k times (the index $j = 1, \ldots, k$ denotes the subsample):

(ii) Select randomly per treatment group at least two data point.

(iii) Estimate with these data points the parameter vector $\hat{\theta}$ according to the underlying model and experimental design situation X leading to estimate $\hat{\theta}_j$.

(iv) Compute the squared residuals $res_i^{2(j)}, i = 1, \ldots, d$ and med$_i res_i^{2(j)}$.

(v) Compare med$_i res_i^{2(j)}$ with med$\{res^*\}$ and if med$_i res_i^{2(j)} <$ med$\{res^*\}$ replace $\hat{\theta}^*$ by $\hat{\theta}_j$ and med$\{res^*\}$ by med$_i res_i^{2(j)}$.

(vi) Return $\hat{\theta}^*$ as final estimate and med$\{res^*\}$ as an approximate minimum of median$_i res_i^2$.

In contrast to general nonlinear regression situations where we generally have more data points, it is often possible to construct all combinations of subsets. The number of possible subset samplings is normally $n_s = \sum_l \binom{d_l}{2}$. If we have n_s too large with too many uninteresting and adverse data constellations the procedure can be improved by introducing a new variable Z with $Z_i = 1$ if $\hat{p}_i = r_i/n_i \geq 0.5$ and $Z_i = 0$ if $\hat{p}_i < 0.5$ and step (ii) of the algorithm has to be modified as follows: (ii,a): Select randomly per treatment group and per Z value at least one data point.

In case of using not all possible subsets a still better estimate can be found by using the Nelder-Mead simplex algorithm as described in [4]. Using $\hat{\theta}^*$ as starting value that algorithm minimizes med$_i res_i^2$. Furthermore the procedure can be improved by searching for the smallest strip that contains $[n/2] + 1$ data points.

3 Quantitative Dose-Response Relations

Many biological experiments with a continuous response variable like the radioimmunoassays, radioligand assays or immunoradiometric assays estimate the effect of a stimulus by the bounded or free fraction of radiation counts of the antigens or antibodies. Most of these experiments are perhaps not strictly bioassays, since they do not depend on responses measured in living organisms or tissues. But they are nevertheless so similar in structure that they need consideration from the viewpoint of bioassay.

We assume that the expectation of the response is given by a four parameter sigmoid function:

$$y = \delta_0 + \varrho G(\beta + \gamma \lg x) + \varepsilon_{ij}, \tag{8}$$

where x denotes the stimulus concentration and the error term ε_{ij} has mean 0 and variance σ_i^2. δ_0 and ϱ are additional parameters to be estimated that stand for the response level at zero dose and the range, respectively. β and γ are curve parameters that as well as the distribution function $G(x)$, $G_i(x)$ and the matrix X have a similar form and interpretation as in the quantal response case. Let be $\Theta = (\delta_0, \varrho, \theta)$ the parameter vector of interest. The asymptotic parameters δ_0 and ϱ are assumed to be independent from X.

In those cases where the observed responses are radioactive decay counts we can model the observed expectations by an overdispersed Poisson process (cf. [1]) with a variance larger than the normal Poisson variance. The marginal moments are of the form

$$y_i = \mathrm{E}[\text{counts} \mid X, \Theta] = \delta_0 + \varrho G_i(X\theta), \tag{9}$$
$$\mathrm{Var}[\text{counts} \mid X, \Theta] = y_i(1 + \kappa y_i) \tag{10}$$

Other variance proposals than that given in (10) are also conceivable. In standard situations using quasi-likelihood methods one can estimate the parameters Θ and the additional parameter κ in the variance by performing a three step generalized least squares (GLS) algorithm, whereas in the case of outliers in Y-direction a good practicable procedure is a winsorized version of the GLS algorithm (cf. [1]).

If we have outliers in both, Y-direction and X-direction we can proceed similar to the procedure in the quantal response case. In nonlinear quantitative response relationships we normally have only few data points per dose subgroup and minor contaminations. A total random selection of at least m data points per treatment group results again in too many uninteresting and adverse data constellations. The following probabilistic method is similar to the algorithm used in the quantal response case:

  (i) Perform the normal GLS estimation procedure to get estimates $\widehat{\Theta}^*$ as initial estimate and let med$\{RES^*\}$ the median of the squared residuals med$_i RES_i^2$, where $RES_i = \widehat{y}_i - y_i$.

Repeat (ii) – (v) for $j = 1, \ldots, k$:

 (ii) Select randomly per treatment group and dose subgroup at least one data point.

(iii) Estimate the parameter $\widehat{\Theta}_j$ according to the underlying model and experimental design situation X.

(iv) Compute the squared residuals $RES_i^{2(j)}$, $i = 1, \ldots, n$ and med$_i RES_i^{2(j)}$.

 (v) Compare med$_i RES_i^{2(j)}$ with med$\{RES^*\}$ and if med$_i RES_i^{2(j)} <$

med$\{RES^*\}$ replace $\hat{\Theta}^*$ by $\hat{\Theta}_j$ and med$\{RES^*\}$ by med$_i RES_i^{2(j)}$.

(vi) Return $\hat{\Theta}^*$ as final estimate and med$\{RES^*\}$ as an approximate minimum of median$_i RES_i^2$.

In case that we are using a random subset of all possible subset samplings of the data point combinations, we can improve the final estimate $\hat{\Theta}^*$ by using the Nelder-Mead simplex algorithm as mentioned above. The winsorized GLS of [1] and the above given probabilistic approach are of about the same computational effort, but sometimes the winsorized GLS converges very slowly and the probabilistic procedure is then the better alternative.

Instead of using a pregiven variance formula as given in (10) we can estimate the variance for each dose subgroup by a robust method as follows: Per treatment group and dose subgroup l the variances of the responses are measured by the estimator $Q_l = 2.2219\{|y_i - y_j|; i < j\}_{(k)}$ as described in [2] or [5]. This estimator has a simple formula and is affine equivariant. The value of the k-th order statistic is approximately $k \approx \binom{n}{2}/4$.

If the estimators Q_l are not in increasing order one can use the isotonic nearest neighbour method of [3] to assure that the variance estimators are proportional to the expectations. The resulting variance is then a nonparametric estimate that does not impose a particular functional model on the relationship between expected response and variance. In some general quantitative dose response relationships weighted least squares or the probabilistic approach above with these variance estimates might be an alternative to GLS or the winsorized GLS.

4 References

[1] H.-P. Altenburg (1992): Estimation of Radioimmunoassay Data Using Robust Nonlinear Regression Methods. In: Y. Dodge and J. Whittaker (eds.): *Computational Statistics I*, Physica, Heidelberg 1992, pp. 367–372

[2] C. Croux and P.J. Rousseeuw (1992): Time-Efficient Algorithms for Two Highly Robust Estimators of Scale. In: Y. Dodge and J. Whittaker (eds.): *Computational Statistics I*, Physica, Heidelberg 1992, pp. 411–428

[3] D.P. Normolle (1993): An Algorithm for Robust Nonlinear Analysis of Radioimmunoassays and other Bioassays. *Statistics in Medicine* 12, 2025–2042

[4] W.H. Press, B.P. Flannery, S.A. Teukolsky and W.T. Vetterling (1986): Numerical Recipes: The Art of Scientific Computing. Cambridge University Press, New York

[5] P.J. Rousseeuw and C. Croux (1993): Alternatives to the Median Absolute Deviation. *Journal of the American Statistical Association* 88, 1273–1283

[6] A.J. Stromberg (1993): Computation of High Breakdown Nonlinear Regression Parameters. *Journal of the American Statistical Association*, 88, 237–244

High Breakdown Regression by Minimization of a Scale Estimator

Christophe Croux, Peter J. Rousseeuw

Department of Mathematics and Computer Science, University of Antwerp (U.I.A.)

Abstract: In this note we discuss some high breakdown regression estimators which are based on the minimization of a scale estimator. We focus on the robustness and the computational aspects of these estimators.

Keywords: Computation, regression analysis, robustness

1 Introduction

In the linear regression model

$$y_i = \mathbf{x_i}'\beta + \sigma e_i \qquad i = 1, \dots, n,$$

where the errors e_i are independent and identically distributed, one wants to estimate the regression parameter β and the nuisance parameter σ. Suppose that the design points $\mathbf{x_i}$ belong to $I\!\!R^p$. Quite often β is determined such that a scale estimator applied to the residuals $r_i(\beta) = y_i - \mathbf{x_i}'\beta$ is minimized:

$$\hat{\beta} = \underset{\beta}{\mathrm{argmin}}\, s_n(r_1(\beta), \dots, r_n(\beta)). \qquad (1)$$

The scale estimator s_n has to be scale equivariant, that is $s_n(ax_1, \dots, ax_n) = |a|s_n(x_1, \dots, x_n)$ for every a. We will call the function $\beta \rightarrow s_n(r_i(\beta))$ the objective function of the minimization problem. A natural estimator of the parameter σ is then given by

$$\hat{\sigma} = s_n(r_1(\hat{\beta}), \dots, r_n(\hat{\beta})). \qquad (2)$$

If we choose $s_n = (\frac{1}{n}\sum_1^n r_i^2)^{1/2}$ we get the least squares estimator. It is well-known that this estimator is not robust, because it is unreliable in the presence of a single outlier. Also the least absolute values estimator, where $s_n = \frac{1}{n}\sum_1^n |r_i|$, does not protect against outliers in the x-direction (called leverage points).

In order to get a robust regression estimator, one has to take a robust scale estimator in (1). One of the first proposals (Rousseeuw 1984) was to take

$$s_n = \{|r_i|;\, 1 \leq i \leq n\}_{(h_p)},$$

where $h_p = [(n+p+1)/2]$ denotes 'half' of the number of observations. (The index $A_{(k)}$ will always denote the k-th order statistic out of the set A.) This yields the *least median of squares estimator* (LMS). Taking

$$s_n = \frac{1}{h_p} \sum_{k=1}^{h_p} \{r_i(\beta)^2; 1 \le i \le n\}_{(k)} \tag{3}$$

yields the *least trimmed squares estimator* (LTS). However, LMS and LTS have a gaussian efficiency not exceeding 8%. Therefore we proposed to use an estimator of Rousseeuw and Croux (1993) as objective function. This yields the *least quartile difference* (LQD) estimator:

$$\hat{\beta}_{LQD} = \underset{\beta}{\operatorname{argmin}} \{|r_i(\beta) - r_j(\beta)|; 1 \le i < j \le n\}_{\left(\binom{h_p}{2}\right)}. \tag{4}$$

One can see that the objective function is now the first quartile of the distances between the residuals. This estimator is discussed in Croux, Rousseeuw and Hössjer (1994). Another recent proposal is the *least trimmed median* (LTM) estimator of Croux, Rousseeuw and Van Bael (1993), defined as

$$\hat{\beta}_{LTM} = \underset{\beta}{\operatorname{argmin}} \frac{1}{h_p} \sum_{k=1}^{h_p} \{\underset{j}{\operatorname{median}} |r_i(\beta) - r_j(\beta)|; 1 \le i \le n\}_{(k)}, \tag{5}$$

where the median stands for the $([n/2] + 1)$-th order statistic out of the n numbers. The objective function is computed as follows: for each residual r_i one computes the median of the distances from the fixed r_i to all the other residuals. Out of these n median distances one computes the average of the h_p smallest ones. The LQD has a gaussian efficiency of 67% and the LTM of 22%.

Note that estimators of the type (1) don't need an initial regression estimator. Therefore they are well-suited to serve as initial estimators for one-step procedures as in Simpson, Ruppert and Caroll (1992) or Coakley and Hettmansperger (1993).

2 Robustness aspects

An important robustness measure of an estimator is its breakdown point. The breakdown point tells us how many observations have to be replaced by arbitrary values before the estimator can become arbitrary large. Formally, the breakdown point of the estimator T at the sample $Z = \{(\mathbf{x}_i, y_i); 1 \le i \le n\}$ is given by

$$\varepsilon_n^*(T, Z) = \min\{m/n; \sup_{Z'} |T(Z) - T(Z')| = \infty\}, \tag{6}$$

where Z' is obtained by replacing m values of Z by arbitrary ones. (Here $|.|$ stands for the euclidean norm.) It turns out that the four regression estimators considered here all have the maximal breakdown point

$$\varepsilon_n^* = \frac{[(n-p)/2]+1}{n},$$

which converges to 50% as $n \to \infty$. In fact, we have chosen h_p such that this maximal value would be reached.

The breakdown point tells us when an estimator can go to infinity, but not how large the estimate can become under a certain amount of contamination. Consider for example an estimator of the type (1) where

$$s_n = \text{average}\{r_i^2; |r_i| \le 10^{10} \text{median}(r_1, \ldots, r_n)\}$$

then this estimator has a 50% breakdown point, but in practice it will be as unsensitive to outliers as the Least Squares estimator. Therefore, it is also necesarry to look at the maxbias curves. It turns out that LMS has the smallest maxbias curve, closely followed by LTM and LQD, while the maxbias of LTS is substantially larger.

3 Computational aspects

To compute estimators of type (1) one has to minimize the objective function, which can be quite hard since this function does not need to be convex and can have local minima. The basic scheme for computing estimators of type (1) is the p-subset algorithm (Rousseeuw and Leroy 1987), which minimizes the objective function over all β_J which correspond to fitting a subset J with p observations (out of the n available points). Therefore we can use

$$\beta^* = \underset{\beta_J}{\text{argmin}}\, s_n(y_i - \beta_J' \mathbf{x}_i), \tag{7}$$

where β_J is determined by the p-subset J.

In practice, we mostly work with a model with intercept, where $\mathbf{x}_i = (\mathbf{u}_i, 1)$ and $\beta = (\gamma, \alpha)$, where γ is the slope parameter and α is the intercept parameter. For a scale estimator which is not location invariant (i.e., which does not have the property that $s_r(x_1 + b, \ldots, x_n + b) = s_n(x_1, \ldots, x_n)$ for all b) one can define the following location estimator:

$$t_s(y_1, \ldots, y_n) = \underset{\lambda}{\text{argmin}}\, s_n(y_i - \lambda).$$

Note that the objective functions of LMS and LTS are not location invariant. Now define the location invariant scale estimator $\tilde{s}$ as

$$\tilde{s}_n(r_1, \ldots, r_n) = s_n(r_i - t_s(r_1, \ldots, r_n)).$$

Table 1: Computation times in seconds for LMS, LTS, LQD, and LTM, where $p = 2$

n	LMS	LTS	LQD	LTM
10	1.0	1.3	2.2	1.3
20	7.4	9.9	21.5	10.5
40	57	85	214	94
60	208	309	807	337
80	493	760	2079	835
100	976	1520	4227	1689

It is now easy to see that the formula(1) in a model with intercept is equivalent with

$$\begin{cases} \hat{\gamma} &= \operatorname{argmin}_\gamma \tilde{s}_n(y_i - \gamma' u_i) \\ \hat{\alpha} &= t_s(y_i - \hat{\gamma}' u_i) \end{cases} \tag{8}$$

If s_n is already scale invariant (which is true for LQD and LTM) then take $\tilde{s}_n = s_n$ and t_s a high breakdown location estimator (e.g. the median) in (8).

When using the p-subset algorithm (7) we can also look at trial values $(\gamma_J, t_s(y_i - \gamma_J u_i))$ instead of (γ_J, α_J). This strategy is called intercept adjustment. It is often argued that intercept adjustment is necessary to get a good approximation for the estimator. A problem is that intercept adjustment needs more computation time, since one has to compute the estimator t_s in every step, which can be quite time consuming. Note that when $s_n = \tilde{s}_n$ we only have to compute t_s once, at the final stage.

If we use the efficient algorithm of Croux and Rousseeuw (1992) to compute the objective function of LQD or LTM then this objective function needs $O(n \log n)$ operations, yielding an overall computation time of $O(n^{p+1} \log n)$ if *all* p-subsets are considered. By comparison, the exhaustive p-subset algorithm for LMS needs $O(n^{p+1})$ time, and also needs $O(n^{p+1} \log n)$ if the intercept is adjusted in every step. In Table 1 we give computation times in seconds for LMS, LTS, LQD, and LTM with the exhaustive p-subset algorithm with intercept adjustment for $p = 2$. We see that although the order of computation time is the same for all the estimators ($O(n^3 \log n)$), there are large differences in computation time. Note that LTM is almost as fast to compute as LTS.

4 Generalizations of LQD and LTM

By generalizing LQD, we could look at the class of *generalized S-estimators* which have an objective function defined by

$$\binom{n}{2}^{-1} \sum_{i<j} \rho(\frac{r_i - r_j}{s_n}) = k, \tag{9}$$

where $k = E_\Phi[\rho(Y_1 - Y_2)]$ to get consistency for gaussian error distributions and $k/\rho(\infty) = 3/4$ to get a 50% breakdown point. The function ρ has to be even and increasing on the positive numbers. A typical example is the Tukey biweight, which is given by $\rho(x) = \min(3x^2/c^2 - 3x^4/c^4 + x^6/c^6, 1)$. We take c=0.9958 and k=0.75. The gaussian efficiency of this biweight GS-estimator is only slightly higher than that of LQD: 68%.

Computing the objective function of the biweight GS takes $O(n^2)$ operations (using a fixed number of iterations to solve equation (9)), which is more time consuming than the LQD. One can reduce the actual computation time, although it remains $O(n^2)$, in the following way. When considering m trial values β_J we don't need to compute $s_n(\beta_J)$ each time. Indeed, suppose that $\tilde{s}_n$ is the currently best scale. Generalizing an observation of Yohai and Zamar (1991), we then have

$$ s_n \leq \tilde{s} \quad \Leftrightarrow \quad \sum_{i<j} \rho(\frac{r_i - r_j}{\tilde{s}_n}) < k\binom{n}{2}. \tag{10} $$

Therefore, we only have to compute a new scale estimate when (10) holds. This happens $O(\log m)$ times. At each new best estimate $\tilde{\beta}$ it is possible to carry out some local improvement as in Ruppert (1992). The smoothness of our objective function indicates that Newton steps can be useful. The number m is obtained by a tradeoff between robustness and speed of computation. When computation time permits, carrying out the Newton steps at each β_J is even more accurate.

A generalization of LTM is the class of *Nested S-Estimators* (Croux, Rousseeuw, and Van Bael). A prototype example of a Nested S-Estimator is given by minimization of the scale estimator defined by

$$ \frac{1}{n} \sum_{i=1}^{n} \rho(\frac{H(r_i)}{s_n}) = k \tag{11} $$

where $H(r_i) = \text{median}_j |r_i - r_j|$. For ρ we can again take the Tukey Biweight, now with $c = 2.1300$ and $k = 0.5$. We can apply as before the "trick"

$$ s_n \leq \tilde{s}_n \quad \Leftrightarrow \quad \frac{1}{n} \sum_i \rho(\frac{r_i}{\tilde{s}_n}) < k. \tag{12} $$

In this way we obtain a 50% breakdown regression estimator with computation time comparable to the LTM, but with about 10% higher efficiency.

Remark: There also exist exact algorithms for LMS and LTS (see e.g. Stromberg 1993) and more sophisticated approximation algorithms for LMS and LTS (e.g. the feasible subset algorithm of Hawkins 1993). We did not include this in our study since we wanted to compare the different estimators using similar algorithms.

References

Coakley, C.W. and Hettmansperger, T.P. (1993), "A Bounded Influence, High Breakdown, Efficient Regression Estimator," *Journal of the American Statistical Association*, 88, 872–880.

Croux, C., and Rousseeuw, P.J. (1992), "Time-Efficient Algorithms for two Highly Robust Estimators of Scale," in *Computational Statistics, Volume 1*, eds. Y. Dodge and J. Whittaker, Heidelberg: Physika-Verlag, 411-428.

Croux, C., Rousseeuw, P.J., and Hössjer, O. (1994), "Generalized S-Estimators," to appear in *Journal of the American Statistical Association*, December issue.

Croux, C., Rousseeuw, P.J., and Van Bael, A. (1993), "Robust Regression by Minimizing Nested Scale Estimators," Technical Report 93–22, Universitaire Instelling Antwerpen. Under revision for *Journal of Statistical Planning and Inference*.

Hawkins, D.W. (1993), "The Feasible Set Algorithm for Least Median of Squares," *Computational Statistics & Data Analysis*, 16, 81–101.

Rousseeuw, P.J. (1984), "Least Median of Squares Regression," *Journal of the American Statistical Association*, 79, 871–880.

Rousseeuw, P.J., and Croux, C. (1993), "Alternatives to the Median Absolute Deviation," *Journal of the American Statistical Association*, 88, 1273-1283.

Rousseeuw, P.J., and Leroy, A.M. (1987), *Robust Regression and Outlier Detection*, New York: John Wiley.

Ruppert, D. (1992), "Computing S-Estimators for Regression and Multivariate Location/Dispersion," *Journal of Computational and Graphical Statistics*, 1, 253-270.

Simpson, D.G., Ruppert, D., and Carroll, R.J. (1992), "On One-Step GM Estimates and Stability of Inferences in Linear Regression," *Journal of the American Statistical Association*, 87, 439–450.

Stromberg, A.J. (1993), "Computing the Exact Least Median of Squares Estimate and Stability Diagnostics in Multiple Linear Regression," *SIAM Journal of Scientific and Statistical Computing*, 14, November issue.

Yohai, V.J., and Zamar, R.H. (1991), "Discussion of 'Least Median of Squares Estimation in Power Systems' by Mili, L., Phaniraj, V., and Rousseeuw, P.J.," *IEEE Transactions on Power Systems*, 6, 520.

Using GLIM4 to Estimate the Tuning Constant for Huber's M-estimate of Location

Robert Gilchrist, **George Portides**

University of North London

Abstract

Probably the best well known technique for the resistant estimation of a location parameter is Huber's (1964) proposal for an M-estimate of location for the ε-contaminated Normal distribution. However, a disadvantage of this technique is the assumption that the amount of contamination, and hence the so called tuning constant is known, a priori. In the present paper the authors present Huber's proposal and how it can be implemented using the iteratively re-weighted least squares (IRLS) procedure used in GLIM4. A proposal for estimating the mentioned tuning constant is presented and illustrated by some simulations using the GLIM4 package. The authors discuss generalisations of this method (i) in the GLM context and (ii) to the joint M-estimation of location and scale.

1 Introduction

Several proposals have been suggested for the resistant estimation of a location parameter, probably the best well known being Huber's proposal for an M-estimate of location for the ε-contaminated Normal distribution; see Huber (1981). However, the use of this technique is disadvantaged by the assumption that the amount of contamination, and hence the so called tuning constant is known, a priori. This is highly unlikely in practice, even though this problem can slightly be overcome through a search for spurious observations in a data set. Note, however, that any model fitting should naturally be accompanied by model checking; see for instance Atkinson (1985), Gilchrist (1993). In the present paper the authors present Huber's proposal (Section 2) and how it can be implemented in the iteratively re-weighted least squares (IRLS) procedure used in GLIM4 (Section 3). In Section 4, a proposal for estimating the tuning constant used in Huber's M-estimate of location is presented and illustrated by some simulations using the GLIM4 package. In Section 5, the authors discuss generalisations of this method (i) in the GLM context and (ii) to the joint M-estimation of location and scale.

2 Huber's M-estimate of location

It is assumed that data $y=\{y_i\}$, i=1,2,...,n, arise from a common distribution $F\big((y_i-\mu)/\sigma\big)$, where μ is the unknown location parameter to be estimated. Here, σ^2 is the scale parameter which we shall assume to be known. An additional problem is that the form of F is partially known. More specifically, the distribution F is assumed to be of the form $F(z_i)=(1-\varepsilon)\Phi(z_i)+\varepsilon H(z_i)$, where $z_i=(y_i-\mu)/\sigma$, Φ is the standard Normal cumulative distribution function, H is an unknown contaminating symmetric distribution and ε, assumed to be known, is the proportion of the y_i's arising from this contaminated distribution. For example, 90% of the observations in a sample are assumed to be Normal with known variance and unknown mean, but the remaining 10% are affected by gross errors. For simplicity, we shall be using z_i as defined above, but where it is applicable we shall use y_i.

Huber (1964), introduced the use of an M-estimate of location for the estimation problem illustrated above. This is defined as a minimisation problem of the form

$$\min_{\mu}\sum\rho(z_i), \quad \text{where} \quad \rho(z_i)=\begin{cases} z_i^2/2 & ,\ z_i^2 \le k^2, \\ k|z_i|-k^2/2 & ,\ \text{otherwise.} \end{cases} \tag{1}$$

The scalar k is the so called *tuning constant*, and it is connected to the amount of contamination, ε, through equation (6) given further down this section. The variance, σ^2, is assumed either to be known, or estimated and fixed prior to any resistant estimation of μ. If σ^2 is not known, it is suggested that either (i) the median absolute deviation from the median (MAD) or (ii) the unweighted least squares estimate under Normal errors be used as an auxiliary estimate of scale. In particular, Huber suggests the use of MAD,

$$\sigma_{MAD}=\text{med}\big\{|y_i-\text{med}\{y_i\}|\big\}, \tag{2}$$

as such an estimate of scale has better breakdown properties under ε-contamination. Having assigned a value to σ^2, we consider only the case where it is kept fixed whatever value of the tuning constant used; i.e. our discussion here deals only with a resistant estimation of location.

Now, the minimisation problem given in (1) can instead be characterised as the solution of

$$\sum\psi(z_i)=0, \quad \text{where} \quad \sigma\psi(z_i)=\sigma\frac{d\rho}{d\mu}=\begin{cases} -k & ,\ \text{if } z_i<-k, \\ z_i & ,\ \text{if } |z_i|<k, \\ k & ,\ \text{if } z_i>k. \end{cases} \tag{3}$$

The solution of (3) is then given analytically as

$$\hat{\mu} = \tilde{y} + \frac{k\sigma(n_3 - n_1)}{n_2} \ , \qquad (4)$$

where n_1, n_2 and n_3 are the number of observations in the intervals $y_i < -k\sigma +\mu$, $-k\sigma +\mu \le y_i \le k\sigma +\mu$ and $y_i > k\sigma +\mu$ respectively, and $\tilde{y}$ is the simple arithmetic mean of the y_i's belonging in the second interval. One can iterate this procedure until convergence by updating the intervals with the new estimates of μ. For proof of convergence, see Huber (1981), pp 179-186.

This procedure actually arises from the maximum likelihood estimate of μ under the ε-contaminated Normal distribution least informative for location. Huber defines this as

$$f(y_i) = \frac{1-\varepsilon}{\sqrt{2\pi\sigma^2}} \exp\left[-\rho\left(\frac{y_i - \mu}{\sigma}\right)\right], \qquad \forall \ y_i \in (-\infty, \infty) \ , \qquad (5)$$

for ρ as given in (1). In other words, the density is assumed to be Normal in the centre and double exponential in the tails. In order for f to integrate to one, the values of k and ε are connected by

$$\frac{2\varphi(k)}{k} - 2\Phi(-k) = \frac{\varepsilon}{1-\varepsilon}, \qquad (6)$$

where $\varphi = \Phi'$ is the standard Normal density.

Hence any arbitrary choice of k implies an assumed proportion of contaminated data in y, e.g. a choice of $k=1.40$ implies 5% contaminated data in a sample. In practice, knowledge of the amount of contamination in a data set is highly unlikely, and a wrong choice of k can result in a misleading estimate of μ. The authors propose a technique for estimating the tuning constant when σ^2 is known. This is illustrated in the following sections.

3 Using IRLS to compute M-estimates of location

An equivalent approach to (3) when iterated to convergence is *iteratively re-weighted least squares* (IRLS). For the problem being dealt with, the IRLS approach uses an assumption of Normality for all data y, but assigns prior weights to the y_i's at each iteration of the fitting cycle of the form

$$w_i^{(j)} = \psi\left(z_i^{(j-1)}\right)\big/ z_i^{(j-1)} = \begin{cases} 1 & , \ \text{if } \left|z_i^{(j-1)}\right| < k, \\ k\big/\left|z_i^{(j-1)}\right| & , \ \text{otherwise}, \end{cases} \qquad (7)$$

where the superscripts correspond to the computed quantities at the jth iteration, i.e. $z_i^{(j-1)} = \left(y_i - \mu^{(j-1)}\right)\big/\sigma$ and $w_i^{(0)} = 1$. In other words the equation given in (3) is expressed as

$$\sum w_i^{(j)}(y_i - \mu) = 0. \qquad (8)$$

When convergence of this procedure occurs, then the estimate of μ is equal to the converged estimate of μ obtained by iterating (3).

This procedure can be implemented in GLIM4 by updating either the iterative or the prior weights at the last stage of the fitting cycle. In both cases this is easily done by the new GLIM4 directive $METHOD * * macro$. The authors suggest the use of the latter way as this will force GLIM to update automatically all other quantities in the fitting cycle.

4 Estimating the Tuning Constant

4.1 Method

It is again emphasised that IRLS is only equivalent to the maximum likelihood estimate of μ given in (3) when iterated to convergence. Now, if it is assumed that convergence of the IRLS procedure occurs in m finite iterations, then the final form of the underlying likelihood maximised in order to evaluate the converged estimate of the mean, is equivalent to the Normal density, $N\left(\mu, \sigma^2/w_i^{(m-1)}\right)$ which we write as f^*, where the $w_i^{(m-1)}$ are the prior weights computed at the end of the $(m-1)^{th}$ iteration as defined in (7). However, convergence of the mean immediately implies convergence of the prior weights. Hence the $w_i^{(m-1)}$ can equivalently be expressed as functions of y_i and μ and for simplicity written as w_i referring only to the prior weights assigned a value less than one. Therefore, f^* should behave as a likelihood where asymptotically the w_i's possess the known functional relation with the y_i and μ but are still treated as prior weights in the maximum likelihood estimation of the mean. In this case f^* does not behave as a proper likelihood with the desired properties necessary for our estimation problem. We therefore define the slightly modified likelihood f^{**}, by

$$f^{**}(y_i) = \begin{cases} \dfrac{1}{\sqrt{2\pi\sigma^2}} \exp\left(-z_i^2/2\right) & , \text{if } \left|z_i^2\right| \le k \\[3mm] \dfrac{1}{2\sqrt{2\pi\sigma^2/w_i}} \exp\left(-w_i z_i^2/2\right) & , \text{otherwise.} \end{cases} \tag{9}$$

We note that this modification does not depend on the tuning constant. We now have $\int_{-\infty}^{\infty} f^{**}(y)dy = 1$ and $E(y) = \int_{-\infty}^{\infty} y f^{**}(y)dy = \mu$.

Thus, since f^{**} is uniquely defined for different values of k, the authors propose the use of $f^{**}(y_i)$ or more precisely the use of $D_k = -2\sum \log f^{**}(y_i)$ as a criterion for choosing the "best" tuning constant. In other words, the profile D_k is computed over a possible range of k, and the tuning constant which gives the minimum D_k is chosen. To allow for possible multiple local minima, it is advisable to compute D_k over a large range of k.

4.2 Simulation

The technique is illustrated by some simulations using the GLIM4 package. The underlying distribution is the standard Normal contaminated by different levels, ε, from a Normal, mean 5 and variance 1, distribution. The results shown in Table 1 are based on the assumption that the variance is known and equal to the true value, i.e. $\sigma^2=1$. Further such simulations are available from the authors.

Table 1. Comparison of estimated with correct tuning constant. n=100

ε	k	$\hat{\varepsilon}$	$\hat{k}$	bias $\hat{k}$	mse $\hat{k}$	$\hat{\mu}_{\hat{k}}$	mse $\hat{\mu}_{\hat{k}}$	$\hat{\mu}_{LS}$	mse $\hat{\mu}_{LS}$
.05	1.40	.046	1.426	.026	.118	.084	.019	.249	.082
			(.015)	(.015	(.017)	(.005)	(.001)	(.006)	(.003)
.10	1.14	.116	1.082	-.058	.057	.177	.046	.515	.296
			(.010)	(.010)	(.003)	(.005)	(.002)	(.008)	(.009)
.20	.86	.277	.720	-.140	.040	.368	.156	1.025	1.095
			(.009)	(.009)	(.003)	(.009)	(.007)	(.023)	(.028)
.30	.69	.385	.568	-.122	.025	.615	.415	1.502	2.307
			(.006)	(.006)	(.001)	(.012)	(.016	(.014)	(.043)

Note: i) k is the true tuning constant given ε amount of contamination, ii) $\hat{\varepsilon}$ is computed using $\hat{k}$ and equation (6), iii) $\hat{\mu}_{\hat{k}}$ (=bias($\hat{\mu}_{\hat{k}}$)) is the resistant estimate of the mean using $\hat{k}$, iv) $\hat{\mu}_{LS}$ (=bias($\hat{\mu}_{LS}$)) is the unweighted least squares estimate of the mean, iv) values in parentheses correspond to standard errors.

5. Generalisations

5.1 Generalised linear models

The method presented in this paper can easily be generalised to the case of Normally distributed y, where the mean is to be estimated using p explanatory variates with any suitable link function. However, in the non-Normal error case as applied in the GLM context, a possible modification to the M-estimate of location criterion can be to replace $\rho\left((y_i - \mu)/\sigma\right)$ given in (1) by $\rho\left(\sqrt{d_i/\phi}\right)$, where the d_i's are the deviance residuals of each observation and ϕ is the known scale parameter; see Pregibon (1979). A possible criterion for estimating the tuning constant could be the maximisation of the extended quasi-likelihood corresponding to the underlying likelihood used in the IRLS procedure; see Gilchrist & Portides (1994).

5.2 Joint M-estimates of Location and Scale

For the case where the underlying distribution is Normal, and the estimation problem is as defined in Huber's proposal for joint M-estimates of location and scale; see Huber (1981), the IRLS procedure can be used for the estimation of the mean only, as illustrated in this paper . Incorporating the estimation of scale in the IRLS approach and the estimation of the tuning constant for this technique is currently being studied.

References

Atkinson, A.C. (1985), *Plots, Transformations and Regression*. Oxford: Oxford University Press.

Gilchrist, R. (1993), In *The GLIM System, Release 4 Manual*, Chapter 10: Guide to Statistical Modelling in GLIM, and Chapter 11: The Theory of Generalised Linear Models. Oxford: Oxford University Press.

Gilchrist, R. and Portides, G. (1994), Resistant fitting in GLIM4, Submitted to *GLIM Newsletter*.

Green, P.J. (1984), Iteratively re-weighted least squares for maximum likelihood estimation, and some robust resistant alternatives, *J. R. Statist. Soc.* **B 46**, 149-192.

Huber, P.J. (1964), Robust Estimation of a location parameter, *Ann. Math. Statist.* **35**, 73-101.

Huber, P.J. (1981), *Robust Statistics*, New York: John Wiley.

Nelder, J.A. and Pregibon, D. (1987), An Extended Quasi-likelihood Function, *Biometrika* **74**, 221-232.

Pregibon, D. (1979), Unpublished PhD. thesis. University of Toronto.

On the Calculation of MSE Minimizing Robust Estimators

Christine H. Müller

Freie Universität Berlin, Fachbereich Mathematik und Informatik, WE 1

Abstract. A conditionally contaminated linear model $y = x^T\beta + z$ is considered where the errors z may have different contaminated normal distributions for different experimental conditions x. At this model one-step M-estimators have an asymptotic bias which should be bounded for robust estimators. For designs which have a support of linearly independent regressors, an algorithm for the calculation of one-step M-estimators which minimize the asymptotic mean squared error (MSE) is presented. The results of this algorithm are given for quadratic regression and for a problem in a one-way lay-out model.

Keywords. Linear model, robust estimation, one-step M-estimator, mean squared error, MSE minimizing estimator, linearly independent regressors.

1   The model

We consider a general linear model

$$y_N = X_N\beta + z_N,$$

where $y_N = (y_{1N}, ..., y_{NN})^T \in I\!\!R^N$ is the vector of observations, $X_N = (x_{1N}, ..., x_{NN})^T \in I\!\!R^{N \times p}$ is the known design matrix, $\beta \in I\!\!R^p$ is the unknown parameter vector and $z_N = (z_{1N}, ..., z_{NN})^T \in I\!\!R^N$ is the vector of errors. Often it is assumed that the errors are normally distributed, i.e. $z_{nN}/\sigma \sim n_{(0,1)}$. But if some outlying observations (gross errors) may appear then this normal linear model is not anymore correct. Then a conditionally contaminated linear model is more adequate. In a conditionally contaminated linear model the errors have a contaminated normal distribution where the proportion of contamination and the form of the contamination may be different for different experimental conditions, i.e. the errors are distributed according to (see Bickel 1984, Rieder 1987, 1993, Müller 1992)

$$\frac{z_{nN}}{\sigma} \sim (1 - N^{-1/2}\epsilon(x_{nN}))n_{(0,1)}(dz) + N^{-1/2}\epsilon(x_{nN})\, k(dz, x_{nN}).$$

2 One-step M-estimators

To estimate a linear aspect $\varphi(\beta) = C\beta$, $C \in \mathbb{R}^{s \times p}$, of β in a contaminated linear model one-step M-estimators can be used. These estimators have the form

$$\hat{\varphi}_N(y_N)$$

$$= C\hat{\beta}_N^0(y_N) + \frac{1}{N}\sum_{n=1}^N \dot{\psi}\left(\frac{y_{nN} - x_{nN}^T\hat{\beta}_N^0(y_N)}{\hat{\sigma}_N(x_{nN})(y_N)}, x_{nN}\right)\hat{\sigma}_N(x_{nN})(y_N),$$

where $\hat{\beta}_N^0$ is some initial estimator and $\hat{\sigma}_N(x_{nN})$ is some scale estimator. Because we will assume only a finite number of different experimental conditions x the variance estimator may be different for different experimental conditions. If C is the identity matrix $I_{p \times p}$ then the ordinary least squares estimator and the M-estimators are one-step M-estimators.

Now assume that the designs $X_N = (x_{1N}, ..., x_{NN})^T$ are converging to an asymptotic design measure δ with finite support $\mathrm{supp}(\delta)$ in the following sense: $\lim_{N \to \infty} \frac{1}{N}\sum_{n=1}^N e_{x_{nN}}(\{x\}) = \delta(\{x\})$ for all $x \in \mathrm{supp}(\delta)$, where e_x denotes the dirac measure on x. Then under some regularity conditions one-step M-estimators are asymptotically normally distributed at the conditionally contaminated linear model, i.e. (see Bickel 1975, 1984, Rieder 1987, 1993, Müller 1992)

$$\mathcal{L}(\sqrt{N}(\hat{\varphi}_N - \varphi(\beta)) \overset{N \to \infty}{\longrightarrow} \mathcal{N}(\sigma\,b(\psi, \epsilon, k), \sigma^2 V(\psi)), \tag{2.1}$$

with $V(\psi) := \int \psi(z, t)\psi(z, t)^T n_{,0,1)}(dz)\,\delta(dx)$ and maximum asymptotic bias

$$\sup_{\epsilon, k} |b(\psi, \epsilon, k)| = \|\psi\|_\infty. \tag{2.2}$$

Here we judge the one-step M-estimators by this asymptotic behaviour which only depends on the score functions ψ and does not depend on the initial estimators and the scale estimators. Hence a one-step M-estimator is called robust if its score function ψ is bounded, and in the following we regard the problem of the calculation of an optimal choice of the score function ψ. The problem of choosing the initial estimator and the scale estimator will be treated elsewhere. Note only that one-step M-estimators with the least squares estimator as initial estimator also can have good robustness properties at finite samples if the score function is bounded (see Müller 1993).

3 MSE minimizing estimators

From (2.1) and (2.2) we get that the maximum asymptotic mean squared error (MSE) of an one-step M-estimator at the contaminated model is (see also Samarov 1985)

$$\sup_{\epsilon, k} \sigma^2\left(b(\psi, \epsilon, k)^2 + \mathrm{tr}\,V(\psi)\right) = \sigma^2\left(\|\psi\|_\infty^2 + \mathrm{tr}\,V(\psi)\right).$$

Hence a one-step M-estimator is called MSE minimizing estimator if its score function satisfies

$$\psi^* = \arg\min\{\|\psi\|_\infty^2 + \operatorname{tr} V(\psi); \ \psi \in \Psi\},$$

where Ψ is the set of all score functions which provide the asymptotic behaviour (2.1). This MSE minimizing score function can be obtained by determing

$$b^* = \arg\min\{b^2 + \operatorname{tr} V(\psi_b); \ b \geq b_{min}\}, \text{ where} \qquad (3.3)$$

$b_{min} = \arg\min\{\|\psi\|_\infty; \ \psi \in \Psi\}$ and
$\psi_b = \arg\min\{\operatorname{tr} V(\psi); \ \psi \in \Psi \text{ and } \|\psi\|_\infty \leq b\}.$

Note for $C = I_{p \times p}$ and $b > b_{min}$ the score function ψ_b is the score function of the Hampel-Krasker estimator (see Hampel 1978, Krasker 1980). Because always $\|\psi_b\|_\infty = b$ we have $\psi^* = \psi_{b^*}$. so that this equality can be used for determing the MSE minimizing score function. In general this is a very difficult task because ψ_b depends on an implicit matrix and also b_{min} is implicitly given (see Müller 1987, Kurotschka and Müller 1992). But for the following special design situation the problem can be solved:

Assume that $\operatorname{supp}(\delta) \subset d = \{x_1, ..., x_I\}$ where $x_1, ..., x_I$ are linearly independent, i.e.

$$X = (x_1, ..., x_I)^T \text{ has rank } I \leq p. \qquad (3.4)$$

Then the linear aspect $\varphi(\beta) = C\beta$ is asymptotically identifiable if $C = K\,I(\delta)$ where $I(\delta) := \int x\,x^T\,\delta(dx)$ and we have (see Müller 1987, Kurotschka and Müller 1992):

$$b_{min} = \max_{x \in d} |C\,I(\delta)^- x| \sqrt{\frac{\pi}{2}} \text{ and}$$

$$\psi_b(z, x) = \begin{cases} C\,I(\delta)^- x \ \operatorname{sgn}(z) \sqrt{\frac{\pi}{2}} & \text{for } b = |C\,I(\delta)^- x| \sqrt{\frac{\pi}{2}} \\ C\,I(\delta)^- x \ \frac{\operatorname{sgn}(z)\,\min\{|z|,b\,y_b(x)\}}{|C\,I(\delta)^- x|\,y_b(x)} & \text{for } b > |C\,I(\delta)^- x| \sqrt{\frac{\pi}{2}} > 0 \quad (3.5) \\ 0 & \text{otherwise} \end{cases}$$

for $b \geq b_{min}$, where

$$y_b(x) = (2\Phi(b\,y_b(x)) - 1)\,\frac{1}{|C\,I(\delta)^- x|} > 0.$$

4 Calculation of ψ_b, in particular of $y_b(x)$

Because under the assumption (3.4) ψ_b depends on the fixed points $y_b(x)$ we need a method for determing the solution of

$$y_b = (2\Phi(b\,y_b) - 1)a > 0 \text{ with } b > \frac{1}{a}\sqrt{\frac{\pi}{2}},$$

i.e. the root $y_b = f(a, b)$ of $F(y) := (2\Phi(b\,y) - 1)a - y$. Because of $F'(0) > 0 \Leftrightarrow b > \frac{1}{a}\sqrt{\frac{\pi}{2}}$, $F'(a) < 0$ and $F''(y) < 0$ for all $y > 0$ this can be done by Newton's method:

Algorithm
- Start: $y_1 = a$.
- Iteration: $y_{n+1} = y_n - \frac{(2\Phi(b\,y_n) - 1)a - y_n}{2\,a\,b\,\Phi'(b\,y_n) - 1}$.

5 Calculation of the optimal bias bound b^*

Because the MSE minimizing score function satisfies $\psi^* = \psi_{b^*}$ where b^* is given by (3.3) the next step is to calculate b^*. Under the assumption (3.4) we have (see Müller 1987, 1994):

$$\operatorname{tr} V(\psi_b) = \sum_{x \in d} \delta(\{x\})\, t(|C\,I(\delta)^- x|^{-1}, b)$$

where

$$t(a, b) := \begin{cases} \frac{1}{a^2}\frac{\pi}{2} & \text{for } b = \frac{1}{a}\sqrt{\frac{\pi}{2}} \\ \frac{g(b\,f(a,b))}{f(a,b)^2} & \text{for } b > \frac{1}{a}\sqrt{\frac{\pi}{2}} \end{cases}, \quad \text{and } g(y) := \int \min\{|z|, y\}^2\, n_{(0,1)}(dz).$$

The first and the second derivative of $t(a, b)$ with respect to b are known (see Müller 1987) and satisfy $\frac{\partial}{\partial b}t(a, b) < 0$, and $\frac{\partial^2}{(\partial b)^2}t(a, b) > 0$. If we set $T(b) := b^2 + \sum_{x \in d} \delta(\{x\})\, t(|C\,I(\delta)^- x|^{-1}, b)$, then b^* is a root of $T'(b)$ and the following algorithm based on Newton's method can be used.

Algorithm
1st step:
- Start: $b_1 = 2\,b_{min}$.
- Do until $(T'(b_l) < 0)$: $b_{l+1} = (1 + (\frac{1}{2})^{l+1})\,b_{min}$.
2nd step:
- Start: $b_1 = b_l$.
- Do until $(T'(b_n) < \epsilon)$: $b_{n+1} = b_n - \frac{T'(b_n)}{T''(b_n)}$.

Because $T'(b_{min}) = -\infty$ in the first step of the algorithm a start value for the Newton algorithm is calculated. Although it is not known if $T'(b)$ is really a convex function the Newton method with this start value - which always is less than b^* - worked in all investigated examples very well where in both steps less than 10 iterations were needed. In particular for the special case of A-optimal designs the algorithm provided the same results as were obtained in Müller (1987, 1993b) by choosing the start value by some trials. In the following we give some applications of the algorithm for designs which are not A-optimal but satisfy condition (3.4). Thereby the algorithm provides not only the optimal b^* but also the mean squared error $MSE(b^*) = (b^*)^2 + \operatorname{tr} V(\psi_{b^*})$ for the optimal b^* and the values of all $y_{b^*}(x)$ so that ψ_{b^*} can be calculated according to (3.5).

6 Examples

Quadratic regression: Consider a quadratic regression model

$$y(t) = \beta_0 + \beta_1 t + \beta_2 t^2 + z = x(t)^T \beta + z, \qquad t \in [-1, 1],$$

where $x(t) = (1, t, t^2)^T$, $\beta = (\beta_0, \beta_1, \beta_2)^T$. If the observations are made at $t = 1, -1, 0$ with equal number of repetitions and β shall be estimated then we have the following input:

Input:

$$X = (x_1, x_2, x_3)^T = \begin{pmatrix} 1 & 1 & 1 \\ 1 & -1 & 1 \\ 1 & 0 & 0 \end{pmatrix}, \qquad \begin{pmatrix} \delta(\{x_1\}) \\ \delta(\{x_2\}) \\ \delta(\{x_3\}) \end{pmatrix} = \frac{1}{3} \begin{pmatrix} 1 \\ 1 \\ 1 \end{pmatrix},$$

$$C = \begin{pmatrix} 1 & 0 & 0 \\ 0 & 1 & 0 \\ 0 & 0 & 1 \end{pmatrix}.$$

Then the algorithm provides the following output:

Output:

$$b^* = 5.3542, \qquad \text{MSE}(b^*) = (b^*)^2 + \text{tr } V(\psi_{b^*}) = 40.199,$$

$$\begin{pmatrix} y_{b^*}(\{x_1\}) \\ y_{b^*}(\{x_2\}) \\ y_{b^*}(\{x_3\}) \end{pmatrix} = \begin{pmatrix} 0.4654 \\ 0.4654 \\ 0.0381 \end{pmatrix}.$$

One-way lay-out with 3 levels and one control level: In a one-way lay-out with 4 levels observations are given by

$$y_i = \mu_i + z = x(i)^T \beta + z, \qquad i \in \{1, 2, 3, 4\},$$

where $x(i) = (1_1(i), 1_2(i), 1_3(i), 1_4(i))^T$, $\beta = (\mu_1, \mu_2, \mu_3, \mu_4)^T$. If the first level is the control level an interesting aspect of β is $C\beta = (\mu_2 - \mu_1, \mu_3 - \mu_1, \mu_4 - \mu_1)^T$ and if the observations are made at $i = 1, 2, 3, 4$ with equal number of repetitions then we have the following input:

Input:

$$X = (x_1, x_2, x_3, x_4)^T = \begin{pmatrix} 1 & 0 & 0 & 0 \\ 0 & 1 & 0 & 0 \\ 0 & 0 & 1 & 0 \\ 0 & 0 & 0 & 1 \end{pmatrix}, \qquad \begin{pmatrix} \delta(\{x_1\}) \\ \delta(\{x_2\}) \\ \delta(\{x_3\}) \\ \delta(\{x_4\}) \end{pmatrix} = \frac{1}{4} \begin{pmatrix} 1 \\ 1 \\ 1 \\ 1 \end{pmatrix},$$

$$C = \begin{pmatrix} -1 & 1 & 0 & 0 \\ -1 & 0 & 1 & 0 \\ -1 & 0 & 0 & 1 \end{pmatrix}.$$

Then the algorithm provides the following output:

Output:

$$b^* = 8.7213, \qquad \text{MSE}(b^*) = (b^*)^2 + \text{tr } V(\psi_{b^*}) = 105.53,$$

262

$$\begin{pmatrix} y_{b^*}(\{x_1\}) \\ y_{b^*}(\{x_2\}) \\ y_{b^*}(\{x_3\}) \\ y_{b^*}(\{x_4\}) \end{pmatrix} = \begin{pmatrix} 0.0186 \\ 0.2411 \\ 0.2411 \\ 0.2411 \end{pmatrix}.$$

References.

BICKEL, P.J. (1975). One-step Huber estimates in the linear model. *J. Amer. Statist. Assoc.* **70**, 428-434.

BICKEL, P.J. (1984). Robust regression based on infinitesimal neighbourhoods. *Ann. Statist.* **12**, 1349-1368.

HAMPEL, F.R. (1978). Optimally bounding the gross-error-sensitivity and the influence of position in factor space. *Proceedings of the ASA Statistical Computing Section*, ASA, Washington, D.C., 59-64.

KRASKER, W.S. (1980). Estimation in linear regression models with disparate data points. *Econometrica* **48**, 1333-1346.

KUROTSCHKA, V. and MÜLLER, Ch.H. (1992). Optimum robust estimation of linear aspects in conditionally contaminated linear models. *Ann. Statist.* **20**, 331-350.

MÜLLER, Ch.H. (1987). Optimale Versuchspläne für robuste Schätzfunktionen in linearen Modellen. *Ph. D. thesis.* Freie Universität Berlin.

MÜLLER, Ch.H. (1992). One-step-M-estimators in conditionally contaminated linear models. *Preprint No. A-92-11*, Freie Universität Berlin, Fachbereich Mathematik. *Submitted to Stat. Decis.*

MÜLLER, Ch.H. (1993a). Behaviour of asymptotically optimal designs for robust estimation at finite sample sizes. In: *Model-Oriented Data Analysis*, eds. W.G. Müller, H.P. Wynn, A.A. Zhigljavsky, 53-62, Physica-Verlag, Heidelberg.

MÜLLER, Ch.H. (1993b). Optimal bias bounds for robust estimation in linear models. *Preprint No. A-35-93*, Freie Universität Berlin, Fachbereich Mathematik und Informatik. *To appear in the Proceedings of the LINSTAT'93 Conference, Poznań.*

MÜLLER, Ch.H. (1994). Optimal designs for robust estimation in conditionally contaminated linear models. *J. Statist. Plann. Inference.* **38**, 125-140.

RIEDER, H. (1987). Robust regression estimators and their least favorable contamination curves. *Stat. Decis.* **5**, 307-336.

RIEDER, H. (1993). *Robust Asymptotic Statistics*, Springer, New York, to appear.

SAMAROV, A.M. (1985). Bounded-influence regression via local minimax mean squared error. *J. Amer. Statist. Assoc.* **80**, 1032-1040.

Robust Fitting of an Additive Model for Variance Heterogeneity

Robert A. Rigby, Mikis D. Stasinopoulos

School of Mathematical Sciences, University of North London

Keywords: Normal model, Variance heterogeneity, Generalised Additive Models, Penalised likelihood, Cubic splines, Robust smoothing.

1 Introduction

Here we consider robust fitting of an Additive model for the variance heterogeneity in a Normal error model. Harvey (1978) and Aitkin (1987) have modelled the variance function explicitly using parametric linear models for both the mean and the variance. The use of a non-parametric function of a single explanatory variable for the variance model was introduced by Silverman (1985) and Muller and Stadtmuller (1987).

Rigby and Stasinopoulos (1993) formalised this to an arbitrary number of regressor variables by modelling the variance using a semi-parametric Additive model (Hastie and Tibshirani (1990)). Our model assumes that the response variable y_i is independently Normally distributed with mean μ_i and variance σ_i^2, for $i = 1,2,...,n$. The mean and variance are known functions of the predictors η_i and ξ_i respectively, and are modelled, for $i = 1,2,...,n$, by :

$$g_1(\mu_i) = \eta_i = \sum_{j=1}^{p} f_j(x_{ij}) \tag{1a}$$

$$g_2(\sigma_i^2) = \xi_i = \sum_{k=1}^{q} h_k(z_{ik}) \tag{1b}$$

where f_j is either a linear or a non-parametric function of explanatory variable x_j for the mean, for $j = 1,2,..,p$ and h_k is either a linear or a non-parametric function of explanatory variable z_k for the variance, for $k = 1,2,..,q$. The x's and the z's are assumed to be fixed, and g_1 and g_2 are monotonic link functions (typically the identity and log functions respectively). Further, let $\mu, \sigma^2, \eta, \xi, x_j, z_k, f_j(x_j)$, and $h_k(z_k)$ be the coresponding vectors of length n. Model (1) is very flexible and allows a wide range of parametric, semi-parametric or non-parametric models for the mean and the variance.

The focus of this paper is robust fitting of the above model (1). To achieve this, the Normal distribution for y_i is replaced by a 'more robust' distribution with heavier tails. Let $y_i = \mu_i + \sigma_i t_i$ for $i = 1,...,n$, where t_i has a standardised robust distribution with probability density function $f(t) = c\exp\{-\rho(t)\}$ and c is the normalising constant for the distribution. Instead of maximising the penalised Normal log-likelihood as in Rigby and Stasinopoulos (1993), model (1) is fitted robustly by maximising the penalised robust log-likelihood l_{pr} (with penalties for lack of smoothness in the additive finctions f_j , for $j = 1,2,..,p$, and h_k for $k = 1,2,..,q$ in the mean and variance models respectively) given by :

$$l_{pr} = -\sum_{i=1}^{n}\rho\{(y_i - \mu_i)/\sigma_i\} - \sum_{i=1}^{n}\log\sigma_i - n\log c$$

$$-\frac{1}{2}\left[\sum_{j=1}^{p}\lambda_{1j}\int_{-\infty}^{+\infty}\{f_j''(t)\}^2 dt + \sum_{k=1}^{q}\lambda_{2k}\int_{-\infty}^{+\infty}\{h_k''(t)\}^2 dt\right]$$

(2)

with respect to the non-parametric additive functions f_j, and h_k in the mean and variance models. The maximising functions $\hat{f}_j$, and $\hat{h}_k$ are cubic splines. (See Hardle and Tsybakov (1988) for a kernel smoothing approach to robustly estimating the mean and variance functions). One possible standardised robust distribution for t_i is the NET distribution decribed in section 3.

A different approach to fitting model (1) robustly is to alternate between separate robust fitting methods for the mean and variance models e.g. estimating the mean model by conditionally maximising l_{pr} given the current fitted variance model and vice-versa but with a different $\rho(t)$, as in Huber 's (1964) proposal 2, (see section 3). Both the above robust fitting approaches can be achieved using the algorithm in section 2. An example is given in section 4 and section 5 contains concluding remarks.

2 The fitting algorithm

The algorithm proposed here extends the algorithm of Rigby and Stasinopoulos (1993) to fit model (1) robustly. The implementation of the algorithm requires a Generalised Additive Model (GAM) procedure which is a Local Scoring algorithm, combining a Generalised Linear Model (GLM) procedure and a Backfitting algorithm (incorporating a smoother). For details see Hastie and Tibshirani (1990, ch 6.3).

The full robust algorithm is as follows:

INITIALISATION : set prior weights $w_1 = 1$, set $f_j = 0$, $h_k = 0$ for $j=1..p, k=1..q$.

REPEAT

STEP 1. Fit a GAM model with response variable y against the x's using Normal errors, scale parameter $\phi_1 = 1$, link function g_1, and prior weights w_1 to give the current fitted $f_j(x_j)$'s and hence $\hat{\mu}$.

For robust fitting of the mean model amend the resulting GLM iterative weights $a_{1i} = w_{1i}(g_1'(\hat{\mu}_i))^{-2}$ to u_{1i}, (see section 3).

STEP 2. *Calculate a robust function s of the residuals* $r = y - \hat{\mu}$ *in order to form the 'response' variable for the next step, (see section 3).*

STEP 3. *Fit a GAM model with response variable s against the z's using Gamma errors, scale parameter* $\phi_2 = 2$, *link function* g_2, *and prior weights* $w_2 = 1$ *to give the current fitted* $\hat{h}_k(z_k)$*'s and hence* $\hat{\sigma}^2$. *A suitable amendment of the GLM iterative weights* $a_{2i} = (\hat{\sigma}_i^2 g_2'(\hat{\sigma}_i^2))^{-2} / 2$ *to* u_{2i} *can be used here if needed (see comments in section 3).*

STEP 4. *Use the fitted values* $\hat{\sigma}^2$ *from step 3 to calculate the new prior weights* $w_1 = 1 / \hat{\sigma}^2$ *for fitting the mean model.*

END

UNTIL : The penalised log-likelihood in equation (2) converges.

Note that linear parametric terms for both the mean and the variance models in equation (1) can be fitted using an extra GLM fit within the backfitting algorithm. In the case where η and ξ are both non-parametric Additive models and where the smoother used in the backfitting is a cubic splines it was shown by Rigby and Stasinopoulos (1993) that the above algorithm, with $s = r^2$ $u_{1i} = a_{1i}$ and $u_{2i} = a_{2i}$ maximises the penalised Normal log-likelihood function of the data. The above algorithm has been implemented by the authors in GLIM4 where the amount of smoothing used for each of the x 's or z 's is determined in advance by fixing the effective degrees of freedom to be used for the variable (Hastie and Tibshirani, 1990, Ch6.8).

3 The NET distribution

This assumes a Normal-Exponential-Student-t (NET) standardised error distribution for the data, i.e. Normal up to k_1, Exponential from k_1 to k_2 and Student-t with $(k_1 k_2 - 1)$ degrees of freedom after k_2, (see Rigby and Stasinopoulos (1994)), giving the following robust function $\rho(t)$ in (2):

$$\rho(t) = \begin{cases} t^2 / 2 & |t| \le k_1 \\ k_1 |t| - k_1^2 / 2 & \text{for} \quad k_1 < |t| \le k_2 \\ k_1 k_2 \log(|t| / k_2) + k_1 k_2 - k_1^2 / 2 & |t| > k_2 \end{cases}$$

This leads to bounded influence functions $\psi(t) = \rho'(t)$ and $\chi(t) = t\psi(t) - 1$ for the mean and variance respectively, since:

$$\psi(t) = \begin{cases} t \\ k_1 \text{sgn}(t) \\ k_1 k_2 / t \end{cases} \quad \text{and} \quad \chi(t) = \begin{cases} t^2 - 1 & |t| \le k_1 \\ k_1 |t| - 1 & \text{for} \quad k_1 < |t| \le k_2 \\ k_1 k_2 - 1 & |t| > k_2 \end{cases}$$

To obtain a consistent estimator of σ^2 for uncontaminated Normal data,
$$\chi_\beta(t) = \chi(t) + 1 - \beta \qquad \text{for} \qquad \beta = 1 - k_1(2/\pi)^{1/2} e^{-k_1^2/2} + 2k_1k_2\Phi(-k_2) - 2\Phi(-k_1)$$
replaces $\chi(t)$.

In the algorithm in section 2, the following iterative weights $u_{1i} = a_{1i}\psi(t_i)/t_i$, where $t_i = r_i/\hat{\sigma}_i$ and $r_i = y_i - \hat{\mu}_i$, are required for fitting the mean model and in the variance model the response variable $s_i = \psi(t_i)r_i^2/\beta t_i$ with iterative weights $u_{2i} = \beta a_{2i}$ is required. For the NET distribution this gives:

$$u_{1i} = \begin{cases} a_{1i} \\ a_{1i}k_1\hat{\sigma}_i / |r_i| \\ a_{1i}k_1k_2\hat{\sigma}_i^2 / r_i^2 \end{cases} \qquad \text{and} \quad s_i = \begin{cases} r_i^2/\beta \\ k_1\hat{\sigma}_i|r_i|/\beta \\ k_1k_2\hat{\sigma}_i^2/\beta \end{cases} \qquad \text{for} \qquad \begin{matrix} |r_i| \le k_1\hat{\sigma}_i \\ k_1\hat{\sigma}_i < |r_i| \le k_2\hat{\sigma}_i \\ |r_i| > k_2\hat{\sigma}_i \end{matrix}$$

The NET method above is related to Huber's (1964) proposal 2 which effectively recommends using a Normal-Exponential (NE) distribution for fitting the mean (i.e. $k_2 = \infty$ in $\psi(t)$ above) and a Normal-Student-t (NT) distribution for fitting the variance (i.e. $k_2 = k_1$ in $\chi(t)$ and $\chi_\beta(t)$ above). Huber's proposal 2 can be fitted using the algorithm in section 2 by using $k_2 = \infty$ in u_{1i} and $k_2 = k_1$ in s and β.

The motivation for using the NET is that it provides an explicit error distribution while maintaining bounded mean and variance influence functions. The model is therefore easier to interpret and the proper likelihood function allows inferences concerning the model e.g. hypothesis tests between different mean-variance models and prediction intervals for future observations. Maximum likelihood estimates of k_1 and k_2 can be obtained and the maximised NET likelihood function compared with the maximised (unrobust) Normal likelihood function.

4 Example

Silverman (1985) used observations of accelerometer readings taken through time in an experiment on the efficacy of crash helmets. Rigby and Stasinopoulos (1993) reanalysed the data by fitting nonparametric mean and variance models with 21 and 14 degrees of freedom respectively. The contaminated data in figure 1a were generated from the original fitted model (with mean and variance functions given by the solid lines in figure 1a and 1b respectively). The ith value y_i in the contaminated data was randomly generated from $N(\mu_i, \sigma_i^2)$ with probability 0.9 and $N(\mu_i + 50, 4\sigma_i^2)$ with probability 0.1, where μ_i and σ_i^2 are the ith values of the original fitted mean and variance functions. The contaminated data were then fitted, (using 21 and 14 degrees of freedom for the mean and the variance respectively), both unrobustly and robustly. The NET model deviance of 1182.95, with $k_1 = 1.5$ and $k_2 = 2$, compares favourably to the unrobust fit of 1232.04 indicating the need for robust analysis. Figure 1a shows the fitted values for the mean model for both the unrobust and the NET model.

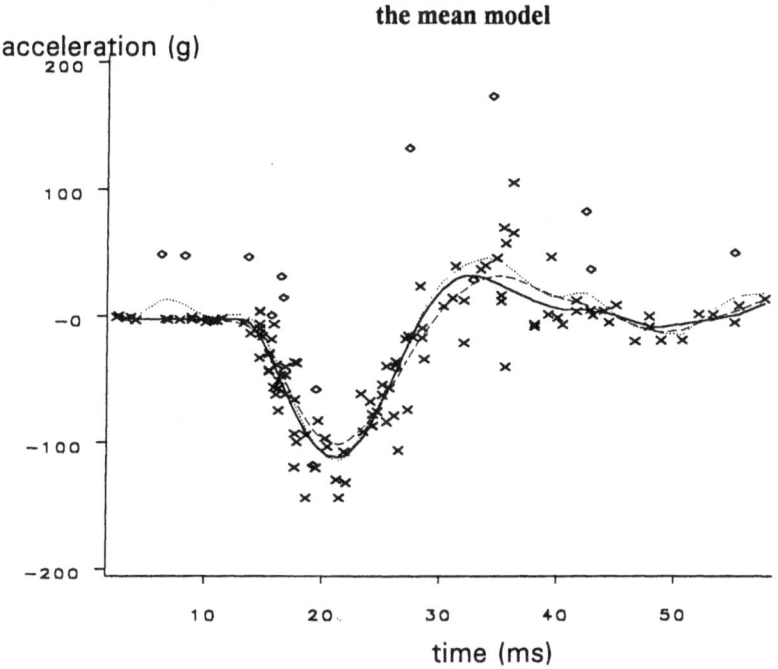

the mean model

acceleration (g)

time (ms)

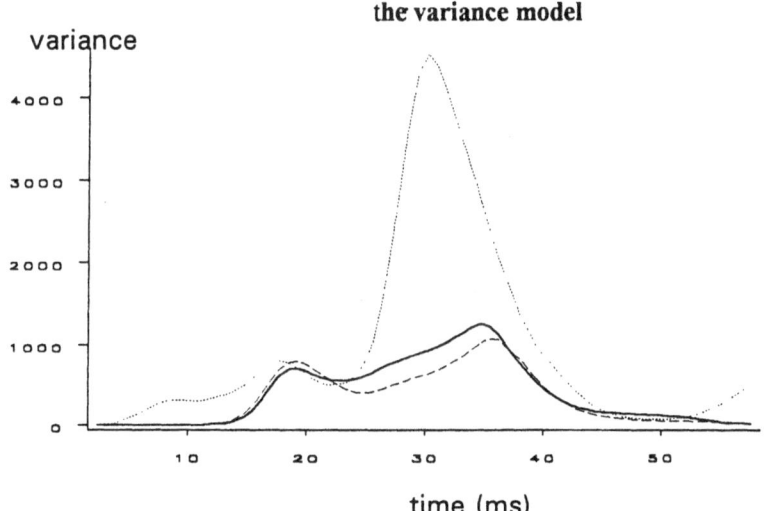

the variance model

variance

time (ms)

Figure 1a: Showing the uncontaminated (x) and the contaminated (◊) observations together with the true mean (solid line), the unrobust fitted mean (dot line) and the NET fitted mean (dash line).

Figure 1b: Showing the true variance (solid line), the unrobust fitted variance (dot line) and the NET fitted variance (dash line).

The NET mean model fit is better than the unrobust fit, since it is on average closer to the true mean (when standardised by the true variance). A striking improvement of the robust method is shown in the fitted variance models in figure 1b, where the NET robust fit for the variance is very close to the true variance.

5 Discussion

The advantage of the NET method is that it provides a proper penalised likelihood for model selection. The method depends on both k_1 and k_2 and can be computationally unstable for low values of k_1 and k_2. The same computational difficulty seems to apply to Huber's method for small k_1. A more thorough assessment is currently in progress.

References

Aitkin, M. (1987) Modelling variance heterogeneity in normal regression using GLIM. Appl. Statist., 36, 332-339.

Hardle, W. and Tsÿbakov, A.B. (1988) Robust nonparametric regression with simultaneous scale curve estimation. *Ann. Statist.*, **16**, 120-135.

Harvey, A.C. (1976) Estimating regression models with multiplicative heteroscedasticity. *Econometrica*. **41**, 461-465.

Hastie, T.J. and Tibshirani, R.J. (1990) *Generalized Additive Models*. London: Chapman and Hall.

Huber, P.J. (1964) Robust estimation of a location parameter. *Ann. of Math. Statist.* , **35**, 73-101.

Muller, H. G. and Stadtmuller, U. (1987) Estimation of heteroscedasticity in regression analysis. *Ann. Statist.*, **15**, 610-625.

Rigby, R.A. and Stasinopoulos, D.M. (1993) A semi-parametric Additive model for variance heterogeneity. (sent for publication).

Rigby, R.A. and Stasinopoulos, D.M. (1994) NET: A usefull distribution for robustly fitting semi-parametric models for the mean and variance. (in preparation).

Silverman, B.W. (1985) Some aspects of the spline smoothing approach to non-parametric regression curve fitting (with discussion). *J. R. Statist. Soc.* B, **47**, 1-52.

Part XIII

Computational Aspects in Multivariate Analysis

Models for Multivariate Data Analysis

Philippe C. Besse

Laboratoire de Statistique et Probabilités, U.A. CNRS 745,
Université Paul Sabatier

Abstract. This paper reviews some models for exploring multivariate data. If a fixed effect model is used to define a linear Principal Components Analysis (PCA), then risk functions can be defined and issues of metric and dimension optimality addressed. The model is then adapted to define a functional PCA which can be used to the study of smooth sampled curves. Finally, this model is generalised, giving a curvilinear PCA, which attempts to build smooth optimal transformations of data for the purpose of dimension reduction.

Keywords. principal component analysis, functional estimation, optimal transformations, spline smoothing

1 Introduction

A large number of statistical methods aim to analyze or to explore multivariate data. This paper is concerned with a class of them where the two main purposes are reduction of dimensionality and graphical presentation of data collected in a $(n \times p)$ matrix. Different frameworks have been proposed. They often lead to a common mathematical core as those based on the spectral analysis of a symmetric positive matrix but the interpretation of results may depend on the adopted framework or the assumed model. On the one hand, observations are considered to come from a population, with certain distributional assumptions, and interest lies more in the population itself than in the statistical units. On the other hand, a descriptive approach aims to investigate possible patterns for the units and for the variables. As it has been noted by several authors (e.g. Gower, 1984), these two aims are somewhat contradictory; strong distributional assumptions seem inconsistent with a descriptive approach. Nevertheless, only the consideration of a model can clarify what kind of arbitrariness is unavoidable; e.g. our perception requires an Euclidean framework for graphical displays. Furthermore, a model allows us to discuss what guidelines can be put forward to help, for example, the choice of metric, dimensionality, smoothing parameter.

This paper is mainly concerned with the analysis of quantitative data and focuses on recent developments of the Principal Component Analysis (PCA)

and its extensions. It aims to emphasize advantages and consequences when introducing the PCA by means of a fixed effect model (Caussinus, 1986) which mainly deals with geometrical assumptions rather than probabilistic ones. The same kind of considerations occur when dealing with categorical data and contingency tables. Baccini et al. (1993) compare different models for Correspondence Analysis and their estimation with Association and Correlation models (Goodman, 1991 with discussion).

Section 2 defines the PCA as the estimation of a fixed effect model and discusses the optimal choices of dimensionality and metric by introducing quadratic risks in order to evaluate the estimation quality. Section 3 makes the fixed effect model suitable for studying sampled curves. This leads to define a new kind of functional PCA featuring spline smoothing. Section 4 generalizes the fixed effect model in order to search for optimal transformations for dimension reduction. This leads to introduce a curvilinear fixed effect model including non-metric PCA's. A Smoothing Alternative Least Squares Algorithm (SALSA) is proposed to achieve the estimation. The lack of space, in this short review, does not allow to clarify all proofs or to give practical examples. These must be found in referenced papers or in a forthcoming book (Besse et al. 1995).

2 Linear PCA

Consider the following framework:

- **Y** denotes the original data matrix displaying observations of p real variables Y^j on n $(n > p)$ statistical units, each weighted by w_i,

- E is the Euclidean p-dimensional space of units where the metric is $\mathbf{M}$,

- F is the Euclidean n-dimensional space of variables where the metric is $\mathbf{D} = \operatorname{diag}(w_1, \ldots, w_n)$.

2.1 A diet model

The data matrix rows are assumed to be the observations of n independent random vectors $\{\mathbf{y}_i; \ i = 1, \ldots, n\}$ with the same covariance matrix $\sigma^2 \mathbf{\Psi}$ but different expectations $\mathbf{z}_i$ belonging to a common q-dimensional affine subspace A_q of E. In this model, $E(\mathbf{y}_i) = \mathbf{z}_i$ is the so-called fixed effect attached to the ith unit; $\mathbf{M}$ is the metric which is introduced to compute distances between units or ultimately $\|\mathbf{z}_i - \mathbf{z}_{i'}\|_{\mathbf{M}}$.

To summarize,

$\{y_i ; i = 1, \ldots, n\}$, n independent random vectors of E,

$y_i = z_i + e_i ; i = 1, \ldots, n$ with $\begin{cases} E(e_i) = 0, \ \text{var}(e_i) = \sigma^2 \Psi, \\ \sigma > 0 \text{ unknown}, \\ \Psi \text{ regular and known}, \end{cases}$

$\exists A_q$, q dimensional affine subspace of E such that $\forall i, z_i \in A_q, (q < p)$.

$$(1)$$

Let $\bar{z} = \sum_{i=1}^{n} w_i z_i$. The model assumptions implies $\bar{z} \in A_q$. Let then E_q be the q dimensional linear subspace of E such that $A_q = \bar{z} + E_q$. The parameters to be estimated are E_q and $z_i, i = 1, \ldots, n$, eventually σ.

2.2 Estimation

Least squares estimation for the problem is formulated as follows

$$\min_{E_q, z_i} \left\{ \sum_{i=1}^{n} w_i \| y_i - z_i \|_M^2 \ ; \ \dim(E_q) = q, z_i - \bar{z} \in E_q \right\}. \qquad (2)$$

Denoting by $X = Y - 1_n \bar{y}'$ the column centered matrix with $\bar{y} = \sum_{i=1}^{n} w_i y_i$ and by Z the $(n \times p)$ matrix whose rows are the $(z_i - \bar{z})'$, as

$$\sum_{i=1}^{n} w_i \| y_i - z_i \|_M^2 = \sum_{i=1}^{n} w_i \| y_i - \bar{y} + \bar{z} - z_i \|_M^2 + \| \bar{y} - \bar{z} \|_M^2,$$

the problem (2) is equivalent to $\widehat{\bar{z}} = \bar{y}$ and

$$\min_{Z} \left\{ \| X - Z \|_{M,D} \ ; Z \in \mathcal{M}_{n,p}, \ \text{rank}(Z) = q \right\}. \qquad (3)$$

This is the approximation of a $(n \times p)$ matrix by another $(n \times p)$ matrix with a lower rank. Estimation of parameters E_q and $\{z_i; i = 1, \ldots, n\}$ of the fixed effect model (1) is obtained by order q generalized PCA of (Y, M, D). This PCA is obtained by considering an order q generalized singular value decomposition (SVD) of (X, M, D)

$$\widehat{Z_q} = \sum_{k=1}^{q} l_k^{1/2} u^k v^{k'} = U_q L^{1/2} V_q'.$$

Finally the estimates are given by

$$\begin{aligned} \widehat{\bar{z}} &= \bar{y}, \\ \widehat{E_q} &= \text{vect} \{ v^1, \ldots, v^q \}, \\ \widehat{P_q} &= V_q V_q' M \text{ is the M-orthogonal} \\ & \qquad \text{projection matrix onto } \widehat{E_q}, \end{aligned}$$

$$\widehat{Z_q} = \mathbf{X}\widehat{\mathbf{P}_q}',$$
$$\widehat{z_i} = \widehat{\mathbf{P}_q}\mathbf{x}_i + \overline{\mathbf{y}}.$$

2.3 Optimal metric

Up until now, the choice of $\mathbf{M}$ seems rather arbitrary. However, under normality assumptions, it is easily seen that maximum likelihood estimation leads to minimize (3) with, in this case, $\mathbf{M} = \mathbf{\Psi}^{-1}$. It is therefore natural to look for an optimal metric choice in the non-normal case.

Since PCA is defined in terms of model estimation, a quality criterion can be investigated in order to optimize this metric choice. Consider the general loss function

$$\mathcal{L}_q = \frac{1}{n}\sum_{i=1}^{n}\|\mathbf{z}_i - \widehat{\mathbf{z}}_i\|_{\mathbf{A}}^2 ,$$

where $\mathbf{A}$ denotes any Euclidean metric in $\mathbb{R}^p$.

The taking of the expectation defines a mean square error

$$R_q(\mathbf{M}, \mathbf{A}) = E(\mathcal{L}_q)$$

which is to be optimized over $\mathbf{M}$ to derive an optimal metric.

Besse et al. (1988) proved, by means of Perturbation Theory, that if the variance error σ^2 is assumed to be small enough, $R_q(\mathbf{M}, \mathbf{A})$ can be expanded as

$$R_q(\mathbf{M}, \mathbf{A}) = \sigma^2\left[(q+1)\mathrm{tr}\mathbf{\Psi}\mathbf{A} + (n-q-1)\mathrm{tr}(\mathbf{P}_q\mathbf{\Psi}\mathbf{P}_q'\mathbf{A})\right] + 0(\sigma^3).$$

Furthermore, for any metric $\mathbf{A}$, the leading term in $R_q(\mathbf{M}, \mathbf{A})$ is minimized for

$$\mathbf{M} = \mathbf{\Psi}^{-1}.$$

This gives a property similar to Gauss Markov for PCA; the optimal metric is given by the inverse of error covariance matrix. Of course, this matrix is generally unknown but it can be estimated under some circumstances as is the case in the next section.

2.3.1 PCA and Discriminant Analysis

Factorial Discriminant Analysis (FDA) is an optimal PCA which displays clusters of units when groups are known. Data consist of p real random variables $Y^1, \ldots, Y^p$ and a *qualitative* variable T which are jointly observed on the same units. It is assumed that T has m categories $\{T_1, \ldots, T_m\}$ and thus defines a partition $\{\Omega_k; k = 1, \ldots, m\}$ of the set Ω of statistical units ω_i. Let $\mathbf{T}$ be the $(n \times m)$ matrix of indicator functions,

$$t_i^k = t^k(\omega_i) = \begin{cases} 1 & \text{if } T(\omega_i) = T_k, \\ 0 & \text{if not.} \end{cases}$$

It is natural to consider that all units, which hold the same k^{th} category of variable T, are n_k repeated measures of a same fixed effect $\mathbf{z}_k$ weighted by $\overline{w_k} = \sum_{i \in \Omega_k} w_i$. We also define $\overline{\mathbf{D}} = \text{diag}(\overline{w_1}, \dots, \overline{w_m})$ and

$$
\overline{\mathbf{Y}} = \overline{\mathbf{D}}^{-1} \mathbf{T}' \mathbf{D} \mathbf{Y} = \begin{bmatrix} \overline{y_1}' \\ \vdots \\ \overline{y_m}' \end{bmatrix} \quad \text{where } \overline{y_k} = \frac{1}{\overline{w_k}} \sum_{i \in \Omega_k} w_i y_i .
$$

This leads to the following model,

$$
\begin{array}{l}
\{\mathbf{y}_i \; ; i = 1, \dots, n\}, \; n \text{ independent random vectors of } E, \\
\forall k, \forall i \in \Omega_k, \mathbf{y}_i = \mathbf{z}_k + \mathbf{e}_i \text{ with } \left\{ \begin{array}{l} E(\mathbf{e}_i) = 0, \; \text{var}(\mathbf{e}_i) = \mathbf{\Psi}, \\ \mathbf{\Psi} \text{ regular and unknown}, \end{array} \right. \quad (4) \\
\exists A_q, \; q\text{-dimensional affine subspace of } E \\
\text{such that } \forall k, \mathbf{z}_k \in A_q, (q < \min(p, m-1)).
\end{array}
$$

Let $\overline{\mathbf{z}} = \sum_{k=1}^{m} \overline{w_k} \mathbf{z}_k$. The model implies $\overline{\mathbf{z}} \in A_q$. Let E_q be the q dimensional linear subspace of E such that $A_q = \overline{\mathbf{z}} + E_q$. The parameters to be estimated are E_q and $\{\mathbf{z}_k \; ; k = 1, \dots, m\}$; $\overline{w_k}$ are nuisance parameters that we will not consider now.

A least squares estimation for this problem can be formulated as follows.

$$
\min_{E_q, \mathbf{z}_k} \left\{ \sum_{k=1}^{m} \sum_{i \in \Omega_k} w_i \|\mathbf{y}_i - \mathbf{z}_k\|_{\mathbf{M}}^2 \; ; \; \dim(E_q) = q, \mathbf{z}_k - \overline{\mathbf{z}} \in E_q \right\} .
$$

We obtain

$$
\sum_{k=1}^{m} \sum_{i \in \Omega_k} w_i \|\mathbf{y}_i - \mathbf{z}_k\|_{\mathbf{M}}^2 = \sum_{k=1}^{m} \sum_{i \in \Omega_k} w_i \|\mathbf{y}_i - \overline{y_k}\|_{\mathbf{M}}^2 + \sum_{k=1}^{m} \overline{w_k} \|\overline{y_k} - \mathbf{z}_k\|_{\mathbf{M}}^2
$$

and we are led to minimize

$$
\min_{E_q, \mathbf{z}_k} \left\{ \sum_{k=1}^{m} \overline{w_k} \|\overline{y_k} - \mathbf{z}_k\|_{\mathbf{M}}^2 \; ; \; \dim(E_q) = q, \mathbf{z}_k - \overline{\mathbf{z}} \in E_q \right\} .
$$

As seen above, an optimal choice for $\mathbf{M}$ is $\mathbf{\Psi}^{-1}$. Since this matrix is unknown it has to be estimated. In this context $\mathbf{\Psi}$ may be estimated by the sample within group covariance matrix $\mathbf{W}$.

The estimation of parameters E_q and $\mathbf{z}_k$ of the model (4) is given by the PCA of $(\overline{\mathbf{Y}}, \mathbf{W}^{-1}, \overline{\mathbf{D}})$. It leads to the computation of the same eigenvectors as the PCA of $(\overline{\mathbf{Y}}, \mathbf{S}^{-1}, \overline{\mathbf{D}})$. It is called the Factorial Discriminant Analysis of $(\mathbf{Y}|\mathbf{T}, \mathbf{D})$.

2.3.2 Projection Pursuit with PCA

In the case of unknown groups and when the within group covariance is "larger" than the between group covariance, a classical PCA, which focuses on maximal dispersion, can be very misleading. Exploratory Projection Pursuit (Friedman, 1987) may overcome such problems. It aims at producing interesting low-dimensional projections by optimizing criteria based on indices of non-normality (entropy) rather than the variance of the projected distributions. Usually, such methods require a very large amount of data which rapidly increases with p and q; this is the "curse of dimensionality". Moreover they are time consuming. In the same context, various kinds of suitably generalized PCA have been proposed by Caussinus and Ruiz (1990) in order to reveal data structures such as outliers or clusters at a very low computational cost. The basic idea is to draw from the data a convenient metric $\mathbf{M}$ on the space E of units.

For instance, a robust estimate of the unknown within group covariance is given by

$$
\widehat{\mathbf{\Psi}} = \frac{\sum_{i=1}^{n-1} \sum_{j=i+1}^{n} k \left(\|\mathbf{y}_i - \mathbf{y}_j\|_{\mathbf{S}-1}^2 \right) (\mathbf{y}_i - \mathbf{y}_j)(\mathbf{y}_i - \mathbf{y}_j)'}{\sum_{i=1}^{n-1} \sum_{j=i+1}^{n} k \left(\|\mathbf{y}_i - \mathbf{y}_j\|_{\mathbf{S}-1}^2 \right)};
$$

where k is a decreasing positive real function (e.g. $k(t) = e^{-\gamma t}$ with $\gamma > 0$). A projection pursuit PCA, aiming to reveal clusters, is then defined as the PCA of $(\mathbf{X}, \widehat{\mathbf{\Psi}}^{-1}, \frac{1}{n}\mathbf{I}_n)$. Another robust estimate aiming to reveal multivariate outliers is defined by

$$
\widehat{\mathbf{\Psi}} = \frac{\sum_{i=1}^{n-1} k \left(\|\mathbf{y}_i - \bar{\mathbf{y}}\|_{\mathbf{S}-1}^2 \right) (\mathbf{y}_i - \bar{\mathbf{y}})(\mathbf{y}_i - \bar{\mathbf{y}})'}{\sum_{i=1}^{n-1} k \left(\|\mathbf{y}_i - \bar{\mathbf{y}}\|_{\mathbf{S}-1}^2 \right)}.
$$

2.4 Optimal dimension

In this section, we set for the sake of clarity, $\mathbf{M} = \mathbf{\Psi} = \mathbf{I}_p$. In other cases of known error covariance matrix, raw data can be transformed by $\mathbf{\Psi}^{-1/2}$.

The quality of the estimates defined in section 2.3 directly depends on the choice of q; how many components need to be retained? Jolliffe (1986) reviews some rules frequently used to reach such a decision. He distinguishes between ad hoc rules that are intuitively plausible and rules based on more formal reasoning. Some of these rules further require strong distributional assumptions to test, for instance, the equality of least eigenvalues. The introduction of a model in order to define PCA allows to consider another framework for discussing this problem.

An optimal dimension could also be searched for by minimizing the risk function as introduced in Section 2. Unfortunately this approach raises a problem of indeterminacy if σ^2 is unknown; it is not possible to estimate simultaneously σ and the optimal dimension q, since the natural estimate for σ^2 is the average of the $(p-q)$ smallest eigenvalues.

This drawback can be overcome by considering the subspace $\widehat{E_q}$, which is an asymptotically consistent estimate (Fine and Pousse, 1992) rather than the vectors z_i. The loss function then becomes

$$\mathcal{L}_q = \frac{1}{2}\left\|\mathbf{P}_q - \widehat{\mathbf{P}_q}\right\|_2^2 = q - \operatorname{tr}\mathbf{P}_q\widehat{\mathbf{P}_q}. \tag{5}$$

It is based on the usual SSQ norm of matrices which is applied to measure distances between projectors, and thus to measure distances between the associated subspaces. In that context, $\operatorname{tr}\mathbf{P}_q\widehat{\mathbf{P}_q}$ is also the sum of the squared canonical correlation coefficients between the component sets which respectively span E_q and $\widehat{E_q}$. This approach follows the principle that, in a PCA, the results are assumed to be reliable if the representation subspace is stable.

A risk function is then defined by taking the expectation,

$$R_q = E(\mathcal{L}_q). \tag{6}$$

R_q being symmetrically defined since its value is invariant under any permutation of the observations y_i. Resampling methods such as the bootstrap and the jackknife are natural candidates to compute estimates for this risk function. Besse and Falguerolles (1993) compare these methods which are all computationally expensive. Furthermore Besse (1992) proposes an approximation of the jackknife which does not require much computational effort.

If n is large enough, it is quite acceptable to consider that any row elimination introduces only a small perturbation in further computations. Perturbation theory enables us to write Taylor expansions of the eigenelements of $\mathbf{S}$ which lead to a Taylor expansion of the jackknife estimate and then to an analytic approximation

$$\widehat{R_{JKq}} = \widehat{R_{Pq}} + O((n-1)^{-2}).$$

This analytic approximation of the jackknife estimate, is given by

$$\widehat{R_{Pq}} = \frac{1}{n-1}\sum_{k=1}^{q}\sum_{j=k+1}^{p}\frac{\tau_{jk}}{(\lambda_j - \lambda_k)^2} \quad \text{with } \tau_{jk} = \frac{1}{n}\sum_{i=1}^{n}c_{ik}^2 c_{ij}^2 \tag{7}$$

where c_{ij} denotes the general entry of the principal component matrix

$$\mathbf{C} = \mathbf{U}\mathbf{L}^{1/2}.$$

This shows the importance of the gap between successive eigenvalues. The leading term in $\widehat{R_{Pq}}$ depends on the difference between the eigenvalues associated with the last selected dimension and the first omitted one. This is consistent with intuitive considerations; if the difference $(\lambda_q - \lambda_{q+1})$ is large enough, data perturbations cannot lead to the swapping of the associated eigenvectors $\mathbf{v}^q$ and $\mathbf{v}^{q+1}$. It is very cheap and easy to plot $\widehat{R_{Pq}}$ versus q and then to choose a convenient value for the dimensionality of the representation space $\widehat{E}_q$.

3   Functional PCA

3.1   Spline regression

In this section, we consider data generated from the observation of sampled curves. The relevant model is then

$$y_i = z_i + e_i, \text{ with } z_i^j = f_i(t_j) \left\{ \begin{array}{l} i = 1, \ldots, n \\ j = 1, \ldots, p \end{array} \right.$$

and where the vectors y_i are noisy observations of n functions f_i for

$$t_1 > \cdots > t_p; t_j \in T = [a, b] \subset \mathbb{R}.$$

Each observation is weighted by $w_i > 0$ with $\sum_{i=1}^n w_i = 1$. The error e_i is considered as a "white noise" with zero mean and whose covariance matrix is $\sigma^2 \mathbf{I}$. The functions f_i are assumed to be regular and smooth. One way to ensure the regularity of f is to impose a condition such as the existence of $(m-1)$ absolutely continuous derivatives and a square integrable mth integrable derivative, $(m \geq 1)$. This means that each f_i belongs to a Sobolev space $W^m(T)$. Furthermore, the smoothness is imposed by a constraint

$$\|f_i\|^2_m = \int_T (f^{(m)})^2(t) dt < c, (c > 0), \tag{8}$$

where $\|.\|_m$ defines a semi-norm on $W^m(T)$. One could generalize this to any equivalent semi-norm of $W^m(T)$.

In the case $n = 1$, this problem is classically solved by spline smoothing or spline non-parametric regression (see Wahba (1990) for a review),

$$\hat{f} = \arg \min_{f \in W^m} \left\{ \sum_{j=1}^p (f(t_j) - y^j)^2 + \rho \|f_i\|^2_m \right\}; \tag{9}$$

ρ is the smoothing parameter; it is also the Lagrange multiplier depending on c. It controls the tradeoff between the roughness of the solution and the lack of fit to the data. The solution is obtained by a polynomial spline interpolation which exactly fits the values that are stored in the vector $\hat{z} = (\hat{f}(t_j) ; j = 1, \ldots, p)$ given by

$$\hat{z} = \mathbf{A}(\rho)\mathbf{y} \text{ with } \mathbf{A}(\rho) = (\mathbf{I} + \rho\mathbf{N})^{-1}; \tag{10}$$

A is the so-called hat matrix. The symmetric semi-definite and positive matrix **N** is the representation in the canonical basis of the semi-norm which is induced by $\|.\|_m$ on a p-dimensional spline subspace of W^m. With matrix notation, it simply amounts to

$$\left\| \hat{f} \right\|_m^2 = \hat{z}'\mathbf{N}\hat{z} = \|\hat{z}\|_\mathbf{N}^2 .$$

The matrix **N** depends on the sequence $\{t_1, \ldots, t_p\}$. It can be built by means of reproducing kernels (Wahba, 1990) or, computationally faster, by using tridiagonal matrices as described for instance by Buja et al. (1989).

For large n, we assume that a data set consists of sampled curves, issued from noisy observations of n regular trajectories $f_i(t)$. The problem then is one of functional estimations of several curves, and means simultaneous non-parametric regressions. This could be achieved by computing n classical non-parametric (kernel or spline) regressions where each smoothing parameter is independently optimized by Generalized Cross Validation (GCV). But this approach does not take any special features of the data into account and simulations show (Besse, 1994) that it induces a wide information loss. This kind of data has been previously studied by Besse and Ramsay (1986), Ramsay and Dalzell (1991), Besse and Pousse (1992) who deal with Principal Components Analysis of curves. These are closely related to works by Kneip and Gasser (1992) and to other papers which aim at approximating a covariance function by smooth eigenfunctions (Rice and Silverman, 1991, Pezzuli and Silverman, 1993).

3.2 Model and estimation

The nature of the data introduces two kinds of constraints in a fixed effect model. On one hand, a dimension constraint assumes that curves only span a subspace of $W^m(T)$. In finite dimension, this means that sampled curves lie in a q dimensional subspace of $\mathbb{R}^p$. On the other hand, each real curve is assumed to be smooth enough and then they obey the constraint (8) with a common constant c.

The fixed effect model becomes

$$\{\mathbf{y}_i \; ; i = 1, \ldots, n\}, \; n \text{ independent random vectors of } \mathbb{R}^p,$$

$$\mathbf{y}_i = \mathbf{z}_i + \mathbf{e}_i \; ; i = 1, \ldots, n \text{ with } \left\{ \begin{array}{l} E(\mathbf{e}_i) = 0, \; \text{var}(\mathbf{e}_i) = \sigma^2 \mathbf{I}, \\ \sigma > 0 \text{ unknown}, \end{array} \right. \tag{11}$$

$$\exists A_q, \; q \text{ dimensional affine subspace of } \mathbb{R}^p \text{ such that}$$
$$\forall i, \mathbf{z}_i \in A_q, (q < p), \|\mathbf{z}_i\|_\mathbf{N} < c, (c > 0).$$

As above, the smoothness constraint is taken into account by introducing the Lagrange multiplier $\rho, (\rho > 0)$ in the least squares criterion; it leads to consider the following problem

$$\min_{\mathbf{z}_i, A_q} \left\{ \sum_{i=1}^n w_i \left(\|\mathbf{y}_i - \mathbf{z}_i\|_\mathbf{I}^2 + \rho \|\mathbf{z}_i\|_\mathbf{N}^2 \right) ; \forall i, \mathbf{z}_i \in A_q, \; \dim(A_q) = q \right\}. \tag{12}$$

Let $\bar{z} = \sum_{i=1}^{n} w_i z_i$, $\bar{y} = \sum_{i=1}^{n} w_i y_i$ and $\mathbf{X}$ $(n \times p)$ be the column centered matrix with rows $(y_i - \bar{y})'$; $\mathbf{Z}$ $(n \times p)$ denotes the matrix with rows $(z_i - \bar{z})'$. It follows that the vectors $z_i - \bar{z}$ belong to the linear subspace E_q $(A_q = \bar{z} + E_q)$ and thus the rank of $\mathbf{Z}$ is q.

Besse (1994), in parallel with Denby and Mallows (1993) proves that the solution of (12) is given by

$$\widehat{\mathbf{Z}_q} = \mathbf{U}_q \mathbf{L}_q^{1/2} \mathbf{V}_q' \mathbf{A}(\rho)^{1/2}, \tag{13}$$

$$\widehat{z}_i = \left[\widehat{\mathbf{Z}_q} \right]_i' + \mathbf{A}(\rho) \bar{y}; \tag{14}$$

where matrices $\mathbf{U}_q(n \times q), \mathbf{L}_q(q \times q)$ and $\mathbf{V}_q(p \times q)$ define the first q leading terms of the generalized SVD of $(\mathbf{XA}(\rho)^{1/2}, \mathbf{I}, \mathbf{D})$; $\mathbf{A}(\rho)$ denotes the hat matrix. $\mathbf{V}$ contains the eigenvectors of $\mathbf{A}(\rho)^{1/2} \mathbf{S} \mathbf{A}(\rho)^{1/2}$ where $\mathbf{S} = \mathbf{X}'\mathbf{DX}$ denotes the empirical covariance matrix. They are stored in the decreasing order of their associated eigenvalues. $\widehat{E}_q$ is spanned by the first q vectors $\tilde{\mathbf{v}}^k$ of $\tilde{\mathbf{V}}$.

As in the case $n = 1$, the estimates of functions f are then obtained by the spline interpolation functions $\widehat{f}_i \in W^m$ that exactly fit the values of the vectors

$$\widehat{z}_i = (\widehat{f}_i(t_j); j = 1, \dots, p)$$

and which minimize $\left\| \widehat{f}_i \right\|_m$. The estimates are derived from the PCA of $\mathbf{XA}(\rho)^{1/2}$ that leads to simultaneously perform spline smoothing and a singular value decomposition. This produces a kind of smooth Karhunen-Loeve expansion of the observed process in that functional context.

3.3 Smoothing and dimension optimality

Both splines and SVD act as smoothing tools. The first is driven by a smoothing parameter ρ whose optimal value depends on data regularity. In contrast, SVD of longitudinal data gives approximations of trajectories whose smoothness usually depends on the number q of selected components; neglected components are often essentially noisy. These parameter values must be jointly optimized in order to obtain better fits of the original functions. GCV is commonly used to optimize the spline smoothing parameter. Unfortunately, its application to the dimension choice in PCA (Krzanowski, 1987) is not convincing (Besse and Ferré, 1993) and never used in practice.

Besse and Pousse (1992) suggest another strategy based on the stability criterion which is described in section 2.4. PCA results are assumed to be reliable if the estimated subspace $\widehat{E}_q$ is stable with respect to data perturbations. It is then proposed to search for a dimension and a smoothing parameter which induce the greatest stability.

Optimality is achieved by minimizing $\widehat{R_{P_q}}$ on both ρ and q. Simulations (Besse, 1994) show that this is not easy but this heuristic approach leads to fairly good working choices of ρ and q values. An asymptotic study still needs to be done.

4 Curvilinear PCA

Different kinds of generalized additive models were studied by Hastie and Tibshirani (1990) in the framework of regression. In the context of PCA, different generalizations, which introduce non linear transformations for each variable, were previously developed. De Leeuw et al. (1981), De Leeuw (1982), Rijckvorsel (1988), Gifi (1990) define different kinds of "Non metric PCAs". Winsberg and Ramsay (1983), Ramsay (1988) search optimal transformations for dimension reduction. A generalization of the fixed effect model is proposed in order to take into account these kind of problems. This leads to a new method of dealing with spline smoothing.

4.1 Smooth transformation

Let Y be a real statistical variable valued within the range T; n realizations of this variable are collected in the vector $\mathbf{y}$ of $\mathbb{R}^n$. Several approaches, in different frameworks have been proposed to define transformations of a statistical variable as power functions, polynomial piecewise functions based on B-splines, with monotony constraint or not.... In this section, we look for functions f belonging to a Sobolev space $W^m(T)$ to define data transformations

$$\tilde{\mathbf{y}} = f(\mathbf{y}) = [f(y_1), \ldots, f(y_n)]'.$$

These are built by using a spline smoother in order to easily control the smoothness. The spline smoothing principle is briefly recalled in section 3.1. There, the smoothing criterion becomes

$$\|f\|^2_m = \tilde{\mathbf{y}}'\mathbf{N}\tilde{\mathbf{y}} = \|\tilde{\mathbf{y}}\|^2_{\mathbf{N}}$$

and $\mathbf{N}$ depends on $\mathbf{y}$ and m.

We denote by $(\mathcal{C})$ the following set of constraints,

$$(\mathcal{C}) \Longleftrightarrow \left\{ \begin{array}{l} f \in W^m(T), \\ \|f\|^2_m \leq c, \\ \text{var}(f(Y)) = 1. \end{array} \right. \tag{15}$$

4.2 The fixed effect curvilinear model

Consider now $\{\mathbf{y}_i; \ i = 1, \ldots, n\}$ n independent random variables in $\mathbb{R}^p$. Their realization is collected in a $(n \times p)$ matrix $\mathbf{Y}$ which is assumed, without loss of generality, to be column centered when each row is weighted by w_i.

This framework leads to consider a set f of p smooth transformation functions f_j operating on the p statistical variables Y^j which are each valued in a range T_j. The columns $\mathbf{y}^j$ of the data matrix become

$$\tilde{\mathbf{Y}} = f(\mathbf{Y}) = [f_1(\mathbf{y}^1), \ldots, f_p(\mathbf{y}^p)],$$

Each function f_j much obey C_j. It belongs to a subspace of $W^m(T_j)$ whose the semi-norm matrix is $\mathbf{N}_j$; the smoothing parameter is c_j; $\mathbf{I}$ denotes the identity matrix in $\mathbb{R}^p$.

The data matrix rows are assumed to be the observations of n random vectors $\{\mathbf{y}_i; \ i = 1, \ldots, n\}$ whose the transformations $\{f(\mathbf{y}_i); \ i = 1, \ldots, n\}$ have the same covariance matrix $\sigma^2 \mathbf{I}$ but different expectations $\mathbf{z}_i$ belonging to a common q-dimensional $(q < p)$ subspace of $\mathbb{R}^p$.

To summarize, the curvilinear fixed effect model is defined by

$$\left.\begin{array}{l} \{\mathbf{y}_i; i = 1, \ldots, n\}\ n \text{ independent random variables in } \mathbb{R}^p \\ f(\mathbf{y}_i) = \mathbf{z}_i + \mathbf{e}_i \ ; i = 1, \ldots, n \text{ with } E(\mathbf{e}_i) = 0, \ \mathrm{var}(\mathbf{e}_i) = \sigma^2 \mathbf{I}, \\ \exists E_q, dim(E_q) = q \\ \exists f, f_j \text{ obeying } (\mathcal{C}_j) \end{array}\right\} \text{ such that } \forall i, \mathbf{z}_i \in E_q. \tag{16}$$

4.3 Estimation

$\mathbf{Z}$ denotes the $(n \times p)$ matrix whose rows are the $\mathbf{z}_i$ and columns the $\mathbf{z}^j$. A least squares estimation of the transformations and the fixed effects leads to the solution of the following problem

$$\min_{\mathbf{z}^j, f^j} \left\{ \sum_{j=1}^p \left\| f^j(\mathbf{y}^j) - \mathbf{z}^j \right\|_{\mathbf{D}}^2 \ ; f_j \text{ holding } (\mathcal{C}_j) \text{ and } \mathrm{rank}(\mathbf{Z}) = q \right\}. \tag{17}$$

Other sets of constraints than (15) lead to equivalent definitions of other methods.

- If f_j is assumed to be a step function depending on a sequence of knots belonging to T_j, it is the non-metric PCA as defined by de Leeuw (1982).

- If f_j is assumed to belong to a B-spline subspace of $W^m(T)$, it gives the approach by de Leeuw et al. (1981).

With matrix notation and when denoting the transformed data columns by $\tilde{\mathbf{y}}^j$, Besse and Ferraty (1994) prove that problem (17) is equivalent to

$$\min_{\mathbf{Q}_q, \tilde{\mathbf{Y}}} \left\{ \left\| \tilde{\mathbf{Y}} - \mathbf{Q}_q \tilde{\mathbf{Y}} \right\|_{\mathbf{I}, \mathbf{D}}^2 + \sum_{j=1}^p \rho_j \left\| \tilde{\mathbf{y}}^j \right\|_{\mathbf{N}_j}^2 \ ; \left\| \tilde{\mathbf{y}}_j \right\|_{\mathbf{D}} = 1 \right\} \tag{18}$$

where $\mathbf{Q}_q$ is a rank q $\mathbf{D}$-orthogonal projection matrix.

4.4 Algorithm

In this class of problem, the solution is achieved through an Alternative Least Squares algorithm (Gifi, 1990). An implementation of the "classical" non metric PCA can be found in a SAS (1989) procedure (PRINQUAL) in which the user must set several prior parameter values;

- number and positions of knots which define the B-spline basis for each variable transformation,

- shape and properties of transformations (degree of B or I-splines),

- the dimension q (note that solutions are not nested).

The introduction of a spline smoother rather than B or I-splines led Besse and Ferraty (1994) to propose a new algorithm to estimate the curvilinear fixed effect model. For this algorithm, only the dimension q and the p smoothing parameters must be set.

SALSA   Smoothing Alternative Least Square Algorithm

1. *Let $\tilde{\mathbf{Y}}_{(0)} = [\mathbf{y}^1| \cdots |\mathbf{y}^p]$ be the original data set. A matrix $\mathbf{N}_j$ is associated to each vector $\tilde{\mathbf{y}}^j$.*

2. k^{th} *iteration*

   - *Set $\tilde{\mathbf{Y}}_{(k)}$ and solve (18) by the first q terms of the singular value decomposition of $\tilde{\mathbf{Y}}_{(k)}$;*

   $$\hat{\tilde{\mathbf{Y}}}^q_{(k)} = \mathbf{U}_q \mathbf{L}_q^{1/2} \mathbf{V}'_q \ and \ \mathbf{Q}_{(k)} = \mathbf{U}_q \mathbf{U}'_q \mathbf{D}.$$

   - *Set $\mathbf{Q}_{(k)}$ and solve for $j = 1, \dots, p$*

   $$\min_{\tilde{\mathbf{y}}} \left\{ \|\mathbf{Q}_{(k)}\tilde{\mathbf{y}} - \tilde{\mathbf{y}}\|_{\mathbf{D}}^2 + \rho_j \|\tilde{\mathbf{y}}\|_{\mathbf{N}_j}^2 ; \|\tilde{\mathbf{y}}_j\|_{\mathbf{D}} = 1 \right\}. \qquad (19)$$

   The solution $\tilde{\mathbf{y}}^j_{(k+1)}$ is the eigenvector of $\mathbf{D}\left(\mathbf{I} - \mathbf{Q}_{(k)}\right) + \rho_j \mathbf{N}_j$ corresponding to the smallest eigenvalue.

As the least squares criterion is positive and monotonously decreasing, SALSA reaches a minimum but this could be a local one. When a minimum is reached, the last SVD provides a rank q estimate of the fixed effect and also an estimate of the graphical representation subspace $\widehat{E}_q$. It is spanned by the first q eigenvectors $\mathbf{v}^1, \dots, \mathbf{v}^q$ of the correlation matrix of the transformed data $\mathbf{R} = \tilde{\mathbf{Y}}'\mathbf{D}\tilde{\mathbf{Y}}$.

References

BACCINI, A., CAUSSINUS, H., and DE FALGUEROLLES, A. (1993). Analyzing Dependence in large Contingency Tables: Dimensionality and Patterns in Scatter-plots, in *Multivariate analysis: future directions 2*, Cuadras, C.M. and Rao, C.R. eds., Elsivier Science Publishers.

BESSE, P.C. (1992). PCA Stability and Choice of Dimensionality, *Statistics & Probability Letters*, North-Holland **13**, 405-410.

BESSE, P.C. (1994). Simultaneous Non-Parametric Regressions, submitted for publication.

BESSE, P.C., CAUSSINUS, H., FERRÉ, L. and FINE, J. (1988). Principal Components Analysis and Optimization of Graphical Displays. *Statistics*, **19**, 301-312.

BESSE, P.C. and DE FALGUEROLLES, A. (1993). Application of Resampling Methods to the Choice of Dimension in Principal Components Analysis, in *Computer Intensive Methods in Statistics*, Härdle W. and Simar L. eds., Physica-Verlag.

BESSE, P.C., DE FALGUEROLLES, A. and FINE, J. (1995). *A Modeling Approach to Principal Component Analysis*, to appear, Springer-Verlag.

BESSE, P.C. and FERRATY, F. (1994). A Fixed Effect Curvilinear Model, submitted for publication.

BESSE, P.C. and FERRÉ, L. (1993). Sur l'usage de la Validation Croisée en Analyse en Composantes Principales, *Rev. Statistique Appliquée*, **XLI**(1), 71-76.

BESSE, P.C. and POUSSE, A. (1992). Extension des Analyses Factorielles, in *Modèles pour l'Analyse des Données Multidimensionnelles*, J.J. Droesbeke et al. eds., Economica, 129-158.

BESSE, P.C. and RAMSAY, J.O. (1986). Principal Component Analysis of Sampled Curves, *Psychometrika*, **51**, 285-311.

BUJA, A., HASTIE, T. and TIBSHIRANI, R. (1989). Linear Smoothers and Additive Models, *The Annals of Statistics*, **17**, 453-555.

CAUSSINUS, H. (1986). Models and Uses of Principal Component Analysis, in *Multidimensional Data Analysis*, J. de Leeuw et al. eds., DSWO Press, Leiden, 149-170.

CAUSSINUS, H. and RUIZ, A. (1990). Interesting Projections of Multidimensional Data by Means of Generalized Principal Components Analysis, *COMPSTAT 90*, Physica-Verlag, Heidelberg, 121-126.

DENBY, L. and MALLOWS, C.L. (1993). Smooth Reduced-Rank Approximations, *I.S.I.*, 49th session, contributed papers, book 1, 355-356.

FINE, J. and POUSSE, A. (1992). Asymptotic study of the Multivariate Functional Model; application to the Metric choice in PCA, *Statistics*, **23**, 63-83.

FRIEDMAN, J.H. (1987). Exploratory Projection Pursuit, *J. Amer. Statist. Assoc.*, **82**, 249-266.

GIFI, A. (1990). *Non Linear Multivariate Analysis*, Wiley & Sons.

GOODMAN, L.A. (1991). Measures, Models, and Graphical Displays in the Analysis of Cross-classified data, with discussion, Journal of the American Statistical Society, **86**, 1085-1123.

GOWER, J.C. (1984). Multivariate Analysis: Ordination, Multidimensional Scaling and Allied Topics. in *Handbook for Applicable Mathematics*: vol. VI. Statistics, (ed. E.H., Lloyd), 727-781, Wiley.

HASTIE, T.J. and TIBSHIRANI, R.J. (1990). *Generalized Additive Models*, Chapman and Hall.

JOLLIFFE, I. (1986). *Principal Component Analysis*, Springer-Verlag, New-York.

KNEIP, A. and GASSER, T. (1992). Statistical Tools to Analyse Data Representing a Sample of Curves, *The Annals of Statistics*, **20**, 1266-1305.

KRZANOWSKI, W.J. (1987). Cross Validation Choice in Principal Components Analysis, *Biometrics*, **43**, 575-584.

DE LEEUW, J. (1982). Non-linear Principal Component Analysis, COMPSTAT 82, H. Caussinus et al. eds., Physica-Verlag, 77-85.

DE LEEUW, J., VAN RIJCKVORSEL, J.L.A. and WONDER H. (1981). Non-linear Principal Component Analysis with B-spline, *Methods of Operations Research* **33**, 379-393.

PEZZULLI, S. and SILVERMAN, B.W. (1993). On Smoothed Principal Components Analysis, *Computational Statistics*, **8**, 1-16.

RAMSAY, J. (1988). Monotone Regression Splines in Action, *Statistical Sciences* (With Discussion) **3**, 425-461.

RAMSAY, J.O. and DALZELL, C.J. (1991). Some Tools for Functional Data Analysis (with discussion), *J.R. Statist. Soc. B*, **53**, 539-572.

RICE, J.A. and SILVERMAN, B.W. (1991). Estimating the Mean and Covariance Structure Nonparametrically when the Data are Curves, *J. R. Statist. Soc. B*, **53**, 233-243.

RIJCKVORSEL, J. (1988). Fuzzy Coding and B-spline, in *Component and Correspondence Analysis*, J.L.A. van Rijckvorsel et J. de Leeuw eds., Wiley & Sons.

SAS INSTITUTE INC. (1989). *SAS/STAT User's Guide, Version 6, Fourth Edition, Volume 2*, Cary, NC : SAS Institute Inc., 846 pp.

WAHBA, G. (1990). *Spline Models for Observational Data*, SIAM.

WINSBERG, S. and RAMSAY J.O. (1983). Monotone Spline Transformations for Dimension Reduction, *Psychometrika* **48**, 575-599.

Comparing Independent Samples of High-dimensional Observation Vectors

Siegfried Kropf, Jürgen Läuter

Institut für Biometrie und Medizinische Informatik, Otto-von-Guericke-Universität Magdeburg

Abstract. Based on an algorithm for accomplishing permutational tests by means of a Monte–Carlo method with sequential stopping rule, stabilized test statistics for the use in high-dimensional data are proposed, and their performance is compared wi th that of common statistics in a simulation study. Furthermore procedures for the treatment of missing values are discussed.

Keywords. Permutational test, Monte-Carlo method, sequential stopping rule, stabilized test statistics, one-factor model, missing values, EM algorithm

1 Introduction

In the context of multivariate tests for the comparison of independent samples, we proposed the following algorithm for deriving the permutational distribution of the test statistic (Kropf, 1993):

1. Compute the test statistic on the basis of the original allocation of the observation vectors to the samples.
2. Carry out a random permutation of all vectors which results in a new allocation of the vectors to the samples.
3. Recalculate the test statistic on the basis of the new allocation.
4. Determine the rate of permutations having a value of the statistic that is greater than or equal to the original one (as an estimator for the 'true p-value' π of the permutational test).
5. Test the hypothesis $\pi \leq \alpha$ in a sequential procedure, where α is the prespecified level of the permutational test. If either $\pi \leq \alpha$ (significant outcome of the permutational test) or $\pi > \alpha$ (nonsignificant o utcome) can be concluded within the sequential test procedure, then stop, else repeat from step 2.

This sequential procedure enables us the practical evaluation of the permutational distribution even in large samples. It has neglectable influence on the error rates and shortens the procedure considerably.

The proposed algorithm does not refer to special tests but is generally applicable in the comparison of independent samples. Moreover, step 3 can include preparing steps such as elimination of outliers and imputation of missing values, provided that outli ers or missing values occur randomly with equal probability in all samples.

On this basis, test statistics for high-dimensional observation vectors are proposed and compared to common test statistics for multiple endpoints in section 2. The treatment of missing values is discussed in the last section.

2 Stabilized statistics

2.1 Statistical model

Though the proposed procedure is distribution-free in nature, we shall focus our consideration onto normal data. But we are free now to use statistics whose distribution is complicated or still unknown.

Restricting to the case of two populations with the same covariance matrix, we have the p-dimensional observation vectors

$$y_{ij} \sim N(\mu_i, \Sigma) , \qquad i = 1, 2; \; j = 1, \ldots, n_i \quad .$$

As an important (more parsimonious) special case we use the so-called one-factor model (Läuter, 1992), which is characterized by the covariance structure

$$\Sigma = K + \omega \mu \mu'$$

with $\mu = \mu_1 - \mu_2$, ω a positive scalar factor, and K a diagonal matrix of nonnegative specific variances. This model implies a parameter restriction with the property of smoothing. It is likely to describe phenomena in medicine or other branches of science where a set of variables is controlled by a latent univariate variable. Each variable has a specific disturbance and an individual linear scaling.

2.2 Test statistics

As basic statistic we shall regard Hotelling's T^2-statistic

$$T^2 = \frac{n_1 n_2}{n(n-2)} y' S^{-1} y$$

where $n = n_1 + n_2$, $y = \bar{y}_1 - \bar{y}_2$ (difference of sample means) and S is the pooled sample covariance matrix. It is used in the common F-test as well as in the framework of the permutational test.

The T^2-statistic can be calculated for $p < n - 2$, but it is known that the statistics becomes instable with increasing p (Läuter, 1992).

As first stabilized statistic we introduce

$$T_U^2 = \frac{n_1 n_2}{n(n-2)} y' (\text{Diag}(S + \frac{n_1 n_2}{n(n-2)} yy'))^{-1} y ,$$

which is based on an independency assumption of the variables.

Exploiting features of the one-factor model we can use the statistic

$$T_{of}^2 = \frac{\frac{n_1 n_2}{n} p y'(G^* + \text{Diag}(\frac{n_1 n_2}{n} yy'))^{-1} y}{\text{tr}(G(G^* + \text{Diag}(\frac{n_1 n_2}{n} yy'))^{-1})}$$

with $G = (n-2)S$ and the modified matrix

$$G^* = \begin{cases} \frac{1}{p-1}(p \text{ Diag}(G) - G) & \text{for} \quad p \geq 2 \\ G & \text{for} \quad p = 1 \end{cases}$$

which is the result of a Bayesian consideration (Läuter, 1992).

By means of the ridge technique as another more general method of stabilization we obtain a statistic

$$T_R^2 = \frac{n_1 n_2}{n} y'(G + \frac{p(n-2)}{(n-4)(n+p-3)} T_0^{-1})^{-1} y ,$$

where

$$T_0^{-1} = \text{Diag}(G (\text{Diag}(G))^{-1} G)$$

is the so-called 'stabilized inverse product sum matrix' (Läuter, 1992).

In the above form, the statistics can be used for undirected alternative hypotheses. A directed version of them is derived when the calculations include only those of the p variables for which $y_i = \bar{y}_{1i} - \bar{y}_{2i} > 0$ and – only for T^2 – the ith component of $S^{-1}y$ is positive, too.

Other competitors for these statistics in their one-sided form are the OLS and GLS statistics of O'Brien (1984) and two variations — the centered statistic (CS) and the approximate likelihood ratio statistic (ALR) — proposed recently by Tang, Geller, and Pocock (1993).

2.3 Simulation results

Simulation experiments have been done for the above statistics with $n_1 = n_2 = 8$, $p = 10$ and three different data structures which all have a Mahalanobis distance of 4.0 between the two populations:

Structure A: One-factor structure with all 10 variables correlated with $\rho = 0.25$ and equal, uniformly directed mean differences.

Structure B: One-factor structure with 5 variables correlated with $\rho = 0.8$ and equal mean differences, and 5 uncorrelated variables with no mean differences.

Structure C: Random covariance structure produced by linear transformation of standard normal vectors by a matrix with random elements (only for undirected tests).

The level of the tests was $\alpha = 0.05$. All simulations have been done with 10,000 replications.

The results are displayed in table 1. Regarding the T^2-test and its variations they show that

Table 1. Error of 2nd kind in simulations with structures A, B, and C

statistic	A undir.	A directed	B undir.	B directed	C undir.
T^2 (param.)	.740		.739		.731
T^2 (perm.t.)	.742	.042	.744	.163	.730
T_U^2	.050	.018	.049	.020	.525
T_{of}^2	.046	.019	.046	.018	.545
T_R^2	.091	.041	.181	.134	.503
OLS		.016		.046	
GLS		.326		.786	
CS		.017		.021	
ALR		.516		.513	

- the results of the T^2-statistic in parametric and permutational framework agree very well,
- the statistics with restricted models are more powerful than T^2, even when the simulated parameter structure doesn't match the restriction completely,
- in the one-factor structures A and B T_{of}^2 has the best results, in the more general structure C the less restricted ridge statistic does best,
- for directed alternatives an essential improvement is possible though the statistics are quadratic forms in mean differences.

With respect to O'Brien's and related statistics we can state that the more sophisticated GLS and ALR method should not be used in high dimensions with small or moderate samples, whereas both OLS and CS method work well and are even better than T_{of}^2 in the uniformly directed structure A, but are less efficient in structure B with noice variables.

Simulations with tests on the basis of the four statistics under utilization of their asymptotic distribution (cf. Tang, Geller ansd Pocock, 1993) show that they do not keep the error of first kind. For the same covariance structures and sample sizes as a bove, OLS and CS have increased error levels of about 0.07, GS and ALR even of above 0.20. So these statistics should be used only in permutational tests when p is large.

3 Treatment of missing values

There are three general possibilities to treat missing values in multivariate tests:

i. cases with missing values are omitted (complete case analysis),
ii. missing values are substituted by estimates before usual test,
iii. tests statistics are used that accept missing values.

The practicability of the complete case analyses is limited in high-dimensional data because the number of complete cases will often be small.

We assume that missing values are of type MCAR (missing completely at random, cf. Little and Rubin, 1987). Then the use of permutation tests ensures the α-level in cases ii and iii, supposed the estimation procedure for missing values in the substitution methods is involved in the permutational process.

Usual estimation methods for missing values are variablewise substitution by total means or by group means, respectively, and regression methods.

Examples for case iii are the proposals of Wei and Lachin (1984) and subsequent work.

We shall combine the statistics of section 2 with the substitution methods of type ii 'total means' and 'group means', respectively. As third method of type iii, we will use the same statistics with $\bar{y}_1$, $\bar{y}_2$ and S replaced by such estimates of mean vectors and covariance matrices where each element is derived on the basis of all available data for the corresponding variable(s).

A more sophisticated estimation method is the EM algorithm (cf. Little and Rubin, 1987) which essentially consists in an iterative regression procedure. To avoid the problems of instability in high-dimensional data, we propose to estimate the mean vectors and the covariance matrix by an EM algorithm assuming a one-factor structure. Unfortunately, even the loglikelihood function for the complete data

$$l_0 = \ln \left(\prod_{j=1}^{2} \prod_{l=1}^{n_j} \frac{1}{(2\pi)^{p/2} |\Sigma|^{1/2}} e^{-\frac{1}{2}(y_{jl}-\mu_j)'\Sigma^{-1}(y_{jl}-\mu_j)} \right)$$

with $\Sigma = K + \omega\mu\mu'$ and $\mu = \mu_1 - \mu_2$ is a rather complicated function in its arguments μ_1, μ_2, K, and ω, so that a solution of the likelihood equations (M-step of EM algorithm) is not obtained in a noniterative way, and can be outside parameter restrictions of the model (Läuter, 1992). That's why it should be better to use a generalized EM algorithm for the maximization such as the EM1 algorithm of Rai and Matthews (1993) which carries out a Newton-Raphson algorthm in the M-step but restricts it to a single iteration under regard of the parameter restrictions. To avoid the tedious evaluation of the information matrix (matrix of second derivatives of l_0 with respect to the model parameters) it should be approximated by the sampling covariance matrix of the score (vector of first derivatives of l_0 with respect to the parameters, cf. Little and Rubin, 1987).

Table 2 shows results of simulation experiments for data with structures B and C from section 2. The columns represent the errors of second kind for the tests with complete data and for the three outlined procedures of handling missing data applied to the same data where in each variable three out of the 16 values were randomly replaced by missing values.

The results for the parametric T^2-test cannot be compared because only the substitution by total means kept the α-level, whereas the other two methods yielded rejection rates greater than 0.10.

Table 2. Error of second kind for procedures with complete data and those with a portion of about 20 % of missing values, treated by three different methods for structures B (undirected and directed) and C (undirected only)

statistics	complete data			total means			group means			avail. cases		
	$B_{und.}$	$B_{dir.}$	$C_{und.}$	$B_{und.}$	$B_{dir.}$	$C_{und.}$	$B_{und.}$	$B_{dir.}$	$C_{und.}$	$B_{und.}$	$B_{dir.}$	$C_{und.}$
T^2-p.t.	.744	.163	.730	.816	.211	.842	.735	.211	.820	.870	.500	.907
T^2_U	.049	.020	.525	.069	.029	.581	.070	.029	.583	.069	.030	.580
T^2_{of}	.046	.018	.545	.062	.025	.599	.062	.025	.597	.061	.024	.603
T^2_R	.181	.134	.503	.199	.139	.585	.227	.158	.603	.264	.219	.612
OLS		.046			.069			.069			.071	
GLS		.786			.770			.766			.858	
CS		.021			.041			.042			.064	
ALR		.513			.608			.506			.999	

Despite of a high rate of missing values the loss of power is only moderate and the differences between the three methods for treatment of missing values are neglectible *in most stabilized statistics*. The substitution of group means was distinctly favourable only for the statistics with poor performance (T^2, GLS, and ALR). In the other cases the substitution of total means was favourable or at least not far below the other methods, but requires less computational effort, because this substitution is invariant of permutations and needs to be performed only once before permutation. Results regarding the EM algorithm are not available yet.

Concluding we can summarize that the proposed permutational procedures are valuable tools to provide high-dimensional multivariate tests under the conditions of instability and under presence of missing values.

References

Kropf, S. (1993). Applicability of Multivariate Rank Tests for the Comparison of Independent Samples. Computational Statistics 8, 71-85.

Läuter, J. (1992). Stabile Multivariate Verfahren. Akademie Verlag, Berlin.

Little, R.J.A.; Rubin, D.B. (1987). Statistical Analysis with Missing Data. John Wiley, New York.

O'Brien, P.C. (1984). Procedures for Comparing Samples with Multiple Endpoints. Biometrics 40, 1079-1087.

Rai, S. N.; Matthews, D.E. (1993). Improving the EM Algorithm. Biometrics 49, 587-591.

Tang, D.I.; Geller, N.L.; Pocock, S.J. (1993). On the Design and Analysis of Randomized Clinical Trials with Multiple Endpoints. Biometrics 49, 23-30.

Wei, L.J.; Lachin, J.M. (1984). Two-Sample Asymptotically Distribution-Free Tests for Incomplete Multivariate Observations. JASA 79, 653-661.

Unmasking Influential Observations in Multivariate Methods

Yutaka Tanaka, Shingo Watadani

Department of Statistics, Okayama University

Abstract. A robust version of general procedure of sensitivity analysis is proposed for detecting influential subsets of observations in multivariate methods in which influence functions are available. A numerical investigation is carried out to illustrate the performance of the proposed procedure for detecting two types of perturbed observations in confirmatory factor analysis.

Keywords. Influence function, influential subset, additivity of influence, sensitivity analysis, robust estimation

1 Introduction

Methods of detecting individually influential observations are well established not only in regression and related methods (see, e.g., Belsley, Kuh and Welsch, 1980; Cook and Weisberg, 1982) but also in other multivariate methods such as principal component analysis, canonical correlation analysis, factor analysis and covariance structure analysis (see, e.g., Critchley, 1985; Radhakrishnan and Kshirsagar, 1981; Tanaka and Odaka, 1989a, b; Tanaka, Watadani and Moon, 1991). Influence measures used in detecting individually influential observations can be directly generalized for evaluating the effect of multiple observations. However, it is not easy to detect influential subsets of observations because of the so-called "masking and swamping effects." which indicate the phenomena that the effect of an influential observation is hidden by other observations. To detect influential subsets in multivariate methods Tanaka and his coworkers have utilized an additivity property that the influence of a set of observations is equal to the sum of the influence of each observation in the set as far as up to the first order term of the perturbation expansion is considered, and have proposed a general procedure of sensitivity analysis shown in section 2. But it still has a possibility to suffer from the masking and swamping effects.

In the present paper we try to robustify their general procedure to protect from those effects. The proposed procedure can be applied to any multivariate method in which influence functions are available. But, here we assume that we are applying one of multivariate methods which are classified into covariance structure analysis.

2 A General Procedure of Sensitivity Analysis

Let us consider the influence of a set of k observations $A = \{\underline{x}_{i_1}, \dots, \underline{x}_{i_k}\}$ on a parameter vector $\underline{\theta}(F)$, which is given as a functional of the cumulative distribution function (cdf). To do this we introduce a perturbation on the cdf from F to $\tilde{F} = (1-\varepsilon)F + \varepsilon G$, where $G = k^{-1}\sum_{\underline{x}_i \in A} \delta_{\underline{x}_i}$, $\delta_{\underline{x}_i}$ being the cdf of a unit point mass at $\underline{x}_i$, and define a generalized theoretical influence function of A as the limit $TIF(A; \underline{\theta}) = \lim_{\varepsilon \to 0}[\underline{\theta}(\tilde{F}) - \underline{\theta}(F)]/\varepsilon$. Then it can be easily verified that $TIF(A; \underline{\theta}) = k^{-1}\sum_{\underline{x}_i \in A} TIF(\underline{x}_i; \underline{\theta})$, where $TIF(\underline{x}_i; \underline{\theta})$ is the ordinary influence function of $\underline{x}_i$. The similar relation holds for the empirical influence function (EIF). Hence the parameter estimate based on the sample with a subset A omitted can be approximated as

$$\hat{\underline{\theta}}_{(A)} \cong \tilde{\underline{\theta}}_{(A)} \equiv \hat{\underline{\theta}} - (n-k)^{-1} \sum_{\underline{x}_i \in A} EIF(\underline{x}_i; \hat{\underline{\theta}}), \qquad (1)$$

where symbol (˜) indicates the linear approximation based on the EIF. Generally in most multivariate methods (see, e.g., Tanaka and Odaka, 1989a, b; Tanaka, Watadani and Moon. 1991) the empirical influence function of $\underline{x}_i$ for $\hat{\underline{\theta}}$, i.e., $EIF(\underline{x}_i; \hat{\underline{\theta}})$, is obtained as a linear function of the elements of $EIF(\underline{x}_i; S)$, where S is the sample covariance matrix defined by $S = n^{-1}\sum_i (\underline{x}_i - \bar{\underline{x}})(\underline{x}_i - \bar{\underline{x}})^T$ and $EIF(\underline{x}_i; S)$ is given by $EIF(\underline{x}_i; S) = (\underline{x}_i - \bar{\underline{x}})(\underline{x}_i - \bar{\underline{x}})^T - S$. Based on the relation (1) we can adopt the following general procedure of sensitivity analysis in any multivariate method, where influence functions are available (Tanaka. Castaño-Tostado and Odaka, 1990; Tanaka, 1992).

Step 1. Compute the EIF vectors, $EIF(\underline{x}_i; \hat{\underline{\theta}})$, $i = 1, \dots, n$.

Step 2. Summarize the EIF vectors into scalar influence measures from various aspects such as the influence on the estimate $\hat{\underline{\theta}}$, on its precision and on the goodness of fit. Find observations which are individually influential.

Step 3. Search for subsets of observations whose members are individually relatively influential and have similar influence patterns using principal component analysis and other multivariate techniques.

Among possible influence measures are the so-called generalized Cook's D defined by $D_i = (\tilde{\underline{\theta}}_{(i)} - \hat{\theta})^T [\widehat{acov}(\hat{\underline{\theta}})]^{-1} (\tilde{\underline{\theta}}_{(i)} - \hat{\theta})$ and the $COVRATIO$-like measure defined by $CVR_i = |\widehat{acov}(\tilde{\underline{\theta}}_{(i)})|/|\widehat{acov}(\hat{\underline{\theta}})|$, where $\tilde{\underline{\theta}}_{(i)}$ is an approximate estimate based on the sample without the i-th observation, and $\widehat{acov}(\cdot)$ is an estimated asymptotic covariance matrix.

3 A Robust Version of the General Procedure

The general procedure in the previous section works well in ordinary circumstances. However, the so-called masking and swamping effects may occur
when the estimated parameters are far from the true values. For example,
when $\bar{x}$ and S are quite different from the true values, the empirical influence
function $EIF(\underline{x}_i; S)$ does not reflect correctly the influence on the covariance
matrix and this fact may cause the masking and/or swamping effects. These
effects can be protected by using an appropriate robust method for parameter
estimation. Among possible procedures we propose the following one.

Step 1. Using the minimum volume ellipsoid estimator (MVE) followed by
the iterative process of one-step improvement (Rousseeuw and Leroy, 1987;
Rousseeuw and van Zomeren, 1990), assign unit weight or zero weight to
each observation. Regard the observations with unit weights as active observations and those with zero weights as supplementary observations.

Step 2. Using only the active observations compute the mean vector $\bar{x}_R$ and
the covariance matrix S_R. Then. obtain the parameter estimate $\hat{\underline{\theta}}_R$.

Step 3. Using only the active observations compute the coefficient matrix in
the relation between $EIF(\underline{x}_i; S_R)$ and $EIF(\underline{x}_i; \hat{\underline{\theta}}_R)$.

Step 4. Compute the EIF for the covariance matrix not only for the active
observations but also for the supplementary observations in such a way that
$EIF(\underline{x}_i; S) = (\underline{x}_i - \bar{x}_R)(\underline{x}_i - \bar{x}_R)^T - S_R$, and compute the corresponding
EIF for $\hat{\underline{\theta}}$ using the linear relation obtained in Step 3.

Step 5. Summarize the obtained EIF vectors into scalar influence measures
and find candidates for influential observations. Also apply principal component analysis or other multivariate methods to the data set of the EIF
vectors for $\hat{\underline{\theta}}$. and find observations which are relatively influential and have
similar influence patterns.

It may seem that a method of M-estimation (see, e.g., Campbell, 1980)
can be used instead of the MVE estimation in Step 1. The major reason
we use the MVE estimation is that the proposed procedure preserves the
property of high breakdown. The important thing is to start from a subset
of observations which does not contain any influential observation. The iterative process of one-step improvement is slightly modified from the original
procedure as follows. The reason of this modification is to take the structure
of the covariance matrix into consideration.

Step 1. Assign a weight w_i to each observation by mean of the rule

$$w_i = \begin{cases} 1, & \text{if } (\underline{x}_i - T(X))^T C(X)^{-1}(\underline{x}_i - T(X)) \leq \chi^2(p, 0.975), \\ 0, & \text{otherwise.} \end{cases}$$

Step 2. Obtain the reweighted mean vector $T(X)$ and covariance matrix
$C(X)$.

Step 3. Apply the method of covariance structure analysis to $C(X)$ and obtain the estimated covariance matrix $\hat{C}(X)$.

Step 4. Go back to Step 1 if not converged.

4 Numerical Investigation

Let us investigate the performance of the proposed procedure in the case of confirmatory factor analysis. To do this three sets of artificial data are generated based on the following factor analysis model:

$$
\begin{bmatrix} x_1 \\ x_2 \\ x_3 \\ x_4 \\ x_5 \end{bmatrix} = \begin{bmatrix} 0.8 & 0 \\ 0.8 & 0 \\ 0 & 0.8 \\ 0 & 0.8 \\ 0 & 0.8 \end{bmatrix} \begin{bmatrix} f_1 \\ f_2 \end{bmatrix} + \begin{bmatrix} 0.6e_1 \\ 0.6e_2 \\ 0.6e_3 \\ 0.6e_4 \\ 0.6e_5 \end{bmatrix} \tag{6}
$$

$$
(f_1, f_2) : NID \left(\underline{0}, \begin{bmatrix} 1 & 0.8 \\ 0.8 & 1 \end{bmatrix} \right) \quad \text{random numbers.}
$$

$e_1, \ldots, e_5 : NID(0, 1)$ random numbers.

Data set A (f-perturbed data set)
A set of 100 observations are generated based on the above factor analysis model. But, in generating observations #41 to #45 the values of (f_1, f_2) are replaced by $(2.4, -2.4)$.

Data set B (e-perturbed data set)
A set of 100 observations are generated based on the above factor analysis model using the same random numbers as in Data set A. But, in generating observations #81 to #85 the values of (e_1, e_2) are replaced by $(3.6, -3.6)$.

Data set C (f and e-perturbed data set)
A set of 100 observations are generated based on the above factor analysis model using the same random numbers as in Data sets A and B. But, both of the two types of perturbation in Data sets A and B are introduced to observations #41 to #45 and #81 to #85.

Before analyzing Data sets A, B, and C, we applied the ordinary general procedure to, say, Data sets A' and B' in which only one observation (#41 or #81) instead of five in each set was replaced by the perturbed observation, and found as the results that both of the perturbed observations were the most influential with the measure of the generalized Cook's D. However, the results of the ordinary general procedure applied to Data sets A, B and C could not reveal the perturbed observations as the most influential. Because of the limitation of space we show only the result for Data set C. Fig. 1 is the index plot of the generalized Cook's D for Data set C. It is noted that

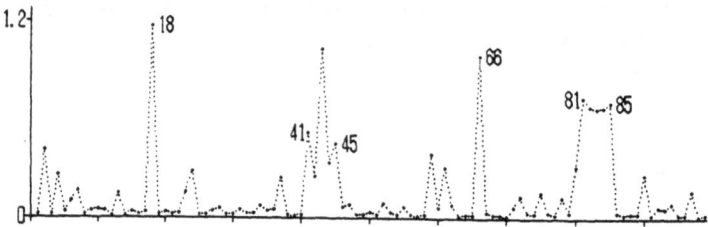

Fig. 1. Index plot of the generalized Cook's D in the ordinary procedure

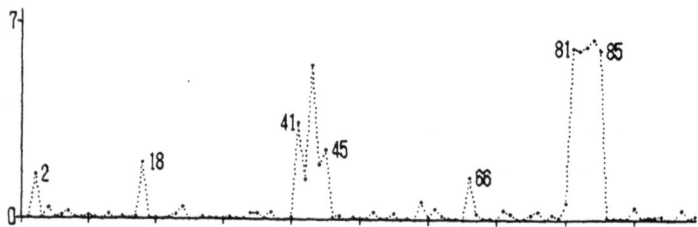

Fig. 2. Index plot of the generalized Cook's D in the robust procedure

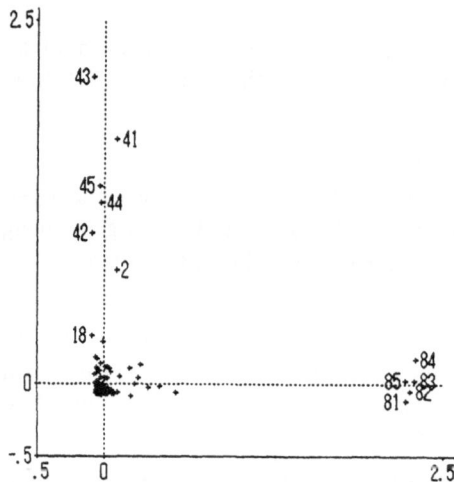

Fig. 3. Scatter plot of the first two canonical variates in the robust procedure

observations #18 and #66 are more influential than most of the perturbed observations #41 to #45 and #81 to #85.

Then, the robust version of the general procedure was applied to Data sets A, B and C. In all cases the perturbed observations were easily found to be influential. We shall discuss the case of Data set C in detail. We randomly drew 1000 subsamples of size 6, searched for the MVE, and then proceeded to the iterative process of one-step improvement. In this iterative process zero weights were assigned to 22 observations including #45 and #81 to #85 at the initial step, and after 4 iterations we obtained the converged solution, in which zero weights were assigned to 15 observations including #41 to #45, #81 to #85, #2, #18 and #66. The index plot of the generalized Cook's D is shown in Fig. 2. Then, to search for observations with similar influence patterns, we applied canonical variate analysis or, in other words, principal component analysis with metric $[\widehat{acov}(\hat{\theta})]^{-1}$. The eigenvalues (and cumulative proportions) obtained were $26.63(51.4\%) > 13.06(76.6\%) > 2.50(81.4\%) > 2.31(85.8\%) > \cdots$ in order of magnitude. Fig. 3 shows the scatter plot of the first two canonical variates. From this scatter plot we can easily find the two sets of influential observations.

References.

Belsley, D. A., Kuh, E. and Welsch, R. E. (1980). *Regression Diagnostics: Identifying Influential Data and Sources of Collinearity*. John Wiley & Sons.

Campbell, N. A. (1980). Robust procedures in multivariate analysis I: Robust covariance estimation. *Appl. Statist.* 29, 231–237.

Cook, R. D. and Weisberg, S. (1982). *Residuals and Influence in Regression*. Chapman and Hall.

Critchley, F. (1985). Influence in principal component analysis. *Biometrika*, 72, 627–636.

Radhakrishnan. R. and Kshirsagar, A. M. (1981). Influence functions for certain parameters in multivariate analysis. *Comm. Statist.*, A10, 515–529.

Rousseeuw, P. J. and Leroy, A. M. (1987). *Robust Regression and Outlier Detection*. John Wiley & Sons.

Rousseeuw, P. J. and van Zomeren, B. C. (1990). Unmasking multivariate outliers and leverage points. *JASA*, 85, 633–639.

Tanaka, Y. (1992). Sensitivity analysis in multivariate methods: Principles, methods and software. *Technical Report of Okayama Statistical Association*. 52.

Tanaka, Y., Castaño-Tostado, E. and Odaka, Y. (1990). Sensitivity analysis in factor analysis: Methods and software. *COMPSTAT 1990* (K. Momirović and V. Mildner, eds.), Physica-Verlag, 205–210.

Tanaka, Y. and Odaka, Y. (1989a). Influential observations in principal factor analysis. *Psychometrika*. 54, 475–485.

Tanaka. Y. and Odaka, Y. (1989b). Sensitivity analysis in maximum likelihood factor analysis. *Comm. Statist.*, A18, 4067–4084.

Tanaka, Y., Watadani, S. and Moon, S. H. (1991). Influence in covariance structure analysis: With an application to confirmatory factor analysis. *Comm. Statist.*, A20, 3805–3821.

Part XIV

Computational Aspects in Time Series Analysis

Application of a Multiagent Distributed Architecture for Time Series Forecasting

Albert Prat, Jesús Lorés, Kanapathipillai Sanjeevan, Ignasi Solé, Josep M. Catot
Universidad Politécnica de Cataluña, Departamento de Estadística e Investigación Operativa

Abstract Designing a software system that incorporates expert knowledge to solve complex problems using sophisticated techniques is a difficult task. Additionally, these systems are required to have on-line help and explanation facilities as well as the flexibility of being utilised by users possessing different levels of expertise (from expert to novice). All this functionality has to be integrated with a suitable Graphical User Interface (GUI). In the last few years, important work has been published that discuss the essential and desirable characteristics of such a system (Dodge et al., 1993) and the architecture and methodologies required to develop them (Prat et al., 1993a). Prototypes (Prat et al., 1993b) and high quality commercial systems have appeared that utilise the work done in this area. These systems are known as Knowledge Based Systems (also known as KBFEs and SKBFEs).

Knowledge Based Systems evolved from simple Rule-based (Expert) Systems to incorporate powerful frame based, object-oriented representation of a domain. Recently, they have been enhanced further by the combination of emerging technologies such as Neural networks, CBR (case based reasoning), Genetic algorithms, Virtual reality and Multimedia (Hedberg et al., 1993). These new systems that integrate off-the-shelf programmes and custom modules are coming to be known as Multiagent Systems.

In this paper we present MEDISA (Multi agEnt DIStributed Architecture) and the tools for the design and development of systems for solving statistical problems. The strategy is to divide a given problem into sub-problems that are worked on by designated "Agents" (soft robot doing tasks in the computational world) that communicate back the results. The architecture and the tools have been developed for the Windows3, Windows-NT platforms. Of the two systems developed using this architecture (DESEXP, FORESEE), we shall be describing the design and functionalities of FORESEE.

FORESEE has the ability to identify, estimate and verify automatically, ARIMA models (univariant, intervention and transfer function).

302

The System has evolved from work carried out by the authors over the last 14 years, especially the methods and tools developed during the FOCUS Project [1].

Keywords. Knowledge Based Systems, Box & Jenkins, Arima models, Multiagent Distributed Systems, Intervention analysis, Transfer function, Genetic Algorithms, Neural Networks.

1 Introduction

Expert Systems were computer programmes that attempted to encode the expertise or the knowledge of an expert in a particular domain. This knowledge was encoded in the form of IF---THEN rules that were consulted by the programme at relevant times. These systems proved to be limited in scope due to the problem of encompassing all relevant knowledge in this form and the inability to model complex interactions and relations between objects in a domain. In the mid-1980's, an improved methodology for representing domain knowledge was proposed that would still retain the rule-based knowledge base and at the same time enhances the modelling of this domain knowledge by linking it with object oriented techniques. The software development tools and applications that follow this methodology are known as Knowledge Based Systems.

The design of knowledge based systems is entering a new phase with the rapid evolution of computer technology. The incorporation of multimedia, neural networks, genetic algorithms, fuzzy logic, etc., can greatly enhance the ability of a KBS designer to develop systems to solve very complex problems. There is a challenge to come up with new architectures that integrate KBSs with the new tools (Huang et al., 1991). In this paper, we propose an overall architecture for one such system and present a set of tools for developing them along with a description of a prototype system that we have developed for forecasting time series.

In the first part of this paper, we give a short description of the proposed architecture that we call MEDISA (Multi agEnt DIStributed Architecture) that has been implemented under the Windows3 / Windows-NT environment. It allows the incorporation of off-the-shelf and/or custom software that interact and co-operate in the problem solving process. We also present some tools that have been constructed to aid in the development of a system. In the second part, we present FORESEE, a prototype system for forecasting time series that has been developed using this architecture and tools. It has the ability to identify, estimate and verify -automatically - univariant ARIMA models along with the ability to do

---

[1] The FOCUS project is an ESPRIT II project, num. 2620, also partially financed by the CICYT (Spanish government, TIC 88-0643 and TIC91-0193-CE.)

Intervention analysis and use Transfer functions. Additionally, the system generates forecasts using Genetic Algorithms and Neural Networks methodology, selecting the best forecast from these methods. This system is also able to determine the underlying trend, seasonality and the inertia corresponding to a given time series.

2 The Multi Agent Distributed Architecture.

Knowledge Based Systems are in general founded on a problem - solving framework for a given application domain. Solving a complex problem entails input from many sources that interact to arrive at a result. The problem solving model that we have developed has its roots in the blackboard model (Engelmore, 1988), which is a framework for partitioning a problem into many sub-problems. In this model, a collection of intelligent agents are gathered around a blackboard, looking at data written on it and subsequently writing their conclusions on the blackboard. This process ultimately leads to the solving of the overall problem in this co-operative way.

The main drawback of this approach is the difficulty of maintaining consistent data values and tracking the state of the problem solving strategy when many agents are concurrently working on the solution. In order to overcome this problem, we have instead followed a modification to this model called the Serial Blackboard Model whose main characteristics are defined as follows: (Engelmore, 1988)

- Only one intelligent agent can be working at any given time.
- In order to co-ordinate the execution of the agents, a scheduler or control mechanism (agent) is needed.
- The blackboard is not globally visible. Each agent has only a limited view of it and has enough information to work on a self-contained sub-problem.
- It is assumed that all agents operate within a valid and consistent context and the scheduled execution of the agents (even when the scheduling is done dynamically by the scheduling/planing agent) preserves the consistency of the global data.

2.1 What Is An Agent?

The idea of an agent was first proposed by researchers at MIT. In our context, an agent could be an object, module, package or programme that has sufficient knowledge to communicate intelligently with other such agents, in order to co-opertively solve a complex problem. The idea is to define a particular agent to be responsible for the solving of a given sub-problem. Figure 1 illustrates the general structure of an agent (Erceau et al., 1993) along with the four basic components that we define:

I.   Communications and Assigning of Tasks.
II.   Action, perception and learning capabilities.
III.   Plans, knowledge and tasks.
IV.   Models, commitments and intentions.

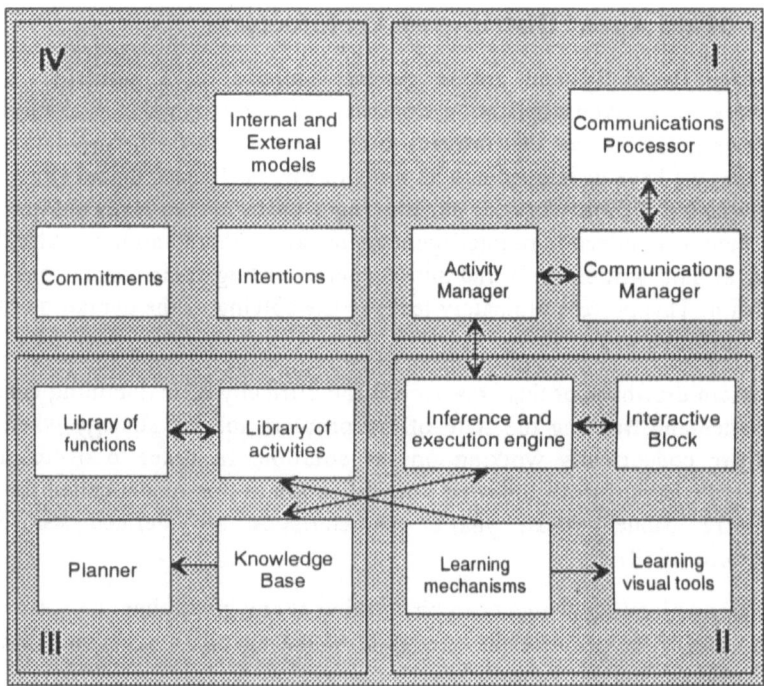

Fig. 1. General structure of an agent.

2.1.1 Communications and Assigning of Tasks

This component of the agent manages the sending and receiving of messages to other agents. Additionally, it also has knowledge about particular tasks that have to be accomplished by its other components and manages the scheduling and execution of these internal tasks.

Agents communicate with each other through messages. All agents are assumed to have implemented a Communication Module (CM) to facilitate this. The CM of an agent is free to choose any protocol (TCP/IP,DDE) at the lower levels while maintaining a uniform set of primitives at the application level.

Messages have a structure that has the following fields:

<Message ID> To identity type of message

<Sender>   To define the sender

<Receiver>   To define the receiver

<Contents>   Contents of the message

An Activity Manager (AM) module within this component monitors the sending and receiving of messages and the control of ativities (the unit of action of an agent).

2.1.2 Action, Perception and Learning Capabilities

This component has a module that is responsible for reasoning and the execution of activities. There is also a user interface module that requests the user for any relevant or required input at a given stage of execution of an activity. A visual tool is also available to the expert of a domain to store knowledge in the agent.

2.1.3 Plans, Knowledge and Activities

This component stores all the domain knowledge for solving a particular problem. This expert knowledge is one that has been entered via the visual tool described above or that which has been learned at run time by the second component of an agent.

The knowledge contained here could be in the form of heuristics or procedural knowledge. It is grouped in the form of "activities" that are the unit of action of agents.

2.1.4 Models, Commitments and Intentions

This last component of an agent has not been implemented in the present system. It is envisaged that it will have special capabilities such as automatic learning and generation of knowledge based on higher level awareness of the problem space and its entities.

2.2 Different Types of Agents

Figure 2 illustrates in detail, the complete set of agents (Lorés, 1994), that participate in the resolution of a problem. This general scheme is defined around the serial blackboard model that we have chosen as the most feasible from the point of view of implementation.

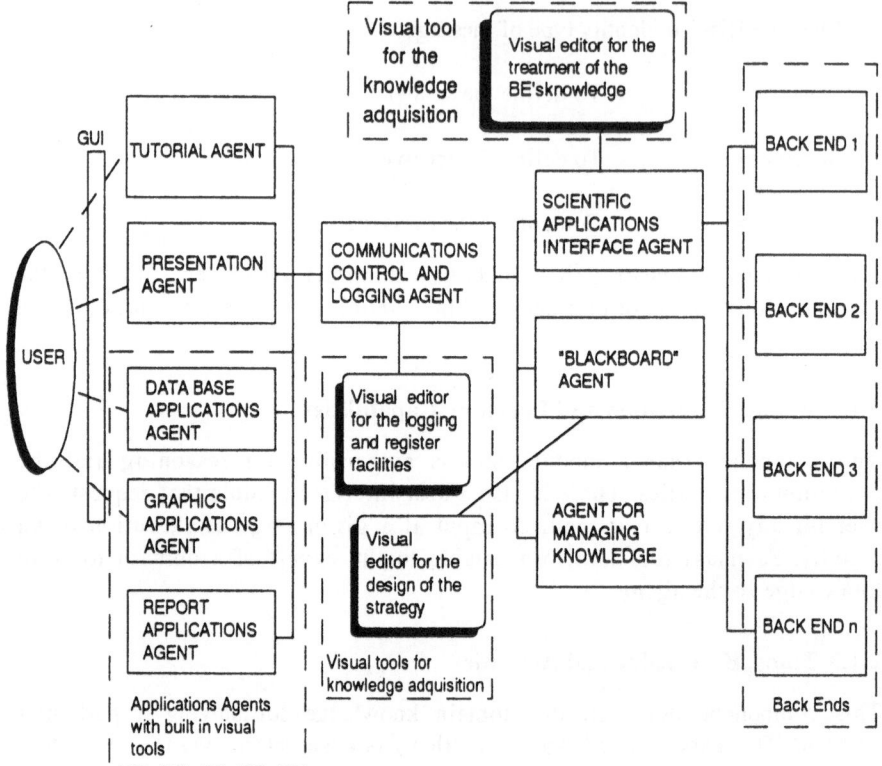

Fig. 2. General components of MEDISA architecture.

From figure 2, the agents of the general architecture are:

- "Blackboard" Agent (Planning and Control)
- Communications Control and Logging Agent
- Presentation Agent
- Agent for Managing Knowledge (Knowledge Based Agent)
- Management of Scientific Applications Agent (Back End Manager)
- Tutorial Agent
- Applications Agents (as for example, Data Base Management, Graphics, Report Writing, ...)

From this general scheme it becomes quite clear that there are two agents that are essential to the over-all architecture: An agent for the planning and control of the problem solving process (i.e representing the blackboard) and an agent for maintaining a registry of other agents and for supervision and management of the communications between them. This last agent is the Communications Control and Logging Agent (CCLA).

In what follows, we present a brief description of each agent.

The "Blackboard" Agent

The Scheduling or control mechanism proposed in the serial blackboard model is implemented here as the "Blackboard" agent. This agent is the repository of the global strategy for solving a complex problem. The domain expert (s) are provided with visual tools to define the problem solving strategy. It is in the form of a state machine diagram where each state corresponds to a pre-defined stage in the solution process and the transitions correspond to one or more activities of an agent.

The "Blackboard" Agent has knowledge about the problem solving strategy and guides the user towards the solution. It also knows at any given moment, the state of the process and receives information from the other agents in order to have a global and complete picture of all the activities.

Communications Control and Logging Agent

The CCLA is one of the essential and important agent of the MEDISA architecutre. It oversees the initiating of all processes that need to be running at system start up. During subsequent invocation(s) of other agents or applications, a connection is established first with the CCLA which thereby maintains the global process state.

A visual tool is available to the system designer to specify the initial configuration to the CCLA. This would entail specifying the host (name, type, etc.) and the communication protocol to be used. Another visual tool enables the designer to enter information about all the agents that might participate in the problem solving process.

The CCLA thus maintains a registry of all the agents and their state (active, idle, etc.). The other important function of this agent is the management of communications between processes.

Presentation Agent

Much work has been done in the area of Human-Computer Interaction (HCI) in the recent past to come up with alternatives to the menu driven or hypertext based user interfaces. From past experience, we had seen that menu-driven interfaces are not well suited for systems solving complex problems where interactions take place among multiple agents and the user. For this reason, we have chosen the story metaphor which consists of scenes and dialogues in which multiple actors (agents) participate.

We define the story as the set of all necessary interactions with the user to solve a given problem. This is divided into subsets called scenes. Scenes contain all necessary interactions to solve a subproblem and consists of a set of dialogues. These (dialogues) become the basic unit of interaction with the user.

The Presentation Agent consists of the following modules:

> Presentation Module (PM)
> On-line help Module
> Test Module
> Communication Module

Tutorial Agent

The Tutorial Agent represents the domain expert. It can be designed to provide high-level information about the problem domain. Multimedia facilities can be incorporated to illustrate important aspects and guide the user on the use of the system.

Applications Agent

An application agent could be any standard software (spreadsheet, database, graphics package, etc.) that could be tailored (by using macros or other means of programming) to suit the needs of the System and be integrated with the rest of the modules.

Knowledge Based Application Agent

This agent contains heuristic knowledge from the domain expert. There is a close interaction between this and the blackboard agent in determing and evaluating rules during the problem solving process. This agent could be designed as a simple rule based module to something much more complex such as one incorporating human reasoning, common sense, opinions and experiences, etc.

Scientific Applications Interface Agent

The class of problems that we are trying to solve always require the use of scientific applications. There exist several software packages, applications and libraries (NAG, IMSL, BMDP, SCA, etc.) that have been written for specific purposes, that a system might wish to utilize in order to solve a sub problem. We have developed the above agent to do this in a clear and simple manner. The Scientific Applications Interface Agent (also known as the Back End Manager) has been described in detail previously (Prat et al., 1992).

3 Description of FORESEE, a Multiagent Distributed System for the Analysis and Forecasting of Time Series

FORESEE, a Multiagent Distributed System for the analysis and forecasting of time series has been developed with the MEDISA architecture. It facilitates the application of the Box-Jenkins methodology, at the univariate, intervention analysis, and transfer function level in a fully automatic or step by step mode, depending on the user's preference. It also offers the option of applying Neural Networks and Genetics Algorithms methods. The KBS compares the results from the different methods and selects the one that offers the best forecasts, applying the criterion of the minimum of the variances of the forecasts. The system has a knowledge base enriched by more than 14 years of application of the methodology of Box and Jenkins by the group that present this work.

Expert domain knowledge is encoded in two locations. The knowledge in the blackboard agent includes the general strategy for solving the problem. The knowledge based agent maintains heuristics about the specific aspects to control and verify in each of the steps inherent in the general methodology. The latter is developed using commercial software (e.g. KAPPA-PC, by Intellicorp).

In order to carry out the required computations for the estimation, diagnostic checking and forecasting, a statistical package is used. The automatic identification of univariate models is done using one of two methods available in the system: using a module that uses the ACF and IACF of the series or one based on genetic algorithms (both developed by our group). In order to automatically identify the transfer function, another module that was also developed by the group is used. BEM, an agent developed during the FOCUS project is used as an interface between the 'Blackboard' Agent and existing statistical software packages. The data base, graphics and report generation are managed by applications selected from those already available in the market.

The link between all the different modules is provided by the agents of the MEDISA architecture, giving the appearance of a single system to the user. In other words, FORESEE is an example of the creation of a sophisticated system from singular and separable elements, using the most appropriate tool available in the market for each component. Developing and/or customizing a component is the responsability of the expert associated with it. It is relatively easy to substitute any component by another that has similar functionalities.

Learning to use FORESEE will be intuitive and interesting, specially due to the interactive multimedia capabilities that are being incorporated in the tutorial and help facilities. The application of sophisticated forecasting techniques with high precision becomes an accessible and productive task.

3.1 FORESEE and the forecasting methods

The Box-Jenkins (Box and Jenkins, 1976) methodology for the analysis and forecasting of time series is widely regarded to be the most efficient forecasting technique, and is used extensively - specially for univariate time series. The three step strategy of identification, estimation and diagnostic checking, requires the person in charge of producing forecasts to have experience and knowledge. When the number of series to forecast is large (like in the case of a bank, an airline or a power utility) the task to model "by hand" becomes practically impossible. The fact that some simple techniques (like exponential smoothing) are still being used along with limited extensions of intervention analysis and transfer function, is basically the consequence of the operational difficulty of applying the methodology of Box and Jenkins.

Two of the basic steps contained in the methodology, identification and diagnostic checking, require a very high level of expertise. Therefore, a way to apply these steps automatically has been studied in order to make the technique more accessible to organisations and industry of all sizes (Tashman et al., 1991).

Knowledge Based Systems provide a powerful way of modelling the skills and kowledge of an expert in the analysis of time series. The difficulty of developing these systems is in acquiring the knowledge of the experts and representing this in an optimal way, in a computer programme.

The area of artificial intelligence provides different methods for modelling time series, like the neural network based systems (ANNs, artificial neural networks)(Cottrell et al., 1993) and genetic algorithms based systems (GAs, genetic algorithms). These are not enclosed in a preprogrammed knowledge base, but instead learn from experience and are capable of increasing the knowledge according to how the problem evolves.

The identification of time series models according to the Box-Jenkins methodology, defines a set of theoretical SARIMA models from which one is tentatively chosen. Parameters of the model are estimated and the model is validated. Future data of the series from the validated model are obtained.

3.2 Components of FORESEE

The following figure shows the FORESEE architecture (specific case of the general MEDISA architecture) and the components of the system.

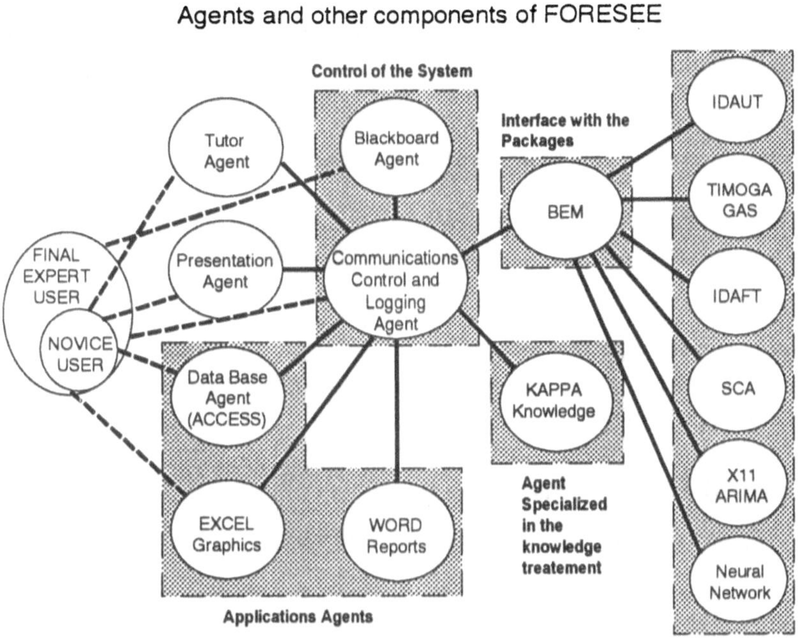

Agents and other components of FORESEE

Fig. 3. FORESEE architecture.

In the figure above, apart from the main agents of the architecture ("blackboard" agent and Communications Control and Logging agent) the following agents can be observed:

*   *Presentation Agent:* Developed with VISUAL BASIC, [*].
*   *Tutor agent of the problem:* Developed with VISUAL BASIC. The recordings are reproduced using the MULTIMEDIA WIEVER, [*].
*   *Data Base Management:* Developed with ACCESS, [*].
*   *Presentation of Graphics:* EXCEL software is used, [*].
*   *Report Producing:* WINWORD software is used, [*].
*   *Knowledge treatment agent:* KAPPA shell is used, (Intellicorp).

[*] These products are by MICROSOFT Corp.

The Scientific Applications Interface agent integrates the following packages:

*   IDAUT: System that does automatic identification of a time series based on a Mahalanobis type of distance between the autocorrelation function (ACF) and the inverse autocorrelation function (IACF) of the given time series and those of the tentative theoretical models.
*   IDAFT: System that allows the identification of the tranfer function based on the Corner Table, computing the standard deviation, determining the

312

elements that are statistically significant, using the algorithm of recognition to determine the transfer function.

- TIMOGA: System to identify a model automatically, based on genetic algorithms that optimize the objective function that takes into account the residual variance and the number of parameters that are included in the model.
- SCA: Statistical System, a powerfull software for forecasting and time series analysis, from Scientific Computing Associates Corporation.
- X11-ARIMA: A software for seasonal adjustment, by Statistics Canada.
- COYUNTURA: A software for the automatic conjunctural analysis of time series, developed by our group, and following the method of Espasa (1993).
- Neural networks: A feedforward neural network for time series forecasting, with back propagation, developed by our group.

3.3 How FORESEE works

Figure 4 shows the problem solving strategy of FORESEE.

The user working with FORESEE has to select the series to be analyzed. The user will be able to graph the selected series, transform it or do a linear combination of them. The next step is the identification, that can be manual or automatic. If the automatic procedure is chosen, it will be performed by the systems IDAUT and TIMOGA. The model or models selected will be estimated and submitted to a verification that will include, among others, the following checks:

- That the mean of the residuals is zero.
- If the ACF and the PACF (or IACF) of the residual lack any structure
- That there is no lag in the ACF and PACF with a significant coefficient
- That the roots of the AR and MA polynomials, both regular and seasonal, are outside the unit circle.
- That all the parameters are statistically significant
- That during the iterations, no model already estimated is repeated
- That the model is consistent over time
- That there are no outliers in the data or identify which observations are suspected to be outliers.
- That in the case of the Transfer Function, it also checks that the residuals are not correlated with the residuals of the univariate series of the input series

If any of the above conditions is not met, the model is specified again in order to correct for the defect.

If there are outliers, the residuals can be graphed in order to mark the possible anomalies (depending on the type). The ACF, PACF, IACF and EACF functions of the residuals can also be graphed. If all the verifications are successful, the model will be taken as the correct one and the forecast will be obtained, providing the user with the more relevant results.

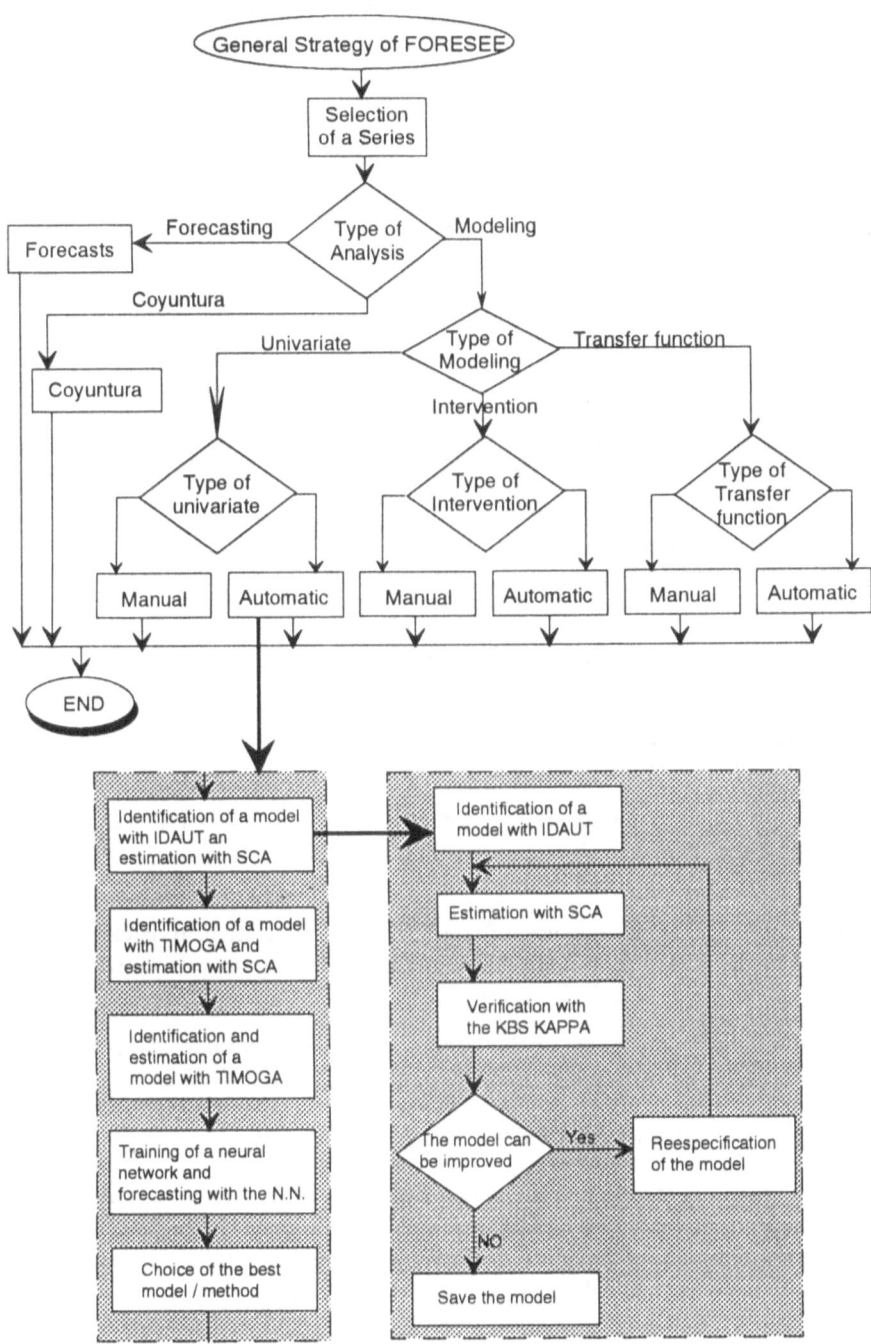

Fig.4. Strategy of FORESEE.

3.4 Types of KNOWLEDGE and the tools to introduce them within FORESEE

FORESEE introduces several types of knowledge:

- Knowledge of a consultant about the general resolution of a forecasting problem
- Knowledge of a teaching professional about forecasting methods to explain them.
- Knowledge of an expert on Box-Jenkins Methodology to identify a model and verify the results of the estimation. Knowledge of this expert to specify again the model according to those verifications. Knowledge to compare the forecast offered by the different methods (e. g., Box and Jenkins and Neural Networks)
- Knowledge of the experts in the use of the different BEs of the system
- Knowledge of the experts in the use of the applications of the system

The separability of the agents and the components enables each expert to concentrate in the introduction and refinement of their own knowledge, using for it the tool asociated with the agent or the component.

Therefore, the knowledge about the general resolution of a problem, managed by the strategy agent and the control of the system, are introduced and refined using the visual editor that incorporates this agent.

The knowledge to teach the methodology of Box-Jenkins are incorporated in the tutor agent and requires multimedia programming.

The knowledge about modeling (identification and verifications of the estimation) are managed by the KAPPA shell and therefore, need to be introduced using the facilities of this shell.

The knowledge about the use of applications and statistical packages are managed by the BEM and are introduced and refined using the own editior of this agent.

The remaining components of the system need to be prepared using the own programming systems of each component (e.g., EXCEL and WORD).

3.5 Validation of the FORESEE system

The MEDISA architecture and the agents and tools associated with it have only been tested internally. Even though it seems to be a powerful architecture that allows the creation of systems for solving complex problems easily and effectively, more methodical testing and study is necessary.

The theoretical methods implemented by FORESEE have all been tested with simulated and real series. For the simulated series the system has had a precision of 98% and for the real series, it has been 86% on 2000 series. At the present time, we have not completed the comparison of results obtained from parametric methods with those from neural networks and genetic algorithms.

Some beta versions of FORESEE are currently being tested by two important organizations: ITEC (Instituto de Tecnologias de la Construcción de Barcelona) and La Caja de Ahorros y Pensiones de Barcelona. In ITEC, forecasts of cement consumption are obtained depending on the licitation, and at the Caja De Ahorros y Pensiones they analyze the evolution of the consumption/price indices and other economic indicators (IPI, car sales, ...). Both organizations have reported to have obtained a very high degree of precision in their forecasts. For instance, the forecasts obtained by ITEC were able to detect strong declines in cement consumption much before it was anticipated by the rest of the sector.

At this early stage, we did not have any reports to indicate that there are any significant problems with learning how to use the system. We anticipate that with more testing and feedback, there will be some changes that will be made to the user interfaces (tutorial and presentation).

4 Conclusions

From our long experience in developing systems to solve complex statistical problems, we have recognized that a generalized framework or architecture was lacking to integrate all relevant components of the problem solving strategy. The MEDISA architecture has evolved from work, done in the past such as STATXPS (Expert Decision Support System)(Prat et al., 1985 and Prat et al., 1988), and the FOCUS project (Prat et al., 1993a). The main motivation was to develop a general architecture that would facilitate rapid and cost effective development of sophisticated systems to solve complex problems.

The FORESEE forecasting system that we have developed using MEDISA integrates state-of-the-art techniques and tools that were previously available only to large organisations. It enables even non-experts who are not concerned with the underlying theory, to do sophisticated statistical analysis. On the other hand, the system also allows an expert user to intervene and analyse all relevant data.

To estimate the parameters of a univariant ARIMA model with the first PC XT´s took between half an hour to 45 minutes. The complete process of automatically identifying and estimating the model, including result verification by different expert systems, done by a current i486 does not take more than 45 seconds. Obviously the problem of capturing the knowledge from the experts and entering them into the system still exists, but this task becomes much easier if we divide the problem into different components and use the most appropriate tool to solve each component as proposed in MEDISA.

The use of BEM (Prat et al., 1992) has allowed us to integrate various statistical packages in an easily interchangeable manner. The distributed architecture has also facilitated the incorporation of new AI techniques (ANNs, GAs, etc.) in order to compare and evaluate against traditional methods.

The process of enriching the knowledge of the system has been continuous and we realize that it most likely will not stop. The application of new methods has allowed us to find aspects that had not been considered before and contributed to the evolution of an accurate and powerful forecasting system.

References

Box, G.E. & Jenkins, G.M. (1976). *Time series analysis, forecasting and control*. 2nd Edition. Holden-Day, San Francisco.

Cottrell, M.; Girard, B.; Girard, Y.; Mangeas, M.; Muller, C. (1993). *"Neural Modeling for Time Series a Statistical Stepwise Method for Weight Elimination"*. Prepublication du SAMOS nº20, Juillet 1993.

Dodge, Y.; Hand, D. J. (1993). *"What Should Future Statistical Software Look Like?"*. Statistical software newsletter.

Engelmore, R. (1988). *"Blackboard systems"*. Addison and Wesley.

Erceau, J.; Ferber, J. (1993). *"La inteligencia artificial distribuida"*. Mundo científico Nº 116, Vol. 11, pp. 850-858.

Espasa, A. (1993). *"Métodos cuantitativos para el análisis de la coyuntura económica"*. Alianza Editorial, Ramón Cancelo (Ed), Madrid.

Hedberg, S.; Liebowitz, J. (Octubre 1993). *"Nuevas Herramientas para el Conocimiento"*, *"Utilice sus propios híbridos"*, *"Ver, oir, aprender"*, Binary, pp. 127-143.

Huang, Y.; Rozenblit, J.W. (1991). *"Architectures for distributed knowledge processing"*. Chapter 17. Neural and Intelligent Systems Integration.

Lorés, J. (1994). *"Diseño y desarrollo de una arquitectura distribuida multiagente para resolución de problemas complejos, Tesis Doctoral.*

Prat. A.; Marti, M.; Catot, J.M. (1985). *"Incorporating Expertise in Time Series Modelling: The STATXPS System"*. Statistical Software Newsletter. Vol. 11, Nº 2, pp 55-62.

Prat, A.; Ginebra, J.; Catot, J.M.; Lores, J. (1988) *"Expert System For Forecasting"*. Elseiver Science Publishers B.V. (North-Holland). IMACS.

Prat, A.; Lorés,J.; Fletcher,P.; Catot, J.M. (1992). *"Back-end manager: An interface between a knowledge-based front end and its application subsystem"*. The Separable User Interface. Academic Press. (ISBN: 0-12-232150-2).

Prat, A.; Catot, J.M.; Fletcher, P.; Lores, J.; Southwick, R. (1993b) *"Using the FOCUS architecture for developing knowledge-based front ends: a KBFE for forecasting"*. Knowledge Based Systems, Vol 6, Nº 1.

Prat, A.; Edmonds, E.; Catot, J.M.; Lores, J.; Galmes, J.; Fletcher, P. (1993a). *"An architecture for knowledge-based statiscal support systems"*. Artificial Intelligence Frontiers in Statistics. Ed. Hand. Chapman & Hall, Cap, 4, pp. 39-45.

Tashman, L.; Leach, M. (1991). *"Automatic forecasting software: A survey and evaluation"*. International Journal of Forecasting Nº 7, pp. 209-230. Elsevier Science Publishers Co. (North-Holland).

Principal Component Analysis on Simulated AR(1) Processes

Lamia Choukair, Bernard Burtschy

Dept. INF - ENST, Télécom-Paris

1   Introduction

When dealing with a set of temporal data, it is possible to describe them using data analysis techniques. Applying these techniques avoids the use of the strong classical hypotheses, such as stationarity, that are usually required in a time series analysis.

We are interested here in Principal Component Analysis (PCA). In fact, when applied to time series, PCA may be considered as an interesting complement to classical time series analyses. In order to modelize and describe some situations, it seems interesting to characterize the properties of the principal components for $ARMA$ processes.

Unlike classical time series analysis methods, the PCA technique does not take into account the chronology underlying the data. However, PCA allows us to obtain a decomposition of the set of data according to the global statistical information inherent in the initial series. This information cannot be obtained through classical methods.

We first describe how to perform a PCA on a time series, and particularily on a stationary time series. This method has the advantage of decomposing a time series into k new time series more homogeneous than the initial one. The remainder part does not contain relevant information and may be considered as a white noise series: $X(t) = \sum_{i=1}^{k} C_i \overline{A_i(t)} + \sum_{i=k+1}^{T} C_i \overline{A_i(t)}$, where the first sum contains all the information inherent in X.

Then, using simulations of $AR(1)$ univariate processes, we seek a relation between the results of the descriptive technique of PCA and the analytical properties of those processes. More precisely, we will try here to find a connection between the $ARMA$ representation of the initial process and that of its principal components.

2 PCA of a stationary time series

[BH79] have used the principal components in order to decompose a univariate time series into trend, cycle and seasonality. According to [BU87], when we are in presence of an MA(1) time series, PCA carry out a partial frequency decomposition of the time series. [BR81] proposed a PCA in the frequency domain.

318

Besides, *PCA* of multivariate time series was often used to reduce the dimensionality of representation of the initial series [PB87]. However, *PCA* has not been often used to investigate the properties of a single time series.

The principal components of a time series are the eigenvectors of its autocorrelation matrix which is a Toeplitz matrix.

Unfortunately, according to [GS84], if X is an $AR(1)$ process, it does not seem feasible to evaluate analytically the zeros of its characteristic equation $\Delta^{(n)}(\lambda)$. Notice however that it is possible to compute the eigenvalues of symmetric Toeplitz matrices using recursive algorithms [CL86; TR89].

Since it is very difficult to establish analytical expressions for the principal components of an $AR(1)$ process, we performed a heuristic study based on simulations.

3   Simulations: principal components of $AR(1)$ processes

The idea of this work is to simplify the representation of a set of data, realizations of an AR process according to the whole information inherent in the data. For that purpose, it is well known that *PCA* method is one of the most useful tools.

The problem is then to find out if the principal components of an AR process may be adequately modelized by an autoregressive representation.

According to [CH93], if $\{Y(t)\}$ is an $AR(p)$ process: $\Phi(B)\,Y = \epsilon$, then it is possible to write it according to the following linear decomposition: $Y_t = \sum_{i=1}^{p}\{\prod_{j\neq i}\frac{1}{z_i-z_j}\}Y_t^i$ The $Y_t^i = \frac{\Phi}{X-z_i}(Y_t)$ are $AR(1)$ processes, B is the Backward shift operator, X is the identity polynomial and the z_i are the zeros of the parameter polynomial Φ.

The latter property allows us to perform simulations only on $AR(1)$ processes and then to deduce results for $AR(p)$ processes.

We have simulated, using the adequate procedures of the IMSL library [IMSL87], several stationary $AR(1)$ time series, *i.e.*: $X(t) = \phi X(t-1) + \epsilon(t)$ which verify $0 < |\phi| < 1$. To cover the definition domain of ϕ, we have used the following three parameters: $|\phi| = 0.2, |\phi| = 0.5$ and $|\phi| = 0.9$.

In order to perform a *PCA*, the autocorrelation matrices of the simulated processes were computed for order 120. We thus obtained 120 eigenvalues and 120 normalized eigenvectors of 120 components each.

For illustration, we will give results obtained for $|\phi| = 0.5$.

3.1   Principal components of $AR(1)$ processes: $X_t = \phi X_{t-1} + \epsilon_t$

For all the 100 simulated series ($|\phi| = 0.2, 0.5, 0.9$), the obtained results are analogous.

On *Figure 1*, we have plotted the eigenvalues and the corresponding cumulated percentage of variance of the autocorrelation matrices, respectively for $\phi = 0.5$ and $\phi = -0.5$.

Figures 2 and 3 represent the first two corresponding eigenvectors with their *ACF* and *PACF*.

For all simulated series, the resulting principal components are sine waves and the ACF also have a very typical periodic behavior.

According to [BJ70], the principal components processes represented *Figures 2 and 3* are probably $AR(2)$ processes: $X_t = \phi_1 X_{t-1} + \phi_2 X_{t-2} + \epsilon_t$. In fact, the behavior of the ACF corresponds to that described by Box and Jenkins when the roots of the characteristic equation $1 - \phi_1 z - \phi_2 z^2 = 0$ are complex numbers ($\phi_1^2 + 4\phi_2 < 0$). In this case, the $AR(2)$ process has a pseudo-periodic behavior that we can notice on the ACF. In all cases, the ACF are damped sine waves, and the $PACF$ are almost zeros beyond the second lag.

So it seems that the principal components of $AR(1)$ processes are also AR processes of small order.

3.2   Periodograms of the series and their principal components

In order to better characterize the resulting series, we have plotted *Figure 4* the periodograms of the initial series and of their first two principal components.

3.2.1   Periodograms of the initial series

According to *Figure 4*, we can notice what the periodograms of the initial series are mainly composed of.

- For $AR(1)$ processes with positive parameter, the series is smooth and this is reflected in an ACF which damps out with lag. It is seen that the corresponding spectrum is large at low frequencies and small (comparable to a noise) at high frequencies. In fact, smooth series are characterized by spectra which have most of their power at low frequencies. Hence, a $AR(1)$ process always provides a large spectrum at low frequencies and small spectrum at high frequencies. Such an $AR(1)$ process is then mainly composed of a superposition of low frequency sinusoids.

- When the stochastic process is $AR(1)$ with negative parameter, $-1 < \phi < 0$, there is a duality with the results obtained when $|\phi|$ is positive.

3.2.2   Periodograms of the principal components

In fact, the periodogram of a time series is not sufficient to represent the whole information contained in the series. Actually, the data are entirely described by the periodogram and the phase of the series. The periodograms are often composed of superimposed rays which are not in phase. Then, the power spectrum represents only a part of the information. Hence, the principal components are useful to obtain complementary information.

If we look at *Figure 4*, we can see that the periodograms of the first two principal components (in both positive and negative parameter cases) have only one significant frequency. This frequency approximately corresponds to the inverse of the period observed on the corresponding correlogram.

So the PCA seems to provide a frequency decomposition in both cases

(AR processes with positive and negative parameters).

We can also notice that the pseudo-periodic behavior of the principal components appears also in their periodogram. The inverse of the pseudo-period p observed on the autocorrelogram is approximately equal to the frequency f_0 corresponding to the dominant peak of the periodogram. According to [BJ70], the principal components seem to be $AR(2)$ processes of which the characteristic equations have complex zeros.

4   Conclusion

The results we have obtained are very interesting for both AR with negative and positive parameter processes. The first principal components show pseudo-periodic behavior, and their autocorrelation functions (ACF) are damped sine waves. On another hand, the partial autocorrelations are almost zeros for orders greater than two.

Then, these first principal components seem to be realizations of $AR(2)$ processes. To confirm this statement, we have computed the estimated spectrum for each of those components. In each case, the periodogram shows only one significant frequency, *i.e.* almost all the variance of the series is accounted for by only one frequency. Then, it seems that PCA provides a spectral decomposition of the initial process in both cases.

References

- [BH79] A. Basilevski and D.P.J. Hum. Karhunen-Loève analysis of historical time series with an application to plantation births in Jamaica. *J. of the Amer. Statist. Assoc., 74, 366, pp. 284-290,* 1979.

- [BJ70] G.E.P. Box and G.M. Jenkins. Time series analysis, forecasting and control. *Holden-Day, San Fransisco, 2nd. ed.,* 1976.

- [BR81] D.R. Brillinger. Time series, data analysis and theory. *Holden-Day, San Fransisco, expanded ed.,* 1981.

- [BU87] B. Burtschy. Factorial analysis of multiple time series. *Proc. of the Joint Statistical Meeting, San Fransisco, pp. 310-314,* 1987.

- [CH93] L. Choukair. Statistical methodology for spatiotemporal series analysis. Application to climatological data. *Ph.D. Thesis, ENST, Paris,* 1993

- [CL86] G. Cybenko and Van Loan. Computing the minimum eigenvalues of a symmetric positive definite Toeplitz matrix. *SIAM J. Scient. Statist. Comput., No 7, pp. 123-131,* 1986.

- [GS84] U. Grenander and G. Szegö. Toeplitz forms and their applications. *Chelsea Publishing Company, New York, 2nd. ed.,* 1987.

- [IMSL87] IMSL Inc. User's Manual, Math Library. *IMSL, Houston, TX,* 1987.

- [PB87] D. Pēna and G.E.P. Box. Identifying a simplifying structure in time series. *J. of the Amer. Statist. Assoc., Vol. 82, No. 399, pp. 836-843,* 1987.

- [TR89] W.F. Trench. Numerical solution of the eigenvalue problem for hermitian Toeplitz matrices. *SIAM J. Matrix Anal. Appl., No 10, pp. 135-146,* 1989.

Annex: Figures

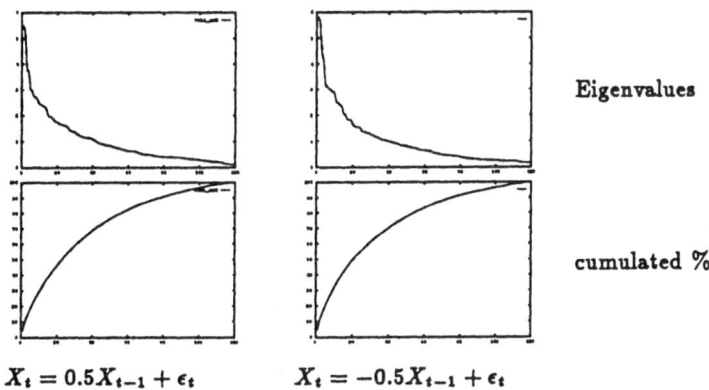

Eigenvalues

cumulated %

$$X_t = 0.5X_{t-1} + \epsilon_t \qquad X_t = -0.5X_{t-1} + \epsilon_t$$

Figure 1: Eigenvalues of the initial series

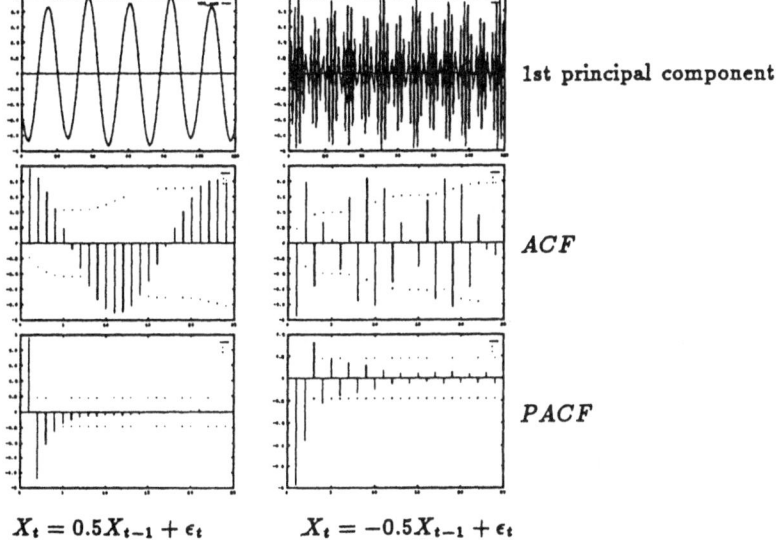

1st principal component

ACF

$PACF$

$$X_t = 0.5X_{t-1} + \epsilon_t \qquad X_t = -0.5X_{t-1} + \epsilon_t$$

Figure 2: 1st Principal component, its ACF and $PACF$

322

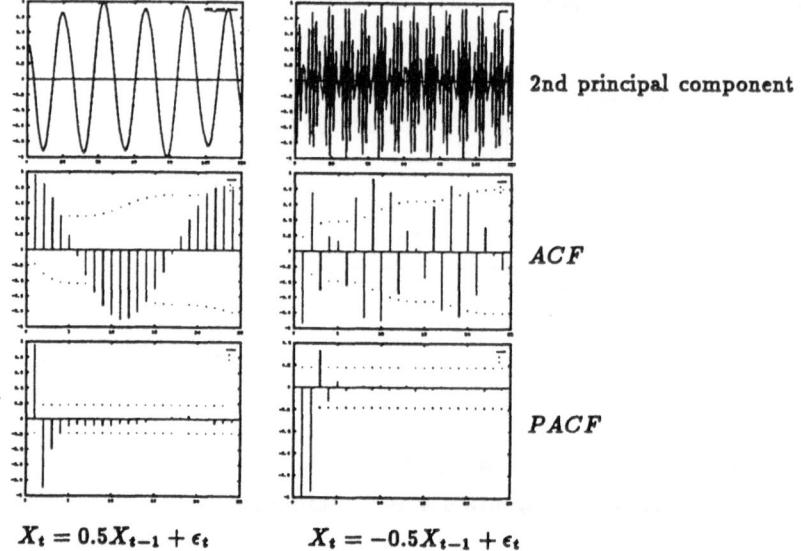

$X_t = 0.5X_{t-1} + \epsilon_t$ $\qquad$ $X_t = -0.5X_{t-1} + \epsilon_t$

Figure 3: 2nd Principal component, its ACF and $PACF$

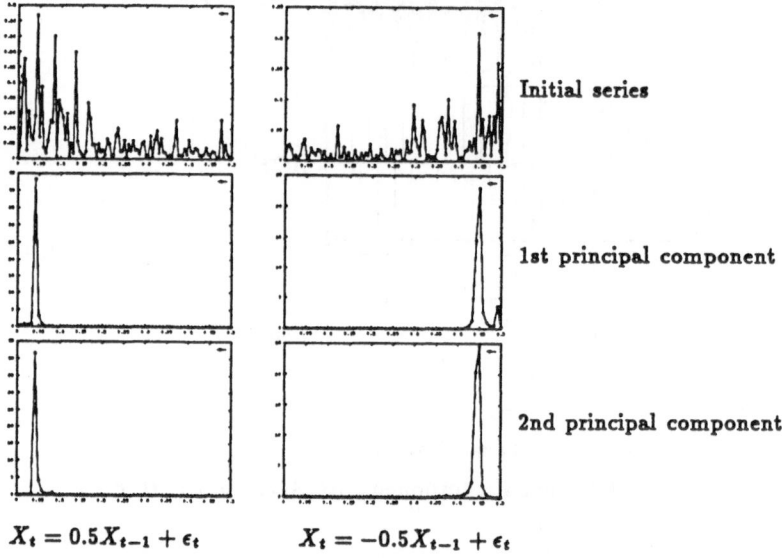

$X_t = 0.5X_{t-1} + \epsilon_t$ $\qquad$ $X_t = -0.5X_{t-1} + \epsilon_t$

Figure 4: Periodograms of the initial series and of its principal components

A Note on Generation, Estimation and Prediction of Stationary Processes

Michael A. Hauser, Wolfgang Hörmann, Robert M. Kunst, Jörg Lenneis

Department of Statistics, University of Economics and Business Administration, Vienna

1 Introduction

Some recently discussed stationary processes like fractionally integrated processes cannot be described by low order autoregressive or moving average (ARMA) models rendering the common algorithms for generation estimation and prediction partly very misleading [cf. Hosking(1981,1984), Sowell(1992), Ray(1993)]. We offer an unified approach based on the Cholesky decomposition of the covariance matrix which makes these problems exactly solvable in an efficient way.

Our starting point are stationary processes with a Wold representation of the form

$$y_t - \mu = \sum_{i=0}^{\infty} \psi_i \epsilon_{t-i}, \tag{1}$$

where ϵ_t is uncorrelated noise with mean zero. The ψ_i are quadratic summable and the (unconditional) variance of the noise, σ_ϵ^2, is greater than zero. We assume for simplicity of the presentation that $\mu = 0$. Y_T denotes the vector $(y_1, \ldots, y_T)'$ and $E_T = (\epsilon_1, \ldots, \epsilon_T)'$. The covariance matrix of Y_T, Σ_T, is positive definite, symmetric and Toeplitz, and thus persymmetric. It may by factorized according to the Cholesky decomposition.

$$\Sigma_T = L_T L_T'. \tag{2}$$

L_T is a lower triangular matrix.

One possibility for the *generation* of a sample of length T of a given process which possesses exactly the same covariance structure is to use the relation

$$Y_T = L_T E_T. \tag{3}$$

Under the assumption of normal distributed noise *estimation* may be performed by maximizing the Gaussian likelihood

$$f(Y_T; \mu, \Sigma_T) = (2\pi)^{-T/2} |\Sigma_T|^{-1/2} \exp[-(Y_T - \mu)' \Sigma_T^{-1} (Y_T - \mu)/2]. \tag{4}$$

For ARMA models there exist computationally simpler presentations of the likelihood. For fractionally integrated models, however, this is the only known exact form [Li and McLeod(1986) or Sowell(1992)].

The implicit *noise* vector may be obtained by

$$E_T = L_T^{-1} Y_T. \tag{5}$$

The *linear prediction* for one step to τ steps ahead may simply be performed by extending the above equation to $T + \tau$ and replacing the future noises by their expectation which is zero. This is

$$Y_{T+\tau} = \begin{bmatrix} Y_T \\ Y_\tau \end{bmatrix} = L_{T+\tau} E_{T+\tau} = \begin{bmatrix} L_T & 0 \\ L_{\tau T} & L_{\tau\tau} \end{bmatrix} \begin{bmatrix} E_T \\ E_\tau \end{bmatrix} \text{ and}$$

$$E[Y_\tau | Y_T] = \begin{bmatrix} L_{\tau T} & L_{\tau\tau} \end{bmatrix} \begin{bmatrix} E_T \\ 0 \end{bmatrix} = L_{\tau T} E_T. \tag{6}$$

The *variance of the linear forecast* Y_τ given Y_T, E_T respectively, is given by means of the covariance matrix $\Sigma_{T+\tau}$, with $\Sigma_{T+\tau} = L_{T+\tau} L'_{T+\tau}$,

$$V[Y_\tau | Y_T] = E[(Y_\tau - E[Y_\tau | Y_T])(Y_\tau - E[Y_\tau | Y_T])' | Y_T] = L_{\tau\tau} L'_{\tau\tau}. \tag{7}$$

If the *innovations* are *conditional heteroscedastic* and Gaussian - i.e. ϵ_t are uncorrelated and normal with non-constant variances, which depend on the past - the process likelihood is given by (4) by replacing the covariance matrix Σ_T by a process dependent covariance matrix [see Hauser and Kunst (1993)]

$$\Sigma_T = L_T H_T L'_T \tag{8}$$

where H_T is diagonal and contains the conditional variances of the normalized ϵ_t. In case of homoscedasticity the H_T matrix reduces to I_T.
Generation and linear prediction is analogous to the homoscedastic case once the heteroscedastic innovations are given. The variance of the linear predictor is, however,

$$V[Y_\tau | Y_T] = L_{\tau\tau} H_\tau L'_{\tau\tau} \tag{9}$$

with $H_{T+\tau} = \begin{bmatrix} H_T & 0 \\ 0 & H_\tau \end{bmatrix}$.

The numerical problems addressed above can be summarized as follows: Generation and prediction require the calculation of the Cholesky factor, the inverse of the Cholesky factor, and the repeated multiplication of the Cholesky factor with an arbitrary vector. Estimation, i.e. the inversion of the covariance matrix, may be implemented by factorizing Σ_T^{-1} in a MDM', M a lower triangular matrix with ones in the diagonal, D a diagonal matrix, via the Levinson algorithm. The determinant of the covariance matrix is then equal $|D|$. The calculation of the variance of the predictor may be obtained by calculating only the lower right $\tau \times \tau$ part of the Cholesky matrix. How the necessary operations can be performed in an efficient way is discussed below.

2 The multiplication of the Cholesky factor with an arbitrary vector

Notation and some properties of Toeplitz matrices:

$$\Sigma_{T+1} = \left[\begin{array}{cc} \Sigma_T & \Sigma'_{1T} \\ \Sigma_{1T} & \Sigma_{11} \end{array} \right], L_{T+1} = \left[\begin{array}{cc} L_T & 0 \\ L_{1T} & L_{11} \end{array} \right], R_{T+1} = \left[\begin{array}{cc} R_T & Er \\ (Er)' & 1 \end{array} \right]$$

E is a square matrix with ones in the secondary diagonal and zeros else. It holds that $EE = I, E^{-1} = E$. R_T is the correlation matrix, $\Sigma_T = \sigma_y^2 R_T$. It is symmetric and Toeplitz, so that $ER_T E = R_T$ and $ER_T = R_T E$ holds. R_T^{-1} is also symmetric and persymmetric.

Lemma 1: [Brockwell and Davis(1991, p.168)]
The best linear 1-step ahead predictor of $\hat{y}_{T+1}$ of y_{T+1} in terms of Y_T and its mean squared error are

$$\hat{y}_{T+1} = \Sigma_{1T}\Sigma_T^{-1}Y_T, \qquad v_T = \Sigma_{11} - \Sigma'_{1T}\Sigma_T^{-1}\Sigma_{1T}. \tag{10}$$

In case of multivariate normal distributed Y_{T+1} this is identical to the moments given by the conditional normal distribution [Johnson(1987, p.50)]. The coefficient in front of y_T may be interpreted as the T-th partial autocovariance.

Proposition 1:
The best linear 1-step ahead predictor of $\hat{y}_{T+1}$ of y_{T+1} in terms of the Cholesky factors and past innovation vector E_T and its mean squared error are

$$\hat{y}_{T+1} = L_{1T}E_T, \qquad v_T = L_{11}L_{11}. \tag{11}$$

Proof:
This may be easily seen by using $\Sigma_{T+1} = L_{T+1}L'_{T+1}$ in the partition representation as given above, multiplying out, and replacing the Σ-matrices by the corresponding expressions in terms of the L-matrices in (10). For Y_T use $Y_T = L_T E_T$. □

The predictor is given by the multiplication of the last line of the Cholesky matrix by the vector $(E_T, 0)'$.

For the generation of samples of a process with given true covariance matrix the best linear predictor can be easily used recursively in the following way starting at $T = 0$ with $v_0 = \sigma_y^2$ [cf. Hosking(1984, p.1900)]:

$$y_{T+1} = \Sigma_{1T} \Sigma_T^{-1} Y_T + \sqrt{v_T} \epsilon_{T+1}, \tag{12}$$

where the ϵ_t are an (possibly heteroscedastic) innovation sequence.

In notation of the Cholesky matrix this amounts to

$$y_{T+1} = L_{1T} E_T + L_{11} \epsilon_{T+1}. \tag{13}$$

This is the multiplication of the last line of L_{T+1} with E_{T+1}, or more compactly for the whole vector Y_{T+1}, $Y_{T+1} = L_{T+1} E_{T+1}$.

An efficient algorithm to compute the best linear predictor and its mean squared error is the Durbin-Levinson algorithm [Brockwell and Davis(1991, p.169)]. Thus the Durbin-Levinson algorithm does multiply the Cholesky matrix with the vector E_{T+1} by requiring $O(T^2)$ flops and $O(T)$ storage. More generally, this algorithm performs the multiplication of the Cholesky matrix of a symmetric Toeplitz matrix with any arbitrary vector. This is remarkable, since there is no procedure known for the simply structured Toeplitz matrices to compute the Cholesky matrix with less than $O(T^3)$ flops and $O(T^2)$ storage. Below we will give a derivation of an equivalent algorithm based on matrix computations and the use of the Durbin algorithm which solves the Yule-Walker equations [Golub and VanLoan(1989, p.185)].

Derivation of the algorithm:
The idea for the algorithm is identical to the first step of the recursion of the Trench algorithm as presented in Golub and VanLoan(1989, p.188). For simplicity we reformulate the problem in correlations instead of covariances, which implies $\sigma_y^2 = 1$, $\Sigma_T = \sigma_y^2 R_T$ respectively. The first two moments of y_{T+1} as given in (10) simplify to $(Er)' R_T^{-1} Y_T$ and $1 - r' R_T^{-1} r$ using the properties of Toeplitz matrices and the matrix E and the notation given above.

$$R_{T+1}^{-1} = \begin{bmatrix} R_T & Er \\ (Er)' & 1 \end{bmatrix}^{-1} = \begin{bmatrix} B & v \\ v' & \gamma \end{bmatrix}. \text{ This implies that}$$

$$\begin{bmatrix} R_T & Er \\ (Er)' & 1 \end{bmatrix} \begin{bmatrix} v \\ \gamma \end{bmatrix} = \begin{bmatrix} 0 \\ 1 \end{bmatrix}.$$

Solving this system for v and γ yields $R_T v = -\gamma Er$ from the first equation. So v can be expressed via the solution y of the Yule-Walker equations, $R_T y = -r$, $y = -R_T^{-1} r$ and $v = \gamma E y$.

By replacing v in the second equation γ can be expressed as $\gamma = 1/(1+r'y) = 1/(1 - r' R_T^{-1} r)$.

The first two moments of y_{T+1} can be then expressed in terms of y. That is: $(Er)' R_T^{-1} Y_T = -(Ey)' Y_T$ and $1 - r' R_T^{-1} r = 1/\gamma = 1 + r'y$. $\square$

The algorithm gives the multiplication of the Cholesky factor with an arbitrary vector. The storage requirements are those of Durbin algorithm which are linear. Its number of flops are $O(T^2)$ which increases by two vector multiplications.

3 The inverse of the Cholesky factor

Proposition 2:
The inverse Cholesky matrix is related to the Cholesky matrix of the inverse by transposing with respect to the secondary diagonal.

Proof:
Σ is positive definite, symmetric and persymmetric. The Cholesky decompositions of Σ and its inverse, which is also symmetric and persymmetric, are $\Sigma = AA'$ and $\Sigma^{-1} = BB'$.
The inversion of the first decomposition is $\Sigma^{-1} = (A')^{-1}A^{-1}$. A, A^{-1} and B are lower triangular matrices. So there is a lower triangular decomposition and an upper triangular decomposition of the same matrix.
$\Sigma^{-1} = E\Sigma^{-1}E = E(A')^{-1}EEA^{-1}E = (E(A')^{-1}E)(EA^{-1}E) = BB'$. Since the Cholesky decomposition is well defined $EA^{-1}E = B'$ and, thus, $A^{-1} = EB'E$ follows. $\square$

4 Computations

As given above *generation* of samples of the process (y_t) may be obtained efficiently in linear storage requirements, once the autocorrelation function is given. [For the calculation of the autocovariance function of fractional integrated processes see Sowell(1992).]

If the *estimation* is performed via the likelihood function given in (4) the Levinson algorithm [see Marple(1987, p.87)] may be used to calculate the Cholesky decomposition of Σ_T^{-1}, $\Sigma_T^{-1} = MDM'$, and thus also the required determinant. This algorithm is $O(T^2)$ in storage and $O(T^2)$ in flops.
The resulting *innovations* may be calculated using the Cholesky decomposition of the last iteration of the optimization procedure, Proposition 2 and (5).

The linear 1- to τ-step *prediction* (forecast) vector given Y_T may be calculated via the (estimated) residual vector and (6) - linear in storage and quadratic in flops - using the (estimated) autocovariance function.

Especially in case of calculating the *variance of the linear predictor*, (7), Proposition 1 is very helpful since τ is typically small. Multiplying $L_{T+\tau}$ by a vector with zeros and a 1 in position $(T+j)$ picks out exactly the $(T+j)$-

th column which is the column j in $L_{\tau\tau}$. Without storing the intermediate results of the multiplication of $L_{T+\tau}$ with the first T zeros the number of flops is $O(\tau T^2)$. The storage is linear if the diagonal elements are needed only.

The procedure can be easily generalized for heteroscedastic innovations. The 1's have to be replaced by the square root of the conditional variances.

5 Summary

An efficient algorithm - $O(T)$ in storage and $O(T^2)$ in flops - for multiplying the Cholesky factor by an arbitrary vector is presented. It may be used for generation of linear processes, linear prediction and calculation of the predictor variance.

It is shown that the Cholesky factor of an inverse symmetric Toeplitz matrix is a simple function of the inverse Cholesky factor of the Toeplitz matrix itself. Thus, given the Cholesky factor of the inverse covariance matrix the noise vector may be easily obtained.

We have outlined that for the simulation of stationary processes, for estimation and prediction two different algorithms are sufficient: the Levinson algorithm for calculating the Cholesky decomposition of the inverse covariance matrix and the algorithm giving a multiplication of a vector with the Cholesky matrix of the covariance matrix. Moreover this way is also very efficient.

References

Brockwell, P.J. and Davis, R.A., 1991, Time series: Theory and methods (Springer, New York).

Golub, G.H. and Van Loan, Ch.F., 1989, Matrix computations, (John Hopkins University Press, Baltimore).

Hauser, M.A. and Kunst, R.M., 1993, Fractionally Integrated Models with ARCH Errors, Paper presented at ESEM, Uppsala.

Hosking, J.R.M., 1981, Fractional differencing, Biometrika, 68, 165-176.

Hosking, J.R.M., 1984, Modelling persistence in hydrological time series using fractional differencing, Water Resources Research, 20, 1898-1908.

Johnson, M.E. 1987, Multivariate Statistical Simulation (John Wiley, New York).

Li, W.K. and McLeod, A.I., 1986, Fractional time series modelling, Biometrika, 73, 217-221.

Marple, S.L. Jr., 1987, Digital Spectral Analysis (Prentice Hall, Englewood Cliffs).

Ray, B.K., 1993, Modeling long-memory processes for optimal long-range prediction, Journal of Time Series Analysis, 14, 511-526.

Sowell, F., 1992, Maximum likelihood estimation of stationary univariate fractionally integrated time series models, Journal of Econometrics, 53, 165-188.

Data Based Selection of "Window Width" for Spectrum Density Estimator Based on Time-averaged Periodograms

Viacheslav Mazur, Alexei Iourovski

Firm "BNK", Moscow

Let x_1 , ... , x_N be the observations of a weak-stationary time series (t.s.) $\{X_t\}$ with zero mean and square-integrable spectrum density (s.d.) $f(x) = \frac{1}{2\pi} \sum c_u e^{iux}$; the Fourier coefficients (F.c.) for f are the autocovariances of the t.s.: $c_u = \mathrm{E} X_t X_{t+u}$. (Here and below: integration limits are always $[-\pi, \pi]$; all functions are periodic with the period of 2π , integrable and even; range for indexies if not specified is either evident or runs over all integers; dependence upon the t.s.' sample length N is sometimes skipped.)

The s.d. estimator

$$f(x) = \int V(x - y) I(y)\, dy \qquad (1)$$

is the averaged periodogram (a.p.) I smoothed with the weight function (w.f.) V in frequency domain. The w.f. depends upon N and concentrates in zero as N grows. The a.p. is the result of the averaging of "T" ordinary periodograms computed over the shifted time segments of length "M" each: $I(x) = \frac{1}{T}\sum_{k=1}^{T} J_k(x)$. (We assume that $N = L(T-1)+M$; parameter "L" $(0 < L \le M)$ is the "shift" value; both segment length and shift value depend upon N.) Each periodogram J_k is computed for the smoothed data points $\{y_t(k) = a_t x_{(k-1)L+t}, \ k = 1, ... , M\}$ of k-th segment: $\quad J_k(x) = \frac{1}{2\pi} |\sum_{t=1}^{M} y_t(k) e^{itx}|^2$; $k = 1, ... , T$. Smoothing coefficients $\{a_j, \ j=1, ... , M\}$ depend upon M (i.e., upon N); usually they are positive and symmetrically tend to zero with respect to the middle of the segment. We shall assume that $\{a_t\}$ are of the form: $\{a_t = C_M h(t/M), \ t = 1, ... , M\}$ with $h \ge 0$ being the function of limited variation on $[0, 1]$, and C_M being such that $\sum a_t{}^2 = 1$. The a.p. is useful when the data contain slight deviations from stationarity as well as for long signals or for real-time computations. If $a_j = (1/M)^{\frac{1}{2}}$, $j=1, ... , M$ (no data smoothing) and $M = N$ (i.e.

$T=1$), the a.p. is the classic Shuster periodogram. (For general theory see, for example, [1]).

We consider the problem of the data-based selection of w.f. by minimising a certain risk function (r.f.) along a given set of w.f. Similar problem for a so called "smoothed periodogram estimator" (s. p. e.), i.e. the estimator of type (1) with I being an ordinary periodogram, was investigated in [2], [3] and [4]. The r.f. in [3] is a kind of a likelihood function; in [2] and [4] r.f. is

$$R(\hat{f}) = E \, ||\hat{f} - f||^2 , \qquad (2)$$

where $|| g || = \int g^2 \, dx$. The solution of the considered problem in [2] is based on minimax approach; in [3] and [4] it is based on minimising an estimator of the r.f. The method presented in [4] is much simplier than these in [2] and [3] from the computational point of view. (The selection of w.f. should be data based because the optimal w.f. appears to depend upon the unknown s.d.) Here we present a result similar to that given in [4], but for the estimator (1) based on the a.p.

Theoretical Result. Suppose that an s.d. f satisfies additional smoothing restriction: $\sum c_u^2 \, |u| < \infty$, and that it is not trigonometric polinomial: $\sum_{u > M} c_u^2 > 0$ for any $m > 0$. Let us define as $\aleph$ the set of all such t.s.

Suppose that the F.c. $\{v_u\}$ for a w.f. $V(x) = \sum v_u e^{-iux}$ are such that $0 \le v_u \le 1$, $v_u \to 1$ for each fixed u . (Here and below all limits are for $N \to \infty$.) Let us define as $\Im$ the set of all such w.f.

For any given s.d. f and any $x > 0$ let us define as $S(x)$ the quantity of points of the set $\{c_u : c_u^2 > 1/x\}$. (It is easy to see that $S(N) \ll N$.)

The a.p. based sample autocovariances $\{\bar{c}_u = \int I(x) e^{-iux} dx\}$ (i.e., the F.c. for the a.p.) are the averages of the F.c. for $\{J_k\}$:

$$\{\bar{c}_u = \frac{1}{T} \sum_{k=1}^{T} \hat{c}_u(k)\}, \text{ where } \{\hat{c}_u(k) = \sum_{t=1}^{M-|u|} y_t(k) \, y_{t+|u|}(k) \text{ for } |u| < M ;$$

$\hat{c}_u(k) = 0$ for $|u| \ge M\}$.

THEOREM. Let t.s. $\{X(t)\} \in \aleph$ and w.f. $V \in \Im$. Let $M \gg \dfrac{N}{S(N)}$. Then the r.f. (2) for the estimator (1) satisfies the asymtotic equation:

$$R(\hat{f}) = ||f||^2 + \frac{1}{2\pi} E\,(r(\hat{f}))\,(1 + o(1)) ,$$

where $\quad r(\hat{f}) = \sum_{|u| \le M} [v_u^2 \, \bar{c}_u^2 - 2v_u (\bar{c}_u^2 - 2\pi \, ||f||^2 A_u)\,] ;$

$$A_u = \frac{1}{T} \sum_{|k| \le \min(T, \frac{MT}{L})} (1 - \frac{|k|}{MT})\, d_u(kL) , \; |u| < M ; \qquad (3)$$

$$d_u(j) = \sum_{t=1}^{M-|u|-|j|} a_t \, a_{t+|u|} \, a_{t+|j|} \, a_{t+|u|+|j|}, \quad |j| < M - |u|$$

Note. For the s. p. e. ($M = N$; $T=1$) the values $\{A_u\}$ appear to be $\{A_u = \dfrac{1}{N}(1 - \dfrac{|u|}{N}), \ |u| < N\};$ together   with $\{c_u = \dfrac{1}{N}\sum_{t=1}^{N-|u|} x_t \, x_{t+|u|}, \ |u| < N\}$ this transforms (3) to the result given in [4].

The proof is based on the four following lemmas.

LEMMA 1. For any $\{X(t)\} \in \aleph$ holds:

$$R(\hat{f}) = \| f \|^2 + \mathrm{E} \, [\textstyle\int \hat{f}^2(x) \, dx - 2\int \hat{f}(x) \, I(x) \, dx \,] +$$

$$+ 2\int\int V(x - z) \, \mathrm{cov}\,(I(x), I(z)) \, dx \, dz -$$

$$- 2\int \mathrm{E}\,\hat{f}(x) \, [f(x) - f * H(x)\,] \, dx,$$

where the "time smoothing" w.f. H is defined as $H(x) = \dfrac{1}{2\pi}\sum_{|u| < M} h_u \, e^{-i u x};$

its F.c. are $h_u = \sum_{t=1}^{M-|u|} a_t \, a_{t+|u|}, \ |u| < M$ $\}$; " $*$ " is for a convolution operation: $f * g(x) = \int f(y) \, g(x - y) \, dy.$

LEMMA 2. For any $\{X(t)\} \in \aleph$ and $V \in \Im$ holds:

$$\int\int V(x-z) \, \mathrm{cov}\,(I(x), I(z)) \, dx \, dz =$$

$$= \| f \|^2 \, (\sum_{|u| < M} v_u \, A_u) \, (1 + o(1)) \, .$$

LEMMA 3. For any $\{X(t)\} \in \aleph$ and $V \in \Im$ holds:

$$\int \mathrm{E}\,\hat{f}(x) \, [f(x) - f * H(x)\,] \, dx = O\,(\dfrac{1}{M})$$

LEMMA 4. For any $\{X(t)\} \in \aleph$ and any w.f. V holds:

$$R(\hat{f}) \geq C \, [\dfrac{S(N)}{N}\,],$$

where $C > 0$ does not depend upon N.

Practical Selection of "Window Width". A widely used one-parametric set of w.f. is determined as $\{V^{(g)}(x) = \frac{1}{2\pi}\sum_{|u|\le W} g(\frac{|u|}{w+1}) e^{-iux}\}$.

Function $g(x)$, called "correlation window", is continious in zero, $g(0)=1$, $g(x)=0$ for $|x|\ge 1$, $0\le g(x)\le 1$ for all x. The most well-known correlation windows are smooth functions descending from maximum in point $x=0$ to zero in $x=1$. The larger values of "window width" (w.w.) w produce w.f. more concentrated in zero. About windows see, for example, [1].

The expression (3) in the theorem statement represents the asymptotically unbiased estimator of $2\pi(R(\hat{f}) - ||f||^2)$. Changing in (3) the unknown value of $||f||^2$ to some estimator of it and using w.f. set $\{V^{(g)}\}$ (for some g) we get the expression which depends only upon the data and w; its minima over some range of $\{w\}$ is proposed as the estimator of the optimal (i.e., minimising r.f. (2)) w.w. It is natural to take the range of w for minimisation being $0\le w\le o(N)$ as the estimator (1) is not consistent for $w = O(N)$.

We have elaborated the algorithm whith implements this idea. The estimator of $||f||^2$ is $\frac{1}{2\pi}\sum_{|u|<m} \bar{c}_u^2 / [\ln(1+\frac{m}{N})]$, m is the round of $N^{\frac{1}{2}}$.

(In practice the observations of each segment are centered and normalised.) As the time smoothing requires too long calculations, the values $\{a_t, t=1,...,M\}$ are all equal to $1/M^{\frac{1}{2}}$. (It is easy to see that the autocovariancies $\{\hat{c}_u\}$ and the values $\{A_u\}$ from the theorem statement can be calculated much quicklier if the $\{a_t\}$ are all equal; on the other hand, it is not difficult to modify the algorithm so that the time smoothing would be included.) All most well-known windows (Blackman's, Hanning's, Blackman-Harris', etc) are available. For the sake of the speed of computations, the w.w. selection is made as a two-step procedure. Firstly we calculate the w.w. $w^{(r)}$ for the rectangular window (i.e., $g(x)=1$ for $0\le x\le 1$); the range for the minimisation of (3) is $0\le w\le W^*$, $W^*=\min(M-1, 10N^{\frac{1}{2}}, N/2)$. (We always use only integer values of w; it is easy to see that (3) is calculated rapidly when all $\{v_u\}$ are equal 1.) Then the window width (for needed window) is searched in the range of $w^{(r)}\le w\le \min(5w^{(r)}, W^*)$. Such procedure gives practically always the same result as the minimisation over the total range $0\le w\le W^*$, but it works much faster. (The theory shows that the order of optimal w.w. does not depend upon the concrete form of window being asymptotically minimal for the rectangular window.)

The computer program implementing the described procedure permits to enter interactively all parameters and display all results. The parameters are the window type, segment length M and "shift" value L (given data set). The graphs of the estimator (1), sample and smoothed autocovariance functions and the estimated risk function are displayed. For the simulated data all graphs contain both estimated and theoretic characteristics.

Simulations. The application of the algorithm shows that the procedure described above surely works in practice. Due to the construction of the estimator (1) (many periodograms averaged, the data segment for each being sufficiently large), the minimal sample value for it can not be small. However, the estimator appears to work even for $N = 200$ ($M = 80$, $L = 40$, $T = 4$). The s.p.e. gives reasonable results even for $N = 100$; for large samples (i.e. $N = 1000$) the both estimators appear to behave in the similar manner. We have performed simulations in order to investigate how close is the estimator f_* using the "adaptive" w.w. to the optimal estimator $\hat{f}^*$ (i.e. using the w.w. minimising the risk (2)). Because the results of such simulations were not published in [4] for the s.p.e., and in order to compare the two estimators, the computations were performed for both of them. The simulated t.s. was the autoregression process of the order 4. Its s.d. represents the smooth curve with two "peaks", one being "sharp" and another - "low" (there is no place for the graphs; our sofware is available from us - see below). Sample value was $N = 1000$; parameters for the estimator (1) were $M = 200$, $L = 100$, $T = 9$. For both estimators the Blackman's window $g(x) = A + B\cos(\pi x) + C\cos(2\pi x)$ (with $A = 7938/18608$, $B = 9240/18608$, $C = 1430/18608$, $0 \le x \le 1$) was selected. The optimal risk $R(\hat{f}^*)$ was 0.062 for the estimator (1) and 0.04 for the s.p.e. Here is the statistics for the relation $\rho(\hat{f}) = R(\hat{f}_*)/R(\hat{f}^*)$ based on ten simulated data samples (first line - for the estimator (1), second line- for the s.p.e.):

Mean± M. S. E. = 1.143± 0.033 ; minimum = 1.0006 ; maximum = 1.27 ;

Mean± M. S. E. = 1.224± 0.041 ; minimum = 1.01   ; maximum = 1.43 .

The media value of the r.f. (2) (i.e. $1.143*0.062 = 0.071$) is larger for the estimator (1) than for the s.p.e. (i.e. $1.224*0.04 = 0.049$), as it should be. On the other hand, the media values of ρ show that the estimator (1) approaches the optimal w.w. better than the s.p.e. Other simulations performed give similar results.

The adaptive w.w. selection show reasonable results even when the data does not satisfy the conditions of the theorem. When the t.s. is "white noise" (i.e. the s.d. is constant), the estimated w.w. almost always equals zero, and so the estimator and the theoretic s.d. are identical. The possibility to recognize noise is important for practice. When the t.s. is harmonical signal, the estimated w.w. always equals its maximal value, and the s.d. estimator shows sharp peaks in the points of the signal frequencies. (All this concernes both the estimator (1) and the s.p.e.)

About our software. The algorithm for the adaptive s.d. estimator (1) and also the similar algorithm for the adaptive s.p.e. are included into our time series' analysis package. The package also includes many other methods (parametric spectrum analysis, cross analysis, digital filtering, etc.); it runs on IBM PC compatible computers (286 and higher) and has modern "friendly" user interface. The current version is non-commercial and is available from our firm.

Referencies. [1]. Marple S.L., Jr. Digital spectral analysis with applications. - Prentice-Hall, Inc., Englewood Cliffs, New Jersey 07632, 1987.

[2]. Beltrao K.I., Bloomfield P. Determining the bandwidth of a kernel spectrum density estimate. - Journal of Time Series Analysis, 1987, N 1, pp.21-38

[3]. Ефроймович М. Ю. Пинскер М. С. Самонастраивающийся алгоритм минимаксного непараметрического оценивания спектральной плотности.- Проблемы передачи информации, 1986, т. 22, вып. 3, с. 62-76.

[4]. Iurovski A. Data based selection of window width for non-parametric spectrum density estimators. - COMPSTAT-90, short communications, pp.137-138.

(Note. Iurovski and Iourovski is the same person.)

Part XVII

Resampling Methods

Bootstrap Confidence Regions for Homogeneity Analysis; the Influence of Rotation on Coverage Percentages

Monica Th. Markus

Department of Data Theory, Leiden University

Abstract. The influence is evaluated of rotation of bootstrap sample homogeneity analysis solutions towards original sample solutions on coverage percentages of the bootstrap constructed confidence regions. The coverage percentages as well as the sizes of the confidence regions tend to decrease.

Keywords. Bootstrap, homogeneity analysis, rotation, coverage

1. Introduction

A serious limitation in the application of non-linear multivariate analysis (NLMVA) is the lack of information on the stability of the parameters (Markus, 1994). NLMVA relates to methods aiming at the representation of the structure of (non-) numerical multivariate data (Gifi, 1990). A specific criterion is optimized by assigning scores to objects and categories of variables. These scores can be used to construct a geometrical representation of the dependencies in the data. In this paper we will focus on homogeneity analysis, also called multiple correspondence analysis, which is a well-known example of NLMVA (Gifi, 1990). A homogeneity analysis solution consists of a series of statistics in several dimensions. We restrict ourselves to category quantifications of two dimensional homogeneity analysis solutions.

In the field of NLMVA, knowledge of the family of population distributions is not always available, and approximations of the sampling distribution of the parameters are hardly ever mathematically manageable. The bootstrap method (Efron, 1979) has been proposed to get insight in the stability of NLMVA parameters because it fits very well in the NLMVA framework; it is possible to apply the bootstrap without making assumptions on the (sampling) distributions of the NLMVA parameters (Heiser, 1981; De Leeuw, 1985; Gifi, 1990; Markus,

1994). The core of the bootstrap method is that from a set of n observations, the original sample, B bootstrap samples of size n are sampled with replacement (Efron, 1979). For each bootstrap sample the values of the parameters of homogeneity analysis are calculated. Because the study is restricted to two-dimensional solutions, it follows that by means of the bootstrap a two-dimensional scatter of bootstrap points is created for each parameter of the multivariate analysis method. We take the scatter of bootstrap points to be the approximation of the sampling distribution of the parameter values. Subsequently, confidence regions are created by peeling convex hulls (Green, 1981).

To evaluate the validity of the bootstrap constructed confidence regions a Monte Carlo study is executed. Initially, we expect that the bootstrap generated 90% confidence regions have actually a chance of $p=.9$ to include the population value of the parameter of interest over repeated sampling. We start with a known population and an NLMVA solution of this finite population is computed. In a Monte Carlo study 100 samples are drawn from the population and for each sample a bootstrap study is applied. Because the population is known, for each NLMVA parameter the 100 constructed 90% confidence regions may be checked by tracing whether the population value of the parameter is included. This empirical confidence coefficient of the constructed confidence regions is called coverage percentage. Coverage percentages below 84% are considered as serious deviations from the target value 90%. In this paper we focus on the question whether the coverage percentages are influenced by rotation of the bootstrap sample solutions towards the original sample solution.

2. Rotation

In our study we want to compare bootstrap NLMVA solutions from a given original sample. A problem in comparing different NLMVA solutions is the indeterminacy of the configurations, in the sense that the solutions are not unique with respect to reflections and rotations round the axes. Suppose several NLMVA solutions are used to generate confidence regions and that no reflection will be applied. This approach will provide unrealistically huge confidence regions for the category quantifications, that are not very informative (cf. Borsboom, 1988). Without any doubt reflection of the bootstrap solutions to one target is necessary, since for each dimension the sign may be changed without affecting the loss function. In order to give all NLMVA solutions the same orientation, a target solution must be chosen. In the bootstrap study we choose the original sample solution as a target. All bootstrap solutions are given the same orientation of this original sample solution. Before this step, the NLMVA solution of the original sample is given the orientation of the population solution. In this study, where reflection is concerned, all bootstrap solutions are given equal orientation.

The indeterminacy of the configurations with respect to rotation implies that in every bootstrap solution the orientation may be different. Specifically if the eigenvalues of the dimensions of interest are very close together, the probability of a shift of emphasis of the explained variance of a variable from one dimension to the other dimension is rather large (cf. Meulman, 1984; Borsboom, 1988). If one is of the opinion that the solution as such must be taken seriously, the position of the axes is considered to be an important aspect of the analysis results. Conversely, one may consider the orientation of the axes of no importance. In the latter case the interest is in the mutual positions of the category points, irrespective of the position of the axis. Consequently, the bootstrap NLMVA solutions may be rotated towards one target configuration, i.e. in our case the original sample representation. In this study the results of rotation with respect to the coverage percentages will be presented and discussed.

Procrustes procedures are applied to realize rotation of the different NLMVA solutions to one target (Borg & Lingoes, 1987). If A is the matrix of the target category quantifications and B is the matrix of category quantifications (weighted by the marginal frequencies) that will be rotated to the target, the problem is solved as follows. The singular value decomposition of $B'A$ is determined, $B'A=P\Phi Q'$, and the rotation matrix T may be found by: $T=PQ'$.

Data III is a population of 4000 observations, consisting of 6 variables, each of 5 categories. From this population 100 samples are drawn of sample size $n=100$, and for each sample a bootstrap study is executed, where the number of bootstrap trials $B=1000$. Detailed information on this data set can be found in Markus (1994).

3. Results

Figure 1 shows that for Data III coverage percentages tend to decrease if rotation is applied. This effect is somewhat harmful because not all of the coverage percentages stay within an acceptable range; quite a number of categories show coverage percentages below 84%.

An explanation of this effect of rotation on the coverage percentages may be found in the influence of rotation on the size of the bootstrap generated confidence regions. Figure 2 shows that rotation of the bootstrap HOMALS solutions towards the original sample solution results in smaller standard deviations for the bootstrap scatters of category quantifications. Smaller standard deviations means a shrinkage of the bootstrap generated confidence regions, which is a nice feature. Especially for the categories with small marginal frequencies this effect is obvious. This effect may be understood in the following way. In most cases categories with small marginal frequencies have peripheral category quantifications. Because of their peripheral location and their low weights these categories have the highest standard deviations under the no-rotation condition. Specially the bootstrap scatters of these categories appear to benefit from rotation.

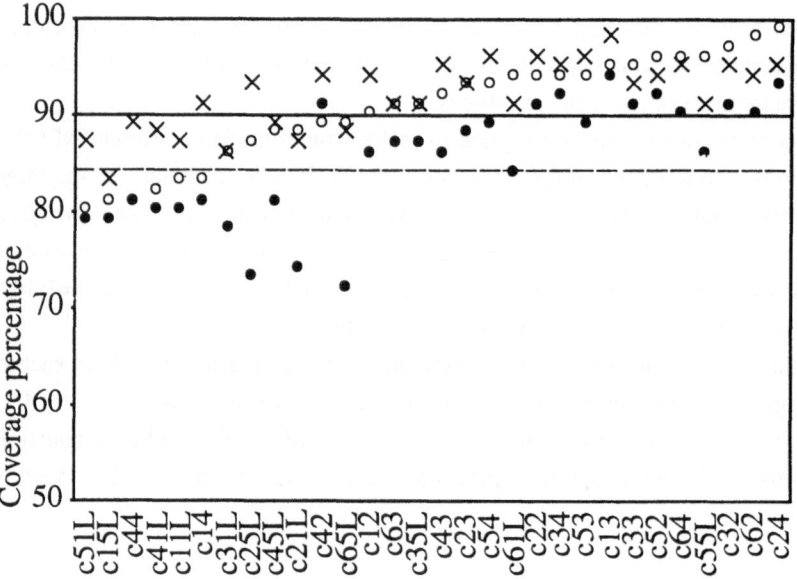

Figure 1 Coverage percentages of the homogeneity analysis category quantifications (horizontal axis) for Data III. Unrotated data: open circles; bootstrap solutions rotated towards the original sample solution: bold circles. The number of bootstrap samples $B=1000$ and the sample size $n=100$. Dashed line at 84%, the additional crosses indicate coverage percentages for rotation, in combination with 95% peeling.

To offer a solution for the possibly large number of coverage percentages below 84%, the following strategy is proposed. Rotation is applied, but the hulls are not peeled off up to a percentage of 90% is reached. Instead, the peeling is stopped as soon as a higher percentage, say 95%, is reached. It appears that this combination, rotation and peeling up to 95% provides satisfying results: the coverage percentages vary within an acceptable range. Figure 1 shows that almost all crosses are higher than 84%, the critical value. To formulate rules of thumb more data have to be gathered. This would be, however, beyond the scope of this paper.

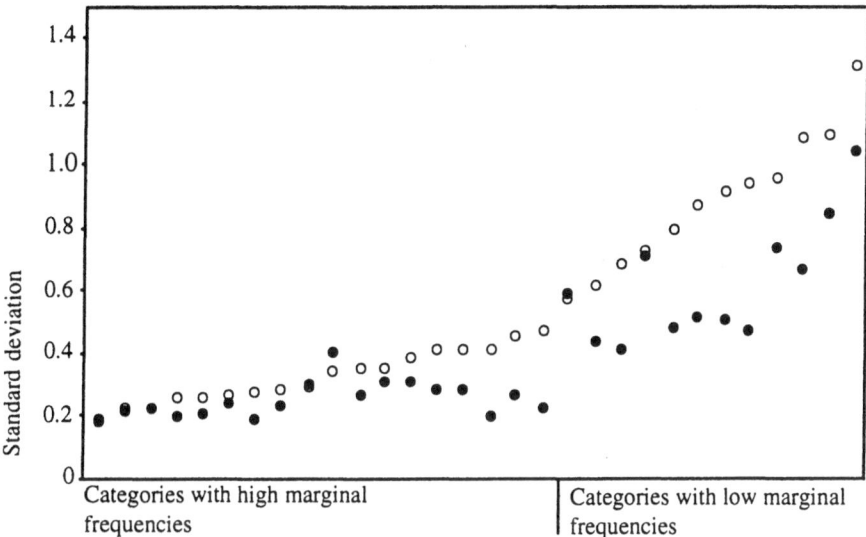

Figure 2 Standard deviations of the bootstrap scatters of the HOMALS category quantifications for Data III, sample size $n=100$. Unrotated data: open circles; bootstrap solutions rotated towards the original sample solution: bold circles. The number of bootstrap samples $B=1000$.

In summary, rotating the bootstrap NLMVA solutions to the original solution seems to have positive effects for the size of the confidence regions. The reverse of this effect is that the coverage percentages of the bootstrap generated confidence regions tend to decrease below the critical value of 84%.

References

Borg, I & J. Lingoes (1987). *Multidimensional similarity structure analysis.* New York: Springer-Verlag.

Borsboom, G. (1988). *Confidence Regions for Multiple Correspondence Analysis Using Bootstrap Methods.* Research Report RR-88-03, University of Leiden: Dept. of Behavioral and Computational Science.

De Leeuw, J. (1985). *Jack-knife and Bootstrap in Multinomial Situations.* Research Report RR-85-16, University of Leiden: Dept. of Data Theory.

Efron, B. (1979), Bootstrap methods: another look at the jack-knife, *Ann. Statist., 7,* 1-26.

Gifi, A. (1990). *Non-linear multivariate analysis.* Chichester: John Wiley & Sons.

Green, P.J. (1981), Peeling bivariate data. In: V. Barnett (ed.), *Interpreting multivariate data.* New York: John Wiley and sons.

Heiser, W.J. (1981). *Unfolding analysis of proximity data.* DSWO Press Leiden University.

Markus, M.Th. (1994). *Bootstrap confidence regions in non-linear multivariate analysis.* Leiden: DSWO Press.

Meulman, J.J. (1984). *Correspondence Analysis and Stability.* Research Report RR-84-01, Leiden University: Dept. of Data Theory.

Bootstrap Confidence Regions for Canonical Variate Analysis

Trevor J. Ringrose

Department of Mathematical Sciences, University of Aberdeen

Abstract. Several papers have recently proposed improvements to the traditional circular confidence regions often placed around sample means in canonical variate analysis. However, the need for improvement over the circles is greatest in small-sample cases where these regions, based on asymptotically correct variances, also perform badly. This paper describes the use of bootstrapping to estimate the required variances and hence produce regions which are clearly superior in such situations, as demonstrated by a simulation study.
Keywords. Canonical variate analysis; Bootstrap; Confidence Region.

1. Introduction

Canonical variate analysis (CVA) is used to provide a descriptive summary of the differences between pre-specified groups of points in a multivariate data set (Krzanowski, 1988, section 11.1). The following is a brief description of the method given largely in order to define notation.

Suppose that we have p observations on each of n individuals falling into g a priori groups, with n_i individuals in the i-th group so that $\sum n_i = n$. Let x_{ij} be the observed p-vector for the j-th individual in the i-th group, and define the group and overall means as $\bar{x}_i = \sum_{j=1}^{n_i} x_{ij}/n_i$ and $\bar{x} = \sum_{i=1}^{g} n_i \bar{x}_i/n$ respectively. Let W and B be the within- and between-groups sum of squares and cross-products matrices, $W = \sum_{i=1}^{g} \sum_{j=1}^{n_i} (x_{ij} - \bar{x}_i)(x_{ij} - \bar{x}_i)^T$ and $B = \sum_{i=1}^{g} n_i (\bar{x}_i - \bar{x})(\bar{x}_i - \bar{x})^T$. Then the $s = \min(g-1, p)$ sample canonical variates are given by $z_k = a_k^T(x - \bar{x})$ for $k = 1 \ldots s$, where the vectors a_k are chosen successively to maximize $a_k^T B a_k / a_k^T W a_k$. Hence the axes $a_1, \ldots, a_s$ are the eigenvectors of $W^{-1}B$ corresponding to the eigenvalues $l_1 \geq \ldots \geq l_s$, with standardisations $A^T B A = (n-g)L$ and $A^T W A = (n-g)I_s$ where the k-th column of A is a_k, I_s is the $s \times s$ identity matrix and $L = \text{diag}(l_1, \ldots, l_s)$. The data points can then be projected onto the new axes in order to examine informally the nature of the differences between the groups. The position of the i-th group mean projected onto the k-th axis is given by $\bar{z}_{ik} = a_k^T(\bar{x}_i - \bar{x})$, which is the k-th element of $\bar{z}_i = A^T(\bar{x}_i - \bar{x})$.

If it is assumed that the i-th group is a sample of size n_i from a $N_p(\mu_i, \Sigma)$ distribution then one-way multivariate analysis of variance (MANOVA) provides a formal statistical test for the equality of the group means and we can also form confidence regions (CRs) around the mean points on the canonical variate display. If we define $\bar{\mu} = \sum_{i=1}^{g} n_i \mu_i/n$ then it is well-known that, for *fixed* axes (eigenvectors) a_k, $\bar{z}_i \sim N_s(\nu_i, n_i^{-1} I_s)$ where $\nu_i = A^T(\mu_i - \bar{\mu})$ is the position of the population group mean relative to the observed sample axes. Hence $n_i(\bar{z}_i - \nu_i)^T(\bar{z}_i - \nu_i) \sim \chi_s^2$. Following previous papers, attention here is restricted to two dimensions only, and the CR for the first 2 elements of ν_i is a circle centred at $\bar{z}_i$ with squared radius $n_i^{-1}\chi_{d,1-\alpha}^2$ in obvious notation.

Krzanowski (1989) and Schott (1990) tried to produce improved regions based on the variance of $\bar{z}_{ik}$ which allowed for the fact that the axes are random variables not constants. However, Ringrose & Krzanowski (1991) and Ringrose (199?) argue that the variance required is in fact that of $a_k^T(\bar{x}_i - \mu_i) \equiv \xi_{ik}$, the difference between the sample and population means when both are projected onto the sample axes. Ringrose (199?) describes theoretical regions based on this variance but finds that, though better than the previous attempts, these are also not wholly satisfactory, and goes on to advocate a form of bootstrapping as the best solution so far. Ringrose (1992) outlined the proposed theoretical regions and gave some brief preliminary results concerning non-parametric bootstrapping. The present paper goes into further detail concerning the bootstrap approach, concentrating in particular on the problems encountered both in constructing the regions and in assessing their performance.

If the sample axes have low variance then, as expected, the circles perform well (Ringrose & Krzanowski, 1991). This is generally the case when n is large, p is small and there is a large degree of separation between the group means. However, if the reverse is true then the sample axes turn out to be far more variable than any of the theoretical methods predict and the circles perform spectacularly badly. The regions described here are, therefore, particularly aimed at cases with large p and small n.

2. Constructing Confidence Regions using Bootstrapping

The regions described below are not true bootstrap CRs but are instead based on bootstrap estimates of the variances of the ξ_{ik}. This is because the former are still rather problematic, in particular because of the difficulty of ordering vectors to define the 'central' resampled points (Owen, 1990). However, the development of genuine bootstrap CRs using, say, density estimation, is clearly a subject worthy of future study. As well as being simpler, the approach used here also allows the calculation of estimated mean square errors for the bootstrap method as described in section 3.

The method adopted is to generate bootstrap replicate data matrices X_b for $b = 1 \ldots B$, a large number, and, by performing the usual CVA calculations on each of these, to produce bootstrap analogues of the sample means and axes, $\bar{x}_{i(b)}$ and $a_{k(b)}$ respectively. Then we can estimate the variance of $\xi_{ik} = a_k^T(\bar{x}_i - \mu_i)$ by that of $\xi_{ik(b)} = a_{k(b)}^T(\bar{x}_{i(b)} - \bar{x}_i)$ over the B replicates.

Putting $\xi_{ik(.)} = B^{-1} \sum_{b=1}^{B} \xi_{ik(b)}$ gives, for $k, h = 1 \ldots s$, $i = 1 \ldots g$

$$\widehat{cov}(\xi_{ik}, \xi_{ih}) = \frac{1}{B-1} \left(\sum_{b=1}^{B} \xi_{ik(b)} \xi_{ih(b)} - B \xi_{ik(.)} \xi_{ih(.)} \right) \tag{1}$$

Krzanowski & Radley (1989) also used bootstrapping in this situation, but their approach was rather different to this. They estimated the variance of $\bar{z}_{ik}$, rather than ξ_{ik}, by jackknifing and by bootstrapping within one group at a time. It seems unlikely that this fully represents the true variability in the sample axes.

There are two main ways of generating the bootstrap replicate matrices. In the non-parametric version the i-th group has n_i vectors randomly drawn with replacement from the data vectors x_{ij}, $j = 1 \ldots n_i$. In the parametric version the n_i vectors are instead generated from the $N_p(\bar{x}_i, \widehat{\Sigma})$ distribution, where $\widehat{\Sigma} = \frac{1}{n-g} W$. The parametric method seems to be preferable in this situation since in cases with small sample sizes, which are of most interest here, non-parametric bootstrapping often performs very badly (the bootstrap version of W is often singular).

Eigenvectors are ordered according to the relative sizes of their corresponding eigenvalues in that particular sample or bootstrap replicate and hence there is no particular reason why the k-th bootstrap eigenvector $a_{k(b)}$ should correspond to the k-th sample eigenvector a_k. If the sample eigenvalues are all well-separated then the ordering of the eigenvectors should be preserved from sample to bootstrap replicate but otherwise we should expect some re-ordering to occur. However, (1) only works if $a_{k(b)}$ corresponds to a_k for every b. If there is reordering of the eigenvectors then the variance along the k-th sample axis (say) may be estimated using some values corresponding to projections onto other sample axes. This will almost certainly lead to an inflated estimate of the variance. This problem is exacerbated by the fact that any eigenvector can be multiplied by -1 (reflected), so that if some of the bootstrap eigenvectors are reflected relative to the sample ones then this too is likely to inflate the estimate. Both of these sources of spurious variation within the bootstrapping must be dealt with.

Krzanowski & Radley (1989) used Procrustes rotations on the sets of group mean points in order to minimise similar indeterminacy. Similarly it would be possible to use Procrustes rotations on each replicate to rotate the bootstrap eigenvectors to a position of best fit to the sample eigenvectors, by treating each set as s points in p-dimensional space. However, this does not seem appropriate, as the variability of the bootstrap eigenvectors must mimic as closely as possible that of the sample eigenvectors. Procrustes rotations would almost certainly lead to the bootstrap eigenvectors consistently being closer to the sample eigenvectors than the latter are to the population eigenvectors, thus making the estimated variance too small. Hence the method used here is to allow the bootstrap eigenvectors just to be reordered and reflected; the modified set chosen is that 'closest' to the sample eigenvectors.

However, even this is likely to remove genuine as well as spurious variance in the cases of most interest where the eigenvectors are very unstable. The bootstrap eigenvector deemed to correspond to the k-th sample eigenvector will usually be the one that is most similar to it, but this does not necessarily mean that it really corresponds to it. In particular this may treat the most 'extreme' bootstrap versions of a particular sample axis as not coming from that axis at all. Experimentation shows that the variation of bootstrap relative to sample (and sample relative to population) eigenvectors can be enormous, often with none of the vectors in one set looking anything like those in the other. This must be at least partly due to the fact that if, say, l_k and l_{k+1} are very close, then $a_{k(b)}$ might not look like either of a_k or a_{k+1} individually but the space spanned by $a_{k(b)}$ and $a_{k+1(b)}$ might be very similar to that spanned by a_k and a_{k+1}. Thus, except in well-behaved cases where population eigenvalues are well separated, $var(\xi_{ik})$ will usually be systematically either over- or underestimated, depending on if and how the reordering is allowed for. It is difficult to see any complete solution to this.

The method and closeness criterion used here is exactly the same as that for Procrustes rotations except that the transformation matrix, instead of being allowed to be any orthogonal matrix (see Krzanowski, 1988, p159, for example), is only allowed to reorder or reflect the eigenvectors. It can easily be shown (Ringrose, 199?) that the optimum ordering in this sense for bootstrap replicate b is that where

$$\sum_{k=1}^{s} c_k \left(\sum_{j=1}^{p} \{a_k\}_j \{a_{\kappa_k(b)}\}_j \right)$$

is maximised, where $\kappa_1 \ldots \kappa_s$ is a permutation of $1 \ldots s$ and $c_k = \pm 1$. This is simple to find if s is 3 or 4, as in section 3.

Ringrose (199?) gives full details of the construction of the CRs but, briefly, if $\widehat{V}_i$ is the 2×2 matrix whose elements are calculated using (1) for $k, h = 1, 2$ then, assuming multivariate normality of the x_{ij}, ξ_{ik} and $\xi_{ik(b)}$, we have $\xi_i^T \widehat{V}_i^{-1} \xi_i \sim \frac{2(B-1)}{B-2} F_{2, B-2}$ which, for large B, is equivalent to χ_2^2. Hence the appropriate CR is centred on $(\bar{z}_{i1}, \bar{z}_{i2})$ and can be drawn using the equation given by Krzanowski (1989, p112).

3. Assessing Confidence Regions

Krzanowski (1989), Schott (1990) and Ringrose & Krzanowski (1991) all assessed the performance of the proposed confidence regions using Monte Carlo simulation. They repeatedly generated samples, calculated the appropriate confidence regions and checked whether the true population means fell within the regions or not. If the rate of inclusion of population means within confidence regions was close to the nominal significance level of the region then it was assumed to be performing well. This is also used here. However, this is effectively only dealing with the bias, or not, of the estimated regions, not the variance. For example, a 95% region which in 5% of samples was a single point and in 95% of samples was a circle of radius 1000 would have an inclusion rate of 95% but would be useless. It might be possible to assess the variability of the regions themselves in some way but the approach adopted here is to estimate the mean square error (MSE) for the estimated variances used to produce the regions, for each combination of method, axis and group. The Monte Carlo trials can be used to estimate the variances of the ξ_{ik} in the obvious way. Call these estimates $MC\widehat{var}(\xi_{ik})$. The estimated MSEs will be the averages over the trials of the squared differences between these and the estimated variances from the methods. However, the sample eigenvectors can be reordered or reflected relative to the population eigenvectors $\alpha_1 \ldots \alpha_s$. The 2-dimensional regions are produced using a_1 and a_2 in each Monte Carlo trial, but these will not necessarily correspond to α_1 and α_2. Hence if, for example, we always use a_1 in the Monte Carlo estimation of $var(\xi_{i1})$, we may again get 'mixing' of axes. Hence $MC\widehat{var}(\xi_{i1})$ will probably be too high. The solution used here is to redefine $MC\widehat{var}(\xi_{ik})$ as being for the k-th *population* axis, using, from each trial, the sample axis most similar to that population axis. This similarity is decided using the same reordering method described above. Thus if, in a particular trial, a_3 is most like α_1, then ξ_{i3} will be this trial's contribution to $MC\widehat{var}(\xi_{i1})$. There are now at least 2 ways of defining the estimated MSE from this. The contribution from each trial to the MSE for axis k will be of the form $(\widehat{var}(\xi_{i?}) - MC\widehat{var}(\xi_{i?}))^2$ with both estimated variances corresponding to the same population axis. However, they could be chosen as either those corresponding to the k-th population axis *or* those corresponding to that population axis which provides the k-th sample axis in this trial. The former has been used here although it could be argued that the latter, though messier, more accurately measures the feature of interest. However, both should give similar results in terms of relative performance of methods.

For brevity, only a few simulation results are reported here, all with $p = 12$ and $g = 4$ since such cases with $p > g$ are usually where the existing methods do worst. Ringrose (199?) conducts a larger (and more fully explained) study while Ringrose & Krzanowski (1991) describe the simulation method in more detail. The results given in table 1 are, for each method, the inclusion rate (IR) for the nominal 95% level and the estimated mean square errors for the variance estimates for the first two axes. These are all averaged over the groups for 1000 simulations. The factor levels are on the right of the table. The n_i are all identical. 'AX' is a measure of the spread of the group means (large AX means they are more spread out) and 'R' is a measure of the closeness of the population eigenvalues ('s' for separated and 'c' for close). The methods tested are the traditional circles, Schott's ellipses and parametric bootstrapping with $B = 100$. Non-parametric bootstrap results have been omitted as it consistently performed more poorly than parametric, especially in terms of MSE. The simulated data can also be used to assess

CIRCLE		ELLIPSE			P BOOT	AX	n_i	R
14.3	11.4.11.3	36.7	80M.99M	88.5	2.84,2.77	1	5	s
14.1	10.0.11.4	36.2	379.1765	89.6	2.38,3.23	1	5	c
21.1	14.3.5.82	58.4	900.601k	90.7	2.75,1.60	4	5	s
19.1	5.57.8.43	64.0	2.4M.997	91.4	2.28,2.10	4	5	c
26.9	14.5.5.15	99.0	122k.126k	92.5	1.74,1.22	100	5	s
25.5	6.62.5.85	99.7	69k.62k	93.0	1.91,2.23	100	5	c
63.8	2.1m.1.6m	76.3	0.36.23	80.4	0.29m,0.57m	1	20	s
63.5	1.1m.1.0m	78.8	9.3.17.6	82.5	0.47m,0.47m	1	20	c
78.3	0.77m.0.75m	92.8	5.7m.9.7	89.2	0.16m,0.18m	4	20	s
77.1	0.56m.0.51m	96.3	1.2k.897	89.3	0.26m,0.25m	4	20	c
88.5	0.37m.0.28m	100	0.88.0.13	94.8	0.13m,0.11m	100	20	s
86.2	0.31m.0.34m	100	305k.335k	94.3	0.15m,0.14m	100	20	c

Table 1: Percentage IR and estimated MSEs for estimates of variance on axes 1,2 (M is millions, k is thousands. m is thousandths).

the use of the F distribution at the end of section 2, and seem to show that the assumption is not too unreasonable. The results clearly show that parametric bootstrapping outperforms the circles and the ellipses under these conditions. in that its inclusion rate is usually closest to 95% and its MSEs are always lowest. Further simulations have shown that this is consistently true except when the axes are very stable ($p < g$ and n_i is large), when the circles are better in terms of MSE.

References

Krzanowski. W.J. (1989). On confidence regions in canonical variate analysis. *Biometrika.* 76, 107–116.

Krzanowski. W.J. & Radley, D. (1989). Nonparametric confidence and tolerance regions in canonical variate analysis. *Biometrics*, 45. 1163–1173.

Owen, A. (1990). Empirical likelihood ratio confidence regions. *Annals of Statistics*, 18. 90–120.

Ringrose. T.J. (1992). Improving confidence regions for canonical variate analysis. Talk given at the 1992 RSS conference, Sheffield.

Ringrose. T.J. (199?). Improved confidence regions for canonical variate analysis. In preparation.

Ringrose, T.J. and Krzanowski. W.J. (1991). Simulation study of confidence regions for canonical variate analysis. *Statistics and Computing*, 1. 41–46.

Schott, J.R. (1990). Canonical mean projections and confidence regions in canonical variate analysis. *Biometrika*, 77, 587–596.

Part XVIII

Metadata and Statistical Information Systems

STORM+: Statistical Data Storage and Manipulation System

Antonia Bezenchek, Fernanda Massari, Maurizio Rafanelli

INFN - Sezione Sanità, Roma

Abstract. In this paper the overall architecture of the statistical data storage and manipulation system STORM$^+$ is described. It basically consists of four components: the Data Model, the Data Definition Language (DDL), the Query Language (QL), and the Statistical Working Environment (SWE). In this paper the data structure, the conceptual and the logical layer of the Object-Oriented STORM$^+$ model is described.

Keywords. Statistical Data Model, Object-Oriented Database, Aggregate Data

1 Introduction

Statisticians often use aggregate-type data as the source of their manipulation. Aggregate-type data are data which have been obtained by applying statistical aggregation (such as sum, count, etc.) and statistical analysis functions over disaggregate-type data, which are also called micro-data. In the last few years many researchers have studied and proposed different data models, operators, query languages and, in general, systems for defining and manipulating aggregate statistical data[1,2]. Many of them based their proposals on the Relational model, other authors on graphical representation and new operators. In this paper the authors take the Storm model [3] as reference and map it on an object-oriented representation, defining the STORM$^+$ object-oriented model. The formal description of the *Statistical Object* data structure is also given. In addition, the layered structure - conceptual and logical - of the Statistical Database System, as well as the type hierarchy and the method inheritance, are described.

2 The System Architecture

Our Statistical data storage and manipulation STORM$^+$ system consists of four components: the Data Model, the Data Definition Language (DDL), the Query Language (QL), and the Statistical Working Environment (SWE). In Fig. 2.1 the overall architecture of the system is represented.

The STORM$^+$ Data Model is the fundamental component on which the system is based. Using the DDL the data structures and its values are defined in the SDB.

The QL uses the data stored in the SDB to compose the answers. The definitions and queries can be written directly as input of the command interpreter. They can also be generated automatically by the visual interface,

352

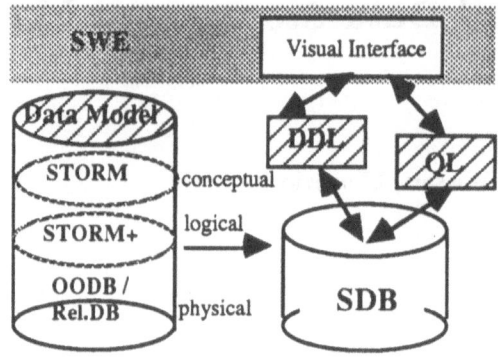

Fig. 2.1. Layout of the Statistical data storage and manipulation system architecture

including icons for the actions, scroll bar and menus for object selection and other elements of the visual interface. This graphical interface is available both for the Data Base Administrator in the database construction phase and for the user in the database query phase. It therefore represents the visual aspect of the Statistical Working Environment. In the rest of the paper the STORM+ data model will be described in more detail.

3 The Statistical Data Model STORM+

In this section the data structure of aggregate data will be proposed and the conceptual model Storm will be briefly discussed. Then the logical layer O-O data model, STORM+, will be proposed.

3.1 The Conceptual Layer: The SO Data Structure and the Storm Model

Data structure used by statisticians are more complex than conventional micro-data, for example, because they have two different attribute types and many different summary types [4], etc. This data structure has been called *Statistical Objects* (SO) [5], and it generally represents a statistical table, even though it can also represent other statistical entities, such as pies, bar-charts, graphics, etc. More formally, a SO is a quadruple $<\mathcal{N}, C, S, f>$ where $\mathcal{N}$ is the *name* which identifies uniquely the SO, C is a set of *category attributes* which describes the summary attribute, S is the *summary attribute* (i.e., the result of the application of a statistical function to the micro data), and f is a *function* which maps from the Cross product of the category attribute values to the summary attribute values of the SO. Each SO has different *properties*. Parts of them are always specified (for example, *summary type* = percentage, or *phenomenon* = Employment), but others do not always appear (for example, *unit of measure* = tons, or *unit of count* = thousands). Moreover, *marginal values* exist for each Statistical Object. They are important in different problems: data protection [6], summarizability of a category attribute, consistency of data, etc.

Recently a new graphical model for summary statistical data, called Storm (an acronym of Statistical Object Representation Model) [3], was proposed. It consists of a graphical representation of SO, in which two *representation spaces*, called respectively *Schema* and *Instance* spaces, and, orthogonally to them, three different *representation levels* (Topics or T-level, Statistical Objects or S-level and Base or B-level) are defined (Fig. 3.1). Only four types of nodes are used in the Storm model: T node, which appears only at the T level (it represents one topic grouping one or more statistical objects, or one sub-topic), S node, which appears only at the S level (it represents the summary data of the SO), A node, which appears at S and B-level (it represents the Cross product of all the category attributes which describe the SO), and C node, which also appears at S and B-level (it represents a category attribute). For *Cross product* we intend a Cartesian product where the operand order is not important.

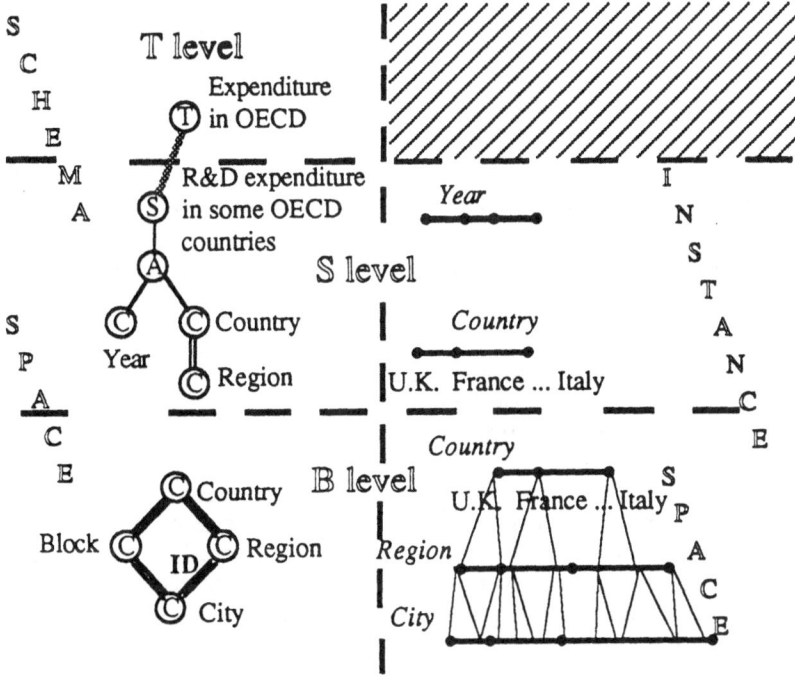

Fig. 3.1. STORM$^+$ conceptual layer

The network of T nodes at T-level represents the way in which different SOs can be grouped in concepts (topics) and these concepts in meta-concepts. This representation can change depending on the point of view of the user.

A set of trees, each of them representing a SO, appears at the S-level. Each SO consists of an *S node* , the root of the SO, which is always linked to an *A node*. The configuration under this A node depends on the complexity of the statistical object, but, in any case, only C nodes and A nodes can appear and C nodes are always the leaves of the tree. A C node may be linked to an A node (with the semantics of *classified by*) by a single solid line, or another C node (with the semantics of *grouping* or *classification hierarchy*) by a double solid line. It is also possible to have a *partitioning* (top-down approach) or *union* (bottom-up approach) of a C node in different C nodes (or, of different C nodes into a C node). In this case a single broken line is used and the SO is called *complex*. If the SO is *composite*, the model uses a notation, a + node, which links two or more S nodes between them.

Finally, the *primitive* or *base* attributes are represented at the B level, without any links to summary data. At this level edges between C and A nodes can exist, depending on the existence of a relation which, in any case, is primitive, i.e., independent from a particular statistical object. Both at the S level and at the B level it is possible to have labelled edges. This case refers to two different situations, i.e., the *Identification dependency* (label ID) and the *Non summarizability* (label NS) of one or more category attributes. The first case happens when, in a hierarchical classification, the lower level is not sufficient to identify uniquely the summary attribute instance. The second case happens, for example, when an overlap exist in the summary attribute, so that the summarization operation [7] gives an incorrect value.

In the *instances space* the domains of all the category attributes (both at S-level, and at B-level), and the eventual relationship tuples (at B-level) are represented.

3.2 The Logical Layer: STORM⁺ - the Object-Oriented Data Model

In this section we discuss the object-oriented typed data model STORM⁺. It derives from the previous model Storm and is its natural mapping in an object-oriented environment, in which the manipulation operations [4] are obtained by the methods defined in the classes of the model.

The object-oriented representation is important for the data manipulation performed as a result of query answer. If the answer to the statistical query doesn't exist as a unique Statistical Object instance, manipulations on the existing data in the database SOs are performed. To reduce the code of the operation definitions the object oriented structure is used. The operations (also called *methods* in the o-o terminology), which can be performed on each SO instance are stored in the respective *classes* (groups of SO instances of the same SO type).

To carry out the link between the STORM⁺ logical model and its conceptual model, we need: 1) a basic element of the o-o model; 2) a mapping tool from the conceptual to the logical representation.

The basic element (the structural part) of the o-o representation is the Statistical Object. The dynamic part consists of methods, describing the operations which can be performed on the SO, in order to obtain new SOs, or to combine it with other ones.

We define *canonical SO* the simplest structure which maintains all the characteristics of the SO to start with, then without classification hierarchy, partitioning, etc. [10].

In the phase of data acquisition and storage the system transforms all the simple, complex and composite SOs in canonical SOs and it maps them in an object oriented (logical, internal) representation. Such a transformation is important because the Query Language will use just the methods defined in this canonical structute, so that they can be simpler, more effective and more efficent.

3.3 Type Hierarchy

More complex real objects can be presented as Functional Objects. For example, we can present the serial number as a function that, given an integer, returns a car (*serial_number: int → car*) [8]. The car-valued function is also vehicle-valued, since every function which returns cars ($\tau \to$ car), also returns vehicles, because a car is also a vehicle. Then, for any type τ we say that $\tau \to$ car is a *subtype* of $\tau \to$ vehicle, i.e. $\tau \to$ car $\preceq \tau \to$ vehicle, because car $\preceq$ vehicle.

Now consider the function: *average_income: Population_in_Italy $\to$ money*.

As "Physicians in Italy" also forms part of "Population in Italy", we can use this function to compute the average income of Physicians in Italy. Hence, *average_income* is also a function from *Physicians_in_Italy* to *money*. In general, every function on "Population in Italy" is also a function on "Physicians in Italy", and we can say that: "average_income: *Population_in_Italy $\to$* money" is a subtype of ($\preceq$) "average_income: *Physicians_in_Italy $\to$* money", because: *Physicians_in_Italy $\preceq$ Population_in_Italy*, and money $\preceq$ money.

Note that the subtype relation is inverted on the left hand side of the arrow. We can conclude that, for a given function f (in the example above the function is *average_income*), we have:

$$f{:}\sigma \to \tau \preceq f{:}\sigma' \to \tau' \quad \text{if and only if} \quad \sigma' \preceq \sigma \text{ and } \tau \preceq \tau' \qquad (1)$$

For a more detailed discussion on the subtyping relation see [8], [9].

As a representation of the object-oriented abstraction, the SO can have its own type hierarchy and then use the type inheritance rules to facilitate the generation of new object types and instances. On the other hand, we have to apply the specific rules for the functional object subtyping.

Applying the rule (1) to the SO type, we can write that one statistical object SO_2 is a subtype of another statistical object SO_1 if and only if the SO_2 subject (its range) is subtype of the SO_1 subject, and the SO_2 category space is supertype of the SO_1 category space. To define the subtyping relations for the subject part, we use the semantics of the subject. The definition of the category space subtyping is more complicated. We have to use the extension to decide whether a type is subtype of another one.

If we have: $SO_2 = \mathcal{F}2{:}\mathcal{A}2 \to S2$ and $SO_1 = \mathcal{F}1{:}\mathcal{A}1 \to S1$, we must have: $\mathcal{A}1 \subseteq \mathcal{A}2$, $S2 \preceq S1$, and $\mathcal{F}2 = \mathcal{F}1 = \mathcal{F}$, to derive that: $SO_2 \preceq SO_1$. This subtyping mechanism will be very useful for the creation of new SOs in the definition phase as well as in the query evaluation phase. In this work we will not describe the new method generation algorithm, which exploits the type hierarchy.

To conclude this paragraph we will give one simple example of SO in subtype relation (Fig. 3.2).

□ *Example*

S2 is subtype of S1, because the semantic type of summary attribute - the subject of S2 (car) is subtype of the subject of S1 (vehicle), and because the extension of the category space of S2, consisting of the Cartesian product of the "region", "year", and "model" extensions is larger than the Cartesian product of "region" and "year" only (for simplicity we assume that "region" and "year" have the same extension for S1 and S2). □

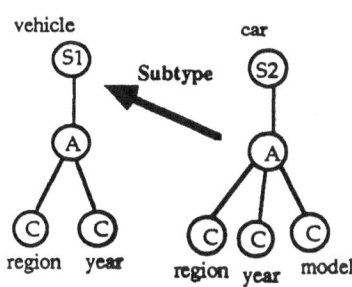

Fig. 3.2. Subtype relationship for the SO types

356

3.4 Methods Inheritance

Using the type hierarchy, based on the subject hierarchy, from the *semantic* point of view, and on the category attribute space hierarchy, from the *structural* point of view, we can manipulate the existing SO types and create the new types, based on the existing SOs. The new types will inherit methods from their parents. This methodology is very useful in the case of statistical objects, because in the query evaluation process we often have to obtain some other statistical object from the existing one by summarizing or cutting some set of category attributes from the statistical object category attribute space. There is an inheritance mechanism which allows the methods to be inherited only for the existing category dimensions and the methods related to the summarised or cut category dimensions in the newly created type to be suppressed. The description of this mechanism is beyond of the scope of this paper and we do not discuss it here.

4 Conclusions and future developments

In this paper we propose an object-oriented type model for the definition of aggregate statistical data, STORM$^+$. The first prototype of the STORM$^+$ system is in progress on a Macintosh computer using the MPW object-oriented environment and the C++ language.

Bibliography

[1]   M.Rafanelli, J.C.Klensin, P.Svensson Ed.s, "Statistical and Scientific Database Management" Proceed. 4th SSDBM, Lecture Notes in Comp. Science, N.339, Springer Verlag Pub., 1989

[2]   Z.Michalewicz Ed., "Statistical and Scientific Database Management" Proc. 5th SSDBM, Lect. Notes in Comp. Science, N.420, Springer Verlag Pub., 1990

[3]   A. Shoshani, M. Rafanelli, "A Model for Representing Statistical Objects", in Advances in Data Management, Eds. P.Sadanandan, T. M. Vijayaraman, McGraw-Hill Pub., 1991, pp. 161-177

[4]   M. Rafanelli, F.L. Ricci, "Mefisto: a functional model for statistical entities" IEEE Transactions on Knowledge and Data Engineering, Vol.5, N.4, pp.670-681, August 1993

[5]   M. Rafanelli, A. Shoshani, "STORM: A STatistical Object Representation Model", Proceed. 5th Intern. Confer. on SSDBM, Charlotte, NC, April 1990, Lecture Notes in Computer Science, Vol. 420, pp. 14-29, Springer Verlag Pub.

[6]   F.M. Malvestuto, M. Moscarini, M. Rafanelli, "Suppressing marginal cells to protect sensitive information in a two-dimensional statistical table" Proc. 10th ACM Symposium on Principles of Database Systems - PODS '91, Denver, Colorado, May 29-31, 1991

[7]   M. Rafanelli, F.L. Ricci, "A functional model for macro-databases" Sigmod Record, Vol.20, No.1, pp. 3-8, March 1991

[8]   L. Cardelli, "A Semantics of Multiple Inheritance", *Information and Computation*, 76: pp.138-164, 1988.

[9]   Cardelli and P. Wegner, "On Understanding Types, Data Abstraction and Polymorphism", *Computing Surveys*, Vol. 17, No. 4, Dec. 1985, pp. 471-522.

[10] M. Rafanelli, A. Shoshani, "A Model for the Graphical Representation of Aggregated Data", Technical Report IASI, R.348, December 1992

A Rigorous Treatment of Microdata, Macrodata, and Metadata

Hans-Joachim Lenz

Dept. of Statistics and Econometrics, Free University of Berlin

Abstract. There is a lot of ambiguities about the semantics of micro-, macro- and metadata which are the main data structures for the data management of statistical packages as well as for statistical databases. We believe that mathematical rigourness can be achieved by introducing abstract data-structures for micro-, macro- and metadata together with the corresponding sets of operators. In such a way we implicitly define what statistical data is like and avoid the usual ambiguities.

Keywords. Datamatrix, multi-way table, metadata, statistical database

1 Introduction

Let us consider an example first. A census was taken in Berlin(-West) in the year 1987. The data set ("Census data") consists of about 2.1 mio. cases (records) and 33 variables (attributes). Such type of data is often called microdata.

The corresponding set of attribute carriers (Berliners) is well defined by statutes with respect to time, region and subject. Frequency distributions (Counts), averages, quotients etc. can be easily derived from this data set using grouping and arithmetic operations. All these figures are meaningless unless there are data available for describing the Census data and the data derived from them. These data are called metadata. They enable the data-analyst to interpret the Census data as well as the derived data in a meaningful way.

It's worthwile introducing three sets which are of importance in the following. A *population* or *class G* is defined as an almost finite, non-empty set of attribute carriers. It's elements (attribute carriers, entities, individuals, objects) are similiar with respect to specific characteristics. The size of the population is given by $N := |G| \in N$. Measurements are taken on each of $0 < n \leq N$ elements of the population G. In order to simplify the notation we assume that the range R_j of the p attributes (variables) X_j (j=1,2,....,p) is the set of reals or coded as a subset of

reals. The cross product $\overset{p}{\underset{j=1}{X}} R(X_j) = R^p$ is called *sample space*.

The measurement can be considered as a mapping (attribute or measurement vector) $F: G \rightarrow R^p$. Consequently, the set T of observations or tupels $(X_{i1}, X_{i2}, ... X_{ip})$ for i=1,2,....,n is a proper subset of the sample space R^p, i.e. $T \subseteq R^p$. Note, that the coding should not be mixed up with the defining of measurement scales for each of the variables X_j (j=1,2,....,p).

The final space we need is the statistics space R^q (q<<p). It is induced by the statistic
$S : R^p \rightarrow R^q$ and is used to represent macrodata.

Generally speaking, the statistic S is composed of a partitioning operation φ and arithmetic operations $\oplus \in \{+, *, /\}$ on the sample space R^p.

In the following chapters we introduce the abstract data structures 'datamatrix', 'multi-way table' and 'metadata' which are the basic structures of statistical databases.

2 Microdata

Microdata are collected by sampling, surveying, running a census or filling forms. The domain of the corresponding variable vector is the population G and the range is the sample space R^p, irrespective of the scales of the single variables (attributes).

2.1 The data structure 'datamatrix'

A natural way to represent data of individuals from a population or universe is to use a datamatrix, cf. Nelder (1974).

A rectangular scheme of numbers $X_{ij} \in R$ ($i = 1, 2, ..., n$; $j = 1, 2, ..., p$) is called matrix of reals. The $n \cdot p$ real numbers are arranged in n rows and p columns. Each row (p-tupel) is identified by a case-id with values s = 1.2,...,n and represents the measurement vector or p-tupel $(X_{s1}, X_{s2}, ..., X_{sp}) \in R^p$ of the sampled attribute carrier $s \in G$.

Def. 1: Datamatrix

A set D of n tupels (rows, measurement vectors) is called *datamatrix* if there is a coding function F' with F':

$$\overset{p}{\underset{j=1}{X}} R(X_j) \to R^P \text{ where F: } G \to \overset{p}{\underset{j=1}{X}} R(X_j).$$

Example 1: Census 1987 data from citizens in Berlin(-West) with p = 33 variables (columns) and $n = N \approx 2.1 \ 10^6$ tupels (rows).

Note, that the term 'datamatrix' doesn't differ between the heading and the set of tupels, i.e. between the mapping F and the set of measurements or tupels.

There are several physical data structures available for storing a datamatrix. One could use a two-dimensional array, a file of records, a grid-file, B-trees, a relation or even an object-oriented data type.

In most cases of modelling the data type 'relation' will be appropriate. It is sometimes called 'flat table'.

Def. 2: Relation ("flat table")

A set D of tupels is called relation if the corresponding mapping (scheme) F is defined by

$$F: G \to \overset{p+1}{\underset{j=1}{X}} R(X_j) \text{ and } D \subseteq \overset{p+1}{\underset{i=1}{X}} R(X_i) \text{ where } X_{p+1} \equiv i, \text{ say:}$$

It is usual in computer science to define a relation by its relation scheme:
R_scheme::= <name> (case_id , X_1, X_2, ... X_p).
We note that

- $R(X_j) \neq R$ for all $j = 1, 2, ... p$.
- One X_j is used as an identifying attribute (primary key), i.e. is used as case identifier (index) : $R(X_{primkey}) \overset{1:1}{\longleftrightarrow} D$.
- D is invariant to permutations of rows and columns implying that this data structure may be inefficient for storing time-series.

2.2 Operations on a Datamatrix

Microdata is modelled in a natural way as a set of p-tupels which are not necessarily numeric. If the non-numerical attributes are coded, a datamatrix can be used for modelling the set of (coded) real p-tupels. In computer science such a data structure is called a relation type. Therefore, the set O_R of relational operators can be applied to the set **D** of datamatrices. Note that O_R is not complete unless recursion and /or looping is included. Recursion and looping are necessary for traversing nested or hierarchical data.

1. Set operations $\cap, \cup, \setminus$ (dyadic operators)

Given $D_1, D_2 \in \mathbf{D}$ then $D := D_1 \otimes D_2$ where $\otimes \in \{\cap, \cup, \setminus\}$ for $n_1 = n_2$ and $p_1 = p_2$.

2. Projection $\prod_R^{A*}$ (monadic operator)

Given a set **A** of attributes (variables) with $|A| = p$ and $A^* \subset A$ then $\prod_R^{A*} D$ is called projection of D on $D_1 := \prod_R^{A*} D$ for all $D \in \mathbf{D}$.

Evidently. the operator $\prod_R^{A*}$ eliminates the set $\mathbf{A} \setminus \mathbf{A}^*$ of variables from D.

3. Selection $\sigma_R{}^C$ (monadic operator)

Given a set $D \in \mathbf{D}$ of tupels with $|D| = n$ and a set C of selection criteria, then the operator σ_R^C selects all tupels which satisfy the selection criterion C,

i.e. $D_1 := \sigma_R^C D$ for all $D \in \mathbf{D}$ and $|D_1| << n$.

4. Join $\gamma_{\mathbf{R}}$ (dyadic operator)

Given $D_1, D_2 \in \mathbf{D}$ then $D := D_1\ \gamma_R\ D_2$ is called (equi) join of D_1, D_2 where $D_1. D_2$ must have at least one attribute in common.

Lemma: The set $\mathbf{O_R}$ of relational operators defined on datamatrices is
1. minimal,
2. closed,
3. not complete with respect to data management because recursion/loop are missing.

3. Macrodata

We consider now macrodata which are, generally speaking, derived from microdata using classification (grouping) and/or defining appropriate statistics like sum, counts (frequencies), average, median, percentage, quotient etc. Such grouped and/or summarized data can be naturally represented as a nested array which is a synonym for a multi-way or p-dimensional table. We firstly define the data structure and secondly describe the set of operators on multi-way tables.

3.1 The data structure 'p-dimensional table'

This data structure was originated by Nelder (1974). An important subclass is defined by contingency tables.

Def. 3: p-dimensional table T

A set of (real) numbers arranged along p dimensions (row. column. block. ...) is called p-dim table T if there is (at least) one summary attribute (statistic X) cross-classified or 'tabulated' on p category attributes $X_1. X_2. ..., X_p$. i.e.
$$X : \mathbf{R}^p \to \mathbf{R}^1 .$$

The datatype of the summary attribute X can be of any statistical type like

- counts (frequencies).
- total (sum).
- average, median. trimmed mean, percentage.
- quotient. index.
- standard deviation etc.

The mapping $X : \mathbf{R}^p \to \mathbf{R}^1$ which is composed of grouping and 'summarizing' or, equivalently, the scheme of a multi-way table can be described by different notations. It should be stressed that these different notational forms don't aim for describing layouts of multi-way tables but the data structure for its own. Shoshani, Chan (1981) introduced a graph-theoretic notation. Pfaltz (1992) used an algebraic notation and Nelder (1974) proposed an operator-oriented notation. For the sake of notational simplicity we follow the last one.

We need only two operators to describe the data structure of any multi-way table. Firstly, we have the dyadic operator x called *crossing*. Given the two ranges R_i of the attributes X_i (i=1.2) the operator x defines the cross-product $R = R_1\ \mathbf{x}\ R_2$. shortly written $X_1\ \mathbf{x}\ X_2$.

As an example consider the 2-dimensional table T_2 representing the number X of inhabitants cross-classified on the category attributes X1=Sex and X2=Age-Group derived from the Census 1987 data. It can simply notated as T_2 (X: X_1 x X_2).

Secondly, we need an operator called (hierarchical) nesting in order to model hierarchical or nested attributes like profession, region, product etc which are defined by specific nomenclatures.The operator nesting (/) represents relationships of the "is-part-of"- type. As an example consider the categorical attribute region. A region is divided into counties, a county may be divided into a different number of cities and so on. A natural form of notation is to use the nesting operator on region, county and city, i.e. (Region/County/City).

Expanding table T_2 by the categorical attribute Region (X_3) with substructure County ($X_{3'}$) and City($X_{3''}$) we get the 3-way table T_3 (X; X_1xX_2x(X_3 / $X_{3'}$ / $X_{3''}$)). Note, that this table is not a 5-dimensional table. Nelder(1974) poposed the following *axioms* for representing multi-way tables:

- A_1: All margins are to be represented explicitly
- A_2: The range of each categorical attribute is expanded by one value to represent the corresponding value of the margin.

3.2 Operators on multi-way tables

We remember ourselves that a multi-way table T can be interpreted as a set of (real) numbers or cells of type real arranged along p dimensions (p categorical attributes). Let **T** be the set of all multi-way tables T. The set O_T of table operators on a multi-way table should fulfill the following *axioms*:

- A_1:   *Closeness* of O_T with respect to **T**.
- A_2:   *Minimality* of O_T in the sense of non-redundancy, i.e. no operator o $\in$ O_T should be derived from a combination $o_1 \otimes o_2 \otimes ... \otimes o_r \equiv o$ with $o_i \in O_T$ (i = 1, 2, ..., r) .
- A_3:   *Completeness* of O_T with respect to data retrieval, but not with respect to data-analysis or modelling.

As the axioms A1, A2 can be easily fulfilled, axiom 3 is rather vaguely defined. The border between retrieval and data-analysis is not sharp. For example, is there a difference at all between the following two queries Q_1 and Q_2 ?

1. Q_1: **select** *avg*(income) **from** Census_87
2. Q_2: **select** *regression* (income on age) **from** Census_87

In the following, we shall exclude ad-hoc queries of the type Q_2. We refer to Nelder (1974) and Meo-Evoli. Ricci. Shoshani (1992, 1993) for further discussions.

The operators presented below are defined on the set **T** of p-dimensional tables:

1. **Arithmetic standard operations**   (dyadic operator)

   $T := T_1 \oplus T_2$ for all T_1. $T_2 \in$ **T** where $\oplus \in$ {+, -, * , /.

2. **Table Union** $\cup_T$ (dyadic operator)
   ("Cellwise linking of (partial) data from two tables")

   $T := T_1 \cup_T T_2$ for all T_1, $T_2 \in$ **T**, where $\cup_T$ is operating cellwise on T_1. $T_2 \in$ **T** .

3. **Table Selection on** σ_T (monadic operator)
   ("Conditioning", i.e. fixing a set of values for some category attribute)
   Let $\sigma_T = (\chi_1, \chi_2, ..., \chi_p)$ where $\chi_j \in$ {0, 1}^{kj} with $k_j :=$ | Range (X_j)| for all j=1,2.....p.
   χ_j is a vector of indicator variables for suppressing single values of the p categorical attributes X_j (j = 1,2,....p). Then $T := \sigma_T T_1 = T_1 \bullet (\chi_1, \chi_2, ..., \chi_p)$ for all $T_1 \in$ **T**.

4. **Table Projection** Π_T (monadic operator)
   ("Marginalization" or eliminating of categorical attributes).
   Let $A \subseteq$ **A** be a set of categorical attributes considered for marginalization.

   Then $T := \prod_T^A T_1$ for all $T_1 \in$ **T** is called T-projection of A in T_1 to A in T.

Note that the marginalization is strongly dependent upon the type of statistics used as summary attribute. If the type is 'counts' then marginalization is nothing else than "summarizing out" a specific attribute except for specific attributes which are non-summarizable. If the type of the summary attribute is 'average', special care must be taken with respect to computing the weighted averages for the margins. However, if quotients, indices or medians etc. are used then the margins can't be computed from multi-way tables. In these cases either the margins are stored and must be retrieved or the margins must be computed, however, using the microdata.

4 Metadata

Metadata are data about data, McCarthy (1982). They are modelled in a metadatabase which includes "nonsequential" text, diagrams, graphics and incorporates other media like voice, video and animation. The relationship between metadatabases, hypertext systems and multi-media systems is apparent.
Metadata describe universes (populations), micro- and macrodata on the semantic, structural, statistical and physical level in such a way that the universe is well defined and micro- and macrodata can be reasonably inputted, stored, accessed, transformed, grouped, summarized, retrieved and disseminated.

4.1 The data structure 'Metadata'

Modelling metadata as a metadatabase enables a formal approach for defining and describing them as an abstract data structure. This can be done using an Entity-Relationship-Model (ERM) approach, cf. Darius (1992), Sundgren (1992), Lenz (1993, 1994), by using an object-oriented database approach, cf. Sato (1991) or by describing conceptually metadatabases using graphs. Of course, the operations on metadata must be defined separately. We shall follow here the graph-theoretic approach.
Let $M = (N, L)$ be a directed graph (digraph) with a non-empty, finite set N of nodes and links L. The nodes are representing flat tables, "documents" which are composed of strings (character sequences, text) and bitmaps (pictures, diagrams) or triggers for audio and video. The links are creating relations between nodes / documents. There exist four types of links in a digraph used for meta- or hypertext databases, cf. Horn (1989):

- *Hierarchical links* expressing nomenclatures and 'a-kind-of'- relationships between documents,
- *Keyword links* connecting an index as a sorted list of keywords and pointers with the corresponding documents,
- *Referential links* among entities or keywords from different documents and
- *Cluster links* cross-referencing various documents and/or keywords within a specified context.

The major types of objects to which metadata may pertain can be briefly summarized as follows, cf. McCarthy (1982), van den Berg et al. (1992), Lenz (1993):

Semantic metadata (including definitions and footnotes)

- Contexts (various views on a database) as a set of entity types and attributes belonging to an universe
- Entity types as types of objects, individuals or events to which data pertain
- Attributes or variables
- Ranges (domains) as sets of values
- Names (identifiers), aliases and labels

Statistical metadata

- Questionnaires
- Surveys/ census/ panels, universes (populations), sampling or report schemes, sampling or report frames, statutes of sampling
- statistical objects, variables together with their meta-attributes like dimension, scale, status, missing value specification, error etc, domains (ranges) and co-domains

Physical metadata

- identifiers of variables including null/not null and key/non key characteristic, datatype, length, privacy lock, access (read/write) rights and
- identifiers of entity types together with a pointer to sub- and superclasses (context)

- We refer to McCarthy(1982) and Lenz (1993) for further details.

Its a specific flavour of databases that there exist different user views defined on subsets of a database. As metadatabases are a kind of databases this feature is inherited.

- A *thesaurus* is part of a document retrieval system. It is a hierarchical structured set of keywords representing an alphabetic or systematic catalogue together with cross-references used for referring to homonymes and synonymes.
- A *data-dictionary* (*DD*) is used to create and maintain entity types and attributes. In its extended form it describes variables on a statistical level.
- A *query panel* is used for an interactive specification of queries operating on a graphical user interface.

4.2 Operators on Metadata

Users of metadatabases need to be able to maintain metadata and search for metadata.

- Editing and visualizing metadata, i.e support for insert, delete and update operations. Note, that integrity constraints on metadatabases are a very hard problem.
- String searching, phonetic searching, scrolling through thesauri as well as range searching for advanced users in addition to arithmetic and logical operators on metadata.
- Fuzzy searching, phonetic searching and navigating (browsing) through the digraph M in order to explore the background of a query along the 'dimensions time, region and subject.

Acknowledgement

The author wishes to thank John McCarthy, LBL, Berkeley, and Bo Sundgren, Statistics Sweden, for a lot of valuable discussions on metadata.

References

van den Berg, G.M., de Feber, E., and de Greef, P. (1992), Analysing Statistical Data Processing, in Eurostat, New Technologies and Techniques for Statistics, Luxembourg, pp 102-111

Chan, P., and Shoshani, A. (1981), SUBJECT: A Directory Driven System for Large Statistical Databases, in Proc. of the LBL Workshop on Statistical Database Management, Lawrence Berkeley Lab, Berkeley, CA

Darius, P. et al. (1993), Modelling Metadata, in Statistical Meta Information Systems, Proc. of the Conference. Office for Official Publications of the European Communities, Brussels, Luxembourg, pp 257-266

Horn, R. E. (1989), Mapping Hypertext, The Lexington Institute

Lenz, H.-J. (1994), M^3-Database Design, Micro-, Macro- and Metadata Modelling, in F. Faulbaum (ed.), SoftStat '93 Advances in Statistical Software 4, Gustav Fischer, Stuttgart etc, forthcoming

Lenz, H.-J. (1993), On the Design of a Statistical Database. Micro-, Macro- and Metadata Modelling, Historical Social Research, vol. 18, No. 4, pp 31-48

McCarthy, J.L. (1982), Metadata Management for Large Statistical Databases, in Proceedings of the Eighth International Conference on Very Large Data Bases, Mexico City

Meo-Evoli, L., Ricci, F.L., and Shoshani, A. (1992), On the semantic completeness of macro-data operators for statistical aggregation, in Hinterberger, H., French, J.C.(eds.), Proceedings of the Sixth International Workshop on Statistical and Scientific Database Management, Department of Informatik, ETH Zürich, Zürich, pp. 239-258

Nelder, J.A.(1974), Genstat - A Statistical System, COMPSTAT 1974, Proc.in computational statistics. Bruckmann, G. et al (eds.),Physica Verlag, Wien, pp.499-506

Pfaltz, J. (1992), A Functional Approach to Scientific Database Implementation, in Hinterberger, H., French, J.C.(eds.), Proceedings of the Sixth International Workshop on Statistical and Scientific Database Management, Department of Informatik, ETH Zürich, Zürich, pp. 239-258

Sato, H. (1991), Statistical Data Models: from a Statistical Table to a Conceptual Approach, in Michalewicz, Z. (ed.), Statistical and Scientific Databases, Ellis Horwood, New York etc., pp. 167-200

Sundgren, B. (1993), Modelling Meta-Information Systems, in Statistical Meta Information Systems, Proc. of the Conference, Office for Official Publications of the European Communities, Brussels, Luxembourg, pp. 59-79

Ullman, J.D. (1988), Principles of Database and Knowledge-Base Systems, Vol. I, Computer Science Press, Rockville

Part XX

Statistical Software, Evaluation and Comparison

Software for the Design of Experiments - a Comparative Overview

Paul L. Darius

Katholieke Universiteit Leuven, Laboratory for Statistics and Experimental Design

Abstract. Although everyone agrees on the importance of the planning or design phase in actual experimentation, and the consequences of poor design decisions for subsequent statistical analysis, this point of view is not adequately reflected in current software. In many of the well-known statistical packages, support for design aspects is lacking or marginal compared to the support for analysis.

On the other hand, many separate (and less well known) programs have been developed that deal specifically with design problems, often in a restricted area (e.g. quality control, variety trials, ...). Recently, the number of these programs has grown considerably, and many of the well-known packages have added modules that specifically support some design aspects. In this paper we try to give a broad overview of what is currently available.

Keywords. experimental design, sample size determination, optimal designs, factorial designs, block designs, mixture designs, variety trials.

1 Introduction

There might be several reasons for the fact that, historically, much of the statistical software has concentrated on analysis aspects, and design aspects have been given relatively little attention.

First, it may be argued that not every experiment needs a design, although even in most observational experiments sample size determination is an important issue. Second , planning an experiment contains many non–statistical components and several of the formal statistical components are non–numerical in nature. Finally, computer software for the design phase is often perceived by the user as being more difficult to use than "default" analysis programs, because design software requires a precise answer to questions about the population to which the results should apply, the precision required and the risks one is willing to take with tests and other decisions. Some of these questions may be hard to answer.

Nevertheless in recent years a substantial amount of software approaching design problems has become available. This paper is meant to give a broad overview. Within the limited space of a proceedings paper it is not feasible to describe all the merits of each program. For a more extensive description of many of the programs we refer to Rasch and Darius (1994). An idea about the analysis capabilities of some of the programs described here can be obtained from Koch and Haag (1993).

The author would be grateful for comments and for information about programs not included here. Unless specified otherwise, all the programs mentioned run on IBM–compatible PC's and are in English.

2 Overview of design software

Designing experiments is an activity with many aspects, and this is reflected in the diversity of the available tools. In this section we try to make a general classification. Each class is then described in more detail in the sections that follow.

A preliminary activity in designing experiments is the recognition of the basic elements of the design : what are the responses to be observed, the experimental units, the factors to be selected, their classification into block and treatment factors, the size of the experimental region... It is a pity that most of the software assumes that these decisions have already been made. Some form of software support (e.g. in the form of a check list, see Jeffers (1984)) would be quite useful.

In a first section we describe programs that assist the user with the choice of an appropriate model, and programs that perform utility functions such as randomization and the preparation of data entry forms.

A second group of programs deals with the actual construction, based on precise user–supplied requirements, of certain types of designs. The designs covered typically include one or more of the following : balanced or partially balanced incomplete block designs, factorial designs, central composite designs, row–column designs and mixture designs. The techniques used are based on combinatorial analysis, finite geometry, Galois theory, difference sets and orthogonal arrays, but also on search procedures.

A third group consists of programs for power analysis and the determination of sample size. They are based on non–central distributions and asymptotical properties of special statistics.

A fourth group deals with the optimal allocation of observations. They are based on D–, A–, G– and E–optimality or related criteria. Some of them (the WACH and REG modules of *CADEMO*) optimize both the size and the structure of the experiment, based on an overall risk function. The techniques used here include functional analysis, integer nonlinear optimization and search procedures.

A fifth group consists of programs that assist in sequential experimentation.

3 Software for model choice and utilities

ECHIP and *EXPERIMENTAL DESIGN* help researchers determine which type of experimental design is most appropriate for their research project. The *CADEMO* package contains several modules that assist in model choice : LPRO in general, WACH for choosing a growth function, NREG for regression models, ANOV for anova models, WIBI for bioassay and LEDA for quality control and survival.

Examples of programs that perform utility functions are *DATACHAIN* (creation of data collection forms, front–end to other statistical packages) and *RANCODE* (preparation of random code lists, sealed envelopes, stick–on labels for large surveys or clinical trials).

4 Software for design construction

A first category of programs is particularly suitable for constructing the type of designs most often used in biometric experimentation in general, and field experiments in particular. *STATITCF* is such a package, available in English, French and Spanish. *DSIGNX* constructs most of the designs used in variety testing. The programs available from *NGUYEN* construct cyclic block designs and optimal or (non)resolvable incomplete block and row–column designs. *ALPHAGEN* constructs alpha and row–column designs. *ALPHA* extends *ALPHAGEN*, constructs randomized field plans for many designs, including alpha and alphaalpha designs with nested treatment structures, and provides facilities to deal with unequal block sizes and restricted treatment randomization.

The module FIEL in *CADEMO* deals with many aspects of field experimentation. In addition to the capabilities to generate the usual designs, there are facilities to assist in specifying the precision requirements and determine the number of experimental units and the size of the plots. Assistance in planning a field trial under practical farming conditions is also provided.

A second category consists of software mainly developed for industrial experimentation. There are many programs that cover some of the typical designs used in industry : two level fractional or full factorials, Plackett–Burman designs, Box–Behnken and central composite designs, mixture designs, orthogonal arrays and Taguchi-type designs. Examples are : *FACTORIAL–DESIGN, BOX–B, DESIGN–EASE, APO, DESIGN–EXPERT, DESIGN–KIT, TAGUCHI–KIT, GENEME* (in French), *MIXTURE–DESIGN*.

Over the years, many of the large analysis packages have added separate modules for experimental design, or included design features in their package. *RS/ DISCOVER* is a module of the RS/1 system and contains an extensive strategy for guiding a user through the design and analysis of an (industrial type) experiment. The packages *STATISTICA, MINITAB* and *STATGRAPHICS* contain facilities for designing (relatively simple) experiments. *SOLO* has a module called Experimental Design, *SYSTAT* has OC/STAT and *SPLUS* has announced a module Industrial Experimental Design. *SAS* has a procedure FACTEX to construct orthogonal designs, which is not part of SAS/STAT but of the quality control module SAS/QC.

There are other programs that can also deal with somewhat more complex design requirements. The ANLA module of *CADEMO* constructs factorial, fractional factorial and several types of central composite designs. *KEYFINDER* is a Prolog program that uses search procedures to generate fractional–replicate and blocked designs meeting complex requirements. *GOSSET* is a general purpose program for constructing designs. It includes a library of several thousand designs and can generate designs based on sophisticated optimality criteria, for special experimental regions and for special models. This program runs under UNIX.

Some programs in this category merely construct a design, others contain an extensive design/analysis strategy and/or a substantial amount of design "wisdom". In the latter category we already mentioned *CADEMO* and *RS/DISCO-VER*. Other examples are *DEXTER*, *SAS/LAB* and *STAVEX*.

5 Software for calculation of power and sample size

Several programs calculate sample sizes or the power of a test for a given size. Most of them are limited to one- or two-sample problems and normal (with possibly some other) distributions. Examples are *N*, *POWER*, *PLANUNG*, *ECHIP*, *STAT–POWER* and *STPLAN* (a public domain program).

Programs with a broader scope (and particularly oriented towards survival problems) are *NSURV*, *POWERPACK*, *DESIGN POWER*, *EX–SAMPLE*, *EGRET–SIZ* and the power analysis add-on to *SOLO*.

The *CADEMO* package has parts for power analysis and sample size for several distributions and underlying models, possibly based on cost considerations. Similar facilities are available for comparing or estimating variances. Another part deals with survival problems, and there is also a part that calculates sample sizes for the selection problem (selection of the t populations with the largest expectations out of k given populations). Another program that deals with the selection problem is *RANKSEL*.

SAS currently has no explicit support for power analysis, although the item ranked high on user suggestions lists. Noncentral distributions are available as functions. The power calculations for general linear multivariate models (including repeated measures) from Muller et al. (1992) is available as an item in the example library distributed with the SAS/IML module (from version 6.09 onwards).

6 Software for optimal allocation

Optimal allocation programs use one out of a variety of optimality criteria, usually to select a set of a given size from a larger set of candidate points. They are most useful when the model or the shape of the experimental region is non-standard.

Examples of such programs are *ACED*, *CAMOS* and the OPTEX procedure in SAS/QC. Some of the programs mentioned in the other sections can also construct optimal designs, e.g. *ECHIP*, *GOSSET* and *CADEMO* (for linear and nonlinear regression).

7 Software for sequential experimentation

In this final class of design software, the roles of design and analysis are very closely linked. After part of the results are available (and based on the preliminary analysis of these results), the design for the next phase is set up or the experiment is stopped. This can be repeated a number of times. Examples of programs that assist in sequential sampling or offer stopping rules are *PEST* and *EAST*. See also Schneider (1992).

8 Contact Addresses

ACED : Prof. W. Welch, Dept of Statistics and Actuarial Sciences, Univ. Waterloo, Waterloo Ontario N2L, 3G1 CANADA

ALPHA : Emlyn R. Williams, CSIRO Division of Forestry, P.O. Box 4008, Queen Victoria Terrace, Canberra ACT 2600, AUSTRALIA

ALPHAGEN : Mike Talbot, Scottish Agricultural Statistics Service, The University of Edinburgh, The King's Buildings, Edinburgh EH9 3JZ, SCOTLAND, UK.

APO : SYSTEGRA, Vertriebsgesellschaft für Automation und Sicherheitstechnik mbH, Frankfurter Str. 63–69, 6236 Eschborn, GERMANY

BOX-B : Statistical Programs, 9941 Rowlett, Suite 6, Houston Texas 77075, USA

CADEMO : BIORAT GmbH, im. Rost. Innov. u. Grüderzentr., Joachim Jungius Str. 9, D-18059 ROSTOCK, GERMANY

CAMOS : see Osyczka, A. (1992) Computer Aided Multicriterion Optimization System. Int. Software Publ. Krakow POLAND.

DATACHAIN : Emlyn R. Williams, CSIRO Division of Forestry, P.O. Box 4008, Queen Victoria Terrace, Canberra ACT 2600, AUSTRALIA

DESIGN-EASE : STATEASE Inc., For Europe: QD Consulting, 68 Station Road, Steeple Morden, Royston, Herts SG8 ONS, UNITED KINGDOM

DESIGN-EXPERT : STATEASE Inc., For Europe: QD Consulting, 68 Station Road, Steeple Morden, Royston, Herts SG8 ONS, UNITED KINGDOM

DESIGN-KIT : M. Hasselaar, Centre for Quantitative Methods, P.O.Box 513, 5600 MB Eindhoven, THE NETHERLANDS

DESIGN POWER : Scientific Software Inc. Neitzel Rd, Mooresville, IN 46158 USA

DEXTER : P. Haaland, Becton Dickinson Research Center, P.O.Box 12016, Research Triangle Park, North Carolina, 27709, USA

DSIGNX : Alec Mann, Scottish Agricultural Statistics Service, The University of Edinburgh, The King's Buildings, Edinburgh EH9 3JZ, SCOTLAND, UK.

EAST : Cytel Software Corp., 675 Massachusetts Ave., Cambridge, MA 02139, USA

ECHIP : ECHIP Inc., 724 Yorklin Road, Hockessin 19707, USA.

EGRET-SIZ : SERC, 909 NE 43d, Suite 202, Seattle, WA 98105, USA

EXPERIMENTAL DESIGN : Statistical Programs, 9941 Rowlett, Suite 6, Houston Texas 77075, USA.

EX-SAMPLE : IDEA Works, Inc., 607 Jackson Street, Columbia, MO 65203, USA

FACTORIAL DESIGN : Statistical Programs, 9941 Rowlett, Suite 6, Houston Texas 77075, USA

GENEME : Mathieu, D., L.P.R.A.I., I.U.T. Avenue Gaston Berger, F-13625 Aix-en-Provence Cedex, FRANCE

GOSSET : Sloane, N.J.A., ATT Bell Labs, Room 2C-376, 600 Mountain Ave, Murray Hill N.J. 07974, USA

KEYFINDER : P.J. Zemroch, Shell Research Ltd, Thornton Research Centre, P.O. Box 1, Chester CH13SH, ENGLAND, UNITED KINGDOM

MINITAB : MINITAB Inc. 3081 Enterprise Drive, State College, PA 16801-3008, USA

MIXTURE-DESIGN : Statistical Programs, 9941 Rowlett, Suite 6, Houston Texas 77075, USA

N : IDV-Datenanalyse und Versuchsplanung, Wessobrunner Str. 6, D-8035 Gauting, GERMANY

NGUYEN : Nam-Ky Nguyen, Biometrics Unit, Inst. of Animal Production and Processing, CSIRO , Private Bag 10, Clayton Vic. 3168, AUSTRALIA

NSURV: IDV–Datenanalyse und Versuchsplanung, Wessobrunner Str. 6, D–8035 Gauting, GERMANY

PEST: Prof. Dr. John Whitehead, University of Reading, Department of Applied Statistics, Whiteknights, P.O. Box 217, Reading, RG6 2AN, ENGLAND, UNITED KINGDOM

PLANUNG : DKFZ, Inst. für Dokumentation, Information und Statistik, Im Neuenheimer Feld 280, D–W–6900 Heidelberg 1, GERMANY

POWER : Epicenter Software, P.O.Box 90073, Padadena, CA 91109, USA

POWERPACK : R.V. Lenth, 3061 Hastings Avenue, Iowa City, IA 52240, USA

RANCODE : IDV–Datenanalyse und Versuchsplanung, Wessobrunner Str. 6, D–8035 Gauting, GERMANY

RANKSEL : H. Edwards, Dept Statistics, Massey University, Palmerston North, NEW ZEALAND.

RS–DISCOVER : BBN Software Products, 10 Fawcett Str, Cambridge, MA 02238, USA

SAS : SAS Institute Inc., SAS Campus Drive, Cary, N.C. 27513, USA

SOLO : BMDP Statistical Software Inc., 1440 Sepulveda Blvd Suite 316, Los Angeles CA 90025, USA

SPLUS: Statistical Sciences Inc., 1700 Westlake Ave N, Suite 500, Seattle WA 98109, USA

STATGRAPHICS : STATGRAPHICS, Manugistics Inc., 2115 East Jefferson Street, Rockville, Maryland 20852, USA

STATPOWER : iec – ProGAMMA, P.O. Box 841, 9700 AV Groningen, THE NETHERLANDS

STATISTICA : StatSoft, 2325 E. 13th St., Tulsa, OK 74104, USA

STATITCF : Jean Tranchefort, I.T.C.F. Service des Etudes Statistiques, Station Experimentale, 91720 Boigneville, FRANCE

STAVEX: H. Flühler, Mathematical Applications, CIBA–GEIGY AG., R–1008 Z 236, CH–4002 Basel, SWITZERLAND

STPLAN : B. W. Brown, The University of Texas, M.D. Anderson Cancer Center, Dept Biomathematics, Box 237, 1515 Holcombe Bld, Houston Texas 77030. ftp from odin.mda.uth.tmc.edu (129.106.3.17)

SYSTAT : SYSTAT, Inc., 1800 Sherman Ave, Evanston, Illinois 60201, USA

TAGUCHI–KIT: M. Hasselaar, Centre for Quantitative Methods, P.O.Box 513, 5600 MB Eindhoven, THE NETHERLANDS

References

JEFFERS, J.N.A. (1984). Statistical Checklist 1 : Design of Experiments. Institute of Terrestrial Ecology, National Environment Research Council.

KOCH, A. and HAAG, U. (1993). The Statistical Software Guide '92/'93. Statistical Software Newsletter, Comp. Statist. & Data Analysis 15, 241–262.

MULLER, K.E., L.M. LaVANGE, S. L. RAMEY, C.T. RAMEY (1992). Power Calculations for General Linear Multivariate Models Including Repeated Measures Applications. Journal of the American Statistical Association, 87, 1209–1226.

RASCH, D. and P. DARIUS (1994). Computer Aided Design of Experiments. in: DIRSCHEDL, P. and OSTERMANN, R. (1994). Computational Statistics. Physika–Verlag.

SCHNEIDER, B. (1992). An Interactive Computer Program for Design and Monitoring of Sequential Clinical Trials. Proc. Int. Biom. Conf. Hamilton, New Zealand. Invited Paper Vol 237–246.

A Distributed LISP-STAT Environment

David C. De Roure, Danius Michaelides

Department of Electronics and Computer Science & Department of Social Statistics, University of Southampton

1   Motivation

With the advent of networking and high-powered workstations, and the rise of end-user computing alongside the traditional centralised computing model, the *heterogeneous network* is emerging as the most significant platform for many computing activities. A heterogeneous network consists of a number of resources (e.g. workstations, fileservers, database engines and computation nodes) interconnected by a fast data network; open systems standards facilitate interoperability in this (possibly multivendor) environment.

This platform offers significant benefits for statistical computing applications. Not only can programs access repositories of online data as required, they can also access other processors in order to exploit specialised features or just to harness idle resources. Hence computationally intensive tasks can be accelerated by farming out subtasks to multiple processors. Further, this platform supports groups of users and hence there is scope for the development of applications which facilitate collaborative working, such as a team of researchers working on a common dataset.

2   LISP-STAT

LISP-STAT [5] is a statistical computing environment based on the Lisp language; it is available on a variety of processors and operating systems. Lisp is a *dynamic language* and as such lends itself to distributed applications— its flexibility, interactive nature and reflective ('code as data') properties make it an effective means of supporting multiple cooperative processes on a heterogeneous network. For these reasons, LISP-STAT is an ideal basis for the development cf a statistical computing environment for heterogeneous networks.

LISP-STAT provides dynamic, interactive graphics. Under the X-windows system, LISP-STAT processes running on one workstation can use another for their display and user interaction. These facilities alone are sufficient to allow one user to run LISP-STAT processes on multiple computers, each having its display set to the user's workstation. However, to harness the full potential of such an environment, mechanisms for communication between those processes must be provided—and these mechanisms need to be supported within the LISP-STAT language so that they are available to a user writing LISP-STAT programs.

3   Distributed Lisp

To establish an environment for distributed statistical computing, it is essential to specify standard interfaces between the components of the system. This will enable the construction of applications from a variety of distributed components complying with the interface conventions.

Our initial experiments have been with LISP-STAT processes communicating with other LISP-STAT processes, but we are now extending this so that they can communicate with programs written in other Lisp dialects and other languages (such as C). In the future, the components of the system may be other application programs which we can harness to comply with the interface. We envisage that LISP-STAT itself will be a front-end to such a hybrid system, and that the Lisp language itself will continue to provide the 'glue' and programmability of our system.

The utility of a *message-passing model* for distributed Lisp applications has been demonstrated [2]; this is one of the many models which we can build using primitive building blocks for concurrency and synchronisation [1]. Tierney [6] has proposed to adopt a broadcasting mechanism in Lisp (called *announcements*) as a future development of LISP-STAT. Like the message-passing models, we are adopting techniques which provide a concept of process location [3], upon which higher-level facilities (such as announcements) can be constructed. It is this which distinguishes our approach from the traditional vision of parallelism solely for speedup—we are able to migrate processes towards data.

4   Communications model

We have investigated a number of different communications models for use in our prototype system. To promote interoperability of components, we anticipate that more than one of these would be supported. The following models have been examined in detail, by building simple implementations and comparing them; we are still evaluating CORBA (the *Common Object Request Broker Architecture*), and the emerging MPI (*Message Passing Interface*) standard.

- RPC or Remote Procedure Call, offers a strict client-server model, i.e. a server publishes the availability of a service and remote clients can then make procedure calls to the server. RPC follows a well defined standard, and has the advantage that it is machine independent and provides a consistent programming interface which, at its highest level, resembles the familiar procedure call model. The RPC model is very strict (the client supplies some data, the server processes it and sends a reply, the transaction terminates) and although it can be relaxed, it is not designed to work with processes which are to behave as both clients and servers.

- **Unix sockets** provide a mechanism for two processes to communicate freely in a bi-directional manner. It requires that a process first set up a known connection point, then act like a server, accepting connections from client processes. Once communications are established, two processes can communicate freely; the connection may be transitory or it may persist for a number of exchanges. Sockets provide independence from the details of particular networking technologies.

- **TLI** or Transport Level Interface, provides a similar model to sockets, but it complies with OSI standards. The programming interface is more sophisticated than sockets and can lead to better engineered solutions. Although TLI is set to replace sockets on UNIX machines, the socket interface is only just becoming widely available on PCs.

- **PVM** or Parallel Virtual Machine aims to harness a collection of heterogeneous machine into a large distributed-memory computer. It provides a message handling system, as well as message parsing, process spawning and configuration library calls. It has emerged as a standard within the parallel processing community.

- **ToolTalk** aims to provide an interface to allow tools (such as editors, compilers, mailers etc) to be aware of what the user is doing in the environment. Tooltalk provides an announcement-like mechanism, as well as point to point communications.

Initially we will build our message passing model on sockets for our prototype; by careful design, the model will migrate to some of the other standard interfaces. The socket model provides a lightweight set of communication primitives, hence is an easy interface to implement, and its simplicity means that we can investigate and implement other communication strategies on top of this interface. Implementations of the socket library exist for both Macs and PCs, and a number of internet tools are based on this interface.

5 Low-level Lisp interface

The model used is that a process sets itself up to listen for connections at a known location, i.e. on a known machine, and a known 'subaddress' called a *port*. In this situation, it is acting like a server. There are a number of routines that set up such a connection. Firstly **socket** creates a new socket, returning a handle. This socket is then bound to a port using **bind**. Finally, the maximum number of queued connection requests is set using **listen**.

At this point the server is ready for clients to call it, which they do by making a socket and then using the **connect** function. On the server, the **accept** function waits until a client makes a call to the specific port, establishes the communication channel, and returns a handle onto the local end of the communication (a new socket). Another client process can then request a connection from this server, by again calling **connect** with the the machine location and the port number. Once the link has been established, the communication can be two way and can persist as long as necessary, i.e. not in the strict client-server model imposed by RPC.

There are three basic data transfer primitives, corresponding to the three basic types integer, float and string. All other types can be transferred by using these functions or combinations of functions and standard conversions.

Although a LISP-STAT process only has a single thread, it can have more than one connection talking to it. Care has to be taken, since if a read activity is initiated, it will not terminate until the data arrives (the read is said to 'block'). A mechanism, based on the UNIX **select** library call, allows you to identify which connections have data ready to be read on them.

6   Encoding data

Data in Lisp takes the form of symbolic expressions (*s-expressions*). In order to transfer data between Lisp processes, the data has to be serialised and transferred over the communication channel. Given the basic primitives to send an integer, string or floating point number, we can build up our protocol to handle any type of Lisp data. For each of the basic Lisp types, we assign an integer code. To send the data we first send the integer code so that the receiver knows what to expect. What follows is dependent on the type of the data. For example to send an integer, all that is sent is the code for an integer, followed by the integer itself.

For non-trivial pieces of data such as **cons**, recursive calls to a sending function are made. So for example a **cons**-pair (the basic unit of list structure in Lisp) consists of sending the code for **cons**-pair followed by sending the **car** expression (head of the list), then the **cdr** expression (tail of the list). This is illustrated in this fragment of code:

```
(defun send-sexp (socket obj)
  (case (class-of object)
    (integer (send-int socket int-code) (send-int socket obj))
    (cons    (send-int socket cons-code) (send-sexp socket (car obj))
             (send-sexp socket (cdr obj)))
    ...))
```

This strategy works fine, in general, but for cyclic data structures the recursive nature causes an infinite loop. This can be avoided by both the transmitting and receiving processes maintaining a list of nodes sent previously. When a node that has been sent before needs to be sent again, a reference to the earlier transmission is sent instead. The cache on the transmission end is based on a hash table to permit fast lookup based on the node itself; on the receive end it is a vector whose elements can be referenced in constant time.

In general, cyclic user data structures are not common, but there are two important exceptions in an object-oriented Lisp interpreter: functions and objects. Typically when we consider transporting functions, we need to consider the environment that the code was defined in (i.e. the values of variables external to a function but in scope when the function was defined), and this may be cyclic; similarly, implementations of objects and classes can contain cyclic structures. The transmission of objects is a similar problem to saving an object to a file, as pointed out by Tierney [5].

7   Example

Having implemented the communications primitives, we were in a position to investigate some form of distribution. As a useful demonstrator and tutorial application, we chose to implement a farm model. A master process is given a list of work that needs to be done. It then waits for requests from *workers*, and delivers each worker an item from the list. The worker then performs the task that is required using the supplied data, and then responds with the result. The master process then collects the results into a list and returns.

This is a relatively straightforward method of distributing work, and for certain applications give useful speedup. The communications overhead is typically very low, since the problem is split up into independent tasks, requiring minimal data transfer at the beginning and end. Course grained computations like this, with little synchronisation, are ideal for distribution.

The example used was the estimation of π by picking random points in a unit square, and testing if the points lie within an enclosed quadrant of a circle of unit radius. The square can be divided up into smaller regions and each region assigned to a process. Hence the only communications are sending the coordinates of the range and returning a ratio of points in the circle to points out of the circle. Another method is to assign a number of iterations to each processor, letting each processor pick random points across the whole square. This example is well-suited to distribution.

It is evident that this system cannot perform well for distributed applications where there is a substantial amount of communication between separate processes. As recognised previously, applications that perform well using this system are those which involve a high degree of computation that can be done in parallel. There are many statistical applications that match this model, typically involving intensive simulation. For example, estimating p-values using Monte Carlo methods, Gibbs sampling using multiple Markov-chains, computer-intensive resampling methods, and jackknife methods.

8 Conclusion and future work

We have established one interface for communicating LISP-STAT processes, based on a message-passing communication model, and demonstrated the use of this to accelerate a simple coarse-grained computation. We are currently generalising this interface, initially to support compiled Lisp systems.

Our prototype implementation will be used to provide support for two statistical applications: the first is the computationally-intensive Monte Carlo approach to the analysis of large sparse social mobility tables proposed by Smith and McDonald [4]; the second is a modified approach to a Markov Chain Monte Carlo (Gibbs Sampler) technique for the analysis of contingency tables. We believe that both applications will gain significant benefit from the distributed LISP-STAT environment.

References

[1] Neil Berrington, Peter A. Broadbery, David C. De Roure, and Julian A. Padget. EuLisp Threads: A Concurrency Toolbox. *Lisp and Symbolic Computation*, 6(1/2):177–200, 1993.

[2] David C. De Roure. Experiences with Lisp and distributed systems. In *Proceedings of the High Performance and Parallel Computing in Lisp Workshop*, Twickenham, London, UK, November 1990.

[3] Christian Queinnec and David C. De Roure. Design of a concurrent and distributed language. In Robert H. Halstead, Jr and Takayasu Ito, editors, *Parallel Symbolic Computing: Languages, Systems, and Applications*, Lecture Notes in Computer Science, pages 234–259. Springer-Verlag, October 1992. LNCS 748.

[4] Peter W. F. Smith and John W. McDonald. Simulate and Reject Monte Carlo Exact Conditional Test for Quasi-independence. In *Proceedings of COMPSTAT 1994*.

[5] Luke Tierney. *LISP-STAT: An Object-Oriented Environment for Statistical Computing and Dynamic Graphics*. J. Wiley and Sons, New York, 1990.

[6] Luke Tierney. Announcements in LISP-STAT. In *Bulletin of the International Statistical Institute, Proceedings of the 49th Session*, pages 117–124, August 2–September 2 1993. Book 3.

S-PLUS and MATLAB: A Comparison of Two Programming Environments

Andreas Futschik, Marcus Hudec

Department of Statistics, Operations Research and Computer Methods, University of Vienna

Abstract: Both S-PLUS and MATLAB provide the statistical researcher with a powerful programming environment covering a wide range of tools for numeric computation and visualization. A comparative evaluation of the systems is given from the viewpoint of a statistician aiming to implement new statistical methods.

1. Introduction

S-PLUS and MATLAB are two popular programming environments, which are both widely used for numerical and statistical computing. As S-PLUS is a package for data analysis it provides the user with powerful tools and convenient features for exploratory data analysis and modern statistical techniques. Thus S-PLUS is clearly superior to MATLAB with respect to availability of statistical methods.

In the context of this paper we will focus on the comparison of the programming environments, which both allow a straightforward implementation of statistical methods and algorithms as well as graphical techniques either newly developed or not yet available in classical statistical software packages. The authors have used both systems in their research and teaching activities.

Our evaluation is based on S-PLUS for Windows version 3.1 and MATLAB for Windows including the statistical and optimization toolbox. While both systems are available on a wider range of computing platforms, we will restrict ourself to the PC-version. Being well aware that any conclusions drawn are therefore of limited scope, we strongly believe that the large number of PC's running under Windows together with the rapidly increasing performance of this type of machines justify our choice. Runtime comparisons are based on a PC with an Intel-486 DX processor with 16 MB RAM.

2. Programming Environment

2.1. Language Elements

In S-PLUS data are organized in so called *data objects*, which can be divided into two main categories of data types. Atomic types cover vectors, matrices, arrays, time series, and factors, while the most important recursive types are lists and data frames.

The basic data element in MATLAB is a matrix. In some situations a special meaning is attached to scalars (1-by-1 matrices) or to vectors (1-

by-n resp. n-by-1 matrices). Besides matrices only strings are available in MATLAB.

Arithmetic operations are confined to mathematically coherent operands in MATLAB, while recycling rules give an additional flexibility to the programmer in S-PLUS. On the other hand unambiguous interpretation of expressions is not always straightforward in the latter case. Furthermore specialized functions available in S-PLUS allow sweeping operations and application of functions to matrices.

Both systems include arithmetic operations as defined by the rules of linear algebra as well as vectorized mathematical operations that are applied element by element to the whole data object.

Various typical language constructs are offered to control the flow of expression evaluation by means of looping and branching. The most remarkable difference seems to us the *ifelse* function available in S-PLUS, which is a vectorized version of the *if* statement that pro ves to be very useful in avoiding loops in the context of vectorized arithmetics.

Concepts of object-oriented programming are available in S-PLUS only. The availability of classes and methods allow the development of generic functions. An efficient implementation of new statistical procedures by means of inheritance and incremental programming is possible, since already in the system available features may be reused directly.

2.2. User Interface Implementation

The most natural way of designing an interface to new written modules is simply to call the function from the command line passing appropriate arguments with the call.

Nevertheless it is desirable in certain applications to offer the user an interactive control. Both systems provide the programmer with simple tools for building a menu-based interface. which allows choice of one alternative out of a small set of items.

A more powerful way of incorporating user input to control the function flow is to parse any user input. which might be requested by *readline* in S-PLUS or the *input* function in MATLAB. In connection with string manipulating functions and the function *eval* a high parametrization of user written functions is possible by means of text macro facilities.

Besides the above mentioned features MATLAB offers various very powerful tools for building much more sophisticated user interfaces by exploiting predefined dialog boxes available under windows. The file dialog boxes give considerably more comfort to the user in the specification of pathnames and filenames. Further dialog elements allow specification of fonts and colors.

Finally MATLAB allows the implementation of professional user interfaces via *uicontrol* (supporting Push buttons, Check boxes, Pop-up menus, Radio buttons, Sliders and Editable text) and *uimenu*. Three demos coming along with the statistics toolbox give a good impression how these techniques may be utilized in preparing statistical teachware.

It has to be noted that similar features are available on UNIX-systems for S-PLUS at additional cost.

An important issue in the distribution of newly developed modules is the availability of an adequate documentation. The creation of help files provides the most efficient way of documentation since it is available to the user interactively. While both systems enable the programmer to generate online help, the concept implemented in S-PLUS is more elaborated than in MATLAB.

In S-PLUS templates containing the standard headings of S-PLUS help files, and entries for each of the function's arguments are created by the *prompt* function. These templates are independent files, which are connected to the corresponding S-PLUS object by a simple naming-convention. In MATLAB, on the other hand, help files are generated from one or more comment lines beginning with the second line of the corresponding M-file.

2.3. Debugging Tools

Both environments allow a very compact denotation of algorithms. While syntax errors occuring during sourcing of S-files and first invocation of M-files are usually straightforward to fix, run-time errors or the occurrence of implausible results might prove to be much more difficult to track down.

Besides the classical remedies like printing intermediate results, various built-in features supports the user in fixing bugs. In S-PLUS a call to the *browser* function stops execution and offers the user a menu based choice of currently available objects to check the assignments. A similar effect may be realized by means of the *keyboard* function in MATLAB.

Furthermore both systems offer a built-in debugger. Although the MATLAB language is less complex than the S language the debugger seems to be more powerful by offering standard debugging features as setting of breakpoints and stepping through the code.

2.4. Interfacing to C and Fortran

Although both systems offer a complete, self-contained environment for programming, it is often very useful to combine the speed and efficiency of compiled code by calling C or Fortran subroutines. Besides the increase of efficiency by avoiding bottlenecks in computations this feature offers a direct access to pre-existing algorithms available in either C or Fortran.

While S-PLUS supports at the moment the Watcom compiler only, MATLAB is additionally compatible with Microway and MetaWare compilers. Furthermore MATLAB gives access to the functionality of the Windows environment, by supporting the standard format for 16-bit Dynamic Link Libraries (DLL) as well as communication with other applications by utilizing the Window's Dynamic Data Exchange (DDE). Creation of libraries according to the DLL format requires the Microsoft C 7.0 compiler along with the Windows Software Development Kit (SDK).

3. Available Numerical Tools

Routines for matrix operations seem to be numerically more accurate in MATLAB. The ill conditioned Hilbert matrices, for example, can be inverted only up to size 6 in S-Plus. Trying to invert Hilbert matrices of higher dimensions leads to a run-time error issuing the message "apparently singular matrix" and causing a dump of the calling function.

In MATLAB even inversion of singular matrices issues a warning message only. In cases when an inverse was found, but is not reliable, a warning displaying the condition number is given.

Important matrix functions available in both systems are solving of linear equations, singular value decomposition, QR-decomposition and calculation of eigenvalues and eigenvectors. A rather surprising omission from the suite of matrix facilities is the lack of a function calculating the determinant of a square matrix in S-PLUS.

Besides the linear algebra capabilities, both procedures have routines for integration, optimization, solution of nonlinear equations, spline interpolation, and fast Fourier transforms.

In MATLAB there are two routines that approximate univariate integrals over finite intervals. Unfortunately both routines allow control only over the relative error, which is not reasonable if the integral evaluates to zero. The integration algorithms are written as M-files that are easily readable but not very efficient. In S-PLUS integrals over infinite regions are allowed and absolute errors may be specified too.

Minimization routines are important e.g. for nonlinear regression and maximum likelihood estimation. Useful tools for one- and higher-dimensional problems are provided in the basic configuration of both packages.

When more specialized optimization procedures are needed, the MATLAB optimization toolbox is valuable. It contains procedures for a larger class of optimization problems, like linear and quadratic programming, goal programming and constrained minimization. The specialized problems that can be solved with S-PLUS are fewer and include polynomial roots and knapsack problems.

Determination of the convex hull is available in S-PLUS for bivariate data, thus providing a good starting point for robust estimation techniques and outlier detection by means of convex hull trimming as indicated by Bebbington [1].

4. Simulation Performance Benchmark

It is popular to test newly developed statistical methods by means of simulation studies. If the investigated methods are computationally expensive, it is of particular importance that the programming environment produces fast running code.

We compare the run time behavior of S-PLUS and MATLAB code for tasks that typically arise in simulation studies.

Simulation is usually based on data produced by random number generators. While both programming languages produce large quantities of normal and uniform random numbers quickly, differences arise for other distributions. Random data from e.g. F, χ^2 and other distributions are produced much faster in S-PLUS, where compiled procedures are used for this task, than in MATLAB, where the random number generating procedures are not built-in but written as M-files.

The situation changes when random numbers from a not yet implemented distribution are needed. We consider the problem of random numbers from a fixed discrete (nonuniform) distribution on $1, \ldots, k$. Table 1 shows the run-time of programs written in S-PLUS and MATLAB for s samples of 1000 observations. Clearly MATLAB performs better in this task.

# of samples samples (s):	10	100	1000
S-PLUS (time in sec.)	8	169	725
MATLAB (time in sec.)	5	34	443

Table 1: Random numbers from discrete nonuniform distribution.

Simulations involving test-statistics require calculation of critical values and p-values. Fast evaluation of distribution functions and their inverses is important. With new methods there might also be a need for evaluating probability distributions not implemented in standard programming packages.

First we look at the normal distribution which is available in MATLAB as M-file only. To compare the performance we calculated the p-values for random $(N(0,1))$ vectors of different lengths. Table 2 shows that the calculations are carried out faster in S-PLUS.

length of vector:	1000	10000	100000
S-PLUS (time in sec.)	0.03	0.33	3
MATLAB (time in sec.)	0.4	4	130

Table 2: Evaluating p-values for large vectors of normal r.v.'s

As an example where a routine has to be programmed by the user, we look at the distribution of $max_{1 \le i \le k} X_i - X_0$ when $X_0, \ldots, X_k$ are standard normal random variables. This distribution is e.g. needed in subset selection problems as investigated by Gupta [2]. Its implementation requires numerical integration of the convolution $\int_{-\infty}^{\infty} \Phi^{k-1}(a+x) \, d\Phi(x)$. Fortran code for tasks of this type has been published by Lund [3]. For presentation it might be desirable to obtain a plot of such a nonstandard distribution. To simulate this task we evaluated Gupta's distribution at 20 points. The calculations took 47 seconds in S-PLUS, but 109 seconds in MATLAB. Also the evaluation of quantiles is done considerably faster in S–Plus than in MATLAB.

Finally, we usually want to calculate test statistics or estimators from the data in a simulation study. As an example we will compute the Euclidean norm of k data vectors, each of length n. This can be done in different ways: Generate a random vector and evaluate its norm k times in a loop; ($M1$) Generate a matrix containing all $k \times n$ random numbers at the beginning and evaluate the vector norms afterwards in a loop ($M2$). Generate the matrix X containing the squared $k \times n$ random numbers, calculate then $v = 1^t X$ and compute the square root of the elements of the vector v ($M3$). Generate the matrix X containing the squared random numbers; evaluate the norms in an implicit loop using the *apply* function of S-PLUS and take the square root.

Table 3 shows that S-PLUS is usually slower, especially if the calculations involve large implicit loops ($k = 10000$, *M4*). Note that in S-PLUS even variants of the most straightforward algorithm *(M1)* can have a considerably longer runtime than the one indicated in the table, depending on what type of operation is used to calculate the norm. Also MATLAB can cope with larger numbers of n and k without running into memory problems.

	$k = 10; n = 10\,000$				$k = 10\,000; n = 10$			
	M1	M2	M3	M4	M1	M2	M3	M4
S-PLUS (time in sec.)	6	6	7	10	21	21	7	624
MATLAB (time in sec.)	5	5	4	—	14	80	5	—

Table 3: Comparison of procedures to evaluate norms of vectors.

The above example shows that the performance of both systems are highly dependent on details of user programming.

5. Discussion

The language concepts of S-PLUS are much richer and more complex than those of MATLAB. While this gives more flexibility to the programmer it simultaneously opens a door for unexpected pitfalls.

While both systems are available under Windows, only MATLAB makes explicit use of its benefits, by providing the programmer access to the functionality of the Windows environment.

In MATLAB the underlying numerical routines are more transparent, since the method of calculation is always indicated and references to the literature are given. Up to our experience the implemented numerical tools seem to have a higher accuracy in MATLAB than in S-PLUS.

User written routines are usually faster in MATLAB than in S-PLUS, especially in cases where they contain large loops. On the other hand many statistically important functions are not built-in in MATLAB. User programs that depend heavily on such M-files functions should better be written in S-PLUS as long as the programmer does not want to include C- or Fortran-subroutines.

References

[1] Bebbington, A.C. (1978). A Method of Bivariate Trimming for Robust Estimation of the Correlation Coefficient, Appl. Statist. 27 221–226.

[2] Gupta, S. S. (1965). On some multiple decision (selection and ranking) rules, Technometrics 7 225–245.

[3] Lund, R. E. (1991). Probabilities and standardized differences for selecting subsets containing the best populations. Appl. Statist. 40 495–515.

A Specification Language for Gibbs Sampling

Peter Hartmann, Song Jin, Andreas Krause

Institute for Statistics and Econometrics, University of Basel

Abstract. A general approach to implement a specification syntax for Markov Chain Monte Carlo Methods, especially Gibbs sampling, is presented.

If one is able to formulate a statistical model in terms of full conditional distributions, a sampling scheme can be derived automatically. The software itself determines automatically the needed functions and data to run the sampling scheme. In this way, one can easily use the functionality of Gibbs sampling schemes without much knowledge of programming. By providing samples from the desired posterior distribution, all further analyses can be based on the sample.

On the other hand, one can also try to derive the sampling scheme automatically. In this case, the task of deriving the setup for obtaining the posterior sample is left to the program.

We present a specification language for time series modelling and demonstrate a flexible way of setting up a variety of models in this context. A first implementation with examples will be demonstrated.

Keywords. Gibbs sampling, Markov Chain Monte Carlo, Simulation

1   Monte Carlo Markov Chains

Introductions to Markov chain Monte Carlo methods, especially Gibbs sampling, are given by various authors (Gelfand and Smith, 1990, Casella and George, 1992; Smith and Gelfand, 1992). We focus on the general idea and derive a framework for implementing a language to formulate models, such that a sampling scheme can be derived more or less automatically.

The Bayesian model setup, such that the Gibbs sampler can be run, is based on the calculation of the full conditional distributions of the parameters of interest.

For example, in the standard normal gamma regression model one obtains the full conditional distributions for the parameters σ^2 and b as

$$y \sim \mathcal{N}(X\mathrm{b}, \sigma^2 I_n)$$
$$(\mathrm{b}, \sigma^{-2}) \sim \mathcal{N}\Gamma\left(\mathrm{b}_*, \sigma^2 H_*, \sigma_*^2, n_*\right)$$

which is equivalent to

$$\sigma^{-2} \sim \Gamma(\sigma_*^2, n_*)$$
$$b|\sigma^{-2} \sim \mathcal{N}(b_*, \sigma^2 H_*).$$

Then the posterior distribution is again a normal gamma distribution.

$$(b, \sigma^{-2})|X, y \sim \mathcal{N}\Gamma (b_{**}, \sigma^2 H_{**}, \sigma_{**}^2, n_{**})$$

with the posterior parameters

$$
\begin{aligned}
b_{**} &= H_{**}(H_*^{-1}b_* + X^T y) \\
H_{**}^{-1} &= H_*^{-1} + X^T X \\
n_{**} &= n_* + n \\
n_{**}\sigma_{**}^2 &= n_*\sigma_*^2 + (y - Xb)^T(y - Xb) + (b - b_*)^T H_*^{-1}(b - b_*)
\end{aligned}
$$

The set of full conditional distributions determines the model completely. Therefore, high dimensional analytical problems like integration or parameter transformation can be reformulated into drawing samples from lower dimensional distributions. All further analyses will then be based on the obtained sample from the posterior distribution.

Based on linear regression models, time series structures are added. Several models like AR, ARX and MA are considered. AR processes can be defined straightforwardly by its conditional distributions, whereas MA processes are more complicated to handle because of their recursive structure.

2   A software approach

There have been some approaches to derive a sampling scheme for Gibbs sampling automatically. Perhaps the best known is the BUGS (Bayesian Inference Using Gibbs Sampling) software, described in Thomas, Spiegelhalter and Gilks (1992). This program covers the class of generalized linear models (GLM), but is on the other hand constrained to formulation of the model in terms of a GLM. Given the GLM specification, the sampling scheme is then derived by using the adaptive rejection algorithm (Gilks, 1992; Gilks and Wild, 1992). The advantage of using this technique is its general applicability without deriving the full conditional distributions, the disadvantage is the non-derivation of the full conditionals and therefore a loss in time, as the program takes longer to run.

In several applications, general algorithms like the adaptive rejection approach are not appropriate, especially if the conditional distributions are available. Using the full information of the conditional distributions is usually faster and easier to control.

In the first step the model is specified exactly, such that the main program picks the corresponding separate module and runs it. For the class of regression models of the normal gamma type, including Tobit type censored observations (Tobin, 1958), a general purpose program is already available by sending email to the authors. The source code is in C with a defined interface to S/S–PLUS (Becker, Chambers, and Wilks, 1988). The list of modules includes simple linear regression, weighted linear regression, linear regression with student distribution error structure, hierarchical regression models (Krause and Polasek, 1992), errors in variables structures, nonlinear regression, switching (two state) regression models, and logistic (binary response) regression.

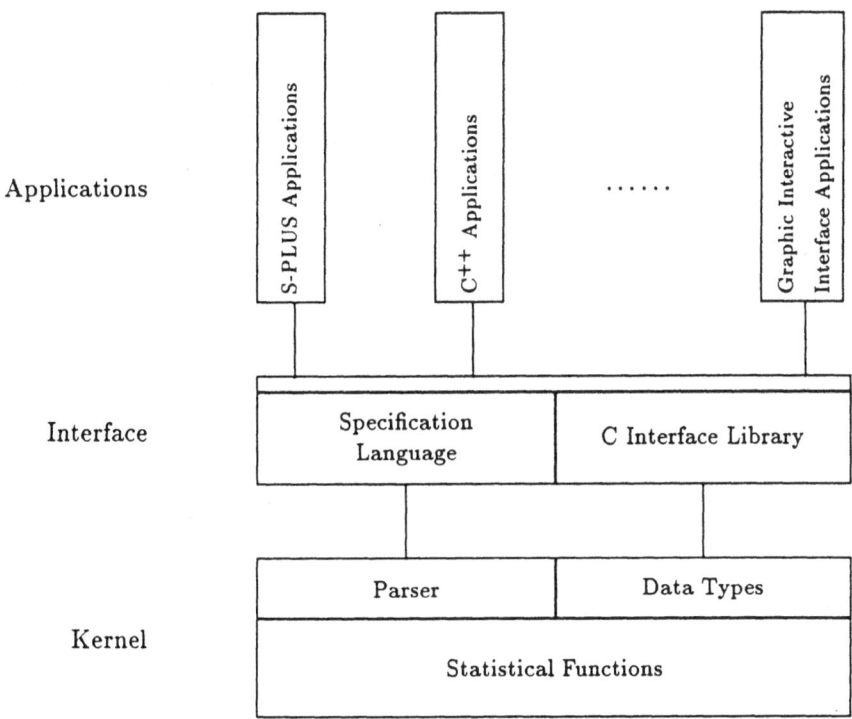

Figure 1: *program structure*

In a further step, as we are developing, the sampling setup is derived by an independent C program which contains a kernel, parser and an open external interface (figure 1). The interface is used for communication between the user application, this approach is flexible and programmable. For the specification language (see Section 3) we need the parser which reads the language formulations and translates them into the internal data structure. The data types and statistical functions build up the kernel.

The third step is the most difficult and ambitious one. Based on the model specification language, one can specify a model with a general data and parameter structure. The program itself tries to determine any conjugate distribution in the model and, if not all the distributions are contained in the set of conjugate distributions, a more general routine like Importance Weighted Sampling or the Metropolis algorithm is applied. Then we have a mixture of a general sampling procedure and a knowledge based system.

3   A general model specification language

The general idea behind our specification language is, that a statistician can specify the prior model and the data structure. The specification is parsed and evaluated by a program, which assembles the different full conditional distributions to a sampling scheme.

Below, an example of a simple linear regression model with Tobit censoring is given.

```
(*******************************************************************)
(* purpose: Syntax example for a linear model. It is written in a *)
(*          language for specifying models for a sampling scheme. *)
(* authors: Andreas Krause, Peter Hartmann, Song Jin.             *)
(* date:    04/02/94                                              *)
(*******************************************************************)

(* parameters in the model *)

OBJECT b = PARAMETER (INTERN b_x; EXTERN h_x) {
  Prior   = NORMAL[b_x,h_x];
  Initial = {0.0, 0.0};
  Returns = {gibbsrun; sample}
};

OBJECT sigmasq = PARAMETER (INTERN sigmasq_x; EXTERN n_x) {
  Prior   = GAMMA[sigmasq_x, n_x];
  Initial = 1.0;
  Returns = {sample; mean; variance}
};

(*******************************************************************)

(* data in the model *)

OBJECT x = DATA (EXTERN xlin) {
```

```
  Scale  = metric;
  Design = fixed;
  Data   = xlin;
};

OBJECT y = DATA (EXTERN y) {
  Scale  = metric;
  Design = random;
  Data   = y;
};

OBJECT cens = DATA (EXTERN status) {
  Scale  = nominal(0,1);
  Design = fixed;
  Data   = status;
};

(************************************************************)

(* model definition, this class will be changed in a further step *)

OBJECT linreg = MODEL () {
  Tobit      = yes;
  Hierarch   = no;
  Eiv        = no;
  Nonlinear  = no;
  Datalist   = {x=x; y=y; censored=cens};
  Parameters = {b=b; sigmasq=sigmasq}
};

(************************************************************)

(* sampling specification *)

OBJECT gibbslr = SAMPLER () {
  Model   = linreg;
  Options = {Chains=10; Iterations=1000};
};

(************************************************************)
```

There are two ways to send OBJECTs of different types to the kernel. Either you transfer them in the kernel's data structure by interface functions, or in the language format where you can load them also from a file using the EXTERN statement. The main class is the SAMPLER which contains the model specification and general sampling options. The MODEL class combines data, parameters and the model type so that the whole statistical information is available. The prior distributions and initial values are saved by the class PARAMETER.

In a future version of the program the class MODEL will be replaced by a general expression list, where one can formulate model equations directly.

4 References

Becker, R.A.; Chambers, J.M.; Wilks, A.R. (1988) : *The New S-Language.* Wadsworth & Brooks Cole, Pacific Grove, California

Casella, G.; George, E.I. (1992) : *Explaining the Gibbs Sampler.* The American Statistician 46, 167-174

Gelfand, A.E.; Smith, A.F.M. (1990) : *Sampling–Based Approaches to Calculating Marginal Densities.* Journal of the American Statistical Association, 85, 398–409

Gilks, W.R.; Wild, P. (1992) : *Adaptive Rejection Sampling for Gibbs Sampling.* Journal of the Royal Statistical Society, Series C, 41, 337–348

Gilks, W.R. (1992) : *Derivative-free Adaptive Rejection Sampling for Gibbs Sampling.* In : Bayesian Statistics 4 (edited by Bernardo, J.M.; Berger, J.O.; Dawid, A.P.; Smith, A.F.M.), Oxford University Press, 641-649

Krause, A. (1994) : *Gibbs Sampling in Regressionsmodellen mit zensierten Daten.* Fischer, Stuttgart

Krause, A.; Polasek, W. (1992) : *Approaches to Tobit Models via Gibbs Sampling.* In : Computational Statistics, Volume 1, Proceedings of the 10th Symposium on Computational Statistics. (Editors : Dodge, Y.; Whittaker, J.). Physica Verlag, Heidelberg

Smith, A.F.M.; Gelfand, A.E. (1992) : *Bayesian Statistics without Tears.* The American Statistician 46, No. 2, 84–88

Thomas, A.; Spiegelhalter, D.J.; Gilks, W.R. (1992) : *Bugs: A Program to Perform Bayesian Inference Using Gibbs Sampling.* In : Bayesian Statistics 4 (edited by Bernardo, J.M.; Berger, J.O.; Dawid, A.P.; Smith, A.F.M.), Oxford University Press, 837–842

Tobin, J. (1958) : *Estimation of Relationships for Limited Dependent Variables.* Econometrica, 26, 24–36

Applying the Object Oriented Paradigm to Statistical Computing

Martin Hitz, Marcus Hudec

Department of Statistics, Operations Research and Computer Methods, University of Vienna

Abstract: After a short summarization of the basic concepts of object oriented programming it is outlined by means of two concrete examples how to utilize this new paradigm of computing in the context of statistical software. The concept of generic functions – as it is available in S-PLUS – is employed for developing specialized printing utilities for classification matrices. A highly parametrized algorithm for computing the minimum volume ellipsoid estimator (MVE) is derived by means of both data and procedural abstraction mechanisms of OOP.

1. Introduction

In the last decade the concept of object oriented programming (OOP) has emerged as a ruling paradigm of computing in general. Although most of the widely used statistical systems neglected this new paradigm for a long time, recent developments in the field of statistical computing indicate that the OOP concept is starting to be of increasing importance for both, development and application of statistical software [2].

LISP-STAT, DataDesk, S-PLUS, and SAS have incorporated OOP concepts to a different degree. However, it seems that the potential benefits of this new approach have not yet been fully realized by many statisticians. Due to the unfamiliarity with the concepts of OOP these packages are still used for data analysis and programming of new statistical procedures in the "classical" way.

In our contribution we emphasize the advantages of object orientation for statistical computing in the context of two concrete examples, which illustrate the potential benefits of these concepts. While the first example is based on the implementation of object oriented methods in S-PLUS, the second example is presented in a pseudo-code to avoid restriction to a particular environment.

2. Basic Concepts of OOP

In object oriented systems, data are packaged in *objects* together with corresponding access and manipulation functions called *methods*. Information hiding thus achieved ensures that the user need not know any details about the data representation. In order to achieve a certain behavior, it is sufficient to send *messages* asking the respective object to select and execute a corresponding method from its method-set (the so-called *protocol*). This concept, also known as *encapsulation*, is complemented by the possibility to arrange objects types (or *classes*) in an inheritance lattice where subordinate classes inherit all data structures and methods from their superordinate ancestors while reserving the ability to add data structures or methods at will and to override selected methods to adapt them to the heir's needs. We will show in our first example how this property may be exploited for a technique called *delta programming* where much of the code implemented is reused from other (superordinate) classes. Last but not least, the term object orientation also demands the ability to dynamically extend the class lattice (*dynamic binding*), providing downward compatibility between superclasses and their subclasses. In effect, operations based on superclasses may well operate on objects belonging to subclasses without even knowing the difference (*polymorphism*).

3. Generic functions in S-PLUS

Together with the underlying inheritance mechanisms the above mentioned class hierarchies support a concept known as delta-programming where the implementation of new modules is based on a maximal reuse of already available features. Moreover, the generation of class hierarchies allows definition of polymorphic methods. As a result, lean interfaces to complex and powerful statistical computing environments may be offered to non-expert users.

The statistical programming environment S-PLUS offers to the user powerful tools for data analysis. While the implementation of these tools employs classes and methods, the underlying mechanism remains invisible to the user, who just utilizes these tools as they are. On the other hand, if a user wants to redefine existing methods or implement new methods S-PLUS allows programming very much closer to the style of object oriented systems than it is available in most other statistical software packages (see for more details [1, 4]).

The most important concept of implementation of the class/method mechanism are *generic functions* in S-PLUS. Functions are called generic, if they adapt their behavior specifically to the class of the object they are applied to by means of the dynamic envocation of class-specific methods.

A typical example of such a generic function is the *print* - function, which has the following form in S-PLUS:

```
print <- function(x, ...)
    UseMethod("print")
```

Thus the call of the print function may be viewed as passing the message *print yourself* to the object x. The actual action performed in response to this message depends on the class to which x belongs. This principle might be exploited by a programmer envolved in the development of new algorithms for classification or discrimination analysis in the following way.

In the above context one often has to deal with matrices (n × k) either indicating group membership of n cases to k groups or containing the posterior probability of group membership for a given observation. While such special matrices should inherit all properties from the class *matrix*, we might be interested in further details when printing such matrices. As a consequence, we introduce two new classes generating the following class hierarchy:

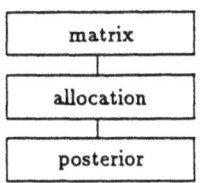

```
# Implementation of class-specific methods
print.allocation <- function(xmat)
{ NextMethod("print")      # delegates to print method of superclass
  disp.membership(xmat)    # prints incremental class-specific details
  invisible(xmat)    }
print.posterior <- function(xmat)
{ NextMethod("print")      # delegates to print method of superclass
  disp.rankgroup(xmat)     # prints incremental class-specific details
  invisible(xmat)    }
```

Note that *NextMethod* in the second algorithm refers to the already refined printing method of class allocation. The following two functions give an implementation of the two class-specific "deltas" of the general printing task:

```
disp.membership <- function(xmat)
{ n.group <- dim(xmat)[2]
  xmat.order <- t(apply(-xmat, 1, order))
  cat("\n", "*** Group - Membership ***", "\n")
  for (i in 1:n.group)
    { print(dimnames(xmat)[[2]][i])
      print(dimnames(xmat)[[1]][xmat.order[,1]==i])
    } }
disp.rankgroup <- function(xmat)
{ n.cases    <- dim(xmat)[1]
  n.group    <- dim(xmat)[2]
  xmat.order <- t(apply(-xmat, 1, order))
  xmat.out   <- matrix(dimnames(xmat)[[2]][xmat.order[,1:n.group]],
                       n.cases, n.group, F)
  dimnames(xmat.out) <-
     list(dimnames(xmat)[[1]], paste("Alloc - ", 1:n.group, sep=""))
  cat("\n", "*** Rows of Matrix Ordered by Probability ***", "\n")
  print(xmat.out)    }
```

An example output as produced by *print(x)* with x belonging to the class posterior is shown below:

```
             Group A Group B Group C
Case No.: 1    0.1     0.1     0.8
Case No.: 2    0.3     0.7     0.0
Case No.: 3    0.1     0.4     0.5
Case No.: 4    0.6     0.2     0.2
Case No.: 5    0.2     0.3     0.5
Case No.: 6    0.1     0.2     0.7
*** Group - Membership ***
[1] "Group A"
[1] "Case No.: 4"
[1] "Group B"
[1] "Case No.: 2"
```

```
[1] "Group C"
[1] "Case No.: 1" "Case No.: 3" "Case No.: 5" "Case No.: 6"
*** Rows of Matrix Ordered by Probability ***
            Alloc - 1 Alloc - 2 Alloc - 3
Case No.: 1 "Group C" "Group A" "Group B"
Case No.: 2 "Group B" "Group A" "Group C"
Case No.: 3 "Group C" "Group B" "Group A"
Case No.: 4 "Group A" "Group B" "Group C"
Case No.: 5 "Group C" "Group B" "Group A"
Case No.: 6 "Group C" "Group B" "Group A"
```

4. Procedural Objects

In the following we will demonstrate how both, data and procedural abstraction mechanisms of OOP can be exploited to implement statistical methods on a very high level of generality, thus allowing for easy maintenance and extendibility: Owing to late binding and polymorphism, specific data types or methods (e.g., auxiliary numerical algorithms) can be exchanged or added at any time without mediation by a systems administrator.

We will base our discussion on the computation of the minimum volume ellipsoid estimator (MVE) for multivariate location and shape, where the location estimator is the center of the minimum volume ellipsoid covering at least half of the points from a multivariate sample, and the covariance matrix is the shape matrix of the ellipse [3]. We may envisage a statistical routine with a protocol defined roughly as follows:

```
MVE(IN PDimSample data, OUT PDimVec location, OUT PxPMat varCov);
```

Of course, some additional parameters might be appropriate to fine-tune the procedure's behaviour, e.g., the kind of optimization method to be used, the termination condition etc.

Focusing on the minimum search procedure, we may note that a number of improvements of the undirected random search originally proposed are currently being discussed [5]. Thus any implementation of MVE should be as open as possible to future enhancements of the underlying optimization procedure. On the other hand, in order to facilitate implementations of new variants, all code pertinent to the problem itself should be factored out and implemented only once.

From an object oriented point of view, we suggest to solve this trade-off by means of what may be called a *procedural object* as it can be viewed as procedural abstraction of correlated methods cast into an object. In this case, the object's instance variables serve an auxiliary purpose, namely, to provide an encapsulated communication platform for the methods thus tied together. Let us restate the protocol of our routine in order to comply with this approach:

```
MVE(IN PDimSample data, OUT PDimVec location, OUT PxPMat varCov,
    IN HeuristicOpt o = randomSearch);
```

The new argument o is defaulted to indicate that a parametric user need not specify this option, thus implicitly employing the standard procedure.

The main method of the class HeuristicOpt might look like this:

```
doOpt():
    Config c, next; real score, nextScore;
    initOpt(c);
    score := eval(c);
    do while not terminated():
        if isBest(score) then c.save();
        next := computeNext(c);
        nextScore := eval(next);
        if acceptable(score, nextScore)
        then:
            c := next;
            score := nextScore;
        nextStep(c);
```

Config specifies an auxiliary class holding a specific configuration in the search space, i.e. a sample of the original data set. Config objects essentially know how to create themselves from a given multivariate data set (initial()), include and exclude data points in order to obtain a new sample, and how to record themselves as temporary result (save()).

The protocol of class HeuristicOpt may now be deduced from the needs of its main procedure doOpt() given above:

initOpt(INOUT Config) Performs all initialization necessary for the optimizing process and defines the initial configuration.

eval(IN Config) → real Implements the objective function.

terminated() → bool Tests if the termination condition ("good enough", "time limit reached", ...) has occurred.

isBest(IN real) → bool Checks if the actual configuration represents a new optimum.

computeNext(IN Config) → Config Computes and returns a successor configuration of the given configuration. The default implementation excludes one data point and includes another one.

acceptable(IN real, IN real) → bool Tests if the new (second) score is "better" than the previous (first) one. This function is predefined as "<", but may be overridden by more complicated criteria.

nextStep(Config) Does some book-keeping job after each iteration.

In addition to the above methods, MVE() will certainly need other auxiliary access methods which are beyond the scope of this presentation.

As already mentioned, the abstract class HeuristicOpt may provide default implementations for many functions, while others must be overridden by the implementor of a specific search algorithm. For instance, considering the algorithms proposed in [5], we get the following class hierarchy:

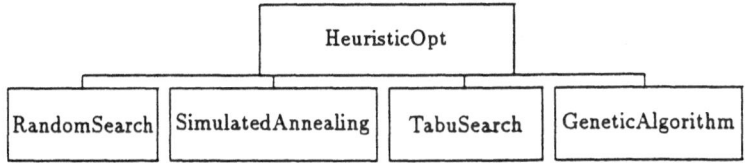

RandomSearch is so trivial that most default methods of HeuristicOpt can directly be reused. Implementation of SimulatedAnnealing, TabuSearch and

GeneticAlgorithm require - depending on the level of sophistication - adequate replacement of the functions outlined above. However, it should be noted that in all cases, MVE() itself need not be changed. It is worth noting that MVE() is indeed open for future improvements in a simple, well-defined manner: the user defines a new subclass of HeuristicOpt (or one of its already existing subclasses), decides which methods can be inherited appropriately and redefines the remaining ones according to his needs. Communication between these methods that is beyond their parameter interface is accomplished via the new class' instance variables in a well encapsulated and thus easily maintainable manner. As an example, we sketch a slightly simplified implementation of SimulatedAnnealing:

```
SimulatedAnnealing, subclass of HeuristicOpt:
    add variables:
        real t, startTemp, finalTemp, tempFactor, step, stepSize;
    override methods:
        initOpt(INOUT Config):
            t := startTemp; step := 1;
            Config := initial();
        terminated() -> bool:
            return t <= finalTemp;
        acceptable(real old, real new) -> bool:
            real delta := new - old;
            return delta<0 or (exp(-delta/t) < uniformRandom(0, 1));
        nextStep(Config):
            if step<stepSize
            then step := step+1
            else:
                step := 1;
                t := t*tempFactor;
```

5. Conclusion

While the object oriented programming paradigm still is rather unfamiliar to many statisticians, we strongly believe that statistical computing environments will increasingly be designed under this new paradigm. As a result lean interfaces to complex scientific computing environments may be developed. Furthermore a maximum of flexibility and efficiency for adapting existing or implementing new statistical procedures will be offered by such systems.

References

[1] Chambers, J.M., Hastie T.J. (1992). Statistical Models in S. Wadsworth & Brooks, Pacific Grove, California.

[2] Galway, L.(1993). Object-Oriented Programming and Statistical Computing. Chance Vol.6, No.1, pp. 52 - 61.

[3] Rousseeuw, P.J. (1985). Multivariate Estimation With High Breakdown Point. In: W. Grossmann et al. (eds.), Mathematical Statistics and Applications, Volume B, Reidel, pp. 283-294.

[4] Statistical Sciences, Inc. (1993). S-PLUS Programmer's Manual, Version 3.1, Seattle: Statistical Sciences, Inc.

[5] Woodruff, D.L. and Rocke, D.M. (1993). Heuristic Search Algorithms for the Minimum Volume Ellipsoid. In: Journal of Computational and Graphical Statistics, Vol. 2, No. 1, pp. 69-95.

Nonlinear Regression on Personal Computers

Jiří Militký

Department of Textile Materials, Textile Faculty, Technical University Liberec

Abstract: This paper describes basic numerical techniques for nonlinear model building by using of regression method. Some procedures for nonlinear least squares minimization are discussed. The procedure MNOPT based on the double dogleg strategy is briefly discussed. Comparison of numerical suitability of well known statistical packages for personal computers is made.

Keywords: Nonlinear regression; numerical techniques for least squares method; comparison of software for personal computers.

1 Introduction

Recently used models in science and engineering are often nonlinear in parameters. Model $f(x, a)$ is obviously complicated function of parameter vector a as for example sums of exponentials. One of the most difficult tasks in building models of such type is estimation of the parameter vector a and subsequent statistical analysis. There are many algorithms and programs for estimation of parameters in general nonlinear regression models that can be used in practice. According to the personal experience of the author, most of the commonly used programs are not much suitable for this task.

In this paper some techniques convenient for analysis of nonlinear regression models (of general type) are described. The structure of new procedure MNOPT designed for personal computers IBM PC-XT/AT is briefly presented. This procedure is a main part of software products TANAL (kinetic analysis of nonisothermal thermoanalytical data) and ADSTAT (general type nonlinear model fitting).

Numerical accuracy of the MNOPT procedure is compared with some well known statistical packages on the base of published test problems.

396

2 Formulation of nonlinear regression problem

The main task in construction of regression models is to estimate a parameter vector **a** (m x 1) in known nonlinear model f(x, **a**). The estimation process is based on experimental data $\{y_i, x_i\}$ i=1,...N. The values x_i (without detriment to generality x_i is supposed to be a scalar) creates nonstochastic set. The measured values y_i are usually supposed to be expressed in frame of additive measurement model

$$y_i = f(x_i, a) + e_i, \quad i = 1,...N \qquad (1)$$

Here e_i are the so-called experimental errors, presumably independent identically distributed random variables having normal distribution $N(0, \sigma^2)$. Considering these assumptions the estimates a^* of the parameter vector **a** may be found by minimizing the least squares (LS) criterion

$$S(a) = \| y - f(a) \| \qquad (2)$$

where **y** is (N x 1) vector of measured quantities, f(**a**) is (N x 1) vector of corresponding model values (the i-th component is f(x_i, **a**)) and $\| . \|$ is Euclidean norm.

Generally, the nonlinear regression problem may be divided into two parts:
- minimization of LS criterion s(**a**),
- statistical analysis of estimates, model and data quality.

In most of the programs the linearization of nonlinear model f(x, **a**) is used for both minimization and statistical analysis. This approach may be observed even in cases where it is not acceptable due to high nonlinearity of model f(x, **a**). In sequel the problem of minimization of LS criterion is discussed. The statistical analysis suitable for highly nonlinear models involved in ADSTAT package is described by Militký (1982)

3 Numerical analysis of the problem

For estimation of parameters **a** minimizing the LS criterion S(**a**) derivative algorithms are profitably being used. The survey of these algorithms gives Nazareth (1980). General scheme of minimization is expressed by the following sequence of steps:
1. Input of first estimates.
2. Finding a suitable directional vector v_j.
3. Determination of the scalar α_j such that the increment $d_j = \alpha_j v_j$ is acceptable. For acceptable increment is obviously valid

$$S(a_j + d_j) < S(a_j) \qquad (3)$$

4. Testing whether the minimum has been reached. If no minimum has been found, the new value $a_{j+1} = a_j + d_j$ is computed and new iteration step (step 2) is started.

Individual algorithms differ by the choice of directional vector v_j and scalar α_j. It may easily be proved that for acceptance of directional vectors v the following expressions should be valid

$$g^T v < 0, \qquad v = -Rv \qquad (4)$$

where R is a regular positive definite matrix and g is a gradient vector of LS criterion having the components $g_j = d\,S(a)\,/\,d\,a_j$. To find an optimum scalar α Taylor expansion up to the quadratic terms is substituted for $S(a)$ in v direction. After analytical minimization with respect to α the well-known Raleigh factor results

$$\alpha = g^T Rg[g^T R^T HRg]^{-1} \qquad (5)$$

where H is the symmetric Hessian matrix having the elements

$$H_{jk} = \frac{d^2 S(a)}{d\,a_j\,d\,a_k}$$

After computation of the above derivatives the gradient and the Hessian may be expressed as follows

$$g = 2\,J^T e \qquad\qquad H = 2\,[J^T J - L^T e] \qquad (6)$$

Here e is the vector of residuals having the components $e_i = y_i - f(x_i, a)$ and J (N x m) is the Jacobian matrix with the components

$$J_{jk} = \frac{d\,f(x_j, a)}{d\,a_k}$$

The field L (N x m x m) is composed from N layers where the i-th one is formed by the matrix L_i having the elements

$$L_{i(j,\,k)} = \frac{d^2 f(x_i, a)}{d\,a_j\,d\,a_k}$$

Accordingly, for assessment of convenient directional vector v the quadratic approximation of $S(a)$ by the Taylor expansion

$$S(a) \approx S(a_j) + d_j{}^T g + 0.5\,d_j{}^T H\,d_j + \ldots \qquad (7)$$

is often employed.

The directional vector v_N defined by the relation

$$v_N = H^{-1} J^T e \qquad (8)$$

corresponds to the well-known Newton method. Here the knowledge of second derivatives of the model function (the field L) is required. The evaluation of the Hessian matrix and the solution of a linear system are required at each step, and problems connected with the singularity or the ill-conditioning of the Hessian matrix may also arise.

The so-called quasi-Newton methods have been introduced in order to eliminate these disadvantages. Along with the sequence a_j a sequence H_j^{-1} approximations to the inverse of Hessian matrix H is also generated. Basic variants of inverse Hessian matrix updates are described by Spedicato and Vespucci (1988).

The notorious Gauss-Newton method is based on the same expansion (7) where the second derivative terms are omitted. The following relationship results

$$v_{GN} = (J^T J)^{-1} J^T e \qquad (9)$$

Numerical experiments (see McKeown (1975)) have shown that the Gauss-Newton method solves the nonlinear LS problem efficiently when the residuals are sufficiently small at the solution, but it can be less efficient if the residuals are significantly nonlinear and their optimal values are large. Basic variants of the Gauss-Newton method are described by Meloun, Militký and Forina (1994).

For seeking of local optimum the v should be chosen as $v_G = -g$ direction, where maximum decrease of criterion condition occurs. The methods of this type are called gradient ones. Some modifications of gradient methods are described by Meloun, Militký and Forina (1994).

Hybrid methods appear very effective making use of fast convergence of the Gauss-Newton method in the neighborhood of minimum and "guaranteed" convergence of gradient methods farther of the stationary point. The best-known representative of these methods is the Marquardt one. The directional vector v_M of the Marquardt method has the form

$$v_M = (J^T J + \lambda E)^{-1} J^T e \qquad (10)$$

where λ is parameter and E is the unity matrix. Some strategies of parameter λ changes during minimization are described by Meloun, Militký and Forina (1994).

It is well known that the Marquardt method converges slowly, if the directional vector v is close to v_G.

In the MNOPT program the modification of the so-called double dogleg strategy of Dennis and Mei (1979) is applied. Optimum directional vector v_D is here sought on the line segment TB of a triangle OTB defined by its vertices

$$O \equiv a_j; \qquad T \equiv a_j + \alpha_1 v_{GN}; \qquad B \equiv a_j + \alpha v_G$$

The vector v_D can be then expressed in the form

$$v_D = \mu \, \alpha_1 \, v_{GN} + (1 - \mu) \, \alpha \, v_G \qquad (11)$$

Parameter μ lies in the interval $(0,1)$ and defines the position on the line segment TB.

Parameter α is computed from equation (5) by substituting $R = E$. For α_1 it holds

$$\alpha_1 = 0.2 + 0.8 \, \|g\|^4 \, [g^T H^{-1} g \, g^T H g]^{-1}$$

Here, the approximation $H \approx J^T J$ is used.

This procedure requires in each iteration only one inverse (in computing $(J^T J)^{-1}$) to be done. Moreover, maximum step length is limited with respect to admissible region where the quadratic expansion of S(a) is acceptable. More detailed description of this algorithm is published by Militký (1982) and Militký and Meloun (1993).

4 Organization of MNOPT

The MNOPT procedure is compiled allowing to do besides effective numerical estimation of regression parameters also their statistical analysis dependent upon the size of bias (model nonlinearity). For the minimization of LS criterion S(a) the double dogleg strategy is carried out.

Instead of inverting the matrix J^TJ the special pseudoinverse technique is employed (a variant of the "rational rank" method). Owing to this minimization is accomplished even in cases where J^TJ is singular or nearly singular. For evaluation of the maximum acceptable step length the heuristic procedure based on the "trust region" is used (see Moré (1983)).
The program is completed by a grid search routine applied if no acceptable directional vector v can be found.

5 Comparative study

The aim of this study is comparison of MNOPT with well known statistical packages for personal computers. The comparison was made by solving the three test problems from literature. The model functions have been in the form: No.1 $y = a_1 + a_2 \exp(a_3 x)$
No.2 $y = a_1 \exp(a_2/(a_3 + x))$ No.3 $y = \exp(a_1 x) + \exp(a_2 x)$
Data and initial parameter values for the test problems No.1 and No.2 are introduced by Meyer and Roth (1972) and for the test problem No.3 by Jennrich and Sampson (1968). The maximum number of iterations equal to 150 was selected. For other special parameters of minimization procedures the default values were used. From results of tests the so-called performance index
   $P = 100*(number\ of\ right\ results)/(3*number\ of\ procedures)$
has been computed.

The following statistical packages and methods has been tested:
BMDP ver.1990 - Gauss-Newton method and nonderivative DUD method.
$P = 16.7\%$

SAS ver.6.03 - Gauss-Newton method, Marquardt method, gradient search and nonderivative DUD method. *P = 16.7%*.
Statgraphics ver. 5.0 - Marquardt method. *P = 33.3%*.
SYSTAT ver. 5.1 -Simplex search and conjugate gradient search. *P = 16.7%*.
BMDP SOLO ver. 4.0 -Marquardt method. *P = 66.6%*.
SPSS PC + ver. 4.0 -Marquardt method. *P = 100%*.
STATISTICA ver. 4.0 - Simplex search, Hooke-Jeeves pattern search, Rosenbrock search and quasi-Newton method. *P = 33.3%*.
S Plus for Win. ver. 3.1 - functions NLS (Gauss-Newton strategy) and MS (Dogleg strategy). *P = 16.7%*.
Sigma Plot ver. 5.1 - Marquardt method. *P = 66.6%*.
ADSTAT ver. 1.25 - MNOPT strategy. *P = 100%*.
PC-Matlab ver. 3.5 - Optimization toolbox. *P = 0%*.
ISP ver. 3.1 - procedure NLWS. *P = 0%*.

Evidently, the most of the tested packages are not able to solve all these test problems. Results of extended test study clearly show that only SPSS and MNOPT perform well, and MNOPT is the best one (see Militký and Meloun (1993)).

6 References

Box M. J.: J. Roy. Stat. Soc. **B33**, 171 (1971)
Dennis J. E., Mei H. W.: J. Opt. Theor. Appl. **28**, 453 (1979)
McKeown J.J.: in Toward Global Optimization (Edited by L.C.W. Dixon and G.R. Szegö) North Holland, Amsterdam (1975)
Jennrich R.I., Sampson P.F.: Technometrics 10, 63 (1968)
Meloun M., Militký J., Forina M.: Chemometrics for Analytical Chemistry, Vol. 2 Chemical Model Building, Ellis Horwood, New York 1994
Meyer R.R., Roth P.M.: J. Inst. Math. Applics 9, 218 (1972)
Militký J.: Proc. Int. Conf. Chemical Engineering Fundamentals, Taormina (1982)
Militký J.: Proc. 2nd Int. Symp. on Statistics, Tampere (1987)
Militký J.: Proc. Conf. ROBUST '90, Prague (1990)
Militký J., Meloun M.: Talanta 40, 269 (1993)
Moré J.J.: In Mathematical Programming (Edited by A. Bachem, M. Groetschel and B. Korte), Springer New York (1983)
Nazareth L.: SIAM Rev. **22**, 1 (1980)
Spedicato E., Vespucci M.T.: J. Opt. Theor. Appl. **57**, 323, (1988)

Part XXI

Using the Computer in Teaching Statistics

Simulation, Modelling & Teaching

Brian P. Murphy

Department of Information Management & Marketing, University of Western Australia

1 Introduction

We have used simulation extensively in courses in this department . The first use is in elementary courses, where we discuss sampling concepts. An advanced use is in a 26 lectures course on simulation modelling to senior undergraduates and MBA students. In that course we show a variety of real-world situations in business, manufacturing and biology in which simulation is currently used with powerful effect. It has been successful in that students not only got the message, but also did, as class exercises, extensive projects we had obtained from consulting clients. Clients have been well pleased, and so have students, who enjoyed the experience, and see enhanced employment chances.

It was always felt necessary that any software used in simulation, either for teaching or modelling, could be effectively used by non-programmer students. For the first years (1986-90) we used **STELLA** [3], on the Mac, even though it depended on formal sequencing of explicit equations. Since 1991 we have been using two more recent programs **Extend** [1] on the Mac, and **SimView** [2], a new PC equivalent. These two programs are alike in being modern graphical user interface (GUI) type systems; the user builds a model with icon blocks (which are really formal subroutines) and connection lines. They are also alike in being generally an order of magnitude more efficient than older systems in development effort. For clarity, we use only Extend in this paper.

2 Simulation in Advanced Courses

General aspects of these systems have been discussed in Murphy & Randolph [4], and the main interest here is in the power of use of simulation tools on students' statistical and modelling development. This is best seen in thepower and maturity of many projects, including models on ecological balances (bay pollution), agriculture (drought effects, pollution runoff), shop-floor dynamics of factories and offices, scheduling (trains/trucks/ships separately and in interaction), inventory control (the classic beer-game problem), finance (workshop economics - extended decision support, replacing spreadsheet approach), to the design of golf courses for various classes of players from weekend thrashers to match professionals. We now have companies giving students interesting problems, as assignments, and actual vacation employment.

404

EX. 1 We reproduce here a basic **Extend** model from [4] just to quickly fix ideas. Fig. 1a represents a simple queuing situation such as arises at any bank. Built by students in minutes, they quickly learn one of the fundamental concepts of queuing : as $\lambda \rightarrow \mu$ then the queue length $\rightarrow \infty$. Only a few more minutes is

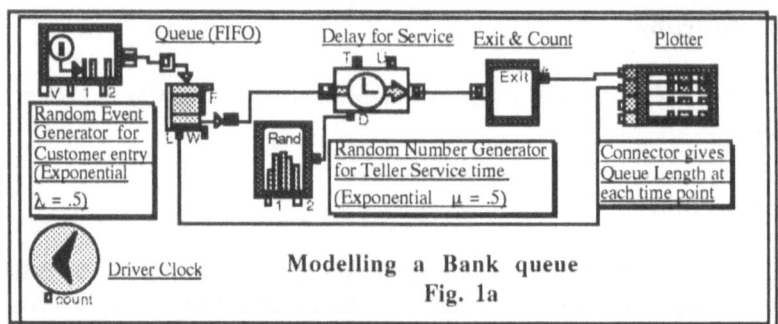

Modelling a Bank queue
Fig. 1a

required to modify it to study the effects of extra servers, multiple $\underline{v}$ single queues etc., as seen in Fig. 1b, which shows the additional icons quickly added to study a two teller situation. This also helps students to understand statistical aspects of the queuing problem, as the question arises: 'how does one compare the two systems, and in particular, how does one combine the variances of the mean delays in system with two servers?' (There is a neat way here - the timer block gives the times for each individual item (i.e. customer) to pass through the system, as in Fig. 1b (&2b), and which has the added advantage of illustrating clearly that this is a proper discrete event simulation system, in which 'next events' drive the time clock and not vice versa).

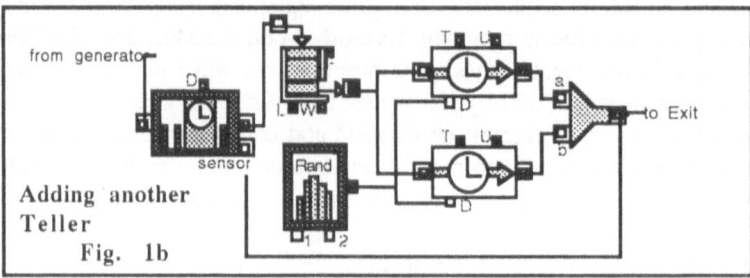

Adding another
Teller
Fig. 1b

This example is the first our students study fully, and is of course mere bagatelle; Extend & SimView are capable of creating very complex real world models in any field. Indeed, we believe that few of the thousands of models annually published could not be developed within either one of these micro-based systems, at a fraction of the cost (in development time) of the mini-mainframe equivalents [5]. Arguments are further detailed in [6] which reports recent research in which these virtues were verified.

EX. 2   Fig. 2a shows a larger real world model produced by students. This investigates the problem of cutting labour losses in a local garage, where random arrival of jobs and types is of the essence. The garage supplied data on profile of job arrivals, their types, estimated times, costs, and stockout delays. The model is really an (large) extension of the bank model above, and introduces submodels ('Hierarchals'). It also has other lessons. Fig. 2b is an hierarchal within the Hoists hierarchal, detailing actions at one hoist.

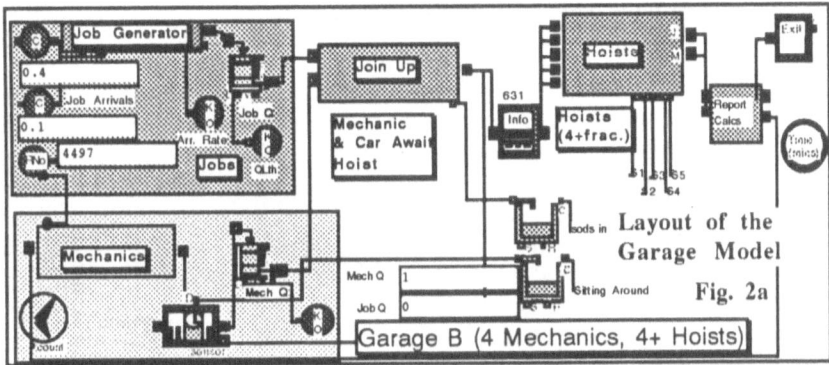

Layout of the Garage Model

Fig. 2a

One major object of the model was to investigate the relationship between the number of mechanics and the number of hoists, because once a car was on the hoist, it could not be removed easily until the job was complete, and so if a part was out of stock, the mechanic became effectively unproductive until it arrived. Would an extra hoist help, if the stockout rate was 1/6?

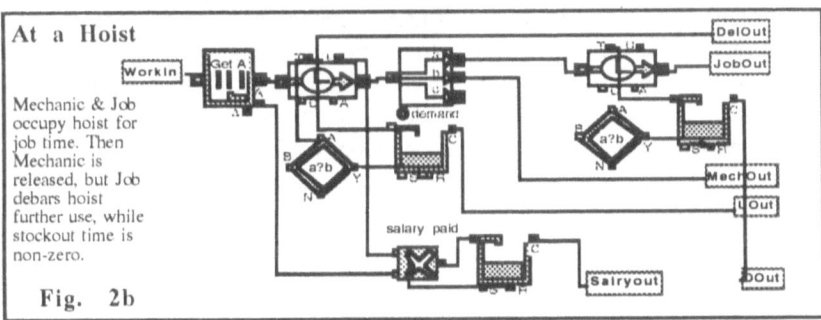

Fig.   2b

Simple statistical calculation shows that a whole extra hoist would be called on .42 of the time, when one or more hoists are tied up with stockout problems. But the lesson was that most managers (and students) did not intuitively accept this initially - they felt the rate should have been simply .1667, and only wanted to allocate .167 of a hoist, which was useless. After being shown the error, a few students then tried a hoist with .50 availability, and were initially surprised to see a total 21% increase in throughput (=.50*.42), but the correct considerations finally emerged.

EX. 3   The above two examples show discrete event models.   Continuous
models can be much more wide ranging. In Fig. 3, our student engineers were
verifying computations of the company who cut a large channel (at a cost of
$70m) to the sea from a polluted estuary, to provide an alternative method of
flushing the algal growth fed by leaching phosphorous. This was causing much
difficulty to the fishing and tourism industries, as well as other farmers. It is an
extensive model, so we only view here the main driver (Fig. 3a) and one River
Hierarchal (Fig. 3b); the other Hierarchals are more complex.

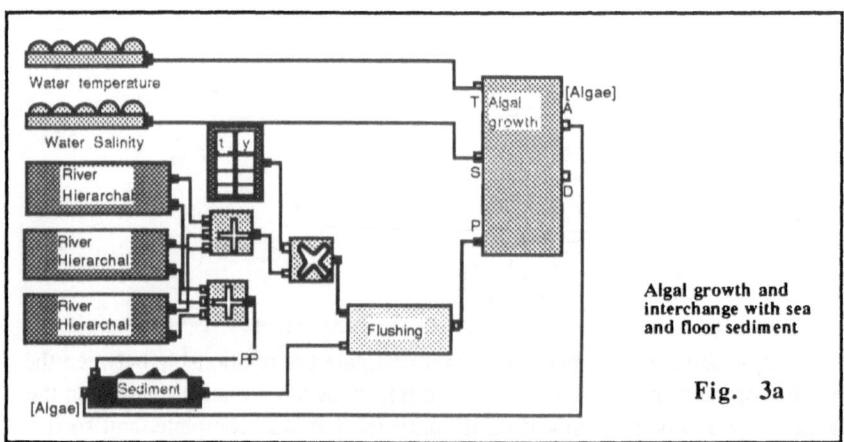

Algal growth and
interchange with sea
and floor sediment

Fig.   3a

In this driving part of the model, the phosphorus is leached by rain into the rivers
from farmlands and carried into the estuary. Depending on water salinity and
temperature, some P is bound or released by the floor sediment . Some P is then
flushed with sea water, and remainder made available to the algae, surplus may be
recombined into the sediment. We thus see two feedback loops here, Re-bound P
(into sediment), and the latest Algae concentration.

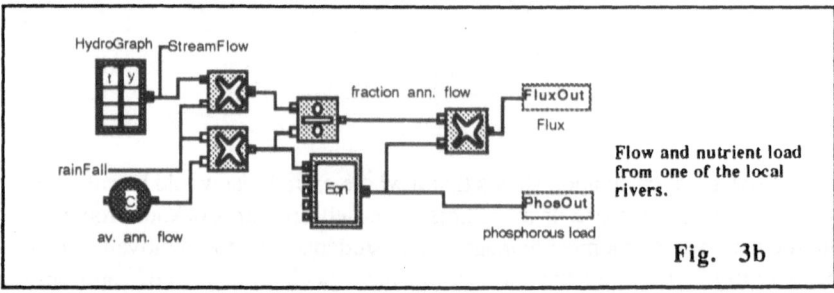

Flow and nutrient load
from one of the local
rivers.

Fig.   3b

The River Hierarchals are submodels of the above, detailing the loads of P in the
respective rivers, as determined by their size and rainfall, known coefficients.
This details of this model will be published in full later; it did verify the
consultants computations, and showed other facts of interest.

3 Simulation in Elementary Courses

We also use simulation extensively in introductory courses for business students, using firstly spreadsheets. The spreadsheet approach to a Central Limit Theorem study seems simple, as in Fig. 4. However, as we all know, elementary courses provide greater difficulties in teaching, as seen in the fact that many students do not seem to get the point of the spreadsheet study as quickly as one might expect (or hope). However, an imaginative new Mac GUI program has arrived to help. **Modls'n'Data** [6] emphasises data as random samples from a model, and quickly engages the students and moves them to a higher level of statistical understanding. Its dynamic simulation, and the facility to get multiple views of the data by rotating it, has had considerable success in cementing statistical concepts of models, random variation, sampling

	A	B	C	D	E	F	G	H	I	J	K
1	Sample	x 1	x 2	x 3	x 4	x 5	Mean	StdDev		Limits	Count
2	1	2.81	3.59	2.76	4.15	3.67	3.40	0.60		0.5	3
3	2	4.98	3.17	4.03	4.68	2.61	3.89	1.00		1	7
4	3	1.78	0.73	4.94	4.46	3.45	3.07	1.78		1.5	20
5	4	2.05	3.58	0.87	3.49	2.06	2.41	1.14		2	26
6	5	0.57	1.86	2.47	1.81	1.13	1.57	0.73		2.5	40
7	6	1.12	3.25	0.37	3.89	1.36	2.00	1.50		3	37

Uniform (0,5) numbers in cols. B to F, means & sds in G , H; Table in J-K

Fig. 4

and sampling distributions, and testing. Of course, there have been many packages before with a simulation facility for the distributions, Central Limit Theorem, confidence intervals etc., but M'n'D is a very powerful extension, looking at tests as outgrowths of (simulation) sampling from models. It covers all main areas of statistics from the univariate distributions to AOV. We have also used it to help honours students trying to perform advanced (e.g. AOV) statistical tests without the usual prior study, as is still, alas, often seen in many places. Some good students do find it boring after a few experiences, but most need the repetition it allows before the principles consolidate; its animated facility for getting multiple views of the data, and for constructing a report sheet, is always welcomed. Unfortunately, it is not something students can use without considerable guidance.

In Fig. 5 we see M'n'D output for three stages of simulating and analysing data from a mixture of bivariate normals.

408

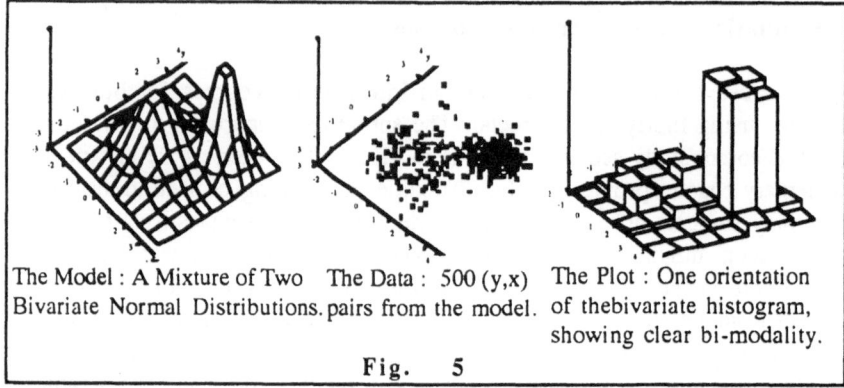

The Model : A Mixture of Two   The Data : 500 (y,x)   The Plot : One orientation
Bivariate Normal Distributions. pairs from the model. of thebivariate histogram,
                                                        showing clear bi-modality.

Fig.   5

4 Conclusions

Simulation is useful in itself in teaching basic concept of statistical inference -
random variation, sampling distributions and tests. It also has large uses in
complex real-world modelling, where mathematical problems seem intractible. In
using simualtion to solve practical problems, one usually needs software that
maximises modeller efficiency rather than computer efficiency, and in this the
advantages of systems like Extend and SimView here discussed over standard
programming languages are vast and well proven.

5   Acknowledgements

Thanks are due to my colleague Mrs Cherry Randolph, who has helped greatly in
the classes, and my daughter Kate who has assembled many waterways models.

6 References

1   Extend (1988,1992): From Imagine That, Inc., San Jose, CA 95115
2   SimView (1992): From InterDynamics Ltd., Adelaide, South Australia 5001
3   STELLA (1986,1991): From High Performance Systems, Inc., Lyme, NH
4   Murphy, B.P. & Randolph, C.M., 1993 : Simulation Packages Today : New
    User Classes and Order-of-Magnitude Breakthroughs in Costs,
    Computational Statistics and Data Analysis, 16, 4, 471-9
5   Murphy, B.P., & Deeley D.M., 1993 : New PC based Simulation Packages
    for More Efficient Development of Statistical Models, **Proc. Intn'l.
    Congress on Modelling and Simulation** (joint host IMACS - Intl.
    Assn. for Maths. in Computer Simulation), Uniprint, Univ. West Australia
6   Modls'n'Data (1993) : By W.D. Stirling; from Intellimation Inc., Santa
    Barbara, CA 93116

Teaching and Learning Through Technology – the Development of Software for Teaching Statistics to Non-Specialist Students

Edwin J. Redfern, Sarah E. Bedford

Department of Statistics, Leeds University

1. Introduction

The increasing number of students with varied mathematical backgrounds entering courses in the applied sciences, geography and economic and social studies is making the teaching of support subjects such as statistics more difficult. Traditionally statistical ideas have been beyond the reach of all but a few, particularly when taught in isolation of the main subject area (Kelly 1992). This is mainly due to lack of motivation, inappropriate backgrounds and single paced presentations. However modern authoring tools make it possible to produce interesting, dynamic software that can be used to support traditional teaching methods, allowing students to work through material at their own pace and placing the statistical ideas more firmly in the context of their principal subject area.

This paper describes the progress of the STEPS (STatistical Education through Problem Solving) consortium of statistics departments, comprising those at the universities of Glasgow, Lancaster, Leeds, Nottingham Trent, Reading, Sheffield and UMIST, funded under the United Kingdom Universities' Teaching and Learning Technology Programme. The consortium's objective is to produce teaching material that introduces statistical ideas as a series of dynamic problems relevant to the students main subject area. Software currently under development targets students in Biology, Psychology, Geography and Business studies. However it will also be useful for specialist statistics students.

The problems are designed to motivate the students need to study statistics by placing each idea in a context that they are more likely to appreciate. Each problem will introduce the need for particular ideas to be studied, and through digressions into the appropriate statistical support material, demonstrate and examine understanding of these concepts. In the modules produced at Leeds this supporting material constitutes a statistical core. The advantage of having a supporting core module is that on completion of the problem the student can be allowed to browse through the related statistical ideas revising and extending his/her knowledge of that area of statistics. Further problems from other subject areas can be set-up around this core making any students approach to the statistical material more relevant to their own subject.

The software is designed to be interactive ensuring that the student actively takes part in the learning process and receives immediate feedback on his/her degree of understanding. It is intended that a record of the student's performance in a series of assessments throughout each module and of their progress through the material will be stored for reference by the tutors.

2. Design Principles

2.1 Selection of Authoring tools and screen design

Much of the early consortium discussion resolved around the selection of an appropriate authoring tool and the basic layout of a screen. Modules are being developed for use on PC's, Macintoshes and under Xwindows. The tool selected for the Macintosh is HYPERCARD while Metacard is being used for XWindows. The PC environment has proved to be much richer for development of this type of software and , based on a minimum of a 386 processor with VGA screen and 4Mb memory using WINDOWS 3.1 and DOS 5, modules are being constructed using TOOLBOOK and VISUAL BASIC. Interfaces with MINITAB and XLISPSTAT have also been developed where interaction with a statistics package and advanced graphics display are required.

The screen design has been based on the standard windows screen appearance together with buttons which permit movement between screens, a map and the ability to exit from the software at any time. A return button is used on a digression so that the user can return to the problem screen at any stage. The principles of this design are illustrated in figures 2 and 3. Each group within the consortium agreed to conform loosely to this design. A menu bar is used for access to items such as *what do I do,* which explains how to proceed if this is not obvious to the user, a *glossary* of statistical terms and *a notebook* for recording information. Within these loose constraints each group in the consortium has the freedom to adapt the software to its particular problem. The approach taken here describes some of the solutions adopted by the Leeds group.

2.2 Problem Module development

The development of the modules follows an iterative path involving paper designs, prototypes and several stages of evaluation within and outside the consortium. The basic steps are

1. *Problem Formulation*

   The first stage is the selection of a problem designed to stimulate the need for a particular statistical concept. These may be based on existing problems and practicals, or a new idea. It is useful to work out the ideas away from the CBL environment so that the relationship between the subject area and the statistical ideas can be thoroughly explored and refined. It is also important to clearly identify the exact statistical concept that is going to be introduced.

2. *Paper Design*

   Next a paper design is constructed in which the structure of the computer module is laid out as a series of screens. At this stage the proposed path or paths through the problem and the links to any supporting statistical core material and subject matter are first identified. Any statistical ideas that can be used in other problems are identified at this stage. These are evaluated and developed as separate units for more general use throughout the consortium.

3. *First Evaluation*

The paper design together with mock-ups of any screens is then evaluated and commented on by members of the consortium and revised and refined through discussion and consultation.

4. *Prototype*

Once the paper design has been agreed a full prototype of the software is developed in which any animation is developed and tested. This enables us to identify potential problems with the initial design and to make revisions and improvements. This is also evaluated by members of the consortium.

5. *Production of final Module*

Once all suggested revisions have been agreed and tested an empty shell is constructed in which only the navigation and map are included. These can then be tested and standardised before the problem component is placed in this shell. The resulting module is then available for full evaluation by a member of a selected panel of people not directly involved in the consortium. Any comments and suggestions made at this stage are taken on board and the module refined.

2.3 Common Statistical Support Material

The emphasis of the modules is placed on the problem, however since many of the statistical concepts are problem independent these have been identified and designed as portable units that can be used in more than one instance. These components have been designed at various levels. They consist of

Screen - a self-contained single screen that explains one idea.

Topic - made up of a set of screens that can be used as a single unit explaining an idea, such as a p value. The component screens could also be used on their own.

Module - this is a group of topics on a related theme. For example a module on chi-square covers aspects of the distribution, contingency tables, goodness of fit, associated p values and a discussion of situations in which it is appropriate.

By grouping supporting statistical ideas in this way they can be used from any problem module as appropriate or used as a stand alone unit to explain and allow revision of the statistical ideas without having to work through the aspects of a problem. Clearly not all ideas in the module would be used in each problem that accesses it, but by allowing the students to browse through the related statistical ideas it might open up other areas that the specific problem would not introduce.

Fig 1 illustrates this for the problem on a salmonella poisoning incident in a hall of residence. The problem path followed by a person using the module for the first time is show on the left hand side and is written as a separate set of linked screens in the authoring language. As appropriate this uses screens or groups of screens from the associated statistical core shown on the right hand side. These are usually used to demonstrate a calculation, statistical principle or graphic. Any demonstrations in

412

the core that require data pick up the current set from the problem. Once the problem has been completed the students are allowed to browse through the related statistical ideas for revision, reinforcement and development. At this stage a variety of data sets may be available to illustrate principles including those considered in the problems. The material on the right hand side can easily be detached from the problem and used in association with other problems in the same statistical area.

Figure 1. *The layout of a module showing the problem path on the left hand side and the core on the right hand side. When working through the problem only the dotted digressions are allowed within the core material.*

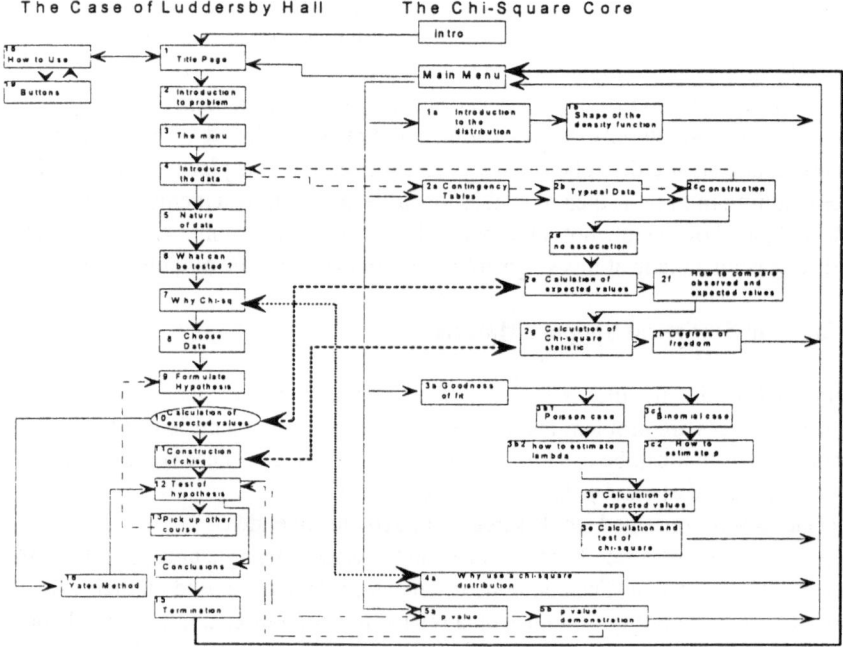

2.4 Navigation and Movement

To help the student use the tutorials effectively navigation has been kept as simple as possible. In the problem modules the path taken through the teaching material is largely dictated by the problem itself. This may be linear, working through the problem step-by-step, or may involve looping around to repeat tests using different data. In addition to the problem structure, there are optional digressions to the statistical support material. When working through the statistical support material in isolation from the problem the student can work through the whole set of material in a basically linear fashion, or they can browse through selected topics.

The user can access a 'roadmap' from most of the pages in the tutorial (by clicking on the compass icon visible in figures 2 and 3). The map provides a pictorial representation of the tutorial and is designed to give the user an overview of what information is available and an indication of how much material there is. The map can also be used to move around within the tutorial, although students are prevented from jumping about in a random manner.

3 The Case of Luddersby Hall - an introduction to the analysis of contingency tables.

Based on a salmonella food poisoning incident the module is designed to introduce the principles of the chi-square test of association. In addition it takes the students through the formulation of the appropriate hypotheses, exploring what information they have available and how it can be used to test that hypotheses. The meal that causes the incident consists of three items, two on the main course plus a desert. The students have to break the problem down into two components, one comparing the main courses, the other the absence or not of a dessert, to identify the culprit. By allowing a free choice of definition of ill (observed ill or salmonella carriers) at the appropriate point, the student who makes the inappropriate selection actually finds no cause for the illness on the first analysis of the desserts and main course. Such a student is then lead through the correct definition of illness and allowed to either repeat the calculations themselves or simply see the correct results. In each instance the student is lead through a discussion of the results. A student who made the correct choice is also shown the other set of results and the discussion as there is always the possibility that the correct choice was made fortuitously.

The structure can be seen in figure 1. The path followed on first entering the module is down the problem on the left hand side. Where appropriate, digressions are made to supporting screens in the chi-square core on the right-hand side. On such a digression the user is only aware of those screens which are allowed on the digression. The data used in the digression is usually that relevant to the problem they are working. These digressions take the form of *show me* support which demonstrate, by animation, how a calculation should be done, e.g. calculation of expected values in a test of association, or the principles behind an idea e.g. p value.

Figure 2. *Screen showing a core screen allowing the student to explore the shape of the density distribution as its parameter changes.*

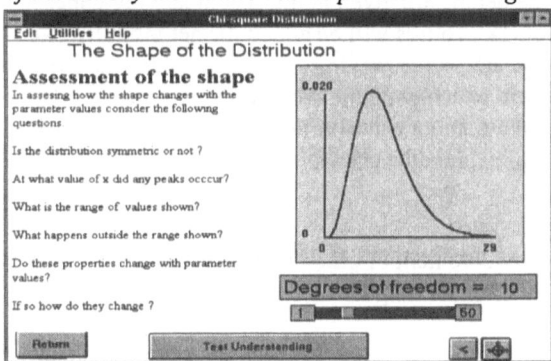

This screen allows the student to explore the shape of the density function by changing the degrees of freedom. This can be done using the scroll bar. Alongside is a set of questions they should be considering while varying the function. Once they feel they have understood the shape of the distribution they can proceed to a set of multiple choice questions that explores their understanding.

414

Figures 2 and 3 illustrate two aspects of this type of software. Figure 2 is a screen from the statistical core that shows the flexibility permitted in allowing a student to explore the shape of a density or probability function. It illustrates this for the chi-square density. It is important that as well as demonstrating the ideas the user is encouraged to think about the processes of problem formulation and development. This can be done using techniques such as multiple choice fields which the student has to correctly complete before being allowed to proceed. The screen in figure 3 illustrates this in the problem at the point where we are asking the student to identify, from the list, the specific questions that they are able to begin answering. Thus the action of reading is done in a more dynamic manner than reading a book.

Figure 3. *Screen in the problem illustrating the use of multiple choice in assisting the student to formulate the correct hypotheses.*

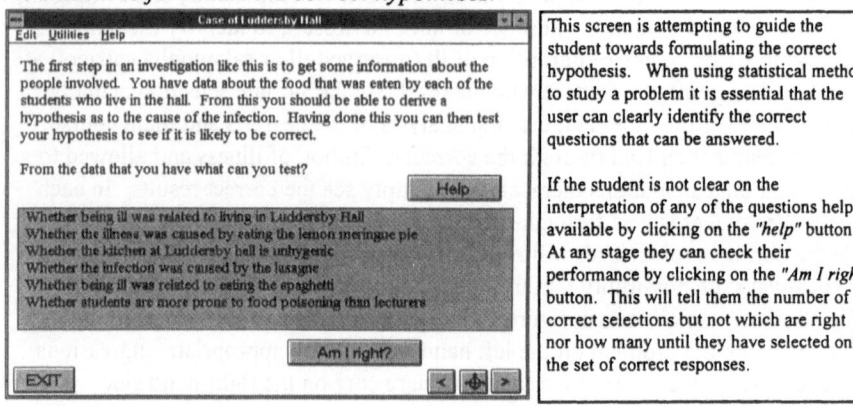

4. Consortium progress - an overview

The consortium is developing problems in four subject areas. Progress has naturally been slow in the first year while many of the design issues have been developed, software evaluated and basic principles developed. The first modules are now entering their final evaluation stage. Despite this, pilot versions have already been tested in some courses at the consortiums own universities. To date 10 modules have reached this stage, the details are given in the accompanying paper in the software section of the proceedings. To this will be added further modules over the next two years to ensure coverage of a wide ranging syllabus of statistical ideas in the four areas. A framework in which to group the modules is also under development and should forge them into a cohesive unit of problem oriented material intended to motivate and explain the principles of statistics.

Acknowledgement
We would like to acknowledge the involvement of other members of the consortium in this work and the UFC and SHEFC for funding.

References
Kelly M.(1992) Teaching statistics to biologists. *Journal of Biol. Education* 26(3) 200-203.

Part XXII

Experimental Designs

An Algorithm for Blocking Response Surface Designs in the Presence of Time Trends

Anthony C. Atkinson, Alexander N. Donev

Department of Statistics, London School of Economics

1 Introduction

Many experiments, especially in science and technology, are carried out in time order. An example is when there is only one apparatus on which several treatments and operating conditions are to be compared. The order in which the treatment combinations are to be observed has then to be decided. A common practice is to randomise. The alternative considered here is to include a smooth time trend as a factor in the experiment. The paper describes algorithms for the construction of optimum designs which accommodate such time variables.

Most of the existing literature on experimental designs in the presence of a time trend is concerned with experiments in which one observation on the response can be made at each of a series of equally spaced time points. Designs are then found which ensure that the parameters describing the time trend are orthogonal to the rest of the model. Such arrangements only exist for a very restricted set of factors, models and trials. For example, the trials of a second-order response surface design cannot be ordered to be orthogonal to a low-order polynomial trend in time. Atkinson and Donev (1994) review the literature and present a general solution based on D-optimality. The optimum design found in this way will not usually be free of the time trend, in that the estimates of the treatment effects will not be orthogonal to those describing the trend. But the designs do allow us to find the best possible estimates of the parameters of interest in the presence of trend. They also allow calculation of the loss in efficiency due to the need to include such time trends in the model.

In nearly all the examples considered by Atkinson and Donev only one trial is allowed per time point. In this paper we are interested in the general case where at each time point a number of experimental trials have to be carried out, each thus forming a conventional block. If there are m blocks, $m - 1$ degrees of freedom are available for the differences between blocks. In the presence of a smooth time trend this number is reduced to 1 or 2, depending on whether the trend is linear or quadratic. In the designs we are looking for, the number of trials N is specified. Finding such exact designs is computationally hard, involving a combinatorial optimization. One algorithm for the conventional blocking of response surface designs is the BLKL algorithm described by Atkinson and Donev (1992), §15.6. But this algorithm cannot be used to construct block designs in the presence of a smooth trend between blocks. The difficulty is caused by the need to give a special status to the time variable. We describe a new algorithm for finding designs balanced for trend which we call the TB algorithm. We give examples of the use of the algorithm to construct block designs where the trend is assumed to be satisfactorily modelled by first or second-order polynomial terms. We compare designs with equal and unequal block sizes and discuss the scope of possible applications of the TB algorithm. Although we are interested in D-optimality, adaptation of the algorithm to other criteria is possible.

The TB algorithm, like the BLKL algorithm, finds optimum designs by searching over a grid of specified design points. A computationally convenient way of searching for better designs off the grid is the Adjustment Algorithm described in §15.7 of Atkinson and Donev (1992).

2   Optimum Experimental Designs

We shall construct D–optimum experimental designs, that is designs which minimise the generalized variance of all, or of a specified subset of, the estimates of the parameters in the model. We write the model as

$$Y = W\gamma = F\beta + G\alpha + \epsilon \tag{1}$$

where the rows of the $N \times p$ extended design matrix F are known functions of m factors and the s columns of G represent the time trend against which balance is required. The formulation is quite general and we only assume that the model is of full rank and that the errors ϵ fulfill the usual second-order assumptions.

When both the treatment and trend parameters are of interest the D–optimum design maximizes the determinant

$$|M(N)| = |W^T W|. \tag{2}$$

The notation in (2) stresses that interest is in exact designs for N trials rather than in the continuous designs introduced by Kiefer (1959) in which the design becomes a measure ξ. An introduction to optimum design theory is given by Atkinson and Donev (1992).

If the s parameters modelling the dependence on time are to be treated as nuisance parameters, the design criterion becomes that of D_s–optimality in which the maximization is of

$$|M_p(N)| = |W^T W|/|G^T G| = |M(N)|/|M_G(N)|. \tag{3}$$

The optimum design makes $|M_p(N)|$ large, or $|M(N)|$ if all parameters are of interest, the case studied in this paper. In the absence of trend, the D-optimum design maximizes $|M_F(N)| = |F^T F|$. The effect of allowing for trend in the analysis of data generated from the design is then the trend factor $TF = |M_p(N)|/|M_F(N)|$. If the design is completely balanced for trend, so that nothing has been lost by allowing for the trend, $TF = 1$, but usually the value will be less.

3   The TB Algorithm

The construction of the design requires an algorithm to select treatment combinations for allocation to the N time points in such a way as to maximize (2). The list of candidate points repeats all factor combinations at each time point. The TB algorithm is based on the BLKL exchange algorithm, modified to construct trend balanced designs. It uses a form of exchange algorithm to move uphill from a random starting point and r random starting points or 'tries' to avoid local maxima. There are two phases.

Phase 1. Construction of an initial design.
    1. Read the number of tries $r, m, N, T, n(i), i = 1, 2, ..., T$ and the model.
    2. Read or generate a list of candidate points for the main factors.
    3. Read or generate a list of candidate time points.
    4. Read and include into the design $j_1, N > j_1 \geq 0$, points for which data are already available: j_1 may be zero.
    5. Include in the design a random number j_2 of points chosen at random from the list of candidate points. A point is included in the design only if this does not lead to a number of points at the ith time point greater than the prespecified number $n(i)$.
    6. The determinant of the information matrix is calculated as $|F^T F + cI|$, with c being a small number, for example 10^{-3}, to guarantee non-singularity.
    7. $N - j_1 - j_2$ points are added to the design sequentially so that each of them gives a maximum increase in the determinant of the information matrix.
Phase 2. Improving the design.
    8. Calculate the determinant of the information matrix $|F^T F|$.
    9. Calculate the effect of exchanging any of the design points with points of the list of candidates provided the constraint $n(i)$ is not violated.
    10. If a beneficial exchange exists, exchange the design point with that from the list of candidates for which the increase in the determinant of the information matrix is greatest.
    11. If a point has been exchanged, go to 8, otherwise go to 12.
    12. If r tries have been carried out print the best design. Otherwise go to 4.

The particular program implementation will clearly depend on the language and the computer used. We ran our FORTRAN program on a SUN work-station and were able to carry out thousands of tries for our examples. Computer time is not an issue.

4 Examples and Discussion

In all our examples we assume that trials can be carried out at equal time intervals with $n(i)$ trials at each time point. Depending on the example, $n(i)$ can take values in the range $N/2 \geq n(i) \geq 0$. In some cases not all time points will be used.

Design Number	m	l	Block Size	Treatment Combinations	TF
1	2	1	1	$(1), ab, ab, (1), a, b, b, a$	1.
2	2	1	2	$\{(1), ab\}, \{a, b\}, \{(1), ab\}, \{a, b\}$	1.
3	2	1	4	$\{1, a, b, ab\}, \{1, a, b, ab\}$	1.
4	3	1	1	$c, abc, ab, a, (1), bc, b, ac$	0.9762
5	3	1	2	$\{1, abc\}, \{b, c\}, \{ab, ac\}, \{a, bc\}$	0.9500
6	3	1	4	$\{c, b, ab, ac\}, \{1, a, bc, abc\}$	$\cdot$ 1.
7	2	2	1	$b, a, (1), ab, (1), ab, b, a$	0.4118
8	2	2	2	$\{(1), ab\}, \{a, b\}, \{a, b\}, \{(1), ab\}$	0.3902
9	3	2	1	$ac, b, ab, c, a, bc, abc, (1)$	0.3922
10	3	2	2	$\{bc, a\}, \{ab, c\}, \{(1), abc\}, \{ac, b\}$	0.392

Table 1: First-order design in m factors, $N = 8$: l is the order of the trend

Example 1. We first illustrate the application of the TB algorithm when the model is first-order in the factors. The design region allows each factor to be at either a high or a low level, so that the D-optimum design in the absence of trend is the 2^m factorial, or equivalently, some of its fractions. When the trend is linear ($s = l = 1$) it is often possible to construct a trend-free design or, failing that, one with a value of TF near one. For example, Designs 1, 2 and 3 of Table 1 for two factors, linear trend ($t = 1$) and $N = 8$ trials, are trend free. For these designs the number of trials at each time point is respectively 1, 2 and 4. Designs 4, 5 and 6 are again for linear trend, but for 3 factors. A completely trend-free design was found only when the points of the 2^3 factorial are concentrated on only two time points, in which case any blocking of the factorial which does not confound the main effects will give the required design. With a smaller number of trials at each time point, the optimum designs are nearly, but not completely, trend free: TF = 0.9762 and 0.9500 for $T = 8$ and 4 respectively. With second-order trend, the loss due to the trend as measured by TF is higher, as is shown by Designs 6 - 10 of Table 1 for 2 and 3 factors and $N = 8$ trials. It is however easy to check how much poorer a random arrangement of the design points would be.

Example 2. We now consider an example in which $n(i)$ is so large that not all time points are used. In order to find out whether equal block sizes are to be preferred we used the TB algorithm to construct D-optimum 16-trial designs with three factors and first or second-order models in both the factors and the trend. For the resulting designs in Table 2 the number of blocks was relatively large, 8, and the maximum block size was variously 2, 3 and 4.

No.	$n(i)$	D-optimum block sizes								Model	$|M(16)|$	$|M_p(16)|$	TF
1	2	2	2	2	2	2	2	2	2	1	.4494E06	.6554E05	1.
2	3	3	3	2	0	0	2	3	3	1	.6420E06	.6554E05	1.
3	4	4	4	0	0	0	0	4	4	1	.7918E06	.6554E05	1.
4	2	2	2	2	2	2	2	2	2	2	.3040E10	.4434E09	0.9855
5	3	3	3	2	0	0	2	3	3	2	.4337E10	.4427E09	0.9840
6	4	4	4	0	0	0	0	4	4	2	.5189E10	.4295E09	0.9546
7	2	2	2	2	2	2	2	2	2	3	.1006E07	.2833E05	0.4323
8	3	3	3	0	3	3	0	1	3	3	.1513E07	.2656E05	0.4053
9	4	4	0	0	4	4	0	0	4	3	.2054E07	.3144E05	0.4797
10	2	2	2	2	2	2	2	2	2	4	.5811E10	.1637E09	0.3638
11	3	3	2	0	3	3	1	1	3	4	.8970E10	.1679E09	0.3732
12	4	4	2	0	3	3	0	0	4	4	.1149E11	.1484E09	0.3298

Table 2: 16 trial D-optimum design for three factors with maximum block size $n(i)$ such that not all time points are used. Model l/k has order l for the trend and order k for the factors. Model 1 1/1, Model 2 1/2, Model 3 2/1, Model 4 2/2

When the model is first order in both the factors and the trend it was easy, as it was in Example 1, to obtain trend-free designs. The resulting Designs 1, 2 and 3 show the effect of increasing the maximum block size $n(i)$, which is to fill the blocks near the edges of the time interval while those in the middle become empty. This is to be expected from optimum design theory when the trend is linear. The three designs give the same value of $|M_p(16)|$ and so are equally good if the trend effect is to be considered a nuisance parameter. However if the trend is also of interest to the experimenter, the best design has the maximum number of trials at the points near the edges and empty blocks in the middle.

If the model is assumed to be of second order in the factors the search for the optimum is over the points of the 3^m factorial. When the trend was not of interest, the design with equal block sizes, Design 4, was slightly better than that with unequal block sizes (Designs 5 and 6). Designs with unequal block sizes were again better when the effect of the time trend was of interest. All of the designs were nearly trend free.

Designs 7-12 were constructed for second-order time trend. As in Example 1, the optimum designs have much lower values of TF than Designs 1-6 for first-order trend. The best designs did not have equal block sizes. Instead, clusters were formed at the edges and at the middle of the time interval, mimicking the D-optimum design for a quadratic in one factor. The best designs, as measured by $|M(16)|$, when the trend effect is of interest were obtained when the block sizes were maximum.

Design	$n(i)$	D-optimum block sizes				Model	$|M(16)|$	$|M_p(16)|$	TF
1	4	4	4	4	4	1	.5825E06	.6554E05	1.
2	8	8	0	0	8	1	.1049E07	.6554E05	1.
3	4	4	4	4	4	3	.1841E07	.2557E05	0.3902
4	8	5	4	2	5	3	.1076E07	.9930E04	0.1515
5	4	4	4	4	4	2	.3969E10	.4466E09	0.9926
6	8	8	0	0	8	2	.6742E10	.3257E08	0.0724
7	4	4	4	4	4	4	.1032E11	.1433E09	0.3186
8	8	6	1	4	5	4	.1259E11	.9943E08	0.2210

Table 3: 16 trial D-optimum design for three factors as in Table 2 but with larger maximum block size $n(i)$. Model l/k has order l for the trend and order k for the factors. Model 1 1/1, Model 2 1/2, Model 3 2/1, Model 4 2/2

The designs in Table 3 were constructed for the same situation as those in Table 2, except that the number of time points was chosen to be relatively small, 4 rather than 8, and the maximum block size $n(i)$ was allowed to increase to N/2, i.e. eight. It can be seen that, when the trend effect is of interest, designs with unequal block sizes are again better than those with equal block sizes. However higher values of TF were generally obtained for designs with equal block sizes.

The examples show that the algorithm is useful both for deciding the optimum block sizes and for calculating the design for the factors in each of the blocks. The algorithm can clearly be used for other models, numbers of factors and design sizes. It may fall short because the search is over a grid of points for the factors, which may not include the global optimum. We used the Adjustment Algorithm to search for the global optimum. Our experience suggests that although the Adjustment Algorithm is effective in finding the exact location of the optimum indicated by the TB algorithm, it often gives only a modest improvement. A disadvantage is that the designs lose their nice property of having only a few levels for each factor.

References

Atkinson, A. C. and A. N. Donev (1992). *Optimum Experimental Designs.* Oxford: Clarendon Press.

Atkinson, A. C. and A. N. Donev (1994). Experimental designs optimally balanced for trend. Submitted.

Kiefer, J. (1959). Optimum experimental designs (with discussion). *J. Roy. Statist. Soc. B 21,* 272–319.

Computational Issues for Cross-over Designs Subject to Dropout

Janice Lorraine Low, S.M. Lewis, B.D. McKay, Philip Prescott

Mathematics Department, University of Southampton

1 Introduction

A cross-over experiment is a form of comparative study in which subjects receive a sequence of $t \geq 2$ treatments, one in each of p successive time periods, with the response of each subject measured at the end of every period. Such experiments are widely used in many areas including clinical and medical research, agriculture and human factors engineering.

For design purposes, a simple model is often assumed for the observations in which the effects of subjects, periods, direct treatments and first-order carry-over treatments are added to the random error, assumed independently and identically distributed. Much work has already been undertaken to develop designs for this situation, see Jones and Kenward (1989).

Frequently in cross-over trials, particularly in medicine, subjects fail to complete a study, most commonly dropping out during the final one or two periods, see Matthews (1988). Herzberg and Andrews (1976) considered an analogous problem in optimal response surface designs. A similar approach can be adopted to examine the effects of fixed period dropouts, unrelated to treatment, on estimating treatment comparisons in cross-over designs. This paper shows how the computational problems produced by such an approach can be reduced through the software application of results from combinatorics.

2 The Problem

Suppose that an experiment is proposed to compare the effects of four treatments using four periods and sixteen subjects in which the model of Section 1 is assumed for the observations. A Williams square design is proposed in which four distinct treatment sequences are employed with four subjects being assigned at random to each sequence. It is believed, a priori, that the probability of subjects dropping out, independently of each other and unrelated to treatment, is negligible during the first two periods but may be non-zero in the last two periods. The Williams square design is proposed because, when the possibility of dropouts is ignored, it is known to be universally optimal, over the class of designs in which $t = p$, for estimating the direct and first-order carry-over treatment effects, under the model. However, when the experiment is performed, there are very many possible outcomes which may result. Two of these are shown in Table 1; numbers in brackets denote the number of subjects present throughout the period; where no number is given all subjects are present throughout the period.

Table 1: Two possible outcomes from a Williams square of side four involving sixteen subjects and two periods of dropout.

		Outcome 1						Outcome 2		
		Period						Period		
	1	2	3	4			1	2	3	4
1	A	B	D(4)	C(4)		1	A	B	D(4)	C(3)
Sequence 2	B	C	A(4)	D(3)	Sequence	2	B	C	A(3)	D(0)
3	C	D	B(4)	A(3)		3	C	D	B(2)	A(0)
4	D	A	C(4)	B(2)		4	D	A	C(3)	B(2)

In outcome 1, four dropouts occur during the final period, namely one on each of sequences 2 and 3 and two on sequence 4. If this outcome were realised from the experiment, the resultant increase in the variance of the estimators of the treatment comparisons, compared with that of the planned design, would not be large. In contrast, the object of the experiment could not be achieved if outcome 2 were realised. The loss of four subjects in the third period and a further seven subjects in the final period results in a *disconnected* design, that is a design from which some pairwise treatment comparisons cannot be estimated.

There are many designs which might result as a consequence of subjects dropping out. Ideally we should like to assess each of these to investigate the robustness to dropouts of the planned design. However this is a daunting task since, even for relatively small studies, the number of designs to be evaluated is considerable; for example, to assess the cross-over experiment described above it is necessary to consider 52,625 different designs. In Section 3 we describe a method of assessing planned designs by considering all possible realisable designs and describe the theory and software for reducing the computational burden in Section 4.

3 Design Assessment

Consider a *planned design*, d, in which a set of n subjects is assigned to each of m treatment sequences. Assume that each subject has a fixed probability θ_i of completing i periods and then dropping out during period $i + 1$, ($i = 0, \ldots, p - 1$). Suppose that, when the experiment is performed, there are $l_{i,j}$ subjects on sequence j, ($j = 1, \ldots, m$), who complete i periods and then drop out during period $i + 1$ resulting in an *implemented design*, d_I, where

$l = (l_{0,1}, \ldots, l_{p,1}, \ldots, l_{0,m}, \ldots, l_{p,m})$ and $l_{p,j}$ denotes the number of subjects on sequence j who complete the study. The probability that d_l is realised is then

$$P(l|\theta_0, \ldots, \theta_{p-1}) = \prod_{j=1}^{m} \frac{n!}{l_{0,j}!, \ldots, l_{p,j}!} \theta_0^{l_{0,j}}, \ldots, \theta_p^{l_{p,j}}, \tag{1}$$

where $\theta_p = 1 - \sum_{i=0}^{p-1} \theta_i$ denotes the probability of a subject completing the study.

For any planned design there is a set, D, of all possible implementable designs. Let $D_0 \subseteq D$ be the set of all disconnected designs. An important requirement on any planned design is that it has a zero, or acceptably small, probability of producing a disconnected implementable design. An additional requirement is that each of the connected implementable designs estimates, as accurately as possible, the contrasts of interest in the direct and first-order carry-over treatment effects. To assess how well a planned design, d, meets the second requirement, probability distributions are considered of random variables X_k $(k = 1, 2)$ defined by

$$X_k(d_l) = \begin{cases} \{\psi[C_k \Omega_k(d_l) C_k']\}^{-1} & \text{for } d_l \in D \backslash D_0, \\ 0 & \text{for } d_l \in D_0, \end{cases} \tag{2}$$

where ψ is an appropriate measure of design performance, C_k $(k = 1, 2)$ are matrices holding the coefficients of the contrasts of interest in the direct and first-order carry-over treatment effects respectively, and $\Omega_k(d_l)$ $(k = 1, 2)$ are generalised inverses of the information matrices for d_l for estimating the direct and carry-over treatment effects respectively. A natural performance measure for such experiments is the A-criterion which is obtained by substituting $\psi[C_k \Omega_k(d_l) C_k'] = tr[C_k \Omega_k(d_l) C_k']$ $(k = 1, 2)$ in equation (2).

The probability distribution of X_k $(k = 1, 2)$ for given values of θ_i $(i = 0, \ldots, p-1)$ is

$$P(X_k = x|\theta_0, \ldots, \theta_{p-1}) = \sum_{l \in L} P(l|\theta_0, \ldots, \theta_{p-1})$$

where $L = \{l : d_l \in D, X_k(d_l) = x\}$.

The mean and variance of X_k $(k = 1, 2)$ will be used to provide *summary measures* for a planned design under repeated use in experiments with fixed θ_i $(i = 0, \ldots, p-1)$. For experiments of a realistic size, the calculations in (2) can involve a prohibitive amount of computation.

4 Computational Reductions

Fortunately it is not necessary to evaluate (2) for every connected implementable design separately, since some of them will give rise to identical random variables and are then equivalent in the following sense:

Definition 1 Consider the planned design, d, and its associated set of implementable designs D. Designs d_{l_1}, $d_{l_2} \in D \backslash D_0$ are *performance equivalent* with respect to direct and first-order carry-over treatment effects if and only if, for $(k = 1, 2)$,

$$\{\psi[C_k \Omega_k(d_{l_1})C_k']\}^{-1} = \{\psi[C_k \Omega_k(d_{l_2})C_k']\}^{-1},$$

where ψ, C_k and Ω_k, $(k = 1, 2)$ are defined in Section 3.

If the performance equivalent implementable designs can be identified, it will suffice to calculate performance measures for one member from each equivalence class of D and then allow for the size of that class when computing the summary statistics.

In order to do this we seek answers to the following questions. Given a planned design, with an associated set of implementable designs: (i) How can we identify the equivalent designs without calculating the performance measures? (ii) How many equivalence classes are there? (iii) What is the size of each equivalence class? (iv) How can we identify one member from each class?

We now define a combinatorial equivalence relationship between designs which is sufficient to establish performance equivalent designs and has the computational advantage that its equivalence classes can be found without calculating Ω_k $(k = 1, 2)$.

Definition 2 Consider the planned design, d, and its associated set of implementable designs D. Designs d_{l_1}, $d_{l_2} \in D$ are *combinatorially equivalent* if d_{l_2} can be obtained from d_{l_1} by permuting firstly the order of the treatment sequences and, secondly, the treatment labels.

A group of permutations which acts on the treatment sequences of the implementable designs may now be established which is sufficient to identify the combinatorially equivalent designs.

The means of identifying the combinatorial equivalence classes and thus of answering the above questions are provided by the theory developed for the colouring problem, which we outline via the following example.

Example 1 Suppose we have a square tray and we place a coloured disc at each of the four corners. If each disc may be either black or white, then the total number of colourings, provided the tray remains fixed, is $2^4 = 16$, see Figure 1. If the position of the tray is fixed all 16 colourings are different. If the tray may be rotated clockwise, then some of the colourings can be obtained from others. For example C2 is equivalent to C3, since we can obtain C3 from C2 by rotating the tray through an angle of $\pi/2$. In this example, there are four distinct rotations of the tray, namely the identity and rotations through $\pi/2$, π and $3\pi/2$. The action of each of these rotations on the four vertices labelled 1, 2, 3 and 4 respectively, can be represented by the following cyclic permutations (1)(2)(3)(4), (1234), (13)(24) and (1432).

Figure 1: Sixteen colourings of a square tray using black and white discs.

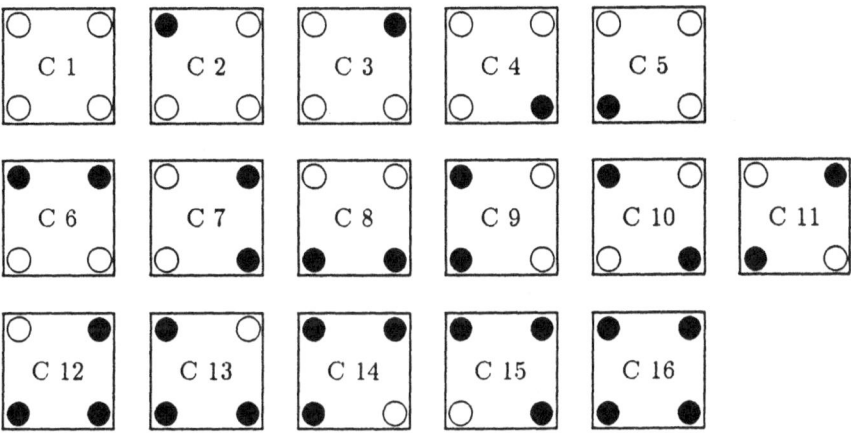

Applying each of these permutations to the colourings we observe that there are only six different patterns namely, C1, C2, C6, C10, C12 and C16. All the remaining colourings may be obtained by applying one of the above permutations to one of these six patterns. We can, therefore, divide the colourings into six equivalence classes in which all the colourings in each class have the same pattern. Using results from combinatorial theory, see for example Slomson(1991), the number of equivalence classes can be found by applying Burnside's Lemma, and the size of each class by applying the Orbit-Stabilizer Theorem.

The underlying theory of the colouring problem concerns the relationship established between a group (in our example the rotational symmetries of a square) and the members of a set (in our example the 16 different colourings). In the example the symmetries of the square interact with each of the colourings to produce another, possibly different, colouring. In general, provided an element belonging to a permutation group can act on any member of a set to give another, not necessarily different, member of the set, then the theory of the colouring problem may be applied.

If, in our design problem, we replace the square tray by the planned design, the set of colourings by the set of implementable designs and the rotational symmetries of a square by the group of permutations, then the problem can be viewed as analogous to the colouring problem. Consequently results from combinatorial theory can be used directly to obtain answers to questions (ii)-(iv) above, and to reduce substantially the computation required to calculate the summary measures. For a Williams square of side four and subject numbers which might realistically be used in practice, Table 2 lists the number of implementable designs and the number of equivalence classes into which D can be partitioned, for two periods of dropout. From Table 2 it may be seen that the computational reduction is approximately three quarters.

Table 2: Number of equivalence classes and implementable designs for a Williams square of side four and two periods of dropout.

Number of subjects.	Number of implementable designs.	Number of combinatorial equivalence classes.
12	10000	2530
16	50625	12720
20	194481	48741
24	614656	153874
28	1679616	420246
32	4100625	1025685
36	9150625	2288440

Software has been developed which obtains one member from each of the combinatorial equivalence classes. It generates each implementable design, d_l, in turn and applies every permutation from the group to d_l to construct designs combinatorially equivalent to d_l. If d_l is the *smallest* of these under some linear ordering, then d_l is used as the class representative. Otherwise d_l is rejected and the next design in D is considered. Furthermore, for each d_l the size of the equivalence class containing d_l is obtained via the Orbit-Stabilizer Theorem. The cost of this search procedure will generally be insignificant compared to the computation of the performance measures, since the permutation group is small.

The program feeds a representative from each equivalence class, together with the class size, into a separate program which calculates the mean and the variance of the X_k, where X_k ($k = 1, 2$) is defined in (2).

This software has been used to assess the performance of particular designs and also to choose between several competing designs for situations when it is anticipated that dropouts may occur. It is available from the first-named author.

References.

Herzberg A. M. and Andrews D. F. (1976), Some considerations in the optimal design of experiments in non-optimal situations. *Journal of the Royal Statistical Society, Series B*, 38, No. 3, pp. 284-289.

Jones B. and Kenward M. (1989), *Design and Analysis of Cross-over Trials*, Chapman and Hall, London.

Matthews, J. N. S. (1988), Recent developments in crossover designs. *International Statistical Review*, 56, pp. 117-127.

Slomson A. (1991), *Introduction to Combinatorics*, Chapman and Hall, London.

Data Structures and Algorithms for an Open System to Design and Analyse Generally Balanced Designs

Roger W. Payne, Michael F. Franklin

Rothamsted Experimental Station, AFRC Institute of Arable Crops Research

Keywords. Experimental design, anova algorithms, design key, general balance

1 Introduction

The design of experiments is now an important facility in many statistical programs and packages. Algorithms are available for constructing partial replicates and designs containing effects confounded with blocks. Alternatively, programs may offer a repertoire of pre-selected designs. Less attention, however, seems to be given to the question of how these confounded designs will eventually be analysed, and to ways of avoiding the constraints on choice of design that arise from menu-based systems. In this paper we focus on the important class of *generally balanced designs*, and describe the information and associated data structures required to form any particular design, and to specify the analysis. We also discuss the tools and algorithms available for deriving these details, and explain how they can be put together to provide an open system in which users can investigate designs and then add them to an available repertoire.

2 General Balance

General balance (Nelder 1965, Payne & Tobias 1992) encompasses a very wide range of designs, with the particular advantage that there can be more than one *block* (or *error*) term. The total sum of squares can then be partitioned up into components known as *strata*, one for each block term. Each stratum contains the sum of squares for the treatment terms estimated between the units of that stratum, and a residual representing the random variability of those units. The properties of a generally balanced design are that (i) the block (or error) terms are mutually orthogonal, (ii) the treatment terms are also mutually orthogonal, and (iii) the contrasts of each treatment term all have equal efficiency factors in each of the strata where they are estimated.

Thus, general balance includes all orthogonal designs and all designs in which there is balanced confounding between treatment and block terms. Designs also occur in which some treatment terms contain several sets of contrasts, each with their own efficiency factor. These too can be accommodated, by allowing such terms to be specified by several *pseudo* terms (one for each set of contrasts). The design is then balanced with respect to the pseudo terms, and the sums of squares, effects and means for the original terms can be obtained by adding together the information from the appropriate pseudo terms (Payne & Wilkinson 1977). This is particularly useful in partially confounded designs, where different sets of treatment contrasts may be confounded with the blocks in each replicate.

Wilkinson (1970) and Payne & Wilkinson (1977) present a very efficient algorithm which can be used for the analysis of generally balanced designs. This operates on a working vector which initially contains the data values, and finally contains the residuals. The terms in the model are fitted by a series of *sweep* operations. Each sweep estimates the effects of a term, and then subtracts them from the current working vector, which then becomes the working vector for the next sweep. The estimated effects are simply the corresponding table of totals for the factor combinations of that term, calculated from the current vector, divided by their replication and efficiency factor. The first sweep is for the grand mean. The block terms are fitted next, to give an initial partitioning into strata. Then the treatments are fitted within each stratum. When there are several strata, each is introduced by a special sweep, known as a *pivot*, in which each value in the working vector is replaced by the corresponding effect for the block term of the stratum; this operation may also need to be repeated after sweeping for any non-orthogonal treatment term estimated in that stratum.

The algorithm has the advantage that it does not involve any matrix inversion, nor the formation of large matrices of sums of squares and products. More importantly perhaps, from the point of view of the user, it automatically allocates each treatment term to the appropriate stratum, allowing sums of squares to be compared with the correct residual and standard errors to be calculated for means.

When treatment effects are estimated in more than one stratum, it is advantageous to present estimates that combine the information from each of the strata where the treatment is estimated. Payne & Welham (1990) discuss the advantages that accrue in computing time and workspace with algorithms specially devised for generally balanced designs (see Payne & Tobias 1992).

Details of the structure of the design (efficiency factors &c) can be obtained by a process known as the *dummy analysis*. This is similar to the analysis of the data, but involves extra sweeps to detect whether each term can be estimated in a particular stratum, and to determine its efficiency factor there (Wilkinson 1970, Payne & Wilkinson 1977). Thus, the information required to specify the analysis of a generally balanced design, consists of the list of block and treatment factors with their values, and the structure formulae to define the block and treatment terms together with any pseudo terms.

3 Design keys

To form a particular generally balanced design requires the block structure formula, the block factors (and their values), and a means of constructing the values of the treatment factors from the block factors, in such a way as to ensure that the design will exhibit all the required confounding and aliasing properties. (The block-factor values generally occur in an easily-constructed lexicographic order, and the block structure formula is necessary to define the randomization of the design, once the treatment factors have been generated, see Nelder 1965.)

The inter-relationship between the treatment and block factors can be represented very conveniently by a matrix known as the *design key* (Patterson 1976, Patterson and Bailey 1978). The construction method requires the factors to have prime numbers of levels, and so the definition may involve defining treatment *pseudo* factors in terms of block *pseudo* factors (usually known as the *plot* factors), and then constructing the original factors from the outer products of the pseudo factors. (So, in a design with 3 blocks each of 9 units, the unit-within-block factor would be represented by two 3-level pseudo factors.) As we explain below, however, these *design* pseudo factors are not necessarily capable of defining the pseudo terms required for the *analysis*.

The design key has a row for each treatment pseudo factor and a column for each plot factor; below k_{ij} represents the element in row i and column j. (This is the transpose of the form used by Patterson 1976, but in statistical packages it seems more convenient to specify the treatments by rows.) There is also a vector, known as the *basevector*, which contains an element b_i for each treatment factor to allow the levels of the factor to be shifted cyclically.

The calculation assumes that the values of the factors are represented by the integers zero upwards. The value $(\alpha_i)_u$ in unit u of treatment factor i is given by

$$(\alpha_i)_u = b_i + k_{i1} \times (\beta_1)_u + k_{i2} \times (\beta_2)_u + \ldots + k_{ic} \times (\beta_c)_u \qquad \text{modulo } t_i$$

where $(\beta_1)_u \ldots (\beta_c)_u$ are the values of the plot factors in unit u, and t_i is the number of levels of treatment factor i. Essentially, the key identifies each treatment factor i with the set of block-factor contrasts (in the usual terminology)

$$\beta_1^{k_{i1}} \quad \beta_2^{k_{i2}} \quad \ldots \quad \beta_c^{k_{in}}$$

To illustrate the process, the treatments to be allocated (before randomization) to the plots of an $n \times n$ Latin Square may be calculated as

$$\text{Latin-factor-value} = \text{Row-factor-value} + \text{Column-factor-value} \quad \text{modulo } n$$

and values of the extra factor in a Graeco-Latin square can be formed as

$$\text{Graeco-factor-value} = \text{Row-factor-value} + 2 \times \text{Column-factor-value}$$
$$\text{modulo } n$$

The design key thus has rows (1,1) and (1,2).

Franklin (1985) gives an algorithm to find defining contrasts and confounded effects in p^{n-m} factorial experiments (for $p \geq 2$ prime), from which the design key can easily be constructed (see, for example, Franklin & Payne 1993). Designs containing sets of factors with several different (prime) numbers of levels can be generated as direct products of designs for each particular prime or, more conveniently, by forming a design key combining the rows and columns from the keys of all the individual designs.

The design key allows treatment factors to be generated that are completely confounded with block factors. Often, there are several ways in which sets of contrasts can be selected to be confounded and, by using different keys for different parts of a design, partially confounded designs can be formed. (Below, these different keys are termed alternative *versions* of the basic design.) Partially confounded designs, however, require treatment terms to be partitioned into pseudo terms for successful analysis. The factors to generate these terms can be formed by inverting the key matrix (using the field of arithmetic modulo p) to obtain a key for defining the block factors in each version in terms of the treatment factors. The block factors identify the treatment contrasts that are confounded in each version. Thus these keys allow the necessary (*analysis*) pseudo factors to be generated from the values of the treatment factors.

4 Data structures

The necessary information to define the construction and analysis of a generally balanced design is thus as follows.

1) description of the design (character string)
2) number of block factors (scalar)
3) number of pseudo factors for each block factor (vector of numbers)
4) number of levels for each block pseudo factor (vector of numbers)
5) block-structure formula for the design (model formula)
6) number of treatment factors (scalar)
7) number of pseudo factors for each treatment factor (vector of numbers)
8) number of levels for each treatment pseudo factor (vector of numbers)
9) number of different "versions" of the design (scalar)
10) description of each version (character string)
    design key for each version (matrix)
11) number of (analysis) pseudo factors per version
    in the treatment formula (scalar)
12) number of levels of each (analysis) pseudo factor (if any)
    (vector of numbers)
13) key for (analysis) pseudo factors (if any) in terms
    of the treatment factors (matrix)
14) treatment-structure formulae (model formula)
    (one formula if there are no analysis pseudo factors;
    otherwise, different formulae may be required depending
    on how many versions of the design are used)

For example

1) '3×3×3 factorial in blocks of size 9'
2) 2

3) 1 2

4) 3 3 3

5) B_1 / B_2

6) 3

7) 1 1 1

8) 3 3 3

9) 4

10) '$T_1+2T_2+2T_3$ confounded with B_1'

    1 1 1

    0 1 0

    0 0 1

    '$T_1+2T_2+T_3$ confounded with B_1'

    1 1 2

    0 1 0

    0 0 1

    '$T_1+T_2+2T_3$ confounded with B_1'

    1 2 1

    0 1 0

    0 0 1

    '$T_1+T_2+T_3$ confounded with B_1'

    1 2 2

    0 1 0

    0 0 1

11) 1

12) 3

13) 1 2 2   (pseudo factor for version 1)

    1 2 1   (pseudo factor for version 2)

    1 1 2   (pseudo factor for version 3)

    1 1 1   (pseudo factor for version 4)

14) $(T_1 * T_2 * T_3) // P_1$ (one version only)

    $(T_1 * T_2 * T_3) // (P_1 + P_2)$ (two different versions)

    $(T_1 * T_2 * T_3) // (P_1 + P_2 + P_3)$ (three different versions)

    $(T_1 * T_2 * T_3)$ (all four versions: design balanced)

For an explanation of the operators in the model formula, including the psuedo-factor operator //, see Payne *et al.* (1993, Chapter 9).

5 Conclusion

The specification above is used in Genstat to define the data base for a suite of procedures, written in the Genstat language and distributed with Release 3[1] of the standard procedure library (Payne, Arnold & Morgan 1993), which allow the user to work through a sequence of pop-up menus to select a design, name the various factors, randomize the design, print the design in a tabular representation, and produce a skeleton analysis of variance showing where each treatment term is

434

estimated and the corresponding efficiency factors. The information is contained in standard Genstat data structures and stored in a backing-store file. The procedures form the menus to list the choices by collating the contents of the file, and a procedure FDESIGNFILE is provided to construct new files or to extend the existing file. The system thus allows users to add new designs as required by their own working environments.

Other design procedures allow the selection and generation of fractional factorials, cyclic designs and alpha designs. There are also procedures to allow designs to be formed as the outer product of other designs or by "adding" designs together, as well as procedures for plotting an experimental plan and generating data forms for the experimenter.

Genstat also has the facilities to allow the properties of potential designs to be assessed (Franklin & Payne 1993), ranging from matrix operations (for example, to study the efficiency factors of unbalanced designs) to methods of analysis for balanced (stratified) designs (Section 2), unbalanced mixed models by residual maximum likelihood (REML), and generalized linear models.

References

Franklin, M.F. (1985). Selecting defining contrasts and confounded effects in p^{n-m} factorial experiments. *Technometrics*, **27**, 165-172.

Franklin, M.F. & Payne, R.W. (1993). Tools for the construction of effective experimental designs. *Proceedings of Applied Statistics in Agriculture*, in press.

Nelder, J.A. (1965). The analysis of randomised experiments with orthogonal block structure. I. Block structure & the null analysis of variance. II. Treatment structure & the general analysis of variance. *Proceedings of the Royal Society of London*, **A283**, 147-178.

Patterson, H.D. (1976). Generation of factorial designs. J. R. Statist. Soc. B, **38**, 175-179.

Patterson, H.D. and Bailey, R.A. (1978). Design keys for factorial experiments. *Applied Statistics* **27**, 335-343.

Payne, R.W. & Wilkinson, G.N. (1977). A general algorithm for analysis of variance. *Applied Statistics*, **26**, 251-260.

Payne, R.W. & Welham, S.J. (1990). A comparison of algorithms for combination of information in generally balanced designs. In: *COMPSTAT90 Proceedings in Computational Statistics*, 297-302. Heidelberg: Physica-Verlag.

Payne, R.W. & Tobias, R.D. (1992). General balance, combination of information and the analysis of covariance. *Scandinavian Journal of Statistics*, **19**, 3-23.

Payne, R.W., Lane, P.W., Digby, P.G.N., Harding, S.A., Leech, P.K., Morgan, G.W., Todd, A.D., Thompson. R., Tunnicliffe Wilson, G., Welham, S.J. & White, R.P. (1993). *Genstat 5 Reference Manual, Release 3*. Oxford: Oxford University Press.

(ed.) R.W. Payne, G.M. Arnold & G.W. Morgan (1993). *Genstat 5 Procedure Library Manual Release 3[1]*. Oxford: Numerical Algorithms Group.

Sequentially Counterbalanced Designs Formed from Two or More Latin Squares

Philip Prescott

Department of Mathematics, University of Southampton

1 Introduction

Experiments in which t treatments are applied to n experimental units or subjects, such that each subject receives a treatment in each of p consecutive time periods, are commonly called cross–over designs or change–over designs. In such experiments, observations may be affected by carry–over effects from previously administered treatments. One model used in such situations to describe the observed response variable includes terms for subjects, periods and direct treatment effects, and an additive term for first–order carry–over treatment effects which depend only on the treatment applied in the preceding period, together with a random error term. See Jones and Kenward (1989) for further details.

For this model, when p=t and n=t or 2t for even or odd t respectively, designs based on Latin squares are optimal, see Williams (1949) or John (1971). These designs are such that each treatment appears equally often in each row and column of the design and each treatment follows every other treatment equally often. Williams' designs for even t are examples of *sequentially counterbalanced addition* Latin squares.

Definition 1. An *addition Latin square* is a txt Latin square with labels 0,1,2,...,t−1, whose ith row is obtained by adding i−1 to the labels in the first row and reducing modulo t, i.e. subtracting t from any label which is greater than or equal to t.

Definition 2. A Latin square in which each treatment label is preceded by every other treatment label exactly once is called a *sequentially counter – balanced Latin square.*

The Latin square with first row given by (0 1 3 2), where columns correspond to periods, and where subsequent rows are formed by cyclic rotation of the treatment labels in columns, is a Williams' design for t=4 treatments.

Definition 3. A design which has the property that each treatment label is preceded by every other treatment label equally often, will be called a *sequentially counterbalanced design.*

Williams designs for odd t, based on two addition Latin squares, are examples of sequentially counterbalanced designs. For t=5, two squares such as (0 4 1 3 2) and (0 1 4 2 3) are required to produce a sequentially counterbalanced design, but neither Latin square is itself sequentially counterbalanced.

For even t, various methods are available for constructing sequentially counterbalanced Latin squares, see Issac and Dean (1993) for a review of these methods. However, computer searches have established that there are no sequentially counterbalanced Latin squares for t=3, 5, or 7, and none has yet been found for t=11. In some cases sequentially counterbalanced Latin squares have been found for odd t. Hedayat and Afsarinejad (1975 and 1978) list examples for t=9, 15, 21 and 27.

Russell (1991), in an investigation of optimal designs when there are fewer subjects than treatments, considered, for odd t, *nearly* sequentially counterbalanced squares in which each treatment is preceded by all other treatments, except for two, exactly once, and by one of the remaining two twice, and by the other not at all. An example of a nearly–balanced square for t=5 has first row (0 3 4 1 2). Russell proposed a general method of constructing addition Latin squares with this property and also listed, for odd t≤19, *complementary* Latin squares, also nearly–balanced, which could be combined with the initial square to produce a sequentially balanced design. Designs formed from these nearly–balanced squares may be used to produce optimal designs when the number of subjects is fewer than t, or for odd t, fewer than 2t. More recently, Newcombe (1992) used a design formed from three nearly–balanced squares to give an overall sequentially counterbalanced design for t=5 using 15 subjects.

In this paper different ways of combining sets of Latin squares to produce sequentially counterbalanced designs are examined. The Williams' designs, Russell's designs and designs based on complete sets of orthogonal squares are identified amongst the many sequentially counterbalanced designs which can be found.

2 The incidence matrix M

The incidence matrix $M=(m_{ij})$ of a design is a txt matrix with elements m_{ij} equal to the number of times that the ith treatment is followed immediately by the jth treatment within the same subject. The off–diagonal elements of M are all the same for a sequentially counterbalanced design. For a Williams' design for even t based on a single Latin square, $m_{ij}=1$ for all $i \neq j$, while for a Williams' design for even t formed from two squares, $m_{ij}=2$, $i \neq j$, where these values are found by combining either 0 or 2 from the first square and, correspondingly, 2 or 0 from the second square. For example, for the Williams' design for t=5 given above, the incidence matrices M_1 and M_2 for the two squares have first rows (0 2 0 2) and (2 0 2 0), omitting m_{11} which is always 0. These may be combined to give M with first row (2 2 2 2).

For the class of cross-over designs with t=p, Hedayat and Afsarinejad (1978) have shown that a cross-over design is universally optimal for estimating direct and first-order carry-over treatment effects, using the model described above, if M is an integer multiple of J-I, where J is the txt matrix with every element equal to 1 and I is the txt identity matrix. Thus sequentially counterbalanced designs for n=kt subjects, formed by combining k addition Latin squares so that the resulting matrix M=k(J-I), will be universally optimal.

3 A complete investigation for t=5

In this section all addition Latin squares for five treatments are investigated to see whether any patterns can be found which will lead to general results for any odd t. Since in practice subjects would be allocated to treatment sequences (rows) at random, the order of the rows within a square is not important and we need consider only the 24 possible squares with first element 0. The full set of squares and their corresponding incidence matrices are shown in Table 1, from which it may be seen that the 24 squares consist of 16 nearly-balanced squares, four of the kind used to form Williams' designs, while the remaining four form the complete set of orthogonal squares. Note that not all arrangements of (2 1 1 0) occur in this list and that only for selected squares can two or more be combined to produce a design with M of the required form. The pairs of squares which may be combined are shown in Table 2.

Table 1 The incidence matrices for all addition squares for t = 5.

Square S_i	Matrix M_i	Type of square
(0 1 2 3 4)	(4 0 0 0)	orthogonal O_1
(0 1 2 4 3)	(2 1 0 1)	nearly-balanced
(0 1 3 2 4)	(1 2 0 1)	nearly-balanced
(0 1 3 4 2)	(2 1 1 0)	nearly-balanced
(0 1 4 2 3)	(2 0 2 0)	Williams W_2
(0 1 4 3 2)	(1 0 1 2)	nearly-balanced
(0 2 1 3 4)	(1 2 0 1)	nearly-balanced
(0 2 1 4 3)	(0 1 1 2)	nearly-balanced
(0 2 3 1 4)	(1 1 2 0)	nearly-balanced
(0 2 3 4 1)	(2 2 0 0)	Williams W_3
(0 2 4 1 3)	(0 4 0 0)	orthogonal O_2
(0 2 4 3 1)	(0 2 1 1)	nearly-balanced
(0 3 1 2 4)	(1 1 2 0)	nearly-balanced
(0 3 1 4 2)	(0 0 4 0)	orthogonal O_3
(0 3 2 1 4)	(0 0 2 2)	Williams W_4
(0 3 2 4 1)	(0 2 1 1)	nearly-balanced
(0 3 4 1 2)	(2 1 1 0)	Russell's nearly-balanced R
(0 3 4 2 1)	(1 0 2 1)	nearly-balanced
(0 4 1 2 3)	(2 1 0 1)	nearly-balanced
(0 4 1 3 2)	(0 2 0 2)	Williams W_1
(0 4 2 1 3)	(0 1 1 2)	Russell's complementary $\bar{R}$
(0 4 2 3 1)	(1 0 2 1)	nearly-balanced
(0 4 3 1 2)	(1 0 1 2)	nearly-balanced
(0 4 3 2 1)	(0 0 0 4)	orthogonal O_4

For groups 1 and 2 in Table 2, either of the squares shown as square 1 may be combined with either of those shown as square 2. There are 10 pairs of squares which will produce a sequentially counterbalanced design with M=2(J-I), eight involving nearly-balanced squares and two Williams' designs. These are not all distinct in that some of these designs may be transformed into other designs by a permutation of the treatment labels and a rearrangement of rows. For example, the two Williams' designs in groups 3 and 4 in Table 2 are not distinct and neither are the left and right pairs as shown in groups 1 and 2 of this table.

Definition 4. A pair of addition Latin squares is called a *reversible complementary pair* if the second square may be obtained by reversing the order of the treatments in the first row of the first square and then rearranging its rows.

Consider the squares (0 1 3 4 2) and (0 2 1 4 3) in group 1. These form a reversible complementary pair since reversing the order of the treatment labels in square 1 gives (2 4 3 1 0), from which (0 2 1 4 3) is obtained by row re-ordering. Russell's squares (0 3 4 1 2) and (0 4 2 1 3) also form a reversible complementary pair as do the other four pairs shown in Table 2.

Similarly, sets of three squares may be used to form sequentially counterbalanced designs with n=3t subjects. An example of one of these, for t=5, is (0 4 1 2 3) with (0 4 2 1 3) and (0 2 3 1 4). Table 3 shows all other combinations of three squares which would result in an incidence matrix M=3(J-I). There are 48 ways of combining sets of three squares to produce a sequentially counterbalanced design with 15 subjects. Those in the first four groups consist of three nearly-balanced squares, while the other groups include a square of the Williams' type. Some of these may be transformed into other designs by label permutation and reordering of rows.

Designs using four squares with n=20 may be formed either by combining two sequentially counterbalanced designs based on nearly-balanced squares or the Williams' designs, or by using the four orthogonal Latin squares.

4 Designs for t=7 treatments

A simple computer search algorithm was employed for t=7 using the 720 Latin squares with first column in standard order. The search revealed that 180 of these are nearly-balanced and that there are either 4, 6 or 8 squares with each of the 30 different arrangements of the elements in their incidence matrices. These may be combined in 15 pairs to give M=2(J-I) in either 4x4, 4x6 or 8x8 ways resulting in 676 sequentially counterbalanced designs formed from two nearly-balanced squares. An example of one group of such squares is shown below where any S_1 may be taken with any S_2. The pairs of squares in the order shown are reversible complementary pairs including R and $\bar{R}$.

S_1 (0 1 3 6 4 5 2) (0 1 5 3 4 6 2) (0 3 5 6 4 1 2) (0 4 5 3 6 1 2)R
S_2 (0 3 2 4 1 6 5) (0 4 2 1 3 6 5) (0 6 2 4 3 1 5) (0 6 4 1 3 2 5)$\bar{R}$

There are many squares of the Williams' type, with incidence matrices containing either 0's or 2's, falling into nine different pairing groups. Three of these groups each consist of six reversible pairs leading to 36 possible pairs, while the remaining six groups each consist of three pairs which are not reversible. The total number of designs of this form is 126 but these are not necessarily distinct.

Sequentially counterbalanced designs for n=21 subjects, formed from three nearly-balanced squares may be found in large numbers. For example, there are 384 designs using three squares with individual incidence matrices (2 1 1 1 0 1), (0 1 2 1 1 1) and (1 1 0 1 2 1), such as (0 1 2 5 4 6 3), (0 3 1 4 6 5 2) and (0 1 6 4 3 5 2).

Table 2. Pairs of squares forming sequentially counterbalanced designs.

Group	Square			Matrix M_i	Number
1	S_1	(0 1 3 4 2)	(0 3 4 1 2) R	(2 1 1 0)	4
	S_2	(0 2 1 4 3)	(0 4 2 1 3) $\bar{R}$	(0 1 1 2)	
2	S_1	(0 1 3 2 4)	(0 2 1 3 4)	(1 2 0 1)	4
	S_2	(0 3 4 2 1)	(0 4 2 3 1)	(1 0 2 1)	
3	S_1	(0 4 1 3 2) W_1		(0 2 0 2)	1
	S_2	(0 1 4 2 3) W_2		(2 0 2 0)	
4	S_1	(0 2 3 4 1) W_3		(2 2 0 0)	1
	S_2	(0 3 2 1 4) W_4		(0 0 2 2)	

Table 3. Three squares forming sequentially counterbalanced designs.

Group	Square			Matrix M_i	Number
1	S_1	(0 1 2 4 3)	(0 4 1 2 3)	(2 1 0 1)	8
	S_2	(0 2 1 4 3)	(0 4 2 1 3) $\bar{R}$	(0 1 1 2)	
	S_3	(0 3 1 2 4)	(0 2 3 1 4)	(1 1 2 0)	
2	S_1	(0 1 2 4 3)	(0 4 1 2 3)	(2 1 0 1)	8
	S_2	(0 3 4 2 1)	(0 4 2 3 1)	(1 0 2 1)	
	S_3	(0 2 4 3 1)	(0 3 2 4 1)	(0 2 1 1)	
3	S_1	(0 1 3 2 4)	(0 2 1 3 4)	(1 2 0 1)	8
	S_2	(0 1 4 3 2)	(0 4 3 1 2)	(1 0 1 2)	
	S_3	(0 2 3 1 4)	(0 3 1 2 4)	(1 1 2 0)	
4	S_1	(0 1 3 4 2)	(0 3 4 1 2) R	(2 1 1 0)	8
	S_2	(0 1 4 3 2)	(0 4 3 1 2)	(1 0 1 2)	
	S_3	(0 2. 4 3 1)	(0 3 2 4 1)	(0 2 1 1)	
5	S_1	(0 1 3 4 2)	(0 3 4 1 2)	(2 1 1 0)	4
	S_2	(0 3 4 2 1)	(0 4 2 3 1)	(1 0 2 1)	
	S_3	(0 4 1 3 2) W_1		(0 2 0 2)	
6	S_1	(0 2 1 4 3)	(0 4 2 1 3)	(0 1 1 2)	4
	S_2	(0 1 3 2 4)	(0 2 1 3 4)	(1 2 0 1)	
	S_3	(0 1 4 2 3) W_2		(2 0 2 0)	
7	S_1	(0 3 4 2 1)	(0 4 2 3 1)	(1 0 2 1)	4
	S_2	(0 2 1 4 3)	(0 4 2 1 3)	(0 1 1 2)	
	S_3	(0 2 3 4 1) W_3		(2 2 0 0)	
8	S_1	(0 1 3 2 4)	(0 2 1 3 4)	(1 2 0 1)	4
	S_2	(0 1 3 4 2)	(0 3 4 1 2)	(2 1 1 0)	
	S_3	(0 3 2 1 4) W_4		(0 0 2 2)	

5 General number of treatments t

Russell's (1991) method of constructing a nearly-balanced square for general t defines $m=(t+1)/2$ and $q=[(m+1)/2]$, where $[x]$ is the greatest integer less than or equal to x. Then the first row of Russell's square has odd–numbered elements $0,m+1,m+2,...,t-1,q$, and even–numbered elements $m,m-1,...,q+1$, $q-1,...,1$. This method gives the following squares for t=5, 7, 9 and 11:

t=5	(0 3 4 1 2)	M = (2 1 1 0)
t=7	(0 4 5 3 6 1 2)	M = (2 1 1 1 1 0)
t=9	(0 5 6 4 7 2 8 1 3)	M = (1 2 1 1 1 1 1 0)
t=11	(0 6 7 5 8 4 9 10 1 3)	M = (1 2 1 1 1 1 1 1 1 0)

The squares for t=5 and 7 may be reversed to form a complementary pair but those for t=9 and 11 cannot.

Rather than use a generating method to find a nearly-balanced square and then a computer search to find a complementary square, we have used a computer search to find a square with M=(2 1 1...0) and then a complementary square may be found be reversing the order of the treatment labels and re–ordering rows. Details for t=5, 7, 9 and 11 are given in Table 4.

Table 4. Reversible pairs of complementary squares for t=5, 7, 9 and 11.

t	No. of squares	No. with M=(211..0)	Reversible pair of complementary squares	
5	24	2	(0 1 3 4 2)	(0 2 1 4 3)
7	720	4	(0 1 3 6 4 5 2)	(0 3 2 4 1 6 5)
9	40320	32	(0 1 3 7 4 5 8 6 2)	(0 4 6 3 2 5 1 8 7)
11	3628800	600	(0 1 3 4 8 5 10 6 9 7 2)	(0 5 7 4 8 3 6 2 1 10 9)

References

Hedayat, A. and Afsarinejad, K. (1975) Repeated measurements designs, I. In A survey of Statistical Design and Linear Models, (Ed. J.N. Strivastava) 229–242, North-Holland, Amsterdam.

Hedayat, A. and Afsarinejad, K. (1978) Repeated measurements designs, II. Annals of Statistics, 6, 619–628.

Isaac, P.D. and Dean, A.M. (1993) Sequentially counterbalanced Latin squares. Technical Report No 512, The Ohio State University, Columbus.

John, P.W.M. (1971) Statistical Design and Analysis of Experiments. New York, Macmillan.

Jones, B and Kenward, M.G. (1989) Design and Analysis of Cross-Over Trials. London, Chapman and Hall.

Newcombe, R.G. (1992) Latin square designs for crossover studies balanced for carryover effects. Statistics in Medicine, 11, 560.

Russell, K.G. (1991) The construction of good change-over designs when there are fewer units than treatments. Biometrika, 78, 305–313.

Williams, E.J. (1949) Experimental designs balanced for the estimation of residual effects of treatments. Aust. J. Sci. Res., A2, 149–68

Part XXIII

Model Fitting

Miscellaneous Procedures Connected with the Change Point Problem

Marie Hušková

Department of Statistics, Charles University

Abstract. The problem to detect and to identify change(s) in statistical models arises in many applications (economic modelling, quality control, biology, medicine, meteorology and ecology among others). It is known as the change point problem, disorder problem or testing the constancy of regression relationship over time. Procedures for changes in univariate and multivariate location models (Section 2, 3) and in linear regression models (Section 4) are presented and their applications are discussed.

Keywords. Change point problem, testing the constancy of the regression relationship over a time

1 Introduction

The problem to detect and to identify changes in statistical models has attracted a host of researchers. They have developed new models incorporating the change(s). Using various principles they have proposed various statistical procedures that are sensitive w.r.t. detection of changes, have studied their properties and, finally, have tried to apply them to real data sets.

Since the list of references is quite extensive we do not bring here the full list of them; we mention only survey papers and books. Zacks (1983) surveyed classical and Bayesian type procedures. Papers by Csörgö and Horváth (1988), Wolfe and Schechtman (1984), Hušková and Sen (1989) concern nonparametric procedures. Miscellaneous types of procedures are described, e.g., in Krishnaiah and Miao (1988), in Antoch and Hušková (1992) and in Deshayes and Picard (1986). Information on sequential procedures is in Zacks (1991). The paper by Brown, Durbin and Evans (1975) is well-known. Concerning the books devoted to this problem: Broemling and Tsurumi (1986) wrote on the Bayesian type procedures, while Hackl (1980) wrote on the recursive ones. Hackl (1989) and Hackl and Westlund (1991) concentrated on the structural changes in econometric models. Recently, the book by Brodsky and Darkhovsky (1993) on nonparametric procedures has appeared.

The objective of the paper is to present procedures for the detection of change(s) in one- and multi- dimensional (Section 2 and 3) and in regression models (Section 4). We focus on the case of independent observations, while in modifications for the case of dependent observations are outlined. The procedures are illustrated on some real data sets and experience with applications is discussed.

As far as it is known to the author the software for procedures connected with the change point problem is not available in any of the currently available statistical packages. However, it can be seen from the description of the procedures that implementation would be relatively simple. Of course, the time needed for the programming will be shorter in the systems which offer basic building tools for the construction of macros like, e.g., Matlab, $S+$ or ISP. These tools should cover not only basic built-in functions, computation of quantiles and classical least squares estimators, but should also appreciate minimization procedures, simple looping facilities, matrix operations etc. Statistical packages like $BMDP$, SAS or $SPSS$ are useful for this purpose. While the methods can be relatively easy implemented by the SAS due to its powerful programming facilities and matrix language, much more time and tricks will be necessary to prepare an analogous imput file for $BMDP$ or an include file for $SPSS$. Algorithms for some procedures can be found in Antoch and Hušková (1992)

2 Change in the mean in location model

The following setup is assumed: $X_1, ... X_n$ are independent observations with mean $\mu_i = EX_i$, $i = 1, ..., n$, common variance σ^2 and the testing problem

$$H_0 : \mu_1 = ... = \mu_n$$

versus

$$H_1 : \mu_1 = ... = \mu_m \neq \mu_{m+1} = ... = \mu_n,$$

where μ_1, μ_n and $m(> 1, < n)$ are unknown, is considered. If the null hypothesis is rejected the problem to find reasonable estimators of the change point m and of the magnitude of the change $\mu_n - \mu_1$ is usually of interest.

If m is known the problem reduces to the classical two-sample problem. In our case, $H_1 = \bigcup_{k=1}^{n-1} H_{1k}$, where $H_{1k} : \mu_1 = ... = \mu_k \neq \mu_{k+1} = ... = \mu_n$, and the testing problem H_0 against H_{1k} (k fixed) is the two-sample problem. The following procedure for testing H_0 against H_1 may serve as a hint: reject H_0 if H_0 is rejected for at least one of the testing problem H_0 against H_{1k}, $k = 1, ..., n - 1$. The test procedure based on T_{n1} introduced below is of this type. It should be noticed that if the level of the test H_0 against H_{1k} is α_k, $k = 1, ..., n - 1$, then for the level α of the resulting procedure the inequality $\max_{1 \leq k < n} \alpha_k < \alpha \leq \sum_{k=1}^{n-1} \alpha_k$ holds true.

As an illustrative example the well known data set – the anual discharges of the Nile river at Aswan in the time period 1871-1970 is used (see Figure 1). These data were examined for a change in the model by many authors. Let us mention Cobb (1978), Hinkley and Schechtman (1987) and recently MacNeil et al. (1991) who gave probably the most sophisticated analysis. In terms of the Nile data we want to test whether the "mean" annual discharges are constant during the whole observational period (100 years) or whether it changes at a certain point(s) and if we detect a change we are interested in estimating when it happened and how big the change is.

The procedures (both for testing and estimation) are typically based on the partial sums

$$S_k = \frac{1}{\hat{\sigma}_n} \sum_{i=1}^{k} (X_i - \overline{X}_n), \quad k = 1, ..., n,$$

where $\hat{\sigma}_n^2$ is an estimator of the variance σ^2 and $\overline{X}_n = \sum_{i=1}^{n} X_i/n$.

The test statistic

$$(1) \qquad T_{n1} = \max_{1 \le k < n} \left\{ \sqrt{\frac{n}{k(n-k)}} |S_k| \right\}$$

is related to the maximum likelihood ratio test statistics when the observations are normally distributed (with σ^2 known). It puts larger weights on S_k with k "small" or "large" (close to n) and it is more sensitive to changes at the beginning or at the end of the observational period than in the middle. Therefore modifications were proposed. James et al. (1987) studied a "trimmed" likelihood ratio type statistic

$$(2) \qquad T_{n2}(a_{n1}, a_{n2}) = \max_{na_{n1} \le k \le n - na_{n2}} \left\{ \sqrt{\frac{n}{k(n-k)}} |S_k| \right\},$$

where $a_{n1} > 0$, $a_{n2} > 0$ are tending to zero and na_{n1} and na_{n2} tend to infinity (e.g., one can put $a_{n1} = a_{n2} = 0.1$). Another modification consists in assigning the same weights to all S_k:

$$(3) \qquad T_{n3} = \max_{1 \le k \le n} \left\{ \frac{1}{\sqrt{n}} |S_k| \right\}.$$

The Bayesian approach leads to the test statistics:

$$(4) \qquad T_{n4} = \frac{1}{n} \sum_{k=1}^{n-1} S_k^2 p_k,$$

where $p_i \ge 0$, $i = 1, ..., n$, ($\sum_{k=1}^{n} p_k = 1$) express the prior probability that the change occurs between the ith and the $(i+1)$st observation. Finally, the nonrecursive MOSUM (moving sum) type test statistic

$$(5) \qquad T_{n5}(G) = \max_{G+1 \le k \le n} \left\{ \frac{1}{\sqrt{G}} |S_k - S_{k-G}| \right\},$$

where G is a suitably chosen number (see Remark 4 below) is sensitive even to multiple changes; for more information see Brodsky and Darkhovsky (1993) and Antoch and Hušková (1994).

Large values of these test statistics indicate that H_0 fails. The exact critical values are not available, only certain approximations based on the large sample behavior. Approximations to the critical values corresponding to T_{n1}, T_{n2} and T_{n3} are in Table 2 at the end of Section 3 (M_{nj} with $p = 1$ corresponds to $T_{nj}, j = 1, 2, 3$). The 5% critical value (approximation) for T_{n4} with uniform prior ($p_i = n^{-1}$, $i = 1, ..., n$) is 0.641. Finally, if T_{n5} exceeds

$$\sqrt{2 \log \frac{n}{G}} + \frac{\log \log \frac{n}{G} + \log \frac{4}{\pi} - 2 \log \log(1 - \alpha)^{-1/2}}{\sqrt{8 \log \frac{n}{G}}},$$

the null hypothesis is rejected (MacNeil (1978) and MacNeil et al. (1991)). Selected critical values are in Table 1.

$G/n = 0.1$	$G/n = 0.15$	$G/n = 0.2$
3.447	3.431	3.414

Table 1. 5% critical values for the tests based on T_{n5}

Worsley (1983) proposed critical values for T_{n1} based on the improved Bonferroni inequality which leads to an asymptotically conservative test, however, it seems to work well for moderate n. For further information see Jarušková and Antoch (1994).

Since under H_0 the test statistics T_{n1} and T_{n2} are asymptotically independent we can introduce a new test (Horváth (1993b)) that rejects H_0 on the (asymptotic) level $1 - (1 - \alpha_1)(1 - \alpha_2)$ if it is rejected by at least one the tests based on T_{n1} or T_{n2} on the (asymptotic) levels α_1 and α_2, respectively. Since the test based on T_{n1} is more sensitive to a change at the beginning and at the end of the observational period and T_{n2} is in the middle part, the newly proposed test is sensitive to a change at any part of the observational period.

The change point m can be estimated in one of the following ways:

$$(6) \qquad \hat{m}_1 = \operatorname{argmax}\{|S_k|, k = 1, ..., n\},$$

$$(7) \qquad \hat{m}_2 = \operatorname{argmax}\left\{|S_k| \frac{n}{\sqrt{k(n - k)}}, k = 1, ..., n - 1\right\},$$

$$(8) \qquad \hat{m}_3(G) = \operatorname{argmax}\{|S_{k+G} - 2S_k + S_{k-G}|, k = G + 1, ..., n - G\},$$

where $1 \le G < n$; for the choice of G see Remark 4 below.

The statistic $\overline{X}_{\hat{m}}^* - \overline{X}_{\hat{m}}$, where $\hat{m}$ is either of the above estimators of the change point m and $\overline{X}_{\hat{m}}^* = \sum_{i=\hat{m}+1}^{n} X_i / n$, provides a reasonable estimator of the magnitude of the change $\mu_n - \mu_1$.

Applying the above procedures to the Nile data we obtain $T_{n1} = T_{n2} = 8.59$, $T_{n3} = 3.91$, $T_{n4} = 9.1$, $T_{n5} = 5.45$, with uniform prior in T_{n4}, which means that all values are significant on the level 0.05 (for the critical values see Table 2 with $p = 1$ and Table 1 with $G/n = 0.15$). Hence using any of the tests we reject the null hypothesis. Concerning estimators of the change point we obtain $\hat{m}_1 = \hat{m}_2 = \hat{m}_3 = 28$. The partial sums $S_k = S - k(+)$, $k =$

1, ..., n, and the 5%-critical functions corresponding to T_{n1}, T_{n2} and T_{n3} are plotted in Figure 2 (the upper and lower quadratic functions correspond to T_{n1} and T_{n2}, repectively, and the horisontal line relates to T_{n3}. Figure 3 shows $S - k(-) = S_k \hat{\sigma}_n / \tilde{\sigma}_n$, $k = 1, ..., n$ and the comparison with Figure 2 gives the evidence of strong influence of the used estimator of σ^2. Figure 4 shows $|S_k - S_{k-15}|(15)^{-1/2}$, $k = 1, ..., n$. All these figures indicate that there is a change point located near $k = 28$.

Remarks and comments

1. All the procedures introduced above (based on S_k's) were originally developed for normally distributed observations, however, it appears that if the number n of observations is "very" large (considerably larger than one needs in order for the central limit theorem to give a reasonable approximation for the distribution of the sample mean) these procedures work reasonably well even when the normality fails. Asymptotically best max-type procedures are related to the maximum likelihood ratio which, unfortunately, depends on the distribution function F of $(X_i - E X_i)/\sigma$. these procedures were studied by Gombay and Horváth (1993).

2. The limit distributions of T_{nj}, $j = 1, ..., 5$, under H_0 are asymptotically distribution free and coincide with the distributions of functionals of the Brownian bridge. For results see James et al. (1987), Jandhyala and MacNeil (1989, 1991, 1992), Gombay and Horváth (1994a, 1994b), Hušková (1993) and Antoch and Hušková (1994) and papers referenced there. Sophisticated comparisons (based on asymptotics) of the max-type and sum-type test procedures were published by Praagman (1988).

3. Simulations contained in the mentioned papers and extensive simulation studies made by Antoch show that none of the tests is uniformly better (more sensitive w.r.t. the change) than others. It appears that the procedures based on T_{n2}, T_{n3} or T_{n4} perform better if the change occurs in the middle of the observational period while T_{n1} is more sensitive to the change at the beginning or at the end. The test statistic T_{n5} is more sensitive in case of more than one change (Antoch and Hušková (1994)).

4. The number G of the summands in the moving sums in T_{n5} should be small w.r.t. the number of observations n, e.g. $G \in [0.1n, 0.2n]$. More precisely, $G \to \infty$, $G/n \to 0$ as $n \to \infty$ (Antoch and Hušková (1994)).

5. A recommended estimator $\hat{\sigma}_n^2$ of σ^2 is of the form:

$$\hat{\sigma}_n^2 = \min_{1 \le k < n} \left\{ \frac{1}{n-2} \left(\sum_{i=1}^{k} (X_i - \overline{X}_k)^2 + \sum_{i=1+k}^{n} (X_i - \overline{X}_k^*)^2 \right), \, k = 1, ..., n-1 \right\},$$

where $\overline{X}_k^* = \sum_{i=k+1}^{n} X_i/(n-k)$, $k = 1, ..., n-1$. Under H_1 the estimator

$$\tilde{\sigma}_n^2 = \frac{1}{n-1} \sum_{i=1}^{n} (X_i - \overline{X}_n)^2$$

usually overestimates σ^2 which may cause H_0 not to be rejected when H_0 is false. In case of the Nile data $\hat{\sigma} = 127.64$, $\tilde{\sigma}_n = 169.19$ and their ratio is $\tilde{\sigma}_n/\hat{\sigma}_n = 1.33$, hence the test statistics are higly effected by the estimator of σ used.

6. Recursive procedures based on the recursive residuals $X_i - \overline{X}_{i-1}$, $i = 2, ..., n$, have been extensively studied, e.g., in Durbin et al. (1975), Hackl (1980) and their robust version by Hušková (1990). The related procedures are known as CUSUM (cumulative sum) and MOSUM (moving sum) procedures.

7. Robust versions of the above procedures can be obtained by replaceing the sample mean $\widehat{X}_n$ and the residuals $X_i - \widehat{X}_n$, $i = 1, ..., n$ by the M-estimators $\widehat{\theta}_n(\psi)$ and by the M-residuals $\psi(X_i - \widehat{\theta}_n(\psi))$, respectively, where ψ is a score generating function. The estimator $\widehat{\sigma}_n^2$ must be replaced by an appropriate one. A survey of the results on such procedures is in Hušková and Sen P.K. (1989) and in Antoch and Hušková (1992).

8. Rank based procedures can be obtained replacing $X_i - \overline{X}_n, i = 1, ..., n$, by $a(R_i)$, $i = 1, ..., n$, where $a(1), ..., a(n)$ are scores and R_i is the rank of X_i among $X_1, ..., X_n$. Also, the estimator $\widehat{\sigma}_n^2$ must be replaced by $\sum_{i=1}^{n}(a(i) - \bar{a})^2/(n-1)$ with $\bar{a} = \sum_{i=1}^{n} a(i)/n$. More information is in Hušková and Sen (1989). The Kolmogorov-Smirnov type tests were also introduced.

9. The procedures considered above can be applied even if the observations are dependent (e.g., form an autoregressive sequence). Of course, a different estimator of σ^2 must be used. More information is in Brodsky and Darkhovsky (1993) and Tang and MacNeil (1993).

10. A long series of simulations made by Antoch show that plotting $\{|S_k|, k = 1, ..., n\}$, $\{|S_k|\frac{n}{\sqrt{k(n-k)}}, k = 1, ..., n-1\}$ and $\{|S_k - S_{k-G}|, k = G+1, ..., n\}$ is extremely useful and that in the final decision these plots together with the results of the tests should be used.

11. The formulation of H_1 admits just one abrupt change. The situation with more than one abrupt change can be met in applications also. The procedures for this situation were treated, e.g., by Lombard (1987), Brodsky and Darkhovsky (1993), Antoch and Hušková (1994). It appears that T_{n5},

$$\max_{G<k<n-G}\{|S_{k+G} - 2S_k + S_{k+G}\}$$

and

$$\max_{1<k<j<n}\{|S_k - S_j|\}$$

are sensitive w.r.t. this type of change. For information on procedures on other types of changes (e.g. gradual) see Brodsky and Darkhovsky (1993) and Antoch and Hušková (1994).·

3 Multivariate location case

In this section we briefly mention procedures for the detection of a change in the mean in location models when the data are multidimensional. We consider the setup: $X_1, ..., X_n$ are independent p-dimensional vectors with the mean vectors $EX_i = \mu_i$, $i = 1, ..., n$ and common nondegenerate variance matrix Σ usually unknown and the problem of interest is to test:

$$H_0 : \mu_1 = ... = \mu_n$$

against

$$H_1 : \mu_1 = ... = \mu_m \neq \mu_{m+1} = ... = \mu_n$$

where μ_1, μ_n and m ($> 1, < n$) are unknown. If the null hypothesis H_0 is rejected then an estimation of the change point m and the magnitude of the change $\mu_n - \mu_1$ is requested.

Quite analogously to the one dimensional case the test procedures are usually based on the vector of the partial sums

$$S_k = \Sigma_n^{-1/2} \sum_{i=1}^{k} (X_i - \overline{X}_n), \quad k = 1, ...n - 1$$

where $\overline{X}_n = \sum_{i=1}^{n} X_i/n$ and Σ_n is an estimator of the variance matrix Σ (a recommendation is at the end of the section). As a test statistics maximum of weighted quadratic forms of the partial sums are mostly used, e.g.,

(9) $$M_{n1} = \max_{1 \leq k < n} \left\{ \sqrt{S_k^T S_k \frac{n}{k(n-k)}} \right\},$$

(10) $$M_{n2}(a_{n1}; a_{n2}) = \max_{a_{n1} \leq k < n - a_{n2}} \left\{ \sqrt{S_k^T S_k \frac{n}{k(n-k)}} \right\},$$

(11) $$M_{n3} = \max_{1 \leq k < n} \left\{ \sqrt{S_k^T S_k/n} \right\},$$

where a_{n1} and a_{n2} have the same properties as in the univariate case.

Most of the remarks and the recommendations formulated in Section 1 take place here, too. Critical values based on approximations for n large can be deduced from the papers by James, James and Siegmund (1987), Horváth (1993a) or the papers referred to there. The critical value (asymptotic level α) for M_{n1} is:

$$\sqrt{2 \log \log n} + \left(\frac{p}{2} \log \log \log n - \log \Gamma(p/2) - \log \log (1-\alpha)^{-1/2} \right) (2 \log \log n)^{-1/2},$$

where $\Gamma(p/2)$ is the gamma function. The critical values for $M_{n3}(a_{n1}, a_{n2})$ can be obtained as the solution (w.r.t. b) of the equation:

$$\frac{b^p \exp\{\frac{-b^2}{2}\}}{2^{\frac{(p-2)}{2}} \Gamma(\frac{p}{2})} \left\{ \frac{2}{b^2} + \frac{1}{2} \left(1 - \frac{p}{b^2} \right) \ln \frac{(1-a_{n1})(1-a_{n2})}{a_{n1} a_{n2}} \right\} = \alpha.$$

The 5% critical values corresponding to M_{n1} for $n = 100$, $M_{n2}(0.1; 0.1)$ and M_{n3} are in the Table 2.

p	M_{n1}	$M_{n2}(0.1;0.1)$	M_{n3}
1	3.64	3.05	1.358
2	4.09	3.52	1.584
3	4.28	3.87	1.748
4	4.33	4.14	1.882

Table 2. 5% critical values for tests based on M_{nj}, $j = 1, 2, 3$.

Since under H_0 the test statistics M_{n1} and $M_{n2}(a_{n1}; a_{n2})$ for a_{n1} and a_{n2} positive close to zero are asymptotically independent and M_{n1} is more sensitive to the change at the begining or at the end of the observational period and $M_{n2}(a_{n1}; a_{n2})$ we construct a new test along the line of Section 1. This test rejects H_0 on (the asymptotic) level $1 - (1 - \alpha_1)(1 - \alpha_2)$ if it is rejected by at least one of the tests based on M_{n1} or M_{n2} on the asymptotic levels α_1 and α_2, respectively. The resulting test is then sensitive to a change in any part of the observational period.

The recommended estimator of the variance matrix Σ is

$$\hat{\Sigma}_n = \frac{1}{n-2}\left\{\sum_{i=1}^{\hat{m}}(X_i - \bar{X}_{\hat{m}})(X_i - \bar{X}_{\hat{m}})^T + \sum_{i=\hat{m}+1}^{n}(X_i - \bar{X}_{\hat{m}}^*)(X_i - \bar{X}_{\hat{m}}^*)^T\right\},$$

where

$$\hat{m} = \operatorname{argmin}\left\{\sum_{i=1}^{k}(X_i - \bar{X}_k)^T(X_i - \bar{X}_k) + \sum_{i=k+1}^{n}(X_i - \bar{X}_k^*)^T(X_i - \bar{X}_k^*),\right.$$
$$\left. k = 1, ..., n-1\right\}$$

with $\bar{X}_k^* = \sum_{i=1+k}^{n} X_i/(n-k)$, $k = 1, ..., n-1$. This $\hat{m}$ can be used as an estimator of m.

The real data sets corresponding to this situation are in the paper by Srivastava and Worsley (1986).

4 Changes in regression models

Standard formulation of the problem assumes that the observations $X_1, ..., X_n$ follow the model:

$$X_i = c_i^T \theta_i + e_i, \quad i = 1, ..., n,$$

where $c_i = (c_{i1}, ..., c_{ip})^T$, $c_{i1} = 1$, $i = 1, ..., n$, are known vectors, $\theta_1, ..., \theta_n$ are unknown regression parameters, $e_1, ..., e_n$ are i.i.d. errors with mean zero and nonzero finite variance $var\, e_i = \sigma^2$ and the problem of prime interest is to test

$$H_0 : \theta_1 = ... = \theta_n$$

against
$$H_1 : \boldsymbol{\theta}_1 = ... = \boldsymbol{\theta}_m \neq \boldsymbol{\theta}_{m+1} = ... = \boldsymbol{\theta}_n$$

where $m(< n)$, $\boldsymbol{\theta}_1$ and $\boldsymbol{\theta}_n$ are unknown. Again as in the previous sections if the null hypothesis is rejected the problem of estimating the change point m and the magnitude of the change $\mu_n - \mu_1$ arises.

Parallel to the location case the tests are usually based either on the vector of the partial sums

$$\boldsymbol{S}_k = (S_{k1}, ..., S_{kp})^T = \frac{1}{\widehat{\sigma}_n} \boldsymbol{C}_n^{-1/2} \sum_{i=1}^{k} \boldsymbol{c}_i (X_i - \boldsymbol{c}_i^T \bar{\boldsymbol{\theta}}_n), \quad k = 1, ..., n,$$

or on the partial sums of residuals

$$S_k^* = \frac{1}{\widehat{\sigma}_n} \sum_{i=1}^{k} (X_i - \boldsymbol{c}_i^T \bar{\boldsymbol{\theta}}_n), \quad k = 1, ..., n,$$

where $\boldsymbol{C}_k = \sum_{i=1}^{k} \boldsymbol{c}_i \boldsymbol{c}_i^T$, $k = 1, ..., n$, and $\bar{\boldsymbol{\theta}}_n$ is the least squares estimator of the regression vector parameter based on $X_1, ... X_n$, i.e., $\bar{\boldsymbol{\theta}}_n = \boldsymbol{C}_n^{-1} \sum_{i=1}^{n} \boldsymbol{c}_i X_i$. The estimator $\widehat{\sigma}_n^2$ of σ^2 (analogously to the location case) is usually of the form:

$$\widehat{\sigma}_n^2 = \frac{1}{n-p} \min_{1 \le k < n} \left\{ \sum_{i=1}^{k} (X_i - \boldsymbol{c}_i^T \bar{\boldsymbol{\theta}}_k)^2 + \sum_{i=k+1}^{n} (X_i - \boldsymbol{c}_i^T \vec{\boldsymbol{\theta}}_k)^2 \right\},$$

where $\bar{\boldsymbol{\theta}}_k$ and $\vec{\boldsymbol{\theta}}_k$ are the least squares estimators of the regression parameters based on $X_1, ..., X_k$ and $X_{k+1}, ..., X_n$, respectively, i.e.,

(12) $$\bar{\boldsymbol{\theta}}_k = \boldsymbol{C}_k^{-1} \sum_{i=1}^{k} \boldsymbol{c}_i X_i. \quad \text{and} \quad \vec{\boldsymbol{\theta}}_k = \boldsymbol{C}_k^{*-1} \sum_{i=k+1}^{n} \boldsymbol{c}_i X_i.$$

with $\boldsymbol{C}_k^* = \boldsymbol{C}_n - \boldsymbol{C}_k$. Motivated by the location case (Section 2, 3) and the maximum likelihood ratio type test procedure when the observations are normally distributed one can develop several classes of tests based either on the partial sums $\boldsymbol{S}_k$, $k = 1, ... n$, or on the partial sums of residuals S_k^*, $k = 1, ..., n$, e.g.,

(13) $$V_{n1} = \max_{1 \le k < n} \left\{ \sqrt{\boldsymbol{S}_k^T \boldsymbol{S}_k \frac{n}{k(n-k)}} \right\},$$

(14) $$V_{n2}(a_{n1}; a_{n2}) = \max_{a_{n1} \le k < n - a_{n2}} \left\{ \sqrt{\boldsymbol{S}_k^T \boldsymbol{S}_k \frac{n}{k(n-k)}} \right\},$$

(15) $$V_{n3} = \max_{1 \le k < n} \{ (\boldsymbol{S}_k^T \boldsymbol{S}_k / n)^{1/2} \},$$

where a_{n1} and a_{n2} have the same properties as in the univariate case. We shall write V_{n1}^*, $V_{n2}^*(a_{n1}, a_{n2})$ and V_{n3}^* when the $\boldsymbol{S}_k$'s in the T_{nj}'s are replaced by S_k^*'s. The test procedures based on the S_k^*'s are computationally more feasible, however, are sensitive only to certain changes. For more information see Ploberger and Krämer (1992).

Generally, under H_0 the limit distributions of the partial sums S_k, $k = 1, ..., n$, and S_k^*, $k = 1, ..., n$, do not depend on the distribution of the error terms e_i's. They depend, however, on the design matrices $(c_1, ..., c_n)^T$ (see Kim and Siegmund (1989) and Jandhyala and MacNeil (1991)) and only for certain design matrices the approximations for the critical values were explicitly calculated.

For a quite large spectrum of the design matrix (C_k and C_k^* for k small should be "smooth"- Horváth (1993a)) the limit distribution (under H_0) of V_{n1} and V_{n1}^* is the same as that of M_{n1} and T_{n1}, respectively.

If the design points $c_1, ..., c_n$ are i.i.d. random variables independent on the error terms and fulfil some more mild conditions (Horvath (1993)) or if the design matrices fulfils

(16)
$$\frac{1}{n} \sum_{i=1}^{[nt]} c_i c_i^T \to tC \quad \text{for } t \in [0, 1],$$

then under H_0 the limit distribution of V_{nj}, $j = 2, 3$ and V_{nj}^*, $j = 2, 3$ coincide with M_{nj}, $j = 2, 3$ and T_{nj}, $j = 2, 3$, respectively. The assumption (16) can be interpreted as: there is no time trend in the design points. This happens e.g. if the design matrix corresponds to a systematic or a purely random repetition of fixed linearly independent design points $c_1^*, ..., c_n^*$.

Kim and Siegmund (1989) showed in case of the polynomial regression with $c_{ij} = (\frac{i}{n})^{j-1}$, $i = 1, ..., n$, $j = 1, 2$, that the critical values corresponding to M_{n2} with $p = 2$ and T_{n2} can be used for the procedure based on V_{n2} and V_{n2}^*, respectively.

MacNeil (1978) and Jandhyala and MacNeil (1991) developed and studied the Bayesian type test statistics of the form

$$V_{n4} = \sum_{k=1}^{n-1} p_k \sum_{j=1}^{p} \left\{ \frac{1}{\sigma \sqrt{n}} \sum_{i=k+1}^{n} c_{ij}(Y_i - c_i^T \bar{\theta}_n) \right\}^2$$

where p_k, $k = 1, ..., n-1$, represent priors. The approximation of 5% critical values for uniform priors and $c_{ij} = (i/n)^{j-1}$ are in Table 3.

p=1	p=2	p=3	p=4
0.461	0.148	0.086	0.06

Table 3. 5% critical values for tests based on $V_{n4}(p)$, $p = 1, ..., 4$

Michels and Trenkler (1990) considered the case when the error terms $(e_1, ..., e_n)$ have normal distribution $N(0, \Sigma)$, where Σ fulfils certain conditions and is unknown otherwise.

The parameters $m, \boldsymbol{\delta}_n = \boldsymbol{\theta}_n - \boldsymbol{\theta}_1$ can be estimated as follows:

$$\hat{m} = \text{argmin} \left\{ \sum_{i=1}^{k} (X_i - \mathbf{c}_i^T \hat{\boldsymbol{\theta}}_k)^2 + \sum_{i=k+1}^{n} (X_i - \mathbf{c}_i^T \vec{\tilde{\boldsymbol{\theta}}}_k)^2, \; k = 1, ..., n \right\}$$

$$\vec{\boldsymbol{\delta}}_n = \vec{\tilde{\boldsymbol{\theta}}}_{\hat{m}} - \bar{\boldsymbol{\theta}}_{\hat{m}},$$

where $\bar{\boldsymbol{\theta}}_{\hat{m}}$ and $\vec{\tilde{\boldsymbol{\theta}}}_{\hat{m}}$ are least squares estimators of the regression parameters based on $X_1, ..., X_{\hat{m}}$ and $X_{\hat{m}+1}, ..., X_n$, respectively.

Remarks and comments

Most of the remarks and comments in Section 2 apply here also. We shall point out only more important items.

1. The recursive procedures are based on the recursive residuals $X_i - \mathbf{c}_i^T \hat{\boldsymbol{\theta}}_{i-1}$, $i = p+1, ..., n$, where $\hat{\boldsymbol{\theta}}_{i-1}$ is defined by (12) and were treated in the same articles and books as in the location case.

2. Robust versions are based on the M-residuals $\psi(X_i - \mathbf{c}_i^T \hat{\boldsymbol{\theta}}_n(\psi))$, $i = 1, ..., n$, where $\hat{\boldsymbol{\theta}}_n(\psi)$ is the M-estimator of the regression parameters.

3. The procedures based on ranks (of $X_i - \mathbf{c}_i^T \hat{\boldsymbol{\theta}}_n(R)$, $i = 1, ..., n$, with $\hat{\boldsymbol{\theta}}_n(R)$ being an estimator of the regression parameter) can be introduced accordingly, however, they are not computationally attractive.

4. The above introduced procedures can be easily modified for the case when the observations are dependent (e.g. forms an autoregressive sequence). The main problem is to substitute $\hat{\sigma}_n$ by a proper estimator related to the scale (Tang and MacNeil (1993)).

5. Similarly, as in the location case one can consider the problem of more than one abrupt changes or some gradual changes. However, here the design matrix plays an important role (Hušková (1993)).

6. Simulations reveal the usefulness of graphs (see the example below).

As an illustrative example data set from Maddala (1977) (see Figure 5,6 and 7) are used. They contain the index of gross domestic product in USA (Y_i), labor input index (L_i) and capital-input index (K_i) between 1929 and 1967. The considered model is:

$$\log Y_i = \theta_{i1} + \theta_{i2} \log L_i + \theta_{i3} \log K_i + e_i, \; i = 1, ..., 39.$$

Applying the introduced procedures to this data set we obtain $\hat{\sigma}_n = 0.025$, $\bar{\sigma}=0.035$, $V_{n1}^* = V_{n2}^*(0.1; 0.1) = 2.89$, $V_{n3}^* = 1.09$, $V_{n1} = V_{n2}(0.1; 0.1) = 5.44$, $V_{n3} = 2.61$. Comparing the values of the test statistics with the corresponding %5 critical values (Table 2 with p=1 for $V_{nj}^*{}'$ and with $p = 3$ for V_{nj}') we find that using V_{nj}' we reject the null hypothesis while using $V_{nj}^*{}'$ we do not. However, Figures 8 and 9 give strong evidence that there is a change in the model.

Acknowledgement. The author wishes to express her sincere thanks to Prof. Antoch for valuable remarks and for preparation of the computational part.

References

Antoch J. and Huškova M. (1989), *Some M-tests for detection of a change in linear models*, Proceedings of the 4th Prague Symposium on Asymptotic Statistics (Huškova M. and Mandl P., eds.), Charles University Press, Praha, pp. 123–136.

Antoch J. and Huškova M. (1992), *Change point problem*, Computational Aspects of model choice (Antoch J., ed.), Physica Verlag, Heideleberg, pp. 11–38.

Antoch J. and Huškova (1994), *Procedures for detection of multiple changes in series of independent observations*, Proceedings of the Fifth Prague Symposium on Asymptotic Statistics (P.Mandl, M.Huškova, eds.), Physica Verlag, Heildelberg, pp. 3–20.

Antoch J., Huškova M. and Veraverbeke N. (1993), *Change-point problem and bootstrap*, submitted.

Basseville M. and Benveniste A. (1986), *Detection of abrupt changes in signals and dynamic systems*, Springer Verlag.

Brodsky B.E. and Darkhovsky B.S. (1993), *Nonprametric Methods in Change-Point Problem*, Kluwer Academic Press.

Broemling L.D. and Tsurumi H. (1987), *Econometrics and Structural change over time (with discussion)*,, Marcel Dekker.

Brown R. L.,Durbin J. and Evans J.M. (1975), *Techniques for testing the constancy of regression relationships over time (with discussion)*, Journal of Royal Statistical Society **B37**, 149–192.

Cobb G.W. (1978), *The problem of Nile; conditional solution to a change point problem*, Biometrika **65**, 243–251.

Csörgő M. and Horváth L.(1987), *Nonparametric tests for the change point problem*, Journal of Statististical Planning and Inference **17**, 1–9.

Csörgő M. and Horváth L. (1988), *Nonparametric methods for the change point problem*, Handbook of Statistics, vol. 7 (P.R. Krishnaiah and C.R. Rao, eds.), J. Wiley, New York, pp. 403–425.

De Long D. (1981), *Crossing probabilities for a square root boundary by a Bessel process*, Communications in Statistics–Theory Methodology **A10**, 2197– 2213.

Deshayes J. and Picard D. (1986), *Off-line statistical analysis of change point models using nonparametric and likelihood methods*, Lecture Notes in Control and Information Sciences (Basseville M. et al., eds.), vol. 77, Physica Verlag, Springer Verlag, pp. 103–168.

Dümbgen L. (1991), *The asymptotic behavior of some nonparametric change-point estimators*, Annals of Statistics **19**, 1471–1475.

Gombay E. and Horváth L. (1994a), *Limit theorems for change in linear regression*, Journal of Multivariate Analysis, to appear.

Gombay E. and Horváth L. (1994b), *An application of the maximum likelihood test to the change-point problem*, Stochast. Proc. Appl., to appear.

Hackl P. (1980), *Testing the constancy of regression relationships over time*, Vandenhoeck and Ruprecht, Göttingen.

Hackl P. ed. (1989), *Statistical Analysis and Forecasting of Economic Structural Change*, Springer Verlag, New York.

Hackl P. and .Westlund A.H. (eds.) (1991), *Economic Structural Change; Analysis and Forecasting*,, IIASA, Springer Verlag.

Hinkley, D.V. and Schechtman, E. (1987), *Conditional bootstrap methods in the mean-shift model*, Biometrika **74**, 85–93.

Horváth L. (1993a), *The maximum likelihood method for testing changes in parameters of normal observations*, Annals of Statistics **21**, 671–680.

Horváth L. (1993b), *Detecting changes in linear regression*, preprint.

Horváth L. and Shao Qi-Man (1993), *Limit theorems for the union intersection test*, preprint.

Hušková M. (1988), *Recursive M-tests for detection of change*, Sequential Analysis 7, 75-90.

Hušková M. (1990a), *Some asymptotic results for robust procedures for testing the constancy of regression models over time*, Kybernetika 26, 392-403.

Hušková M. (1990b), *Asymptotics for robust MOSUM*, Commentationes Mathematicæ Universitatis Carolinæ 31, 345-356.

Hušková M.(1993), *Test and estimators for the change point problem based on M-statistics*, submitted.

Hušková M.(1994a), *Nonrecursive procedures for detecting change in simple linear regression models*, Proceedings of IMSIBAC 4, in print.

Hušková M. (1994b), *Nonparametric procedures for a change in simple linear regression models*, Applied Statistical Sciences, in print.

Hušková M. (1994c), *Estimation of a hange in linear models*, submitted.

Hušková M. and Sen. P. K. (1989), *Nonparametric tests for shift and change in regression at an unknown time point*, Statistical Analysis and Forecasting of Economic Structural Change (Hackl P., ed.), Springer-Verlag, New York, pp. 71-85.

James B., James K. L. and Siegmund D. (1987), *Tests for a change-point*, Biometrika 74, 71-84.

Jandhyala V.K. and MacNeil I.B. (1989), *Residual partial sum limit process for regression models with applications to detecting parameter changes at unknown times*, Stochastic Processes and their Applications 33, 309-323.

Jandhyala V.K. and MacNeil I.B. (1991), *Tests for parameter changes at unknown times in linear regression models*, Journal of Statistical Planning and Inference 27, 291-316.

Jandhyala V.K. and MacNeil I.B. (1992), *On testing for the constancy of regression coefficients under random walk and change-point alternatives*, Econometric Theory 8, 501-517.

Jarušková and Antoch, J. (1994), *Detection of change in variance*, Proceedings of the Fifth Prague Symposium on Asymptotic Statistics (P.Mandl, M.Hušková, eds.), Physica Verlag, Heildelberg, pp. 297-302.

Kim Hyune-Ju and Siegmund D. (1989), *The likelihood ratio test for a change point in simple linear regression*, Biometrika 76, 409-423.

Krishnaiah P.R. and Miao B.Q. (1988), *Review about estimation of change points*, in Handbook of Statistics ,eds. Krishnaiah P.R. and Rao R.C., vol. 7, 375-402.

Lee Tze-San (1994), *Estimating coefficients of two-phase linear regression model with autocorrelated errors*, Statistics and Probability Letters 18, 113- 120.

Lombard F. (1987), *Rank tests for change point problem*, Biometrika 74, 615-624.

Longley J.W. (1967), *An apraisal of least squares programs for the electronic computer from the point of view of the user*, Journ. Amer. Statist. Assoc. 62, 819- 841.

MacNeil I.B. (1978), *Properties of sequences of partial sums of polynomial regression residuals with applications to tests for change of regression at unknown times*, Annals of Statist. 6, 422- 433.

MacNeil I.B., Tang S.M. and Jandhyala V.K. (1991), *A search for the Source of the Nile's change points*, Envirometrics 2, 341- 375.

Maddala G.S. (1977), *Econometrics*, McGraw-Hill, New York.

Michels P. and Trenkler G. (1990), *Testing the stability of regression coefficients using generalized recursive residuals*, Austr. J. Statist. 22, 293- 312.

Page E. S. (1955), *A test for a change in a parameter occuring at an unknown time point*, Biometrika 42, 523-527.

Page E. S. (1957), *On problems in which a change in a parameter occurs at an unknown point*, Biometrika 44, 248-252.

Pettitt A. N. (1979), *A nonparametric approach to the change point problem*, Applied Statistics 28, 126-135.

Ploberger W. and Krämer W. (1992), *The CUSUM test with OLS residuals*, Econometrica **60**, 271–285.

Praagman J. (1988), *Bahadur efficiency of rank tests for the change point problem*, Annals of Statistics **16**, 198–217.

Sen P. K. (1977), *Tied-down Wiener process approxiamtions for aligned rank order statistics and some applcations*, Annals of Statistics **5**, 1107–1123.

Sen P. K. (1980), *Asymptotic theory of some tests for a possible change in the regression slope occuring at an unknown time point*, Zeitschrift für Wahrscheinlichkeitsteorie und verw. Gebiete **52**, 203–218.

Sen P. K. (1982), *Asymptotic theory of some tests for constancy of regression relationships over time*, Statistics **13**, 21–31.

Sen P. K. (1984), *Recursive M-tests for the constancy of multivariate regression relationships over time*, Sequential Analysis **3**, 191–211.

Sen P. K. (1985), *Theory and applications of sequential nonparametrics*, SIAM CBMS **49**.

Sen P. K. (1977), *Tied-down Wiener process approximations for aligned rank order tatistics and some applications*, Annals of Statistics **5**, 1107–1123.

Sen P. K. (1982a), *An invariance principles for rank statistics revisited*, Sankhya ser.A **40**, 215–236.

Sen P. K. (1982b), *An invariance principles for recursive residuals*, Annals of Statistics **10**, 307–312.

Siegmund D. (1988), *Confidence sets in change-point problems*, International Statistical Review **56**, 31–38.

Srivastava M.S. and Worsley K.J. (1986), *Likelihood ratio tests for a change in the multivarite normal mean*, Jour. Amer. Statist. Assoc. **81**, 199–204.

Tang S.M. and MacNeil I.B. (1993), *The effect of serial correlation on tests for parameter change at unknown time*, Annals of Statistics **21**, 552–575.

Wolfe D. A. and Schechtman E. (1984), *Nonparmetric statistical procedures for the change point problem*, Journal of Statistical Planning and Inference **9**, 389–396.

Worsley K. J. (1983), *Testing for a two-phase multiple regression*, Technometrics **25**, 35–42.

Zacks S. (1983), *Survey of classical and Bayesian approach to the change point problem: Fixed sample and sequential procedures of testing and estimation.*, Recent Advances in Statistics. Paper in Honor of Herman Chernoff's Sixtieth Birthday (Rizvi M.H., ed.) Academic Press , New York, 245- 269.

Zacks S. (1991), *Detection and Change-Point Problem*, Handbook of Sequential Analysis (Ghosh, B.K. and Sen, P.K., eds.), Serie Statistics (M.Dekker, New York) **118**, 531- 562.

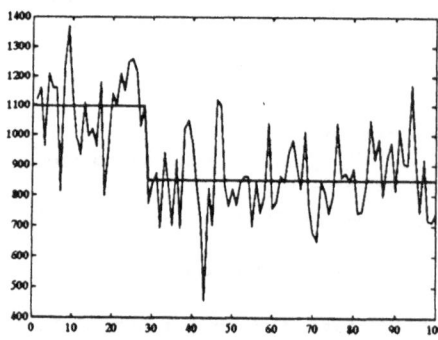

Fig. 1. The Nile data with estimated means.

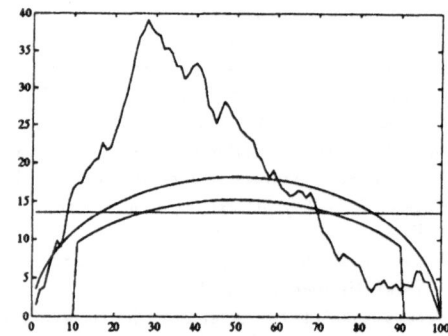

Fig. 2. Plot of S_k(+), k=1,...,100, and the 5% critical regions.

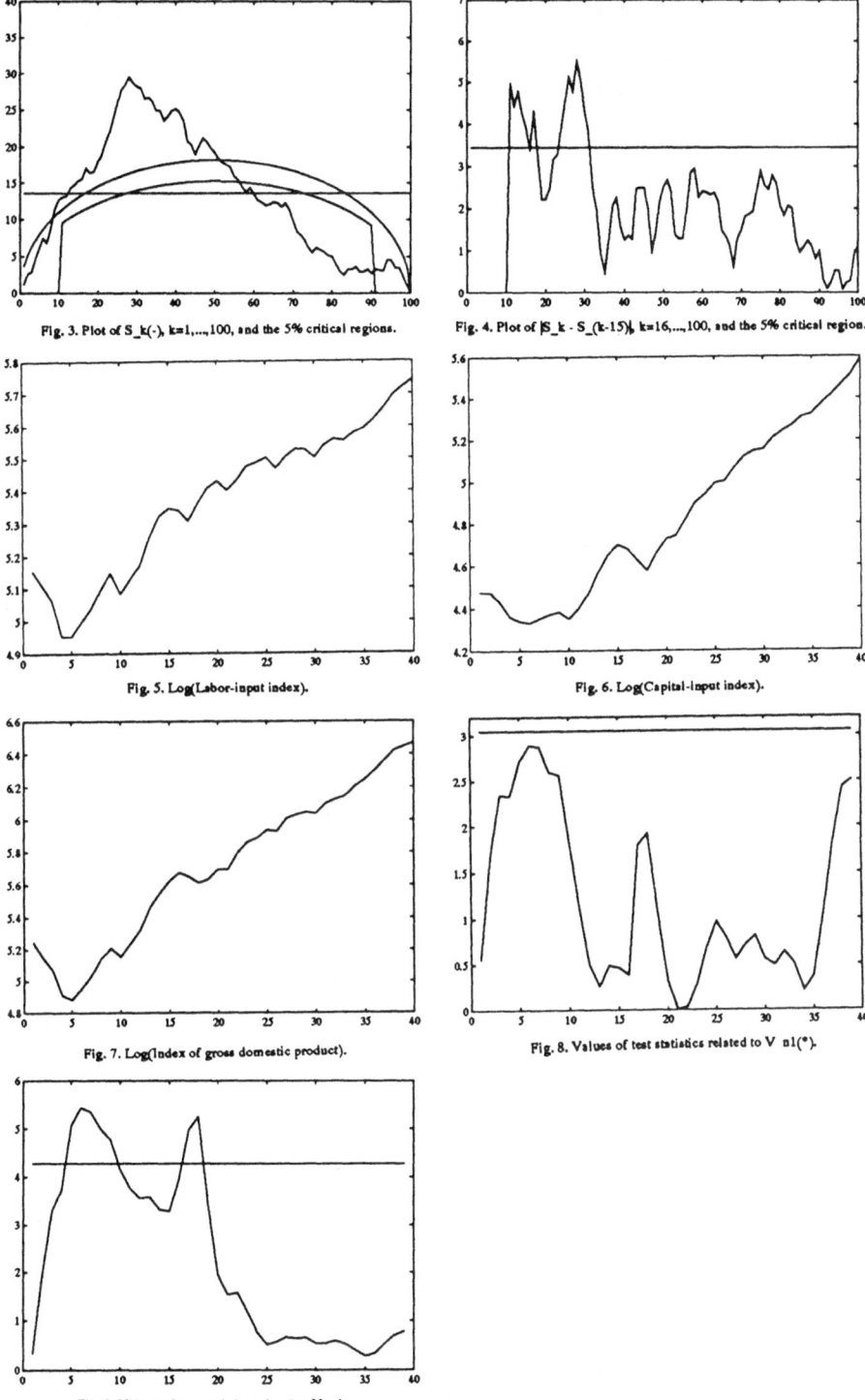

Fig. 3. Plot of S_k(-), k=1,...,100, and the 5% critical regions.

Fig. 4. Plot of |S_k - S_(k-15)|, k=16,...,100, and the 5% critical region.

Fig. 5. Log(Labor-input index).

Fig. 6. Log(Capital-input index).

Fig. 7. Log(Index of gross domestic product).

Fig. 8. Values of test statistics related to V n1(•).

Fig. 9. Values of test statistics related to V n1.

Randomization Methods for Detecting Changepoints

Andrew J. Baczkowski

Department of Statistics, University of Leeds

Abstract. We describe a randomization procedure for detecting change-points in a sequence of values which attempts to incorporate auto-correlation between the data values. This is illustrated using several examples.

1 Introduction

The problem of locating change-points in a sequence of data values has been examined by many researchers. Shaban (1980) gives an extensive bibliography; recent proposals are given by Henderson and Matthews (1993), Stephens (1994), and Basseville and Nikiforov (1993). Unfortunately, most of these approaches are only of use with small single sets of data. In many practical situations large amounts of multivariable data are collected and change-points need to be located.

One example which leads to large multivariable sequences of data is the collection of borehole geophysical logging data. As a logging device is lowered down a borehole it generally records several variables, which will be a function of the lithology and other physical characteristics of the borehole. The problem of interest is to locate change-points in the data, where either the mean level changes or the gradient of the recorded variables alters. These change-points are then used to indicate the position of different strata.

This paper proposes a non-parametric method for locating change-points which tries to allow for the presence of auto-correlation between data values.

2 A Randomization Method

Suppose we have a sequence of n data values, $y_1, y_2, ..., y_n$. Consider a moving window of width $2\Delta T$ centred on each candidate change-point t, where $t \in \{\Delta T + 1, \Delta T + 2, ..., n - \Delta T + 1\}$. This gives two subsets of data values, $\mathbf{y}_1 = \{y_{t-1}, y_{t-2}, ..., y_{t-\Delta T}\}$ and $\mathbf{y}_2 = \{y_t, y_{t+1}, ..., y_{t+\Delta T-1}\}$. Local estimates of the mean level, $\bar{y}_1$ and $\bar{y}_2$, and gradient, β_1 and β_2, about each candidate change-point can be obtained from these two subsets. For example, Redfern and Sawyer (1990, 1992) estimated the mean level and local gradient about each candidate change-point iteratively, incorporating estimation of the auto-correlation between neighbouring values, but assuming multivariate normality.

A randomization procedure can instead be used to test the significance of any detected mean level or slope difference; see Edgington (1987) for a general discussion. Such tests are attractive because of their ability to handle widely different data sets; no distributional assumptions need be made.

A simple randomization procedure would randomize the order of the $2\Delta T$ values $(\mathbf{y}_1, \mathbf{y}_2)$ to yield the values $z_1, z_2, ..., z_{2\Delta T}$, and then construct the appropriate test statistics for a slope change or a mean level change based upon $\mathbf{z}_1 = (z_1, z_2, ..., z_{\Delta T})$, and $\mathbf{z}_2 = (z_{\Delta T+1}, z_{\Delta T+2}, ..., z_{2\Delta T})$. It will be apparent that such a randomization test will not use any of the autocorrelation information which might be present in the data. However, by randomization under suitable restrictions it is possible to allow some of the auto-correlation structure, at least between neighbouring data values, to be retained.

Suppose that in constructing the randomized subsets $(\mathbf{z}_1, \mathbf{z}_2)$ we have so far obtained k values, and that $z_k = y_j$. If the next value in the sequence, z_{k+1}, was to be chosen only from the neighbouring values (y_{j-1}, y_{j+1}) then it is clear that the final subsets $(\mathbf{z}_1, \mathbf{z}_2)$ will retain some of the nearest-neighbour correlation between the y values. Baczkowski (1990) used such a procedure to construct a test for spatial isotropy. For spatially distributed data this procedure produces satisfactory results. However in one dimension it is clear that this randomization method, akin to a bootstrapping procedure, would not produce many useful subsets. Suppose instead that z_{k+1} is chosen from the set of values $Y_\delta = \{y : ((\mid y - y_{j-1} \mid < \delta) \cup (\mid y - y_{j+1} \mid < \delta))\backslash y_j\}$. The smaller the neighbourhood search parameter δ the more faithful we would be to the underlying correlation structure. The justification for this procedure is that if two values y have similar magnitude then we would expect their neighbours to be similar under the null hypothesis of no change-point.

Note that setting $\delta \rightarrow \infty$ is equivalent to choosing z_{k+1} randomly from the set $\{(\mathbf{y}_1, \mathbf{y}_2)\backslash y_j\}$. Including y_j would yield a bootstrapping statistic. Putting $\delta = 0$ is equivalent to moving only to the nearest neighbour with the possibility of jumps to other values y equal to y_{j-1} or y_{j+1}.

We now present several examples.

3   Asthma Cases near Open Cast Mining

Temple and Sykes (1992) examined the weekly number of asthma cases presenting to a surgery at Glynneath, Wales, over 130 weeks between 1989 and

1992. They plotted a cusum chart of the weekly counts, taking as a reference value the average number of cases for the initial 25 week period of their database. Their method showed a clear change at about week 53, when mining commenced. However other changes, though unreported, are apparent in the data. Since a retrospective analysis of the data is required it is better to use a cusum reference value based upon the whole data set, as suggested by Chatfield (1983, p.310), and illustrated in Figure 1. Chatfield also describes a method, due to Ewan and Kemp, of detecting change-points using the cusum chart which searches along both directions of time.

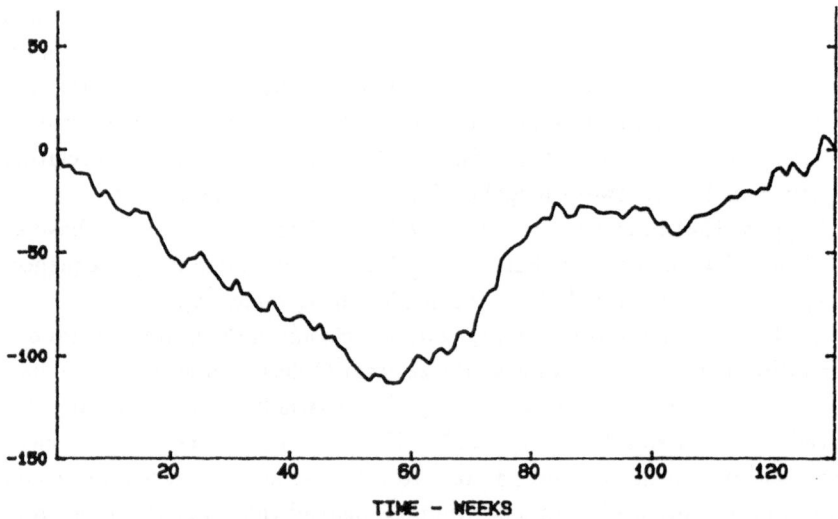

Figure 1. *Cusum chart of asthma data with end-points equated to zero*

With a search window $\Delta T = 5$, Chatfield's method suggested mean shifts at weeks 70, 84, and 104 with $P < 0.05$. Weeks 53, 57, and 76 were found to be significant in only one direction. With $P < 0.01$, only week 70 was indicated as a change-point. Similar results were obtained if the values on each side of the suggested change-point had separately estimated variances. With $\Delta T = 25$, weeks 53, 84, and 58 are significant.

With $\Delta T = 10$ weeks a randomization test approach had significant ($P < 0.01$) mean shifts at weeks 52/53, 57/58, 85, 103, 104, and 105. A slope shift was detected at weeks 49/50. The method of §2 with $\delta = 1$ gave significant weeks at 49, 80, and 84. With $\delta = 0$ only week 49 was significant. With δ very large we detected significant changes at weeks 53, 58, 85, 105, and 110.

Table 1: *Number of homicides in England and Wales 1946-1991*

1946	347	1954	311	1962	299	1970	339	1979	546	1987	601
1947	371	1955	279	1963	307	1971	407	1980	549	1988	551
1948	341	1956	315	1964	296	1972	409	1981	499	1989	526
1949	298	1957	325	1965	325	1973	391	1982	557	1990	572
1950	346	1958	261	1966	364	1974	526	1983	482	1991	675
1951	328	1959	266	1967	354	1975	433	1984	537		
1952	400	1960	282	1968	360	1976	488	1985	537		
1953	327	1961	265	1969	332	1977	418	1986	568		

4   Annual Homicides in England and Wales

In England and Wales homicide includes murder, manslaughter and infanticide. Table 1, (HMSO, 1991), gives the number of known homicides for each year from 1946 to 1991 inclusive.

A non-homogeneous Poisson process is not a good model here. There is more variability within the data than predicted from this model.

With $\Delta T = 10$ years and $\delta = 50$ a change of slope was detected at 1965 ($P = 0.045$). Several changes of mean level were detected after 1965 due to the increasing trend for these years. With small moving windows, $\Delta T \cong 4$ say, the only slope change detected is at 1982 ($P = 0.045$), though significant changes of mean were detected for several other dates. Interestingly, capital punishment was suspended in the United Kingdom in 1965 on a trial basis before its permanent abolition in 1969.

5   Wolverine Patrick Data

As an illustration of handling multivariable continuous data consider the well-logging example described in §1. Data is available from a Michigan well, the Wolverine Patrick, with measurements on ten variables at depths between 7800' and 10214'.

Various approaches have been suggested for such data. For example, Moline *et al.* (1992) take the first principal component of the measured variables and then use a maximum likelihood approach due to Mehta *et al.* (1990). In their study, this principal component accounted for over 80% of the variation between variables, and so this reduction to a single variable was an attractive procedure. Their model did not incorporate the auto-correlation between values at different depths.

For simplicity of presentation we only consider one hundred depth readings between 8332' and 8382'. The first principal component PC1 accounts for 48% of the variation in the data, and is plotted in Figure 2.

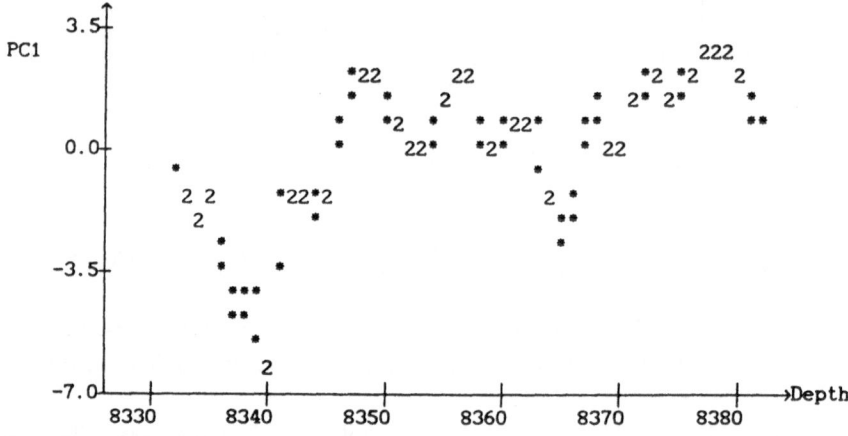

Figure 2.   *Plot of first principal component against depth for Patrick data*

The change of slope near 8340′ was not detected with δ too small, typically δ < 0.5. Larger values of δ allowed this change-point to be detected as a change of slope.

6   Discussion

With several variables we can proceed as in the single variable case. If neighbouring values are uncorrelated then a randomization procedure can be set up for the vector of variables, or for a summary of the variables, such as the first principal component. If values are correlated then a restricted randomization test can be constructed.

The methods used by Redfern and Sawyer (1990, 1992) may be used to assess the significance of the located change-points. They observe that searches with different window widths ΔT are necessary.

Example §5 shows that the parameter δ is important in allowing change-points to be located and should not be too small. One possible approach would be to allow δ to vary along the data sequence.

Acknowledgements

I should like to thank Dr.J.M.F.Temple for providing the set of asthma data in §3. I should also like to thank Dr.G.R.Moline of Oak Ridge National Laboratory for providing the set of well-logging data used in §5.

References

Baczkowski, A.J. (1990) Testing for isotropy through the semi-variogram, in P.A. Dowd and J.J. Royer (eds.), *2nd CODATA Conference on Geomathematics and Geostatistics*, Sci. de la Terre, Sér. Inf., Nancy, **31**, 1-10.

Basseville, M. and Nikiforov, I.V. (1993) *Detection of Abrupt Changes: Theory and Application*, Prentice Hall, Englewood Cliffs, 528 pp..

Chatfield, C. (1983) *Statistics for Technology: 3rd edition*, Chapman and Hall, London, 381 pp..

Edgington, E.S. (1987) *Randomization Tests*, Marcel Dekker, New York, 341 pp..

Henderson, R. and Matthews, J.N.S. (1993) An investigation of changepoints in the annual number of cases of haemolytic uraemic syndrome. *Applied Statistics*, **42**, 461-471.

Her Majesty's Stationery Office, (1991) Home Office Criminal Statistics in England and Wales 1991, Cm2134. HMSO.

Mehta, C.H., Radhakrishnan, S. and Srikanth, G. (1990) Segmentation of well logs by maximum-likelihood estimation. *Mathematical Geology*, **22**, 853-869.

Moline, G.R., Bahr, J.M., Drzewiecki, P.A. and Shepherd, L.D. (1992) Identification and characterization of pressure seals through the use of wireline logs: a multivariate statistical approach. *The Log Analyst*.

Redfern, E.J. and Sawyer, A.W. (1990) Automatic interpretation of geologs in open case mining. *COMPSTAT90, Short Communications*.

Redfern, E.J. and Sawyer, A.W. (1992) An approach to automatic interpretation of borehole logs in open case mining using changes in gradient, in P.A. Dowd and J.J. Royer (eds.), *2nd CODATA Conference on Geomathematics and Geostatistics*, Sci. de la Terre, Sér. Inf., Nancy, **31**, 285-297.

Shaban, S.A. (1980) Change point problem and two phase regression: an annotated bibliography. *International Statistical Review*, **48**, 83-93.

Stephens, D.A. (1994) Bayesian retrospective multiple changepoint identification. *Applied Statistics*, **43**, 159-178.

PLS Regression via Additive Splines

Jean-François Durand, Robert Sabatier

ENSAM-INRA-UM II, Unité de Biométrie, Montpellier

1 Introduction

PLS (Partial Least Squares) regression is a model for situations where a low observation/variable ratio comes with highly collinear predictors. A comparison with other statistical methods can be found in Frank and Friedman (1993). The PLS method, very popular in chemometrics, has been generalized in several ways in order to extend PLS into nonlinearity. The first one (Wold et al., 1989) replaces the standard regression of the Y response matrix on the latent variable t with a quadratic model. The second one (Frank, 1990) uses a smooth nonlinear function of the latent variable through the SMART regression (Friedman, 1984).

Our approach, called ASPLS, consists in performing a nonlinear transformation of the original predictor matrix X by normalized B-splines functions. This way of doing is similar to the Principal Component Analysis with Instrumental Variables extended to the nonlinear case by using B-spline transformations (Durand, 1993). An example is processed thus showing how ASPLS works in comparison with usual algorithms.

2 The ASPLS method

Let Y be the $n \times q$ response matrix. Each column X^j of the $n \times p$ predictor matrix X is transformed in $T^j = B^j a_j$, where B^j is the $n \times r_j$ fuzzy coding matrix of X^j obtained by using a basis of r_j normalized B-splines. T^j linearly depends on the vector a_j whose r_j components are called the spline coefficients. We only note that there exist particular coefficients noted ξ_j, very easy to compute with the knots sequence, that keeps the jth predictor invariant. Those coefficients are called the nodal coefficients. The whole predictor matrix is then replaced with the matrix T and may be considered as a particular T associated with the nodal coefficients. All the a_j vectors are column-binded into a. Without loss of generality we suppose that Y and the B^j matrices are D-centered, the $n \times n$ weighting matrix D being diagonal with positive diagonal elements that sum to 1.

2.1 The step k

Let us use now the standard notation in PLS for the latent variables: $t_k = Tw_k$, $u_k = F_{k-1}c_k$, with $F_0 = Y$. The objective function

$$f(a, w_k, c_k) = w'_k T' D F_{k-1} c_k = cov(t_k, u_k)$$

is to be maximized under the constraints

$$\|w_k\|^2 = \|c_k\|^2 = 1, \tag{1}$$

$$\|B^j a_j\|_D^2 = \|X^j\|_D^2, \quad \text{for } j = 1, \ldots, p, \tag{2}$$

$$t_k' D \bar{t}_j = 0, \quad \text{for } j = 1, \ldots, k-1, \tag{3}$$

where $\|.\|$ is the usual euclidian norm. The D-scalar product being defined by $(x, y)_D = x' D y$, the associated norm is $\|x\|_D^2 = x' D x$. The variance of any $\mathbf{T}$ column is imposed by (2) to be constant and equal to the variance of the corresponding column in $\mathbf{X}$. The constraints (3) mean that the k-latent variable is to be D-orthogonal with the optimal preceding ones that have already been computed. Conditions (3) are to be omitted when $k = 1$.

If we note $\bar{t}_k = \overline{T} \overline{w}_k$ an optimal component, the regression of F_{k-1} on $\bar{t}_k$ is processed at the end of the step k and the updated F_k is defined by

$$F_k = F_{k-1} - P_{\bar{t}_k} F_{k-1}, \tag{4}$$

where $P_{\bar{t}_k}$ is the D-orthogonal projector on $\bar{t}_k$. Thanks to (3) we have the additive decomposition for the inertia of the response

$$\|Y\|_D^2 = \sum_{j=1}^{k} \|P_{\bar{t}_j} F_{j-1}\|_D^2 + \|F_k\|_D^2.$$

Let us examine necessary conditions of optimality by using the Lagrangian function

$$L = f + \frac{1}{2}\lambda(1 - \|c_k\|^2) + \frac{1}{2}\mu(1 - \|w_k\|^2) + \frac{1}{2}\sum_{j=1}^{p} \alpha_j (\|X^j\|_D^2 - \|B^j a_j\|_D^2) - \sum_{j=1}^{k-1} \beta_j w_k' T' D \bar{t}_j.$$

The normal equations are

$$
\begin{cases}
\nabla_{a_i} L = B^{i\,\prime} D F_{k-1} \bar{c}_k [\overline{w}_k]_i - \bar{\alpha}_i B^{i\,\prime} D B^i \bar{a}_i - \sum_{j=1}^{k-1} \bar{\beta}_j B^{i\,\prime} D \bar{t}_j [\overline{w}_k]_i = 0, \text{ for } i = 1, \ldots, p, \\
\nabla_{w_k} L = \overline{T}' D F_{k-1} \bar{c}_k - \bar{\mu}\,\overline{w}_k - \sum_{j=1}^{k-1} \bar{\beta}_j T' D \bar{t}_j = 0, \\
\nabla_{c_k} L = F_{k-1}' D \overline{T} \overline{w}_k - \bar{\lambda} \bar{c}_k = 0, \\
\nabla_{\mu} L = 1 - \|\overline{w}_k\|^2 = 0, \\
\nabla_{\lambda} L = 1 - \|\bar{c}_k\|^2 = 0, \\
\nabla_{\alpha_j} L = \|X^j\|_D^2 - \|B^j \bar{a}_j\|_D^2 = 0, \quad \text{for } j = 1, \ldots, p, \\
\nabla_{\beta_j} L = \overline{w}_k' T' D \bar{t}_j = 0, \quad \text{for } j = 1, \ldots, k-1.
\end{cases}
$$

The optimal Lagrange multipliers are useful for the interpretation of the results. If we note $g_i = \nabla_{a_i} f = B^{i\,\prime} D F_{k-1} c_k [w_k]_i$, the $\bar{\alpha}_i$ coefficients given by

$$\bar{\alpha}_i = \frac{\|g^i\|_{(B^{ii} D B^i)^+}}{\|X^i\|_D}, \tag{5}$$

are positive and represent the part of the optimal covariance explained by the ith predictor

$$\lambda = \bar{\mu} = cov(\bar{t}_k, \bar{u}_k) = \sum_{i=1}^{p} \bar{\alpha}_i \| X^i \|_D^2 . \tag{6}$$

The $\bar{\beta}_j$ multiplier associated with the jth previous latent variable $\bar{t}_j$ is given by

$$(\bar{t}_j, \bar{u}_k)_D - \sum_{i=1}^{p} \bar{\alpha}_i \frac{[\bar{w}_k]_i}{[\bar{w}_j]_i} (\bar{T}_j^i, \bar{T}_k^i)_D = \bar{\beta}_j \| \bar{t}_j \|_D^2 , \tag{7}$$

where $\bar{T}_k^i$ and $\bar{T}_j^i$ represent the optimal transformations of the ith variable for the current step k and the previous j.

2.2 The algorithm for the step 1

The algorithm is based on the use of the normal equations without any β. The iterative method starts with $a_j = \xi_j$ that gives $T = X$. For a fixed T, the second and the third normal equations induce an algorithm very similar to the power method of determining the largest eigenvalue for a matrix. So the first iteration for the first step leads to the first step solution of the usual PLS method. The first step of ASPLS works at least as well as PLS does.

Once w_1, c_1 and $\lambda = \mu$ have been computed, the equation (5) gives the α_i coefficient and the first normal equation supplies the expression of the ith vector of the spline coefficients

$$a_i = \frac{1}{\alpha_i} (B^{i\prime} D B^i)^+ g_i . \tag{8}$$

The p columns of the array T are then updated with $T^i = B^i a_i$ and the iterations continue until a stopping criteria is used.

The values of the objective function constitute an increasing convergent sequence due to the positiveness of the Moore-Penrose inverse matrix in (8) and the positive coefficient α_i.

2.3 The algorithm for the step k $(k > 1)$

The step one algorithm based on F_{k-1} is firstly processed as an initializing procedure $(\beta_j = 0$, for $j = 1, \ldots, k - 1)$. Once the stopping rule is fulfilled, the next part of the algorithm takes into account the orthogonality of the latent variable t with the $k - 1$ optimal previous ones. As an aside we note that the optimal latent variable t obtained at the end of the first step is not far from being orthogonal with the previous ones: equation (4) implies that it is quite orthogonal with the immediate preceding one.

The second part of the algorithm differs from the first in the updating of the β coefficients with the equation (7) before computing the u component and before actualizing the spline coefficients with the first normal equation. On the other hand a new t is forced to be D-orthogonal with all the previous $\bar{t}_j$ after any computation of T or w.

3 Some additive properties replacing the linear ones

For any given step the optimal component t may be written as an additive function of the predictors

$$t = \sum_{i=1}^{p} w_i T^i \,, \tag{9}$$

where T^i is the best transformation of the ith variable. The constraint (2) forces the norm of T^i to be constant so that the coefficients w_i may be interpreted as weights for the optimal tranformations.

Cross-validation can be used in the same way as in usual PLS to choose A, the significant number of components. The equation (4) leads to the additive decomposition of the Y response matrix

$$Y = F_0 = \hat{Y} + F_A \,,$$

where

$$\hat{Y} = \sum_{k=1}^{A} t_k r_k' \,.$$

The vector r_k gives the coefficients of the linear regression of F_{k-1} onto t_k

$$r_k = \frac{F_{k-1}' D t_k}{\|t_k\|_D^2}$$

4 A comparison between ASPLS and other PLS algorithms

The data set (Wright & Gambino, 1984) has a relatively high observation/variable ratio and is a good example for nonlinearities in the structure-activity model (Frank, 1990). A sample of $n = 37$ compounds was collected to examine the quantitative structure-activity relationships (QSAR) of a series of 6-anilinouracil derivatives. The predictors are four chemical parameters π_m, π_p, MR_m and MR_p. The two response variables are the logarithm of the inverse concentrations of 6-anilinouracil required to achieve 50% inhibition of the enzyme and the mutant enzyme activities.

The invoked B-splines are of degree two with three equidistant interior knots and the weighting matrix D is $\frac{1}{n} I_n$. The X and Y data matrices have been D-standardized and the B^i have been D-centered. The comparison between linear PLS, nonlinear NLPLS (Frank, 1990) and ASPLS is presented in Table 1. The goodness of fit of the three models is reflected by the R2 value computed for the two responses with four components.

Table 1. Goodness of fit of PLS, NLPLS and ASPLS.

Comp.	PLS		NLPLS		ASPLS	
	Y1	Y2	Y1	Y2	Y1	Y2
1	0.28	0.22	0.62	0.50	0.89	0.63
2	0.28	0.38	0.65	0.84	0.91	0.85
3	0.36	0.46	0.81	0.86	0.92	0.92
4	0.36	0.46	0.86	0.88	0.95	0.93

Table 1 clearly shows that the two first components are fairly strong for the ASPLS method, the first explaining 75.64% and the second 12.15% of the **Y** variance. The optimal values of the objective function for the two first axes are 1.68 and 0.57.

Plotting the Y-scores (u) against the X-scores (t) for the first two components reveals a slight curvature (Fig. 1). Figure 2 displays the additive contributions of the four predictors concerning the first latent variable t_1.

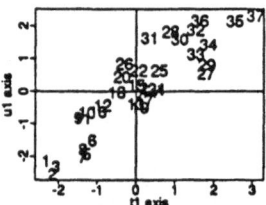

Figure 1: First dimension **Y**-scores plotted against the first dimension **X**-scores.

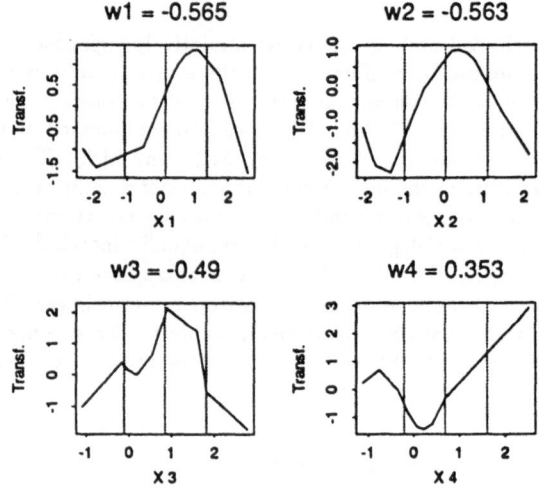

Figure 2: Additive contributions of the four predictors to the first **X**-component t_1. The weights w_i measure the influence of each variable. The dotted vertical lines give the position of the interior knots.

References

Durand, J.F. (1993). Generalized principal component analysis with respect to instrumental variables via univariate spline transformations. *Computational Statistics & Data Analysis*, **16**, 423-440.

Frank, I.E. (1990). A nonlinear PLS model. *Chemometrics and Intelligent Laboratory Systems*, **8**, 109-119.

Frank, I.E., Friedman, J.H. (1993). A statistical view of some Chemometrics regression tools (with discussion). *Technometrics*, **2**, 109-148.

Friedman, J.H. (1984). SMART users guide. *Tech. Rept. 1, LCS.* Dept. of Statistics, Stanford University, Stanford, CA, USA.

Wold, S., Kettaneh-Wold, N. and Skagerberg, B. (1989). Nonlinear PLS modeling. *Chemometrics and Intelligent Laboratory Systems*, **7**, 53-65.

Wright, G.E., Gambino, J.J. (1984). Quantitative structure-activity relationships of 6-anilinouracils as inhibitors of *Bacillus subtilis* DNA polymerase III. *Journal of Medical Chemistry*, **27**, 181-185.

Fitting Power Models to Two-Way Contingency Tables

Antoine de Falguerolles, Brian Francis

Laboratoire de Statistique et Probabilités, U.A. CNRS D0745, Université Paul Sabatier

Abstract

We consider an algorithm for fitting power models to two-way contingency tables. Two approaches are considered: a fast exploratory procedure based on singular value decomposition and a modelling approach which can be implemented in GLM software packages.

1   Introduction

In analysing a two-way cross-classification table, non-independence between the rows and the columns is of prime interest. Departure from independence can be investigated in various ways. Well-known approaches are correspondence analysis (Benzécri, 1973), which is also called the correlation model, and the Goodman $R \times C$ association model (Goodman, 1991). The similarities between the two models have long been recognized and used to define algorithmic procedures of estimation for both models (*e.g.* Escoufier (1988) for a least-square approach, Falguerolles and Francis (1992) for a maximum likelihood approach). Both models can be considered as special cases of a more general *"power model"* (Baccini et al., 1993). For this class of models, departure from independence is modelled by a *bilinear* (or *bi-additive*) term, that is to say, a weighted sum of row by column score products. The scores are used to construct various plots or Gabriel's biplots.

Assuming a theoretical distribution for the cell counts with unknown mean parameter λ_{rs}, these models are formulated as follows:

$$\lambda_{rc} = \gamma \alpha_{r,0} \beta_{c,0} (1 + \frac{1}{\psi} \sum_{k=1}^{K} \sigma_k \alpha_{r,k} \beta_{c,k})^{\psi} \tag{1}$$

where the *"singular values"* are ordered ($\sigma_1 \geq \sigma_2 \geq \ldots \geq \sigma_K > 0$). In the bilinear term of rank K, the vectors of row and column scores ($\underline{\alpha}_k$ and $\underline{\beta}_k$, $k = 1, \ldots, K$) are orthonormalised with respect to given sets of weights. In the independence term, $\gamma \alpha_{r,0} \beta_{c,0}$, standard identification constraints are imposed. The parameter ψ is assumed to be non-negative. Obvious special cases are $\psi = 0$ (independence), $\psi = 1$ (correspondence analysis) and $\psi = \infty$ (Goodman $R \times C$ association).

Small values of K are of practical interest and it is hoped that a proper value of ψ achieves this aim. The relationship with the Box-Cox transform is emphasized by noting that, for known values of the main effect parameters γ, $\underline{\alpha}_0$ and $\underline{\beta}_0$, $\frac{(\frac{\lambda_{rc}}{\gamma^{\alpha_r,0}\beta_{c,0}})^{1/\psi} - 1}{1/\psi} = \sum_{k=1}^{K} \sigma_k \alpha_{r,k} \beta_{c,k}$. For a related approach in the context of ordinal contingency tables see Rom and Sarkar (1992).

In this paper, we consider two approaches for the fitting of the parameters in the bilinear term: exploratory investigation (section 2) and explicit modelling (section 3). In the former the singular value decomposition (SVD) of pre-processed counts is briefly reviewed. In the latter a Poisson distribution for the cell counts is assumed. Then, as in standard Nelder and Wedderburn's Generalized Linear Models where a link function relates each mean to its predictor, we analyse the relationship between the λ_{rc} and the combined terms for main effect and bilinear interaction. Assuming known ψ, an algorithm for maximum likelihood fitting of such models is considered in section 4. The proposed procedure is quite general and, although the models considered are not strictly speaking GLMs, can be implemented in software packages which can fit GLMs. Examples and practical details of implementation will be discussed during the presentation.

2   Fast Exploratory Analysis

We adopt here the point of view of exploratory data analysis: few explicit assumptions and fast computations (even for large data sets). The observed entries in the table, denoted by y_{rc}, are non-negative numbers. In practice, they may be pre-processed before submission to SVD. Well known transformations are the *square root* and, for positive entries, the *log* (Escoufier, 1988). In the context of power models, the former corresponds to $\psi = 2$ and the latter to $\psi = \infty$. The case of no transformation ($\psi = 1$) is standard correspondence analysis. The case where $\psi = 2$ is also known as "spherical correspondence analysis" (Domenges and Volles, 1979).

2.1   Singular Value Decomposition

We assume that the value of parameter ψ is known beforehand. We consider the *general root* transformation which is outlined in Baccini et al. in the discussion of Goodman (1991). SVD of the ψth *root* of the counts gives:

$$(y_{rc})^{\frac{1}{\psi}} = \sum_{k=0}^{M} \tilde{s}_k \widetilde{u_{r,k}} \widetilde{v_{c,k}} \tag{2}$$

where $\tilde{s}_0, \widetilde{u_{r,0}}, \widetilde{v_{c,0}}$ can be chosen to be positive (Perron - Frobenius positivity property) and $\sum_{r=1}^{R} \widetilde{u_{r,0}}^2 = \sum_{c=1}^{C} \widetilde{v_{c,0}}^2 = 1$. Then $(y_{rc})^{\frac{1}{\psi}} = \tilde{s}_0 \widetilde{u_{r,0}} \widetilde{v_{c,0}}[1 +$

$\frac{1}{\psi}\sum_{k=1}^{M}(\frac{\widetilde{\psi s_k}}{s_0})(\frac{\widetilde{u_{r,k}}}{u_{r,0}})(\frac{\widetilde{v_{c,k}}}{v_{c,0}})]$. It is easily seen that the row and column scores $u_{r,k} = \frac{\widetilde{u_{r,k}}}{u_{r,0}}$ and $v_{c,k} = \frac{\widetilde{v_{c,k}}}{v_{c,0}}$ are centered and orthonormed with respect to the "weights" $\widetilde{u_{r,0}}^2$ and $\widetilde{v_{c,0}}^2$. For the generalized SVD of the doubly centered *log* counts see Baccini and Khoudraji (1992).

2.2 Low Dimensional Approximations

The SVD above leads to straightforward reconstitution formulae from which low-dimensional approximations, $K = 1, 2$ or 3 say, are easily derived:

$$\widehat{y_{rc}^{(K)}} = c^{(K)}a_r^{(K)}b_c^{(K)}[1 + \frac{1}{\psi}\sum_{k=1}^{K} s_k u_{r,k} v_{c,k}]^{\psi}. \qquad (3)$$

Correspondence Analysis ($\widehat{y_{rc}^{(K)}} = ca_r b_c[1 + \sum_{k=1}^{K} s_k u_{r,k} v_{c,k}]$) and $R \times C$ Association model ($\widehat{y_{rc}^{(K)}} = c^{(K)}a_r^{(K)}b_c^{(K)}\exp(\sum_{k=1}^{K} s_k u_{r,k} v_{c,k})$) are particular cases. Note that, with the exception of correspondence analysis, in the above formulae the constant $c^{(K)}$ and possibly the main effects $a_r^{(K)}$ and $b_c^{(K)}$ may depend on K; it is usual but not essential to constrain the approximated entries so that they satisfy the observed marginal counts. When required, this is achieved by iterative proportional fitting.

2.3 Heuristic Choice of ψ and K

In all the techniques above it is assumed that there exist some low values for ψ and K, for which the entries in the two-way table are well: approximated. A heuristic determination of ψ and K consists in sequentially ($L = 1, 2, 3$ and $\psi \in [.5, 3.5]$, say) evaluating pseudo chi-squared statistics $\sum_{r=1}^{R}\sum_{c=1}^{C} \frac{(y_{rc}-\widehat{y_{rc}^{(\psi,L)}})^2}{y_{++}f_{r+}f_{+c}}$ where $\widehat{y_{rc}^{(\psi,L)}}$ denotes the inverse transform of y_{rc}. Examination of the pattern of the singular values is also of great help.

3 Modelling Approach

In this section we adopt a modelling point of view which aims to preserve two essential features of the exploratory approach, namely an explanatory term (*bilinear interaction*) and a monotonic function (or scale) relating the (expected) counts to the corresponding associations (*link function*). The counts y_{rc} are now the observed values of random variables Y_{rc} with Poisson distribution and expected values λ_{rc}. (Note that other distributions could as well be considered.) A link function is to be derived from formula (1) relating the expected values λ_{rc} to the parameters in the model. The rank K of the bilinear component and the power parameter ψ are assumed to be known.

3.1 Link Functions for Particular Cases

We consider particular forms of the model with well known link functions. In Goodman's $R \times C$ association model, a *log* link function relates the λ_{rc} to the parameters in the model: $g(\lambda_{rc}) = \log \lambda_{rc} = \log \gamma + \log \alpha_{r,0} + \log \beta_{c,0} + \sum_{k=1}^{K} \sigma_k \alpha_{r,k} \beta_{c,k}$.

In Correspondence analysis, the parameter λ_{rc} relates to the bilinear term in a more subtle way: $\lambda_{rc} = \gamma \alpha_{r,0} \beta_{c,0} [1 + \sum_{k=1}^{K} \sigma_k \alpha_{rk} \beta_{ck}]$. It can be proved (in the absence of missing or structural values for some cells) that this amounts to defining an *"inverse-independence"* class of link function: $g_{rc}(\lambda_{rc}) = \frac{1}{y_{++} f_{r+} f_{+c}} \lambda_{rc} - 1 = \sum_{k=1}^{K} \sigma_k \alpha_{r,k} \beta_{c,k}$ where f_{r+} and f_{+c} denote the marginal observed frequencies and y_{++} the observed total count.

3.2 Link Functions for the General Case

As defined in formula (1), the relationship between the parameters and the expected counts λ_{rc} may be seen as combining two types of link functions. If γ, $\alpha_{r,0}$ and $\beta_{c,0}$ are known, a first class of link function relates the expected counts to the *bilinear* component:

$$g_{\psi,rc}^{(1)}(\lambda_{rc}) = \frac{(\frac{\lambda_{rc}}{\gamma \alpha_{r,0} \beta_{c,0}})^{1/\psi} - 1}{1/\psi} = \sum_{k=1}^{K} \sigma_k \alpha_{r,k} \beta_{c,k}. \qquad (4)$$

If σ_k, $\alpha_{r,k}$ and $\beta_{c,k}$ $(k = 1, \ldots, K)$ are known, a second link function relates the expected counts to the *independence* component $\log \gamma + \log \alpha_r + \log \beta_c$:

$$g^{(2)}(\lambda_{rc}) = \log \lambda_{rc} = \log \gamma + \log \alpha_r + \log \beta_c + \psi \log(1 + \frac{1}{\psi} \sum_{k=1}^{K} \sigma_k \alpha_{r,k} \beta_{c,k}). \quad (5)$$

Note that in formula above the logarithm of the *interaction* component, $\psi \log(1 + \frac{1}{\psi} \sum_{k=1}^{K} \sigma_k \alpha_{r,k} \beta_{c,k})$, behaves as an *offset* (Francis et al., 1993).

3.3 Identification Constraints

While in the exploratory approach the weights used for identification are derived from the data (see section 2), in the modelling approach the vectors of row and column scores (α_k and β_k, $k = 1, \ldots, K$) are orthonormalised with respect to sets of weights imposed by the modeller. For a related discussion see Lauro and D'Ambra (1984).

4 Algorithmic Considerations

We propose an algorithm to obtain maximum likelihood estimates of the parameters when K and ψ are known. When this is not the case, we suggest the approach of *profile likelihood*: the likelihood (or the deviance) is computed as a function of K and ψ ($k \in 1, 2, 3$ and $\psi \in [.7, 3.5]$, say).

4.1 Preliminary Remark

Assume that the *independence* component and the column scores $\beta_{c,k}$ are known. Then $g^{(1)}_{\psi,rc}(\lambda_{rc}) = \sum_{k=1}^{K}(\sigma_k\alpha_{r,k})\beta_{c,k}$. It appears that the unknown parameters $(\sigma_k\alpha_{r,k})$ are standard parameters in a Poisson regression. Given the $\alpha_{r,k}$, the same property holds for $(\sigma_k\beta_{c,k})$. If it is assumed that the row and column scores are both known then $g^{(1)}_{\psi,rc}(\lambda_{rc}) = \sum_{k=1}^{K}\sigma_k(\alpha_{r,k}\beta_{c,k})$. Again the unknown parameters σ_k are provided by ordinary Poisson regression.

Now assume that the *interaction* component, $(1 + \frac{1}{\psi}\sum_{k=1}^{K}\sigma_k\alpha_{r,k}\beta_{c,k})^{\psi}$, is known. Then $g^{(2)}(\lambda_{rc}) = \log\gamma + \log\alpha_r + \log\beta_c + \log(interaction)$. Here the unknown parameters γ, α_r, β_c are provided by fitting the independence model with the logarithm of the *interaction* component as an offset.

4.2 Orthonormalisation of Scores

Gram-Schmidt orthogonalisation of the score vectors is easily performed by repeated ordinary regressions: in stage 1.1 of the algorithm, the kth current row (column) score vector is equal, up to a normalising factor, to the residuals of the ordinary regression of the kth block in the vector of regression parameters onto the first $(k-1)$ score vectors.

4.3 Algorithm

The discussion above suggests the following two stage algorithm.

alternate stage 1 and stage 2 until global convergence:
   stage 1: use current value for the *independence* component (link $g^{(1)}_{\psi,rc}$),
      stage 1.1: use initial **column** scores $\beta_{c,k}$,
      alternate until convergence of scores:
         estimate the $(\sigma_k\alpha_{r,k})$,
         extract revised (centered and orthonormed) **row** scores $\alpha_{r,k}$;
         estimate the $(\sigma_k\beta_{c,k})$,
         extract revised (centered and orthonormed) **column** scores $\beta_{r,k}$;
      stage 1.2:
      estimate the "**singular values**" σ_k ;
      extract revised *interaction* component;
   stage 2: use current value for the *interaction* component (link $g^{(2)}$),
      extract revised *independence* component;

4.4 Initial Values

Initial value for the independence component is given by the fitted values of the standard model of independence. Initial column scores can be either random or provided by the SVD discussed in section (2).

5 Concluding Remarks

GLIM 4 (Francis et al., 1993)) is particularly well suited for the implementation of this algorithm: great flexibility is offered to specify an *own* link function, an *own* probability distribution and an *ad hoc* offset. Profile likelihood estimation of K and ψ is simple. Structural zeros or missing entries may be handled. Note that the power model readily extends to other situations as described in Falguerolles and Francis (1992).

Nevertheless, our approach suffers from two drawbacks: the convergence is slow when using random starting values in the algorithm; and the (co)variance of the estimates cannot be produced. The first drawback is less of a problem if the algorithm instead recycles from fitted values produced from a model with ψ close to the value required.

References

Baccini, A., Caussinus, H., and Falguerolles, A. de (1993), Analysing Dependence in Large Contingency Tables: Dimensionality and Patterns in Scatter Plots. In: C.M. Cuadras and C.R. Rao (eds.): *Multivariate Analysis: Future Directions 2*, 245-63, Elsevier B.V. (North-Holland).

Baccini, A., and Khoudraji, A. (1992), A Least Square Procedure for Estimating the Parameters in the $R \times C$-Association Model for Contingency Tables, *Computational Statistics*, 7, 287-300.

Benzécri, J.P. (1973), *L'analyse des données*, vol.2, Dunod, Paris.

Domenges, D. and Volle, M. (1979), Analyse Factorielle Sphérique : une Exploration, *Annales de l'INSEE*, 35, 3-84.

Escoufier, Y. (1988), Beyond Correspondence Analysis. In: H.H. Bock (eds.): *Classification and Related Methods of Data Analysis*, 505-14, Elsevier Science Publishers B.V. (North-Holland).

Falguerolles, A. de, and Francis, B. (1992), Algorithmic Approaches for Fitting Bilinear Models. In: Y. Dodge and J. Whittaker (eds.): *COMPSTAT 92, Computational Statistics*, Vol. 1, 77-82, Physica-Verlag, Heidelberg.

Francis, B., Green, M., Payne, C. (eds.) (1993), *The GLIM System, Release 4 Manual*, Clarendon Press, Oxford.

Goodman, L. (1991), Measures, Models, and Graphical Displays in the Analysis of Cross-Classified Data (with discussion), *J. Amer. Statist. Ass.*, 86, 1085-138.

Lauro, N. C., and D'Ambra, L. (1984), L'Analyse Non Symétrique des Correspondances, In: E. Diday et al. (eds.): *Data Analysis and Informatics*, III, 433-446, North-Holland, Amsterdam.

Rom, D., and Sarkar, S.K. (1992), A Generalized Model for the Analysis of Association in Ordinal Contingency Tables, *J. Statist. Planning Infer.*, 33, 205-212.

Optimized Local Regression: Computational Methods for the Moving Average Case

Valery Fedorov[4], Peter Hackl[1], Werner G. Müller[3]

University of Minnesota, Dept. of Applied Statistics

[1] on leave at the Department of Applied statistics, University of Minnesota, St. Paul, MN 55108, USA; on leave from Laboratory of Social and Economic Measurement, USSR Academy of Sciences, Moscow, Russia

[2] Department of Statistics, Wirtschaftsuniversität, A-1090 Vienna, Austria

[3] Department of Statistics, University of Iowa, Iowa City, IA 52242, USA; on leave from Department of Statistics, Wirtschaftsuniversität, A-1090 Vienna, Austria

Abstract. Optimized moving local regression is an extension of Cleveland's *loess* technique that takes a suspected misspecification of the model into account. The weights are chosen so that the effect of the misspecification is minimized. The derivation of optimal weights is shown to be similar to that of an optimal design for an experiment.

Keywords. Nonparametric regression, moving local regression, moving averages, optimal experimental design

1   Introduction

Moving regression analysis techniques are intuitively simple statistical methods for smoothing, interpolating, and forecasting. The main idea is to calculate such estimates for a certain point by weighting down the observations so that the weights reflect the "distance" of the observations from the point of interest. The model behind these techniques is specified by smoothness assumptions, leading to polynomial functions that are used as local models; a corresponding estimation procedure is the *loess* technique (Cleveland, 1979). Optimized moving local regression (Fedorov *et al.*, 1993a) is an extension that takes a suspected misspecification of the model into account. The additional flexibility is achieved by supplementing the model with a remainder term that represents the suspected misspecification; the weights are chosen so that the effect of a possible misspecification of the suspected form is minimized. Applications of that idea in time series analysis and discriminant analysis are discussed by Fedorov *et al.* (1993b) and Müller (1992), respectively.

We state the problem in Section 2 and suggest in Section 3 an estimating procedure for the model parameter. The derivation of optimal weights are discussed in Section 4, and an appropriate algorithm is suggested. Concluding remarks are given in Section 5.

2 Statement of the Problem

Let X_p be the set of points of interest, where an interpolated, smoothed or predicted value is desired, and $X_l = \{x_1, \ldots, x_n\}$ the "supporting set", i.e., the set of points where observations are available. The points of interest are assumed to be reasonably close to the points in X_l, and in the best case surrounded by or mixed with those points. In the following, we will have in mind just one point of interest, x; of course, this point can be any element of X_p.

We will consider models for the response y_i at $x_i \in X_l$, $i = 1, \ldots, n$, that are obtained by means of a Taylor series or another approximation of the true model at the point of interest $x \in X_p$

$$y_i = \theta^T f(d_i) + \delta^T \varphi(d_i) + \varepsilon_i . \tag{1}$$

The m-component vector $f(d_i)$ of known functions, $f(d_i) = [f_1(d_i), \ldots, f_m(d_i)]^T$, has the argument $d_i = x_i - x$; the m-vector θ contains unknown parameters. In addition, we allow for the *remainder term* $\delta^T \varphi(d_i)$ in the systematic part of the model. For the disturbances ε_i we assume that $\mathrm{E}\{\varepsilon_i\} = 0$, $\mathrm{E}\{\varepsilon_i^2\} = \sigma^2$, and $\mathrm{E}\{\varepsilon_i \varepsilon_{i'}\} = 0$ for $i = 1, \ldots, n$ and $i \neq i'$.

This model should be flexible enough to approximate the response function of the regression model in the vicinity of x. For interpolating, smoothing, or forecasting, a weighted least squares regression analysis of the model will be performed, where the weights decrease with increasing distance between the points of interest and observation. It is assumed that $\theta^T f(d_i)$ is an adequate specification. If, however, $\theta^T f(d_i)$ deviates systematically from y_i for the various i, these deviations are covered by $\delta^T \varphi(d_i)$ and will be taken into account via the weights.

The discussion of the following will be based on the simplified and the probably most frequently applied version of (1), viz. the model

$$y_i = \theta + \delta \varphi_i + \varepsilon_i , \tag{2}$$

where $\theta^T f(d_i)$ degenerates to a constant. This form of the model allows to expose the ideas in the simplest way.

3   Estimation of θ

As the functions φ are unknown in applications, a reasonable approach for estimating θ is to apply the weighted least squares method to the reduced model $E\{y_i\} = \theta^T f(d_i)$:

$$\hat{\theta} = \arg\min_{\theta} \sum_{i=1}^{n} \lambda_i [y_i - \theta^T f(d_i)]^2 ; \tag{3}$$

the use of (nonnegative) weights λ_i gives us the opportunity to take the distance $d_i = x_i - x$ of x_i from the point of interest $x \in X_p$ into account. For the model (2), the estimator $\hat{\theta}$ for the point of interest x is simply the weighted average of the y_i:

$$\hat{\theta} = (\sum_{i=1}^{n} \lambda_i)^{-1} \sum_{i=1}^{n} \lambda_i y_i = (l^T \lambda)^{-1} \lambda^T y , \tag{4}$$

with $\lambda = (\lambda_1, \ldots, \lambda_n)^T$ and $l = (1, \ldots, 1)^T$. Calculating this estimator for the various points of interest corresponds to a *moving average* approach. The mean squared error (m.s.e.) matrix of $\hat{\theta}$ is the scalar function

$$\begin{aligned} R(\lambda) &= E\{(\hat{\theta} - \theta)(\hat{\theta} - \theta)^T\} \\ &= [\gamma^2 (\lambda^T \varphi)^2 + \lambda^T \lambda](l^T \lambda)^{-2} . \end{aligned} \tag{5}$$

Multiplication of λ by a real constant does change neither $\hat{\theta}$ nor $R(\lambda)$. Therefore, we assume in the following, besides $\lambda_i \geq 0$, that $l^T \lambda = 1$.

4   Optimal Weights

We define the optimal weights – for a given x – as a weighting structure which minimizes the m.s.e. R, taking into account the location of the supporting points x_i, $i = 1, \ldots, n$, and the structure of the remainder term $\delta^T \varphi(d_i)$. For a given point x of interest and a given supporting set X_n, the m.s.e. R of $\hat{\theta}$ is a function $R(\lambda)$ of the weights $\lambda_1, \ldots, \lambda_n$. Hence, improving the quality of $\hat{\theta}$, i.e., reducing $R(\lambda)$, implies an appropriate choice of the weights $\lambda_1, \ldots, \lambda_n$.

For the weights to be optimal, the following condition has to be fulfilled.

Assertion 1: The necessary and sufficient condition for λ^* to be optimal is the inequality

$$\min_{i}(A\lambda^*)_i \geq R(\lambda^*) , \tag{6}$$

where $A = I_n + \gamma^2 \varphi \varphi^T$.

The proof of the assertion is given by Fedorov *et al.* (1993a) and is based on the fact that R and the set λ are convex.

In some situations the equivalent form

$$\min_i[\lambda_i^* + \gamma^2\varphi_i(\varphi^T\lambda^*)] \geq R(\lambda^*)$$

of (6) is computationally more convenient.

The weights can be obtained in closed form when we require that the condition

$$A\lambda^* = R(\lambda^*)l \tag{7}$$

is fulfilled, i.e., each component of $A\lambda^*$ should attain the minimum value of condition (6). Under the constraint $\sum_{i=1}^n \lambda^* = 1$, we find

$$\lambda_i^* = n^{-1}\left[1 - \frac{\gamma^2(\varphi_i - \overline{\varphi})\overline{\varphi}}{n^{-1} + \gamma^2 s(\varphi)}\right] \tag{8}$$

for all i and

$$R(\lambda^*) = n^{-1}\left[1 + \frac{\gamma^2\overline{\varphi}^2}{n^{-1} + \gamma^2 s(\varphi)}\right], \tag{9}$$

where $\overline{\varphi} = n^{-1}\sum_{i=1}^n \varphi_i$ and $s(\varphi) = n^{-1}\sum_{i=1}^n (\varphi_i - \overline{\varphi})^2$. If γ^2 is sufficiently small, i.e., δ^2 is small or σ^2 is large, then $\lambda^* > 0$ for all i and (8) defines the optimal λ. The weights (8) can be found directly by differentiation of (5): All λ_i are positive and the constraints $\lambda_i \geq 0$ are not active.

The analysis is more complicated if some of the λ_i^* defined by (8) are negative. This can happen if either γ^2 or the corresponding φ_i (or both) are large. It is typical in time series analysis when we face a long series of x_i. In such cases the following iterative procedure can be recommended:

Start with a set of n_0 points $x_i \in X_l$ that are neighboring the given point of interest, and use (7) to construct $\lambda^*(n_0)$.

At stage s of the procedure,

(a) remove all points with $\lambda^*(n_s) < 0$; construct for the set of n'_{s+1} remaining points the weights $\lambda^*(n'_{s+1})$;

(b) if the new weights fulfill $\lambda^*(n'_{s+1}) > 0$, add Δn nearest points and repeat (a) with $n_{s+1} = n'_{s+1} + \Delta n$.

Stop at stage s if $R[\lambda^*(n_s)]$ cannot be further decreased.

When x_i is one-dimensional we have no problem to choose points to be added or removed. In multi-dimensional cases, it is not a simple question to define what a "nearest point" is. In this case, (6) can guide the process. For instance, to add a point we have to find

$$i^*(s) = \arg\min_i \varphi_i[\varphi^T \lambda^*(n_s)],$$

where the minimization is taken over the set of points that are candidates for being added.

5   Concluding Remarks

From condition (6) it is clear that the weights depend on the unknown functions φ. We suggest to investigate the structure of λ^* for reasonable choices for the unknown φ like $\varphi(d_i) = d_i$ or $\varphi(d_i) = d_i^2$. In addition, good prior information is needed about γ^2.

As a closed form derivation of λ^* is not possible in general, the first step is to look for a suitable candidate for λ^* on the basis of intuition or common sense ideas and to check subsequently whether the condition (6) is fulfilled. In general, for a fixed set X_n the structure of λ^* is given by

$$\lambda_i^* = a + b\varphi(d_i),$$

where a and b are defined by (8).

For the general case of interpolation when the relative position of the points of interest and the supporting points is constant, the following cross-validation procedure can be applied: Given the structure of $\lambda_i = a + b\varphi(d_i)$, find

$$a^*, b^* = \arg\min_{a,b} \sum_i (y_i - \sum_{j \neq i} \lambda_j y_j)^2$$

such that $\lambda_i \geq 0$ for all i and $\sum_i \lambda = 1$.

A general recommendation is to investigate the structure of λ^* for a given set $X_n = \{x_1, \ldots, x_n\}$. As the corresponding values $\varphi_1, \ldots, \varphi_n$ are usually unknown, prior knowledge from theory or past experience must be used to make assumptions on the form of the weight function. Then, for a special application, the so obtained general structure can be normalized and adapted to the features of the situation in mind.

References

Cleveland, W.S. (1979). Robust Locally Weighted Regression and Smoothing Scatterplots. *Journal of the American Statistical Association*, **74**, 829-836.

Fedorov, V.V., Hackl, P., and Müller, W.G. (1993a). Moving Local Regression. The Weight Function. *Journal of Nonparametric Statistics*, **3**, 355-368.

Fedorov, V.V., Hackl, P., and Müller, W.G. (1993b). Optimized Moving Local Regression. Another Approach to Forecasting, pp. 137-144 in Müller, W.G., Wynn, H.P., and Zhigljavsky, A.A. (eds.), *Model-Oriented Data Analysis*. Heidelberg: Physica-Verlag.

Müller, W.G. (1992). The Evaluation of Bank Accounts Using Optimized Moving Local Regressions, pp. 145-150 in Fahrmeir, L., Francis, B., Gilchrist, R., and Tutz, G. (eds.), *Advances in GLIM and Statistical Modelling*. Berlin: Springer-Verlag.

Equivalence in Non-Recursive Structural Equation Models

Thomas Richardson

Philosophy Department, Carnegie-Mellon University

Introduction
In the last decade, there has been considerable progress in understanding a certain class of statistical models, known as directed acyclic graph (DAG) models, which encode independence, and conditional independence constraints. (See Pearl, 1988). This research has had fruitful results in many areas: there is now a relatively clear causal interpretation of these models, there are efficient procedures for determining the statistical indistinguishability of DAG's, reliable algorithms for generating a class of DAG models from sample data and background knowledge, etc. Two important elements in these investigations were: First, a purely graphical condition for calculating the conditional independence relations entailed by a DAG. Second, a 'local' characterization of equivalence between two graphs, in the sense that all of the same conditional independencies are entailed by each graph. Such a local characterization was essential in allowing the construction of efficient algorithms which could search the whole class of DAG models and to find those which fitted the given data.

1 Non-Recursive Structural Equations and Cyclic Graphs

1.1 The limitations of DAG models
The DAG formalism is very general: A gamut of more familiar constructs such as recursive linear structural equation models with independent errors, regression models, factor analytic models, path models, and discrete latent variable models can be represented as DAG models. However, as you might suspect from the name, directed **acyclic** graphs, DAG models do exclude a kind of model familiar in engineering, economics, and the social sciences: those in which variable A influences variable B, and at the same time, B influences A. Economic and physical processes are often modelled by linear systems of this sort; so-called non-recursive structural equation models. Another use for such systems is modelling time series in which feedback is present, as well as structures in which causal influences propagate in different directions in certain subpopulations.

---

[1]I thank P. Spirtes, C. Glymour, R. Scheines and C. Meek for helpful conversations. Research for this paper was supported by the Office of Naval Research through contract number N00014-93-1-0568.

This paper will give a survey of some recent research aimed at filling this gap; developing a theory of **cyclic** graphical models, that would allow the generalization of acyclic techniques and methods to the cyclic case.

1.2 Linear Structural Equation Models (linear SEMs)

In an SEM, the variables are divided into two disjoint sets, the error terms, and the non-error terms. Associated to each non-error variable V there is a unique error term ε_V. A linear SEM contains a set of equations in which each non-error random variable V is written as a linear function of other non-error random variables and ε_V. A linear SEM also specifies a joint distribution over the error terms. In our discussion we will consider only linear SEMs with error terms that are jointly independent, but as we shall see, in an important sense, at least within the context of our discussion, nothing is lost by this restriction.

The following is an example of such a model:

$$X=\varepsilon_X \qquad Y=\varepsilon_Y$$
$$A= \alpha_1 \cdot X + \alpha_2 \cdot B + \varepsilon_A$$
$$B= \beta_1 \cdot Y + \beta_2 \cdot A + \varepsilon_B$$

The ε_V's are jointly independent standard normal error terms.

A structural equation model in which, for some ordering of the variables, the matrix of coefficients is in lower triangular form, is said to be *recursive*.

1.3 Graphs

There is a directed graph, naturally associated with a given linear SEM, by the rule that there is an edge from X to Y (X→Y) if and only if the coefficient of Y, in the equation for X, is non-zero. By convention we do not include error terms in the graph. Hence the graph relating to the model above is (here the error terms are omitted, being assumed jointly independent):

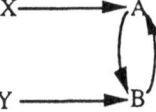

A linear SEM with a jointly independent distribution over the error terms constitutes a parameterization of its associated graph. It is easy to see that the linear SEM associated with an **acyclic** graph will be a recursive structural equation model.

1.4 Linear Entailment

A directed graph containing disjoint sets of variables **X**, **Y**, and **Z**,[2] *linearly entails* that **X** is independent of **Y** given **Z** if and only if **X** is independent of **Y** given **Z** for all values

---

[2] We use bold face letters (**X**) to denote sets of variables.

of the non-zero linear coefficients and all distributions of the exogenous variables in which they have positive variances and are jointly independent. It is important to note that in any particular SEM with directed graph G, there may be conditional independencies which hold even though they are not linearly entailed by G. However, if a zero-correlation holds for some but not all parameterizations of G, then the set of parameterizations in which this 'extra' conditional independence holds, is of zero Lesbesgue measure over the set of all parameter value assignments to the non-zero linear coefficents.

1.5 Conditional Independencies and Equivalence in a graph

In an **acyclic** graph G, there is a graphical 'path' condition which holds between disjoint vertex sets X, Y and Z in the graph if and only if G linearly entails that $X \perp\!\!\!\perp Y \mid Z$.[3] Similarly, the same graphical 'path' condition holds between X, Y and a set Z, not containing X or Y, if and only if G the partial correlation between X and Y controlling for Z, vanishes: $\rho_{XY.z} = 0$. We can calculate the partial correlations that are zero in all linear parametizations of G in which X and Y have correlated errors in the following way. First, form a directed graph G' in which X and Y are the effects of a latent common cause T. The same graphical path condition holds in G' iff in every parameter assignment to G in which X and Y have correlated errors, $\rho_{XY.z} = 0$.

This observation is central to the usefulness of the graphical method. The task of generalizing this result to the cyclic case has already been accomplished: Building on the work of Haavelmo(1943), Spirtes (1993) showed that the same graphical condition, the Geiger-Pearl-Verma d-separation criterion (defined in the Appendix) which determines whether a particular conditional independence relation or zero partial correlation is linearly entailed by a *recursive* structural equation model, can also be used with linear *non-recursive* models. Or equivalently, that the same technique used for reading conditional independencies from an acyclic graph can be applied in the cyclic case.

We say that two graphs are equivalent if they both linearly entail the same set of conditional independencies. It is important to be clear what we are establishing when we work out the conditional independencies entailed by a given model: we are calculating the conditional independence consequences of having a certain **form** of linear equations, i.e. having linear equations in which certain coefficients are zero. We are **not** trying to estimate parameters, we are **not** making any distinction between latent and measured variables, and we are **not** constructing a model from data; though the development of efficient procedures for determining the equivalence of cyclic models will facilitate the construction of computer aids for model specification and updating.

---

[3] '$X \perp\!\!\!\perp Y \mid Z$' means that 'X is independent of Y given Z'.

2 Partial Results about Equivalence for Cyclic Graphs

Although there is an $O(n^3)$ algorithm which can determine when two **acyclic** graphs are equivalent in our sense, a feasible algorithm which will establish this equivalence for **cyclic** graphs has not yet been obtained. Several preliminary results indicate that there is considerable heterogeneity in the equivalence class - much more than in the acyclic case.

2.1 Contrast with the Acyclic case

In the acyclic case, if A and B are dependent conditional on *any* subset of the other variables, then either A→B, or B→A, i.e. either B is a direct cause of A or A is a direct cause of B. However, this implication fails in the cyclic case.[4] For example:

$$A \longrightarrow B \rightleftarrows C$$

There are no independencies, conditional or otherwise entailed by this graph, and yet there is no edge between A and C. This marks a significant difference between the cyclic and acyclic equivalence classes.

2.2 Non-locality property

In the acyclic case, if two graphs are not equivalent then there will be some conditional independence between variables separated by at most two edges, entailed by one graph, and not by the other. This means that we need only look at the structure of triples of adjacent vertices in order to establish that two graphs are equivalent. This is not true for the cyclic case, as the following example shows:

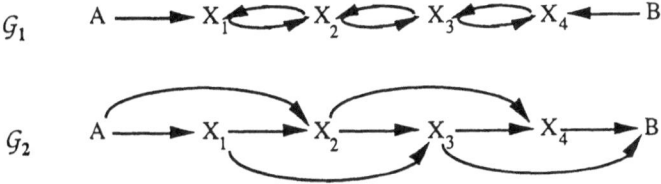

G_1 and G_2 are **not** equivalent. Although every independence implied by G_2, is also implied by G_2, in G_1, A⊥̸ B | ∅, while in G_2, A⊥ B | ∅. But A and B are separated by more than two edges (in both graphs). It is simple to see that we could extend these graphs so that they continued to entail the same conditional independencies, with the exception of A ⊥ B | ∅, while A and B were separated by arbitrarily many edges. This result, however leaves open whether a polynomial time algorithm exists for deciding equivalence.

---

[4]A similar point is made in Whittaker (1989).

2.3 Criteria for detecting feedback:

We give a set of conditions which are sufficient, though not necessary conditions for cyclicity. The following four conditional independence conditions can only be linearly entailed (simultaneously) by a cyclic graph.

For some triple of vertices, the following hold:

(i) For any set S, not containing A or B, $A \perp\!\!\!\perp B \mid S$

(ii) For any set S, not containing B or C, $B \perp\!\!\!\perp C \mid S$

(iii) There is a set T, with $B \in T$ and $A \perp\!\!\!\perp C \mid T$

(iv) There is a set T^*, with $B \notin T^*$ and $A \perp\!\!\!\perp C \mid T^*$.

As an illustration of this criterion, the following model, presented by Whittaker (1989), is not equivalent to any acyclic graph:

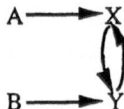

Here <A,X,B> and <A,Y,B> satisfy the conditions given above.

2.4 The Orientation of cycles

Our last and possibly most interesting result says that given a graph G with a cycle C, there is a graph G^*, in which C is replaced by another cycle C^*, having the opposite orientation to C. Thus if C is clockwise, C^* is anti-clockwise, and vice-versa.

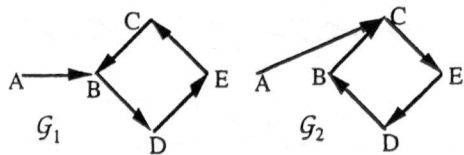

In the above example, in G_1, the cycle <C,B,D,E> has anti-clockwise orientation, while in G_2, the corresponding cycle <C,B,D,E> has clockwise orientation.

One important consequence of this result is that it is not possible to orient a cycle merely using conditional independence information.

Appendix

Definition of d-separation:

If there is an arrow from A to B or from B to A is called an edge between A and B.

Given three vertices A, B and C such that there is an edge between A and B, and between B and C, then if the edges 'collide' at B, (i.e. $A \rightarrow B \leftarrow C$) then we say B is a *collider* between A and C, relative to these edges. Otherwise we will say that B is a non-collider between A and C, relative to these edges. e.g. in the following cases A is a non-collider: A→B→C, A←B→C, A←B←C.

If there is an arrow from A to B (A→B), then we say that A is a parent of B, and B is a child of A. We define 'descendant' relation as the transitive reflexive closure of 'child', and similarly, 'ancestor' as the transitive reflexive closure of 'parent'. A sequence of distinct edges $<E_1,...,E_n>$ in G is an *undirected path* if and only if there exists a sequences of vertices $<V_1,...V_{n+1}>$ such that for $1 \leq i \leq n$ either $<V_{i+1},V_i>=E_i$ or $<V_i,V_{i+1}>=E_i$. We are now in a position to define d-connection:

For disjoint sets, **X**, **Y** and **Z**, **X** is *d-connected to* **Y** *given* **Z** if for some X∈ **X**, and Y∈ **Y**, there is a path from X to Y, satisfying the following conditions:

(i) if A, B and C are adjacent vertices on the path, and B∈ **Z**, then B is a collider between A and C.

(ii) If B is a collider between A and C, then there is a descendant D, of C, and D∈ **Z**.

If **X** and **Y** are not d-connected given **Z** then **X** and **Y** are said to be d-separated by **Z**.

The following important theorems gives the relationship between d-separation and the linear entailment of conditional independencies, and partial correlations. It was proved for the cyclic case by Spirtes (1993).

Theorem (Spirtes): In a (cylic or acyclic) graph G, for disjoint sets, **X**, **Y**, **Z**, **X** and **Y** are d-separated given **Z**, if and only if G linearly entails **X** ⊥⊥ **Y** | **Z**.

Theorem (Spirtes): In a (cylic or acyclic) graph G, for any set **Z**, not containing X or Y, X and Y are d-separated given **Z**, if and only if G linearly entails $\rho_{XY.Z} = 0$.

REFERENCES:

COX, D.R., and WERMUTH, N. (1993) Linear dependencies represented by chain graphs. In *Statistical Science, 1993*, 8 No.3 , 204-283

GEIGER, D. (1990). Graphoids: a qualitative framework for probabilistic inference. PhD dissertation, Univ. California, Los Angeles.

HAAVELMO,T.(1943). The statistical implications of a system of simultaneous equations.

HEISE D.(1975). Causal Analysis. Wiley, New York.

KIIVERI, H. and SPEED, T.P. (1982). Structural Analysis of multivariate data: A review. In *Sociological Methodology, 1982* (S. Leinhardt, ed.) 209-289. Jossey Bass, San Francisco.

PEARL, J. (1988) Probabilistic Reasoning in Intelligent Systems. Morgan Kaufman, San Mateo, CA.

SPIRTES, P. (1993) Directed Cyclic graphs, Conditional Independence and Non-Recursive Linear Structural Equation Models. Philosophy, Methodology and Logic Technical Report 35, Carnegie Mellon University

SPIRTES, P., GLYMOUR C. and SCHEINES R. (1993), Causation, Prediction and Search. Lecture Notes in Statistics, Springer-Verlag.

WHITTAKER, J. (1989) Graphical Models in Applied Multivariate Statistics, Wiley.

Automatic Stable Parameters for Nonlinear Regression

Gavin J.S. Ross

Statistics Department, IACR Rothamsted Experimental Station

Abstract. Stable parameters allow nonlinear regression models to be fitted efficiently and accurately. Ideally the parameters are defined by the user and the transformations are programmed explicitly. It is shown that for some models and data sets the parameters may be chosen automatically and the transformations solved numerically. The method may be adapted for use with separable linear parameters.

Keywords. Nonlinear models, stable parameters, optimization

1. Introduction

Stable parameters (Ross, 1970,1990) are parameter systems expected to provide rapid and efficient solution of nonlinear optimization problems. If they are estimators of contrasting data statistics they will have relatively small variance and low mutual correlations. Preliminary summarization of the data is necessary to ensure that they are appropriate, and they may need to be chosen carefully to enable the defining parameters to be calculated. The outstanding problem is to devise an algorithm which, given a data set and a user-defined procedure for evaluating expectations as functions of one set of parameters, will provide automatically an alternative set of stable parameters, fit the model using the stable parameters, and report the equivalent values of the defining parameters.

In this paper it is shown that automatic stable parameters may be found for a certain class of models and parameter systems, but that some structural difficulties cannot be avoided. There is an additional computing cost in solving the transformation equations, but this is partially offset by the reduced number of iterations required in optimization.

2. Monotone nonlinear curve fitting

Given n (x,y) pairs and a model with expectation function $E(y) = f(x,\theta)$ with p parameters, a set $\mathbf{X}$ of p specific values of x may be chosen such that the values of $\phi = f(\mathbf{X},\theta)$ are each supported by neighbouring data points, and correlations between them are low. Then the quantities ϕ may be used as stable parameters, to be estimated instead of θ, guaranteeing rapid convergence of optimization routines and a dispersion matrix from which the required estimates of θ and its dispersion matrix may be obtained.

The main difficulty, apart from choice of $\mathbf{X}$, is to solve the p simultaneous equations for θ given ϕ. If f is a fairly simple function and $\mathbf{X}$ is chosen in some regular manner it is possible to solve the equations explicitly. Exponential, logistic and hyperbolic curves may be treated in this way, as described in Ross (1990). The range of values of ϕ leading to acceptable values of θ may have to be restricted (for example to avoid complex roots of polynomials), and the stability of ϕ depends on the distribution of data points and their relative weighting.

The numerical solution of the equations requires an initial estimate $\theta_{(0)}$ of θ with which we can compute the corresponding values of $\phi_{(0)}$ and the Jacobian matrix $J = (\partial\phi_i/\partial\theta_j)$ by differencing over a suitable short interval of θ_j. We now have a locally linear approximation to the transformation which can be solved by inverting J and computing a new estimate

$$\theta_{(1)} = \theta_{(0)} + J^{-1}.(\phi - \phi_{(0)})$$

and testing for convergence in terms of the modulus of $(\phi - \phi_{(1)})$ which should decrease towards zero. The values of θ are then used to compute the vector of expectations $E(y)$ for the whole data set. The main difficulty occurs when the solution does not exist, or when some parameters tend to infinity and then change sign, and the numerical procedure does not converge.

For numerical optimization of the residual sum of squares or likelihood function it is important that the solution of θ is performed in a consistent manner. It is better to choose a constant starting value and a constant number of iterations than to rely on a convergence criterion which may produce an abrupt change in θ for small changes in ϕ. Too many iterations will slow down the algorithm, and exact solutions are unimportant during optimization, since they merely affect the form of the transformation, not the function being fitted.

2.1 Numerical example

Suppose that we wish to fit the two-compartment model

$$E(y) = (p.\exp(-qx) - q.\exp(-px))/(p-q)$$

to data for which we choose stable ordinates at $x = 1$ and at $x = 2.6$, where the expected values ϕ are required to be .92 and .75 respectively. Given initial estimates $p = 2$ and $q = .2$ we obtain in successive iterations:

p	q	ϕ_1	ϕ_2	J		J^{-1}	
2.0000	0.2000	.8947	.6600	-.0271	-.0346	-68.44	24.70
				-.4880	-1.352	1.752	-1.372
2.4899	0.1209	.9271	.7675	-.0130	-.0155	-130.28	44.75
				-.5740	-1.671	1.208	-1.014
2.6335	0.1301	.9199	.7500				

showing rapid convergence in this simple example. However for different initial values of p and q it may be necessary to modify the steps to ensure convergence. For some values of ϕ there is no real solution, and the iterations are terminated at some point which is not a true solution.

2.2 The automatic transformation algorithm

Given the data set $(\mathbf{x}, \mathbf{y})$ and a function evaluation routine, the simplest method of automating the evaluation of $f(\mathbf{X})$ is to choose as $\mathbf{X}$ p values from the n values of x in the data. It is then only necessary to modify the function evaluation routine to operate on the subset $\mathbf{X}$, thereby saving much time if p<<n, but avoiding special programming. The parameters are individually stable because they are direct estimates of at least one observation.

   Choice of the subset $\mathbf{X}$ may be manual or automatic. If the function is monotone increasing or decreasing it is sensible to take the x values corresponding to the extreme values of y, and to find p-2 intermediate values of y approximately equally spaced, and use their corresponding x values. The values of x must be different, so checks are needed to ensure that is so. Thus if there are four parameters to estimate it is necessary to find y_{min}, y_{max} and the values closest to $(2y_{min}+y_{max})/3$ and $(y_{min}+2y_{max})/3$. Expectations at equally spaced x values may not be jointly stable, because if the slope of the curve is close to zero they may be highly correlated, for example as an asymptote is approached. If the function is not monotone the solution may not be unique, and so a manual selection of points may be necessary.

   Initial estimates of θ are required, and these must be sufficiently close to the solution to allow the transformation equations to be solved, at least approximately. Starting with estimates of ϕ obtained from the observations, the values of θ are estimated and the remaining expectations computed, from which the log likelihood or sum of squares is obtained as the objective function for optimization. If possible the initial estimate of θ should be periodically updated, between iterations (to avoid numerical instability in differencing algorithms), to improve precision of the transformation. This may require communication back from the optimization routine, or counting of function evaluations.

   When the algorithm converges and the estimates of ϕ and the dispersion matrix are obtained, the original parameters θ are estimated as computed at the final function evaluation, and the inverse Jacobian matrix used to convert from the dispersion matrix of ϕ to that of θ.

2.2.1 Numerical Example

The annual numbers (in millions) of passengers travelling by sea from UK ports between 1964 and 1983 is given in Table 1 (adapted from Ross, 1990). It is not difficult to fit a sigmoid curve to these data, with asymptotes about 4 and 14, and median date about 1977. The four-parameter logistic curve,

$$E(y) = a + c/(1 + \exp(-b(x-m)))$$

may be fitted with x = date - 1963, and starting values (b,m,c,a) = (.5,14,10,4). The algorithm then selects as parameters the expectations at x=1,11,15 and 20 respectively, using the observed data as starting values. A simple Newton-Raphson optimization converges in four iterations, as follows:

Iteration	ϕ_1	ϕ_2	ϕ_3	ϕ_4	Residual S.S.
0	4.1000	6.8700	9.9700	13.3800	2.33205
1	4.1806	7.3009	9.9298	13.7234	0.97757
2	4.3135	7.1949	9.9730	13.6910	0.84336
3	4.3156	7.1853	9.9790	13.6832	0.84210
4	4.3157	7.1852	9.9791	13.6832	0.84210

The estimated standard errors of the stable parameters are (.134, .093, .100, .176) and the maximum correlation (between the first two) is - .40. The corresponding estimates of the defining parameters, their standard errors and correlations are as follows:

Parameter		S.E.	Correlations
b	0.2346	0.0305	1.000
m	15.8403	0.8809	-0.891 1.000
c	13.4514	1.6542	-0.964 0.962 1.000
a	3.9141	0.2714	0.898 -0.656 -0.827 1.000

Direct optimization of the same problem using the defining parameters required 19 iterations from the same initial values. The overall time taken was similar because the transformation was not required, but with a longer series of data the advantage would tend to be greater because only four expectations have to be fitted each time.

Table 1. Millions of sea passengers leaving UK ports.

Year	1964	1965	1966	1967	1968
Number	4.10	4.44	4.46	4.63	4.76
Year	1969	1970	1971	1972	1973
Number	5.39	5.S8	5.99	6.13	6.44
Year	1974	1975	1976	1977	1978
Number	6.87	7.94	8.28	9.15	9.97
Year	1979	1980	1981	1982	1983
Number	10.66	11.S0	12.51	13.17	13.38

The method was tried on a five-parameter generalised logistic, but the power parameter in the model was too close to infinity for the method to work unless the initial estimates were very close to the solution. This is because of the extreme nonlinearity of the transformation, which in practical algorithms requires a different type of transformation.

3. Stable transformations for separable linear parameters

Separable linear parameters are parameters that may be estimated by linear methods if the remaining parameters are fixed. The simplest linear parameters are the overall scale factor (necessary to model a change of units) and a parameter to represent a change of origin. For example, the four-parameter logistic curve in the previous section may be written

$$E(y) = a + c.z(x,b,m)$$

where $z(x,b,m) = 1/(1 + \exp(-b(x-m)))$.

The parameters a and c are estimated separately given b and m, and are not evaluated until the optimization has converged.

If we eliminate the parameters a and c from the simultaneous equations, the remaining equations for b and m become

$$(\phi_2-\phi_1)/(\phi_4-\phi_1) = (z(x_2) - z(x_1)) / (z(x_2) - z(x_1))$$
$$(\phi_3-\phi_1)/(\phi_4-\phi_1) = (z(x_3) - z(x_1)) / (z(x_2) - z(x_1))$$

and we can optimise with respect to the two ratios as stable parameters. Given starting values of b and m, the Jacobian algorithm (using numerical differencing) provides estimates corresponding to a pair of ratios, and the function z is then worked out for all observations, and linear regression used to determine a and c and the residual sum of squares.

In the example of the sea passengers (2.1) the iterations were as follows:

Iteration	Ratio 1	Ratio 2	Residual S.S
0	.29849	.63254	1.07232
1	.31007	.60029	0.85443
2	.30650	.60439	0.84212
3	.30632	.60458	0.84210

and the final results were almost identical to those obtained in the four-parameter optimization. The direct optimization of b and m using separable linear parameters required seven iterations.

In general there may be more than two separable linear parameters, as in the fitting of mixtures of components, but there is unfortunately no simple analogue of differences and ratios that allows the nonlinear parameters to be expressed as transformations of functions of expectations. However the method described is extremely useful as in curve fitting there is nearly always a scale factor and very often a variable origin parameter.

4. Programming the algorithm

For ease of use the algorithm should be provided as a simple modification to existing facilities for fitting general models. How this is done depends on the structure of the program into which it is incorporated. The implementation in MLP (Ross, 1987) will be described.

To specify a general model fitting problem in MLP it is necessary to define the parameters of the model (by writing NAMES p1=b,m,c,a), the data variates and expectation vector (v1=x,y,ey) and a model definition: MODEL (2) ey = a + c/(1+exp(b*(m-x)));
together with options to specify the number of parameters, the number of data variates, the

observation and expectation variates, and the method of calculating the sum of squares or log likelihood (normal least squares, separable normal, Poisson, Binomial, multinomial, gamma etc.). Initial estimates, step lengths and parameter bounds may also be specified.

The transformation algorithm is invoked by a directive TRP, after the model has been specified. TRP has several actions, including the insertion of an explicit transformation into the model, and restoration of the original model. The automatic method requires a working array to store the original defining parameters, and a list of unit numbers which correspond to the expectation parameters. If the list is absent the unit numbers are chosen by the algorithm described. Thus if the working array starts at constant 11 in the fixed store, the directive

TRP c11;;

is all that is required to organise the transformation equations, the rewriting of the model code, the replacement of initial values, steps and bounds and parameter names, and modification of the output to report back the estimates with respect to both sets of parameters. The directive TRP ;; restores the original setup.

The model code is modified so that it calculates either the expectation at the selected units only, or the complete expectation variate, according to a switch. The likelihood calculation incorporates the transformation algorithm, so that when TRP is invoked the parameters are assumed to be the expectation parameters, and the model calculations first operate on selected units only, and after a fixed number of iterations, on the complete variate. The transformation algorithm also communicates with the optimization algorithm to decide when the provisional estimates may be updated, which is essential in order to ensure numerically differentiable log likelihoods without being handicapped by poor initial estimates.

It is hoped that a version of the algorithm may be incorporated into a future release of GENSTAT.

5 Conclusion

The feasibility of automatic transformation to stable parameters has been shown for a restricted class of problems. The algorithm cannot be expected to handle non-existent solutions or to overcome the numerical problems of singularities, infinite values or multiple roots. The success rate decreases with the complexity of the model. But the method provides an introduction to the power of the stable parameter method, and can encourage the user to write a more stable parameterisation of the model.

References

Ross, G.J.S. (1970)   The efficient use of function minimization in non-linear maximum likelihood estimation. Appl. Statist. vol. 19, 205-221

Ross, G.J.S. (1987)   Maximum Likelihood Program 3.08. Numerical Algorithms Group, Oxford.

Ross, G.J.S. (1990)   Nonlinear Estimation. Springer Verlag, New York.

Computationally Intensive Variable Selection Criteria

Johannes H. Venter, **J. Louise J. Snyman**

Department of Statistics, Potchefstroom University

Keywords. multiple linear regression, variable selection, nearest subspace, risk estimation, Mallows C_p, bootstrap, empirical Bayes estimation, simulation

1 Model and approach

Consider the standard multiple linear regression model in which we observe the n-vector Y having the form

$$Y = X\beta + e = \sum_{j=1}^{m} \beta_j x_j + e \tag{1}$$

where e is $N_n(0, \sigma^2 I_n)$-distributed, β is an m-vector of unknown regression coefficients and $X = [x_1\ x_2\ ...\ x_m]$ is a given fixed $n \times m$ design matrix of full column rank m with columns representing the values of the explanatory variables. Our inference objective is that of accurate model description and prediction, that is, accurate estimation of $\mu = X\beta$ in a parsimonious way. For this purpose some of the explanatory variables may be redundant and the goal of variable selection is to select a subset of these explanatory variables for inclusion in the model. Assume that all the explanatory variables are open for selection. If the inclusion of some variables such as a constant term is mandatory, the problem can easily be transformed to the present form.

Denote the column space of X by M with dimension $\dim(M) = m$ and note that $\mu \in M$. Further let $\mathcal{L}$ denote the family of 2^m linear subspaces of M with typical member L, $\dim(L) = l$, spanned by a subset of l of the columns of X. The problem of variable selection amounts to writing μ in terms of only some of the columns of X. So, we wish to data-dependently select $\hat{L} = \hat{L}(Y) \in \mathcal{L}$ to which we think μ belongs and estimate μ accordingly by some estimator $\hat{\mu} = \hat{\mu}(Y) \in \hat{L}$.

Under the assumption that $\mu \in M$ the optimal estimator of μ is the ordinary least squares estimator, that is, the orthogonal projection of Y on the full space

*This research was supported in part by grants from the FRD of South Africa. Computations were carried out on IBM RS6000 workstations using IMSL routines.

M, denoted $P_M Y$ (with many attractive properties amongst which being minimax). Therefore, if we thought that $\mu \in L \in \mathcal{L}$, $P_L Y$ seems the relevant estimator to use. The residual sum of squares associated with L is

$$\text{RSS}_L = \|Y - P_L Y\|^2 = \|P_{L_\perp} Y\|^2 \qquad (2)$$

where $L_\perp$ denotes the orthogonal complement of L in $\mathfrak{R}^n$ and $\|z\|$ is the Euclidean norm of $z \in \mathfrak{R}^n$. RSS_L measures the prediction error/lack of fit of the model corresponding to L. Geometrically, it may be interpreted as the orthogonal distance between Y and the subspace L. The predictive risk in using $P_L Y$ to estimate μ is given by

$$E\|P_L Y - \mu\|^2 = \|P_{L_\perp} \mu\|^2 + \sigma^2 \dim(L), \qquad (3)$$

provided that L does not depend on the data Y.

2 Mallows C_p and the nearest subspace estimators

The risk of $P_L Y$ in (3) is estimated unbiasedly by

$$C_L = \text{RSS}_L + \left(2\dim(L) - n\right)\hat{\sigma}^2 = \|P_{L_\perp} Y\|^2 + \left(2\dim(L) - n\right)\hat{\sigma}^2 \qquad (4)$$

where

$$\hat{\sigma}^2 = \text{RSS}_M \big/ \left[n - \dim(M)\right] = \|Y - P_M Y\|^2 \big/ (n - m) = \|P_{M_\perp} Y\|^2 \big/ \dim(M_\perp) \qquad (5)$$

is the usual estimator of σ^2 from the full model. This suggests selecting $L = \hat{L}$ to minimize (4). This risk estimator (4) is essentially the well-known Mallows C_p - criterion, more appropriately called C_L here. Mallows just uses the scaled form of this by dividing with $\hat{\sigma}^2$ throughout.

The C_L criterion in (4) is a monotone function of RSS_L as is most other variable selection criteria in the literature. Hence, for fixed $\dim(L) = l$, C_L and the others will select the subspace with smallest RSS_L (that is, the nearest subspace of dimension l to Y). Therefore it seems reasonable to limit our choice to only the nearest subspaces of a given dimension to Y. To put this more formally, let $\hat{L}(Y,l) \in \mathcal{L}$ be the nearest subspace (NS) of dimension l to Y, i.e. $\hat{L}(Y,l)$ minimizes RSS_L among all $L \in \mathcal{L}$ with $\dim(L) = l$. Then the sequence $P_{\hat{L}(Y,l)} Y$ are the NS estimators with l varying over the allowable dimensions. The problem of selecting a subspace from $\mathcal{L}$ has thus now reduced to the problem of selecting a subspace dimension l. But how do we select l ? The predictive risk of a typical NS estimator for μ is given by

$$R_l(\mu, \sigma^2) = E\|P_{\hat{L}(Y,l)} Y - \mu\|^2. \qquad (6)$$

Suppose we possess an estimator $r(Y,l)$ for this unknown risk. Then we may consider $r(Y,l)$ as a variable selection criterion and select $l = \hat{l}(Y)$ based on the data Y to minimize $r(Y,l)$. The subspace selected by this criterion is then given by $\hat{L}(Y, \hat{l}(Y))$ with final estimator $P_{\hat{L}(Y, \hat{l}(Y))} Y$ for μ. Therefore the only remaining issue is how to estimate $R_l(\mu, \sigma^2)$. This is not an easy matter, especially in view of the complexity of $P_{\hat{L}(Y,l)}Y$. For the sake of simplicity, we assume σ^2 known and equal to 1 and drop it from notation henceforth.

3 Breiman's little bootstrap criterion

Breiman (1992) points out that $C_{\hat{L}(Y,l)}$, obtained by replacing L with $\hat{L}(Y,l)$ in (4), is a severely downwardly biased estimator of the risk (6). He introduces a resampling type criterion which he calls the little bootstrap. A brief derivation of his proposal follows: Rewrite the risk of the NS estimator as

$$R_l(\mu) = E\|P_{\hat{L}(Y,l)}Y - \mu\|^2 = E\|Y - \mu - P_{\hat{L}(Y,l)}Y\|^2 = n + E\|P_{\hat{L}(Y,l)}Y\|^2 - 2\psi(\mu) \quad (1)$$

with $\psi(\mu) = E(Y - \mu)' P_{\hat{L}(Y,l)}Y$. The first term of (7) is known while $\|P_{\hat{L}(Y,l)}Y\|$ estimates its expectation unbiasedly. To estimate $\psi(\mu)$ Breiman introduces the new n-vector $\tilde{Y} = Y + tU$ with $t > 0$ and U $N_n(0, I_n)$-distributed independently of Y and defines

$$\hat{\psi}_t(Y) = \frac{1}{t^2} E\left[(\tilde{Y} - Y)' P_{\hat{L}(\tilde{Y},l)}\tilde{Y} \mid Y \right]. \quad (2)$$

Then, by manipulating conditional expectations, it can be shown that

$$E\,\hat{\psi}_t(Y) = \psi\left(\mu/\sqrt{1+t^2}\right) \approx \psi(\mu) \text{ if } t \approx 0. \quad (3)$$

Hence for t small, $\hat{\psi}_t(Y)$ estimates $\psi(\mu)$ approximately unbiasedly and

$$r_{LB}(Y, l) = n + \|P_{\hat{L}(Y,l)}Y\|^2 - 2\hat{\psi}_t(Y) \quad (4)$$

is an approximately unbiased estimator for the risk of the NS estimator. $\hat{\psi}_t(Y)$ must be evaluated by Monte Carlo simulation.

How do we choose t? It can be shown that $\mathrm{Var}(\hat{\psi}_t(Y))$ becomes large if $t \to 0$. So, whereas small t reduces the bias of the estimator, it inflates the variance; thus a compromise choice of t is needed. Breiman suggests choosing t in the range 0.6 to 0.8 or even as large as 1. From (8) and the definition of $\psi(\mu)$ it follows that $\hat{\psi}_t(Y) = \psi(Y/t)$. Thus, for $t = 1$ the little bootstrap actually becomes the simple substitution estimator $\psi(Y)$ for $\psi(\mu)$.

4 Empirical Bayes approach

We now introduce an empirical Bayes approach towards estimating the risk of the NS estimators. Typical β-configurations relevant in the variable selection context have some components equal (or close) to zero with the remaining components different from zero. Thus, a reasonable yet tractable prior for β is

$$\beta \sim N_m(0, D) \text{ where } D = \mathrm{diag}\{d_1, \dots, d_m\}, \ d_j \geq 0 \ \forall \ j. \quad (5)$$

By taking d_j equal (or close) to zero or different from zero, provision is made for β_j to be equal (or close) to zero or different from zero respectively. This prior on

β translates into the prior $N_n(0, XDX')$ for $\mu = X\beta$. With this choice of prior, the posterior for μ given Y follows as $N_n(CY, C)$ with $C = I_n - (I_n + XDX')^{-1}$. The Bayes estimator of the risk $R_l(\mu)$ of the NS estimators $P_{\hat{L}(Y,l)}Y$ is given by its posterior expected value $E[R_l(\mu)|Y]$ which can be shown to be equal to

$$r_{EB}(Y,l) = E[R_l(\mu)|Y] = \text{tr}(I+C)^{-1}C + E\left[\left\|P_{\hat{L}(\tilde{Y},l)}\tilde{Y} - (I+C)^{-1}C(\tilde{Y}+Y)\right\|^2 \Big| Y\right] \quad (1)$$

where $\tilde{Y}$ is a new n-vector such that $\tilde{Y}|Y \sim N_n(CY, I_n + C)$. Again evaluation by simulation is needed.

The prior hyperparameter D must be determined by empirical Bayes means. We choose D to maximize the $N_n(0, I_n + XDX')$ marginal likelihood of Y. Unfortunately it does not seem possible to obtain an explicit expression for D and we have to proceed in an iterative fashion. For given j, it can readily be shown that this likelihood is unimodal in d_j and the maximizing choice of d_j can be expressed as a function of the d_i's with $i \neq j$. We therefore start the iteration at some initial guess of D. An improved value of D is obtained by replacing d_j by its maximizing choice in terms of the remaining d_i's with $i \neq j$, successively for $j = 1, \ldots, k$. In this process, the likelihood continually increases and the process of improving D is repeated until convergence of D as whole is obtained. In practice we found convergence to be quite rapid, but we have not been able to proof that convergence is to the global maximum.

5 Examples

We demonstrate the proposed procedures by applying them to two data sets. The first is the artificial data set of Abt (1967) which is only mildly ill-conditioned. The second data set is the well-known highly collinear Hald data (Draper and Smith (1981)) on cement hardening. In both cases the original designs include a mandatory intercept. These were then transformed to no-intercept designs in correlation form. For the transformed designs the correlation matrices $X'X$ and their eigenvalues are shown in Tables 1 and 2. Tables 3 and 4 provide the NS variables and the $C_{\hat{L}(Y,l)}$, $r_{LB}(Y,l)$ and $r_{EB}(Y,l)$ risk estimates for each of the possible subdimensions l. In the case of $r_{LB}(Y,l)$ three t-choices were considered. For example, for the Abt data the $C_{\hat{L}(Y,l)}$ procedure selects the NS model of dimension $l = 2$ since it has lowest risk estimate here. Note that the downward bias of $C_{\hat{L}(Y,l)}$ becomes more apparent as m increases and that $C_{\hat{L}(Y,l)}$ and $r_{LB}(Y,l)$ are identical for $l = 0$. Looking at the risk estimates $r_{LB}(Y,l)$ it is clear that the choice of t is not critical as it quite often results in the same choice of model. From these examples it might appear that $C_{\hat{L}(Y,l)}$ selects a more parsimonious model, but this is not generally true.

Table 1. Abt (1967) data:

Correlation matrix $X'X$			Eigenvalues
1.000			1.921
0.017	1.000		0.983
-0.580	-0.705	1.000	0.096

Table 2. Hald data (Draper and Smith (1981)):

Correlation matrix $X'X$				Eigenvalues
1.000				2.236
0.229	1.000			1.576
-0.824	-0.139	1.000		0.187
-0.245	-0.973	0.030	1.000	0.002

Table 3. Abt (1967) data:

# variables l	0	1	2	3
NS variables	-	3	1,2	1,2,3
$C_{\hat{L}(Y,l)}$	830.33	122.62	2.80	3.00
r_{LB} $(t=0.25)$	830.33	120.63	2.95	3.00
r_{LB} $(t=0.50)$	830.33	122.00	2.64	3.00
r_{LB} $(t=1.00)$	830.33	122.94	2.75	3.00
r_{EB}	830.52	136.12	3.23	3.00

Table 4. Hald data (Draper and Smith (1981)):

# variables l	0	1	2	3	4
NS variables	-	4	1,2	1,2,4	1,2,3,4
$C_{\hat{L}(Y,l)}$	441.92	137.73	1.68	2.02	4.00
r_{LB} $(t=0.25)$	441.92	140.26	4.36	3.48	4.00
r_{LB} $(t=0.50)$	441.92	142.77	5.46	3.16	4.00
r_{LB} $(t=1.00)$	441.92	144.88	4.11	3.05	4.00
r_{EB}	443.28	143.76	4.08	3.58	4.00

6 Risk comparisons

One possible way to compare the procedures is in terms of their risk behavior over μ. The risk of any of the final estimators $P_{\hat{L}(Y,\hat{i}(Y))}Y$ relative to the minimax value m is defined as

$$E\left\| P_{\hat{L}(Y,\hat{i}(Y))}Y - \mu \right\|^2 / m \tag{13}$$

In general this is too complicated to calculate analytically and simulation must be used. The minimum and maximum values attained by the risk (13) over μ are of particular interest. We conjecture that the most favorable configuration (MFC) at which the minimum occurs is at $\mu = 0$ and all the simulation studies carried out so far bears this out. Regarding the least favorable configuration (LFC) at which the maximum occurs, the curse of dimensionality and the numerical intensity of the procedures make it virtually impossible to determine the precise location by simulation. In order to make some progress we determined the maximal risks along a number of lines in μ-space which may be considered on intuitive grounds to be candidates for containing the LFC. Lower bounds for the true maximal risks obtained in this way are reported in Table 5 along with the minimal risks for the Abt and Hald design matrices of Tables 1 and 2 and an orthogonal design with 3 regressors.

Table 5. Extreme relative risks

Data set	Maximum relative risk			Minimum relative risk		
	C_L	LB (t=1)	EB	C_L	LB (t=1)	EB
Orthogonal	1.65	1.48	1.51	0.59	0.41	0.42
Abt	1.85	1.65	2.27	0.63	0.52	0.36
Hald	1.68	1.50	2.44	0.56	0.45	0.31

From Table 5 we conclude that the LB and EB procedures both achieve lower risk than Mallows's procedure in the MFC. For the non-orthogonal designs, the EB procedure does this at the cost of a substantially higher maximal risk whereas no such penalty is involved with the LB procedure.

The extreme relative risk behavior of the procedures is not the only feature on which to base a recommendation. Other relevant configurations are those containing many zero β-components (i.e. μ in a low dimensional subspace of M). In such favorable configurations, the estimators should be *adaptive* in the sense of automatically achieving small risk. Our impression at this stage is that the LB and EB procedures both perform very well on this count and clearly better than the C_p procedure.

7 Concluding remarks

Overall the LB procedure with $t = 1$ seems most recommendable but more extensive simulations are needed to clarify the behavior of the proposed procedures, especially for larger m. The computational burden required for this is extremely heavy but since the general approach outlined has application far beyond variable selection only, this effort may be well spent.

8 References

Abt, K. (1967). On the identification of the significant independent variables in linear models. *Metrika*, **12**, 2-15.

Breiman, L. (1992). The little bootstrap and other methods for dimensionality selection in regression: X-fixed prediction error. *Journal of the American Statistical Association*, **87**, Theory and Methods, 738-754.

Draper, N.R. and Smith, H. (1981). *Applied Regression Analysis*. New York: John Wiley.

Mallows, C.L. (1973). Some comments on C_p. *Technometrics*, **15**, 661-675.

Part XXIV

Statistical Inference

A Nonparametric Testing Procedure for Missing Data

Anna Giraldo, Andrea Pallini, **Fortunato Pesarin**

Dipartimento di Scienze Statistiche, University of Padua

Abstract. This paper deals with treatment of missing data in some multidimensional testing problems. Cases in which missing data are not missing at random are studied. In this context, to obtain valid parametric inferences, the process that causes missing data must be specified. Through a nonparametric approach some problems of missing data not missing at random are discussed and characterized. The proposed solution is based on a nonparametric combination of dependent permutation tests, not requiring any specification of non-response model.

1 Introduction

In statistical analyses the problem of missing data is quite common and various solutions have been proposed. See Rubin (1976), Little and Rubin (1987), Dempster *et al.* (1977), Grenlees *et al.* (1982), Wei and Lachin (1984), Servy and Sen (1987), Barton and Cramer (1989), Kim-Hung Li *et al.* (1991), and Giraldo and Pesarin (1992, 1993). The parametric treatment of missing values in multidimensional testing problems strongly depends on the nature of the missingness mechanism. If the probability of one datum being missing does not depend on the value it would assume, then the missing data are said to be missing at random (MAR) and usually inferences can be made conditionally by ignoring the mechanism causing missingness. Otherwise, for data not missing at random (not MAR) this mechanism must be specified and inferences become more complicated. For a more complete introduction see Little and Rubin (1987).

In this paper, for both cases, we examine a solution, based on a nonparametric combination of several dependent permutation tests. See Pesarin (1992) and Pallini and Pesarin (1992). This solution allows a characterization for some not MAR models, in the sense that for such models it gives an asymptotically efficient solution and their treatment is substantially the same as that of MAR models. The data layout we are referring to is the One-Way $MANOVA$; i.e. the equality of c non-degenerate k-variate distributions according to c levels of a symbolic treatment, $c \geq 2$ and $k \geq 1$. We indicate by $\{(\mathbf{X}, \mathbf{O})_1, \ldots, (\mathbf{X}, \mathbf{O})_c\} = (\mathbf{X}, \mathbf{O})$ the observed data, where the elements of observational configuration $\mathbf{O}$ are such that $O_{hji} = 1$ if X_{hji} is observed and $O_{hji} = 0$ if it is missing, $(1 \leq h \leq k; 1 \leq i \leq n_j; 1 \leq j \leq c)$.

In section 2 some notions and definitions are introduced and main conditions for nonparametric combination of several dependent tests are recalled. In section 3 the structure of MAR testing problems and a characterization of not MAR models admitting a suitable solution are discussed. Treatment of missing values is discussed in section 4. Section 5 gives an illustrative example.

2   A nonparametric approach based on a resampling procedure

Let us first suppose the multidimensional testing problem is the equality of c, $c \geq 2$, non-degenerate k-variate distributions without missing values; i.e. the standard layout of a One-Way MANOVA. Given a set of independent k-dimensional random vectors $\mathbf{X}_{ji}$, $i = 1, \ldots, n_j$, $j = 1, \ldots, c$, with unknown distribution $P_j(\mathbf{x})$, the null hypothesis is set as $H_0 = \{P_1(\mathbf{x}) = \cdots = P_c(\mathbf{x})\} = \{\mathbf{X}_1 \overset{d}{=} \cdots \overset{d}{=} \mathbf{X}_c\}$, while $H_1 = \{at\ least\ one\ of\ the\ equalities\ stated\ by\ H_0\ is\ not\ satisfied\}$. The family $\mathcal{P}$ does not require special conditions except the existence of non-degenerate probability distributions $p \in \mathcal{P}$ generating the data. We further suppose the treatment effects are additive with respect to a suitable function Ψ of data. So H_0 is set equivalent to the equality of the mean vectors $\mu_j = E[\Psi(\mathbf{X}_{ji})]$, $i = 1, \ldots, n_j$, $i = 1, \ldots, c$, i.e. $H_0 : \{\mathbf{X}_1 \overset{d}{=} \cdots \overset{d}{=} \mathbf{X}_c\} = \{\mu_1 = \cdots = \mu_c\}$. Here, H_0 is implicitely assumed to be a subclass of a location family of possible distributions of Ψ.

Observe that hypotheses and set of assumptions are such that a permutation test may be defined and applied (cf. Romano (1989,1990)). In fact, null hypothesis H_0 implies that data are exchangeable (with respect to the c samples) under the action of all permutations of their labels (Bell and Sen (1984)). Note further that hypotheses H_0 and H_1 may be conveniently written as

$$H_0 = \bigcap_{h=1}^{k} \{\mu_{h1} = \ldots = \mu_{hc}\} = \bigcap_{h=1}^{k} H_{0h}, \quad H_1 = \bigcup_{h=1}^{k} H_{1h}, \tag{2.1}$$

where specific decompositions of H_0 and H_1 is emphasized.

The problem may be conveniently solved by a nonparametric combination of k dependent univariate tests of H_{0h} against H_{1h}, $h = 1, \ldots, k$. To be specific, we can use a permutation counterpart of the well known Fisher combining statistic, where unknown dependencies between univariate tests are captured by its permutational distribution. We write $\mathbf{T}$ to indicate the k-dimensional vector of real valued statistics which define the k tests of H_{0h} against H_{1h}, $h = 1, \ldots, k$. The Fisher combining statistic is a nondecreasing function in each component of $\mathbf{T}$ which satisfies nice optimality properties. See Pesarin (1991a,1992) for additional references and more details about its use in combining dependent tests.

Here, we assume that all univariate tests are defined by test statistics significant for large values and Positively Upper Orthant Dependent ($PUOD$) (cf. Dharmadhikari and Joag-Dev (1988), chapter 6). This latter condition may be easily shown to be equivalent to unbiasedness of univariate tests to combine. It plays a central role in the characterization of not MAR problems for which a suitable testing solution can be found. See Giraldo and Pesarin (1992,1993) for more on this aspect.

In principle, the permutation test critical values can be determined exactly. In practice, perumtation tests can be generally used by approximating the exact permutation distributions of their test statistics. In the nonparametric combination of dependent tests a convenient solution is that of estimating permutational distributions by a Without Replacement Resampling Procedure ($WRRP$) carried out with S replicates (cf. Dwass (1957)). For more details on nonparametric combination and on the $WRRP$ see Pesarin (1991a,1992). We will use this approach in the next sections.

3 Structure of testing problems

Let us first consider a MAR model for missing data. In this case we can fix $\mathbf{O}$ at its observed value and proceed conditionally. Thus, $H_0 = \{(\mathbf{X},\mathbf{O}) \overset{d}{=} \cdots \overset{d}{=} (\mathbf{X},\mathbf{O})\}$ is set equivalent to $\{(\mathbf{X}_1 \overset{d}{=} \cdots \overset{d}{=} \mathbf{X}_c) \,|\, \mathbf{O}\}$, since symbolic treatment does not affect the selection model. Hence, taking into consideration that the data layout is a $MANOVA$ one, we must take into account the configuration of missing data $\mathbf{O}^*$ in the set of all permutations of the data $(\mathbf{X},\mathbf{O})$. Observe that the number of effective valid data is a permutational variable: it varies according to the random attribution of the missing values to the c samples. In fact, its distribution is multidimensional hypergeometric. A solution for the problem might be found if we are provided with test statistics whose permutational distribution is invariant with respect to the permutational configuration of missing data. In such a case, these test statistics do behave as if they were independent of $\mathbf{O}^*$.

If the missing data are not MAR, i.e. the probability of non-response depends upon the unobserved outcomes, the hypotheses H_0 and H_1 must take into account also the mechanism producing missing data and not only conditioning on it. In fact, symbolic treatment can affect both the distributions of variable $\mathbf{X}$ and of missingness indicators $\mathbf{O}$. Hence, in the case of missing data not MAR, H_0 requires the joint distributional equality of the variable $(\mathbf{X}\,|\,\mathbf{O})$ and of the mechanism of nonresponse in the c samples:

$$H_0 = (\{\mathbf{O}_1 \overset{d}{=} \cdots \overset{d}{=} \mathbf{O}_c\}) \bigcap (\{\mathbf{X}_1 \overset{d}{=} \cdots \overset{d}{=} \mathbf{X}_c \,|\, \mathbf{O}\}). \tag{3.1}$$

This means that the permutational principle must be applied to the whole data $(\mathbf{X},\mathbf{O})$. Hypothesis (3.1) can be decomposed into $2 \cdot k$ sub-hypotheses:

$$H_0 = \left(\bigcap_{h=1}^{k} \{\mathbf{O}_{1h} \overset{d}{=} \cdots \overset{d}{=} \mathbf{O}_{ch}\} \right) \bigcap \left(\bigcap_{h=1}^{k} \{\mathbf{X}_{1h} \overset{d}{=} \cdots \overset{d}{=} \mathbf{X}_{ch} \,|\, \mathbf{O}\} \right)$$

$$= \left(\bigcap_{h=1}^{k} {}_{\mathbf{O}}H_{0h} \right) \bigcap \left(\bigcap_{h=1}^{k} {}_{\mathbf{X}}H_{0h} \right), \tag{3.2}$$

where ${}_{\mathbf{O}}H_{0h} = \{\mathbf{O}_{1h} \overset{d}{=} \cdots \overset{d}{=} \mathbf{O}_{ch}\}$ indicates the equality of the h-th component of the process producing the missing values, and ${}_{\mathbf{X}}H_{0h} = \{(\mathbf{X}_{1h} \overset{d}{=} \cdots \overset{d}{=} \mathbf{X}_{ch})\}$ indicates the distributional equality conditional on the observed configuration of missing values of the h-th component of $(\mathbf{X}\,|\,\mathbf{O})$. Note that for each of the k sub-hypotheses ${}_{\mathbf{O}}H_{0h}$, permutational statistics such as Pearson's χ^2 or other suitable statistics for proper testing categorical data could be appropriate. Moreover, for each of the k sub-hypotheses ${}_{\mathbf{X}}H_{0h}$ the mentioned permutational tests $\mathbf{O}^*$-invariant are appropriate. Hence, the nonparametrical combination of the $2 \cdot k$ univariate tests provides a suitable solution.

Of course, we are able to solve this testing problem if we are provided with univariate test statistics which are unbiased or $PUOD$ for H_0 against H_1. In particular: *the two sets of univariate tests* $\mathbf{T}_1$ *on the* $\mathbf{O}$ *components and* $\mathbf{T}_2$ *on the* $(\mathbf{X} \,|\, \mathbf{O})$ *components need to be PUOD.* This characterizes the not MAR models for which a suitable solution does exist. A general case in which this characteristc condition is achieved includes situations where the probability of non-response is monotonically related to treatment levels.

4 Treatment of missing values

The key for a suitable solution is then to use test statistics whose permutational distribution is invariant with respect to the permutational sample sizes of valid data. Such a test can be seen in Pesarin (1991b) and Giraldo and Pesarin (1992). Here, we will summarize arguments for such a solution. For this task, let us first consider a MAR model. Consider then a test statistics $\mathbf{T}$, based on linear functions of the data and observe that realizations $\mathbf{o}^*$ of $\mathbf{O}^*$, i.e. the random attribuition of valid data to the c samples, induce a partition on the whole orbit associated to $(\mathbf{X}, \mathbf{O})$ into sub-orbits. If the permutational distribution of $\mathbf{T}$ is invariant (independent) with respect to the sub-orbits induced by $\mathbf{O}^*$ we may estimate $F(\mathbf{t} \,|\, (\mathbf{X}, \mathbf{O}))$ by a $WRRP$, i.e. by ignoring the sub-orbits. Let us indicate by $f_{hj} = \sum_i O_{hjl}$ the number of valid data for the h-th variable in the j-th sample; these are represented in the $k \times c$ matrix $\mathbf{f}$. Under the assumptions for the permutational principle, the $\mathbf{o}^*$-invariance (equivalent to the $\mathbf{f}^*$-invariance) implies $F(\mathbf{t} \,|\, (\mathbf{X}, \mathbf{O})) = F(\mathbf{t} \,|\, (\mathbf{X}, \mathbf{f}^*))$ where $\mathbf{f}^* = (\mathbf{f}_1^*, \dots, \mathbf{f}_c^*)$. This distributional invariance is quite difficult to be exactly achieved when $k > 1$ or n is not sufficiently large. So we have to look at approximate solutions. Our way here is to consider the r.v's $(\mathbf{T} \,|\, (\mathbf{X}, \mathbf{f}^*))$ as having mean values and variances not dependent on the realizations of $\mathbf{O}^*$.

Without loss of generality, let us consider $k = 1$, and suppose the test statistic is based on linear functions of the data, $W_j^* = \sum_{i=1}^{n_j} \Psi(X_{ji}^*) \cdot O_{ji}^*$ say, $j = 1, \dots, c$, where Ψ is a suitable function. We first consider the case of $c = 2$ and suppose the test statistics has the form $T^*(a, b) = a \cdot W_1^* - b \cdot W_2^*$, where a and b are two coefficients to be determined assuming that, under H_0, the variance $Var[T^*(\cdot)] = \delta$ is invariant with respect to permutational sample sizes f_j^*, $j = 1, 2$, and the mean value $E[T^*(\cdot)] = 0$. The solutions are $a = (f_1^*/f_2^*)^{1/2}$ and $b = (f_2^*/f_1^*)^{1/2}$. Hence the test, permutationally invariant in mean value and variance has the form:

$$T^* = W_1^* \cdot (f_2^*/f_1^*) - W_2^* \cdot (f_1^*/f_2^*)^{1/2}. \tag{4.1}$$

In the case of $c > 2$, the solution becomes

$$T^* = \sum_{j=1}^{c} \{W_j^* \cdot [(f^0 - f_j^*)/f_j^*]^{1/2} - (W - W_j^*) \cdot [f_j^*/(f^0 - f_j^*)]^{1/2}\}^2, \tag{4.2}$$

where $W = \sum_{j=1}^{c} W_j = \sum_{j=1}^{c} W_j^*$, and $f^0 = \sum_{j=1}^{c} f_j = \sum_{j=1}^{c} \sum_{i=1}^{n_j} O_{ji}$. In case of no missing values, this last test is equivalent to the permutational test for $ANOVA$.

Of course, if $k > 1$ a nonparametrical combination will follow. For not MAR models we have to combine nonparametrically also the k test statistics on the $\mathbf{O}$ components. This combined permutational test has general asymptotic properties. In particular, *if in some sense best univariate tests are used then it is asymptotically best in the same sense.* See Pesarin (1992) for theoretical details. Some simulation experiments carried out in order to evaluate the effects of the missing values on the distributional invariance in a general k-dimensional problem showed that the number of valid data in the observed configuration as well as in each resampling replicate have to satisfy the condition $\min_{j,h}\{f_r^*\} > \min\{3; (3n/2n)^{1/2}\}$, $r = 0, 1, \ldots, S$, where f_r^* indicates the permuted configuration of r-th resample, $r = 0, 1, \ldots, S$, and $r = 0$ indicates the observed configuration.

5 An illustrative example

This example comes from Pesarin (1991b). Let us imagine a situation in which 20 seeds of a given plant are sown in soil without any treatment and 20 others in a fertilized soil. The treatment expected effects are a) improved frequency of germination (r.v. O); b) improved index of production (r.v. X); c) improved index of leaf development (r.v. Y). Variable O is typically categorical, while X and Y are positive reals, subject to $O = 1$. The underlying distribution P is supposed to be unknown. Vector $\mathbf{O}$ can be thought of as playing the role of the missing values indicators with respect to variables $(\mathbf{X}, \mathbf{Y})$: $(X, Y)_{ji}$ are observed on ji-th individual if $O_{ji} = 1$. $O_{ji} = 0$ means seed ji-th is not germinated, $i = 1, \ldots, 20$, $j = 1, 2$. Hence, this example represents a typical not MAR situation.

Suppose the data are:

(first sample) $f_1 = \sum_{i=1}^{20} O_{1i} = 12$ germinated; valid data $(\dot{X}, Y)_1$ are $\{(6.03, 12.54)$, $(4.20, 14.81)$, $(4.49, 16.71)$, $(2., 7.53)$, $(2.84, 7.02)$, $(3.88, 8.09)$, $(2.04, 5.76)$, $(5.48, 18.01)$, $(2.31, 8.81)$, $(1.90, 8.17)$, $(1.75, 6.62)$, $(3.02, 7.69)\}$;

(second sample) $f_2 = \sum_{i=1}^{20} O_{2i} = 17$ germinated; valid data $(X, Y)_2$ are $\{(3.31, 18.49)$, $(6.56, 19.20)$, $(3.16, 9.85)$, $(4.07, 15.8)$, $(2.09, 6.16)$, $(6.7, 17.58)$, $(3.9, 19.29)$, $(2.56, 10.77)$, $(8.3, 18.8)$, $(4.21, 10.56)$, $(1.86, 9.48)$, $(3.09, 12.54)$, $(5.09, 18.35)$, $(4.08, 11.8)$, $(3.6, 11.44)$, $(2.6, 7.66)$, $(5.2, 12.)\}$.

Suppose the hypotheses we want to test are:

$$H_0 = \{(O, X, Y)_1 \stackrel{d}{=} (O, X, Y)_2\}$$
$$= [O_1 = O_2] \cap \{[(E(X_1^2) = E(X_2^2)) \cap (E(Y_1) = E(Y_2))] \mid \mathbf{O}\},$$

$$H_1 = [O_1 < O_2] \cup \{[(E(X_1^2) < E(X_2^2)) \cup (E(Y_1) - E(Y_2))] \mid \mathbf{O}\}.$$

This is a triplet of directional hypotheses, for which univariate tests could be based on the statistics $T_1^* = (f_2^*/n_2) - (f_1^*/n_1)$, $T_2^* = \gamma_2^* \cdot \sum_j (X_{2j}^*)^2 - \gamma_1^* \cdot \sum_i (X_{1i}^*)^2$, $T_3^* = \gamma_2^* \cdot \sum_j Y_{2j}^* - \gamma_1^* \cdot \sum_i Y_{1i}^*$, where $\gamma_i^* = (f_j^*/f_i^*)^{1/2}$, $j \neq i = 1, 2$, all significant for large values and $PUOD$ because treatment effects are positive on all variables. The role of T_1^* is clear. T_2^* is chosen presuming that for variable X increments in both mean value and variance are expected under H_1. This role is interpreted by second moments. T_3^* is a standard comparison between means and chosen assuming an additive effects model for variable Y.

The complexity implied by the directionality of alternatives and the unusual structure of the data is such that ordinary methods do not apply. Using a $WRRP$ with $S = 1000$ we obtain the p-values: $\lambda_1^* = 0.079$, $\lambda_2^* = 0.132$ and $\lambda_3^* = 0.071$ each one not significant at $\alpha = 0.05$, while their nonparametric combination (based on the Fisher combining algorithm) gives: $\lambda^C = 0.032$, which on the contrary is significant.

508

References

BARTON, C.N. and CRAMER, E.C. (1989). Hypothesis testing in multivariate linear models with randomly missing data. *Commun. Statist. -Sim. Comput.* **18**, 875-895.

DHARMADHIKARI. S. and JOAG-DEV, K. (1988). *Unimodality, Convexity, and Applications.* Academic Press, Boston.

DEMPSTER, A.P., LAIRD, N.M. and RUBIN, D.B. (1977). Maximum likelihood from incomplete data via the EM algorith. *J. R. Statist. Soc. B* **39**, 1-38.

DWASS, M. (1957). Modified randomization tests for nonparametric hypotheses. *Ann. Math. Statist.* **6**, 181-187.

GIRALDO, A. and PESARIN, F. (1992). Verifica d'ipotesi in presenza di dati mancanti e tecniche di ricampionamento. *Proc. Riun. Scient. S.I.S.* **2**, 271-278.

GIRALDO, A. and PESARIN, F. (1993). A resampling procedure for missing data in testing problems. In: *Proc. Conf. "Due Temi di Metodologia Statistica"*, G. Diana, L. Pace and A. Salvan, Eds.. R. Curto, Napoli, 115-122.

GRENLEES, J.S., REECE, W.S. and ZIESCHANG. K.D. (1982). Imputation of missing values when the probability of response depends on the variable being imputed. *J. Amer. Statist. Assoc.* **77**, 251-261.

LI. K.-H., MENG, X.-L., RAGHUNATHANT, E. and RUBIN, D.B. (1991). Significance levels from repeated p-values with multiply-imputed data. *Statistica Sinica* **1**, 65-92.

LITTLE, R.J.A. and RUBIN, D.B. (1987). *Statistical Analysis with Missing Data.* Wiley and Sons, New York.

PALLINI, A. and PESARIN, F. (1992). A class of combinations of dependent tests by a resampling procedure. In: *Proc. Conf. "Bootstrapping and Related Techniques"*. Jöckel, K.-H., Rothe, G. and Sendler, W. Eds.. Lecture Notes in Economics and Mathematical Systems **376**, Springer-Verlag, Berlin, 93-97.

PESARIN, F. (1991a). Tecniche di ricampionamento e verifica d'ipotesi multidimensionale. *Statistica* **50**, 483-501.

PESARIN, F. (1991b). Some multidimensional testing problems for missing values via a resampling procedure. *Statistica Applicata* **3**. 569-577.

PESARIN, F. (1992). A resampling procedure for nonparametric combination of several dependent tests. *J. It. Statist. Soc.* **1**, 87-101.

ROMANO, J.P. (1989). Boostrap and randomization tests of some nonparametric hypotheses. *Ann Statist.* **17**, 141-159.

ROMANO, J.P. (1990). On the behavior of randomization tests without a group invariance assumption. *J. Amer. Statist. Assoc.* **85**. 686-692.

RUBIN, D.B. (1976). Inference and missing data. *Biometrika* **63**. 581-592.

SERVY, E.C. and SEN, P.K. (1987). Missing variables in multi-sample rank permutation tests for MANOVA and MANCOVA. *Sankhya A* **49**, 78-95.

WEI. L.J. and LACHIN, J.M. (1984). Two-sample asymptotically distribution-free test for incomplete multivariate observations. *J. Amer. Statist. Assoc.* **79**, 653-661.

Simulate and Reject Monte Carlo Exact Conditional Tests for Quasi-independence

Peter W.F. Smith, John W. McDonald

Department of Social Statistics, University of Southampton

1 Introduction

In a two-way contingency table, the null hypothesis of quasi-independence (QI) usually arises for two main reasons: 1) some cells involve structural zeros or 2) interest is focused on part of the table, e.g., the off-diagonal cells. Consider Table 1, analyzed by Becker (1990), which cross-classifies two independent interpretations of sputum cytology slides for lung cancer. Since the two interpretations tend to agree, most of the observations lie on the main diagonal and the hypothesis of independence is rejected. The hypothesis of QI for the off-diagonal cells, i.e., that the interpretations are independent given that they differ, is considered. However, the sparseness of the off-diagonal cells causes concern about the validity of using asymptotic tests, and an exact conditional test is used.

Table 1: Cross-classification of first and second independent interpretations of sputum cytology slides for lung cancer (Source: Archer *et al.*, 1966)

First interpretation	Second interpretation					
	N	A	S	P	T	
Negative	26	19	1	0	7	53
Ambiguous cells	2	11	5	3	4	25
Suspect	0	1	6	6	0	13
Positive	0	0	0	4	1	5
Technically unsatisfactory	1	1	0	0	2	4
	29	32	12	13	14	100

In order to perform an exact test of quasi-independence the null distribution of an appropriate test statistic must be calculated or simulated. For both independence and quasi-independence calculating the required distribution is often computationally infeasible. So simulation is used and a Monte Carlo exact conditional test is performed.

A Monte Carlo exact conditional test for independence is described by Agresti, Wackerly and Boyett (1979), Kreiner (1987) and Whittaker (1990). Briefly, one generates a random sample of tables according to the conditional distribution of the table counts given the marginal totals. For each generated table, an appropriate test statistic is calculated and the exact conditional p-value is estimated by the proportion of generated tables which are at least as discrepant from the null as the observed. The accuracy of this unbiased estimate may be evaluated using binomial confidence intervals.

The problem, when using this approach to test for quasi-independence, is how to generate a random sample of tables from the null distribution. Since, as shown by Smith and McDonald (1993), the null distribution has a normalizing constant which is very difficult to evaluate. In the next section, we introduce a simulate and reject procedure based on simulating tables under independence. We then suggest some modifications which dramatically reduce the rejection rate and so make the procedure viable.

2 Simulate and Reject Procedure

Let $X = \{X_{ij} : ij \in I = (1, \ldots, r) \times (1, \ldots, c)\}$ be a $r \times c$ contingency table, and let I^* be a proper subset of the index set I. We call the cells in I^* the cells of interest and the cells not in I^* fixed. For the 5×5 Table 1, I^* refers to the off-diagonal cells. The saturated log-linear model for $m_{ij} = \mathrm{E}(X_{ij})$ has the form

$$\log m_{ij} = \lambda + \lambda_i^1 + \lambda_j^2 + \lambda_{ij}^{12}.$$

The hypothesis of quasi-independence over I^* corresponds to $\lambda_{ij}^{12} = 0$ for $ij \in I^*$. Now λ_{ij}^{12} for $ij \notin I^*$ are nuisance parameters with sufficient statistics x_{ij}, $ij \notin I^*$. Therefore, an exact conditional test for QI is constructed using the conditional distribution of the table counts, given the margins and the observed counts in the fixed cells. Hence, tables under QI can be generated by simulating tables under independence and only retaining those where the counts in the fixed cells match the observed values. For Table 1, we simulate tables from a multivariate hypergeometric distribution, thus maintaining the margins, and reject all tables which do not match the diagonal (26,11,6,4,2). Methods for simulating from a multivariate hypergeometric distribution are given by Agresti, Wackerly and Boyett (1979) and Patefield (1981). Alas, this naive simulate and reject procedure is not computationally viable, since we failed to simulate under independence a table with a matching diagonal in over one billion attempts!

All is not lost. Smith and McDonald (1993) show that the distribution of the cells of interest under quasi-independence does not depend on the observed values of the fixed cells. By replacing the values in the fixed cells with any counts and adjusting the margins accordingly, the simulate and reject procedure yields the correct null distribution. Therefore, the rejection rate can be significantly reduced by replacing the counts in the fixed cells by tho-se closest to independence, based on the adjusted margins. For Table 1 we replace the diagonal with $(3,13,1,0,1)$. Note that the row and column margins for this adjusted tables are $(x_{i+}) = (30, 27, 8, 1, 3)$ and $(x_{+i}) = (6, 34, 7, 9, 13)$, respectively, and that now x_{ii} equals the nearest integer to $x_{i+}x_{+i}/x_{++}$, whe-re $+$ denotes summation over a subscript. Using this adjusted table, in order to obtain 2000 tables with matching diagonal, 234,595 tables were simula-ted under independence. The rejection rate of 99.15% is very large, but this adjusted-margins simulate and reject procedure is now computationally feasible.

Patefield (1981) simulates the required multivariate hypergeometric distri-bution by simulating cell by cell and row by row from univariate hypergeomet-rics, based on a factorization of the multiple hypergeometric mass function. Note that each $r \times c$ table requires $(r - 1) \times (c - 1)$ simulated counts (the others obtained by subtraction). For Table 1, $234, 595 \times 16 = 3, 753, 520$ simulations were required to obtain 2000 tables with matching diagonal, i.e., an average of 1877 simulations per retained table. We now propose various ways of reducing the average number of simulated cell counts required per retained table, by modifying Patefield's algorithm.

2.1 Rejecting Partly Simulated Tables

Patefield's algorithm simulates tables cell by cell, so a mismatch can be iden-tified immediately after the count for a fixed cell has been simulated, thus eliminating unnecessary simulation of the remaining cell counts in the table.

For Table 1, after adjusting the diagonal and margins, we would repeatedly simulate the (1,1) cell count until a match of 3 occurs, then simulate the (1,2) to (1,4) cell counts and obtain the (1,5) cell count by subtraction. However, since the number of rejections does not affect the distribution of the tables retained, the (1,1) cell count can be set at its observed value and the rest of the row obtained as described. Next the (2,1) and (2,2) cell counts are simulated. If the simulated (2,2) cell count matches the observed value of 13, the rest of the row can be simulated. If not, the whole table must be rejected and a new table started. Once we have a successful match for the (2,2) cell count, we can continue simulating the table until the count in the next fixed cell is simulated, the (3,3) cell here. Again, if we have a match, we continue simulating the table; a mismatch means that we must reject the table and start again. We continue in this manner until we have simulated a table with the required matching counts for all fixed cells, remembering to check for matches where the count is obtained by subtraction.

Partly simulated tables are now rejected, so efficiency is measured by the number of cell counts simulated per retained table. By fixing the first cell count and rejecting partly simulated tables, 481,605 simulations were required to obtain 2000 tables, an average 241 per retained table (versus 1877 without these improvements).

2.2   Changing the Order of Cell Count Simulation

A further improvement is to permute the rows and columns of the table in order to attempt to match the counts in the fixed cells as early as possible. Hence, on average, reducing the number of wasted simulations.

For example, if the only fixed cell in a $r \times c$ table is the (r, c) cell, we must simulate the whole table before checking for a match for the last cell. By permuting the rows and columns so the fixed cell becomes the $(1,1)$ cell, we can set the count in the $(1,1)$ cell at its observed value and simulate the rest of the table. Therefore, no rejection is required. McDonald and Smith (1994) extend this idea to triangular tables and propose an algorithm where no rejection is necessary. However, when simulating cell counts row by row, no such permutation is possible for tables with only diagonal fixed cells.

We now discuss the important and common situation of testing for off-diagonal QI in a $r \times r$ table. Recall that Patefield's algorithm simulates cell by cell, row by row. However, one can show that in order to simulate the (i, j) cell only cells above and to the left need to have been simulated, i.e., the cells $(k, l), k = 1, \ldots, i; l = 1, \ldots, j, k \neq l$. Note that these cells plus the cell whose count is being simulated form a rectangle. Therefore, we can change the order in which the cells are simulated so as to attempt to match the counts in the fixed cells as early as possible. When matching on the diagonal, we can set the $(1,1)$ cell count to the (adjusted) observed value. The next fixed count to match on is in the $(2,2)$ cell, so we need only simulate cells counts above and to the left before checking that the simulated count for the $(2,2)$ cell equals the (adjusted) observed value. Here we have only simulated 3 cell counts before checking for a match. If we have a mismatch, we have saved $r - 3$ unnecessary simulations for the first row. If we have a match, we continue by simulating the counts of the cells above and to the left of the $(3,3)$ cell, which reduces the number of simulations required before the second match is attempted. After each successful match we continue through the table in this manner. We call this the expanding-rectangle algorithm. For Table 1, this algorithm reduced the average number of simulations per retained table to 172 (from 241 when simulating row by row).

For a $r \times r$ table with fixed diagonal, the counts in the fixed cells can be reordered by permuting the rows and columns using the same permutation. For the $r!$ possible reorderings, the average number of simulations per retained table varies. In our experience, attempting the "hardest" matches first reduces the average number of simulations per retained table. For example if the (r, r) cell count is the "hardest", we would simulate the whole table only to have to reject the table frequently because the final match is the "hardest". On the other hand, if the "hardest" match is the $(1,1)$ cell count, this count is set to the (adjusted) observed value and the "hardest" match never attempted. Our measure of hardness of match for the (i, i) cell count is the conditional probability of a match, given that we have matched on the $(k, k), k = 1, \ldots, i$, cell counts.

When trying to determine the optimum permutation of the diagonal, the problem is how to calculate the conditional probability of a match, i.e., the hardness of a match. However, our experience suggests that the conditional probability of a match is approximately equal to the "marginal" probability of a match, i.e., the probability of a match if we were simulating the whole table before checking for matches. This is easily calculated for each diagonal cell since, as shown by Patefield's factorization and used in his algorithm, the marginal probability of a match is hypergeometric. For Table 1 with diagonal (3,13,1,0,1), the marginal probabilities of a match are 0.3095, 0.1925, 0.4117, 0.8696, 0.3821, respectively. We permute the rows (and columns) using the permutation (2,1,5,3,4) so that these probabilities are in increasing order for the rearranged table. Now using the expanding-rectangle algorithm on the permuted table, the average number of simulations per retained table is reduced to 104 (from 172 before permuting).

2.3   Estimated P-values

The likelihood ratio test statistic for quasi-independence for Table 1 is 37.194 with estimated exact p-value of 0.00005 and associated 99% confidence interval of (0.00000, 0.00018), based on 20,000 tables generated under QI. While the observed test statistic and associated p-value are extreme, note that the rejection rate does not depend on their values.

3   Discussion

In this paper, we propose improvements to a naive simulate and reject procedure for generating $r \times c$ tables under quasi-independence for an arbitrary pattern of fixed cells. Although some of the algorithmic improvements are described for generating under QI for the off-diagonal cells of a square table, the ideas are applicable to other patterns of fixed cells. Apart from complete enumeration, which is only viable for small tables, the simulate and reject procedure is currently the only method for generating independent tables from the exact null distribution under QI. Our improvements to the naive procedure greatly increase its efficiency.

Smith, McDonald and Forster (1994) discuss another method for generating tables under QI using a Gibbs sampling approach, based on theoretical results in Forster, McDonald and Smith (1994). However, the generated tables are not necessarily independent and are only realizations from an approximation to the exact null distribution. When using a single Markov chain, the observed table is the obvious starting value. For multiple chains, obtaining other starting values with the same sufficient statistics for the nuisance parameters as the observed data is problematic. A possible solution is to generate a small number of independent starting values using the simulate and reject algorithms proposed.

514

Acknowledgements

This work was supported by Economic and Social Research Council award H519255005 as part of the Analysis of Large and Complex Datasets Programme.

References

Agresti, A., Wackerly, D. and Boyett, J. M. (1979). Exact conditional tests for cross-classifications: approximation of attained significance levels. *Psychometrika*, 44, 75–83.

Archer, P. G., Koprowska, I., McDonald, J. R., Naylor, B., Papanicolaou, G. N. and Umiker, W. O. (1966). A study of variability in the interpretation of sputum cytology slides. *Cancer Res.*, 26, 2122–2144.

Becker, M. P. (1990). Quasisymmetric models for the analysis of square contingency tables. *J. R. Statist. Soc. B*, 52, 369–378.

Forster, J. J., McDonald, J. W. and Smith, P. W. F. (1994). Monte Carlo exact conditional tests for log-linear and logistic models. *Working Paper*, University of Southampton.

Kreiner, S. (1987). Analysis of multi-dimensional contingency tables by exact conditional tests: techniques and strategies. *Scand. J. Statist.*, 14, 97–112.

McDonald, J. W. and Smith, P. W. F. (1994). Exact conditional tests of quasi-independence for triangular contingency tables: estimating attained significance levels. *Appl. Statist.*, (to appear).

Patefield, W. M. (1981). Algorithm AS 159: An efficient method of generating random $R \times C$ tables with given row and column totals. *Appl. Statist.*, 30, 91–97.

Smith, P. W. F. and McDonald, J. W. (1993). Exact conditional tests for incomplete contingency tables: estimating attained significance levels. *Working Paper*, University of Southampton.

Smith, P. W. F., McDonald, J. W. and Forster, J. J. (1994). Monte Carlo exact conditional tests for quasi-independence using Gibbs sampling. *Working Paper*, University of Southampton.

Whittaker J. (1990). *Graphical Models in Applied Multivariate Statistics*. Chichester: Wiley.

Empirical Powers of MRPP Rank Tests for Symmetric Kappa Distributions

Derrick S. Tracy, Khushnood A. Khan

Department of Mathematics and Statistics, University of Windsor

Abstract. Two MRPP rank tests are applied to simulated data from symmetric kappa populations with three different values of the parameters for two and three equal sized subgroups. Based on four moment approximations of their permutation distributions, empirical powers of these tests are computed and compared.

Keywords. MRPP tests; empirical powers; symmetric kappa distribution

1 Introduction

Multiresponse permutation procedures (MRPP) tests were introduced by Mielke, Berry and Johnson (1976) to test randomness of classification, where the only assumptions required are that the data is at ordinal or higher levels and that responses have the same range via rank order transformations. They are used to test H_o: Classification of a set of data into g subgroups is random versus H_a: Classification is done in accordance with an *a priori* classification scheme. Such tests have a wide range of applications in atmospheric, biological, physical and social sciences.

A population of N objects $X_1,...,X_N$ is classified into g subgroups of sizes $n_1,...n_g$, with $N = \sum_1^g n_i$

and $n_i \geq 2$. For a distance measure Δ_{IJ} between X_I and X_J, the MRPP test statistic

$$\delta = \sum_1^g c_i \xi_i, \ c_i > 0, \ \sum_1^g c_i = 1, \text{ and}$$

$$\xi_i = \binom{n_i}{2}^{-1} \sum_{I<J}^N \Delta_{IJ} S_i(X_I) S_i(X_J),$$

where S is an indicator function with $S_i(X_I) = 1$ if X_I is in the ith subgroup and 0 otherwise. For classification according to an *a priori* scheme, there is a cluster pattern, causing δ to take a small value and H_o is rejected.

With large N, the number $N!/\Pi_1^g(n_i!)$ of permutations is very large. So approximate

Pearson type distributions are suggested by Mielke, Berry and Johnson (1976) using three moments, and by Tracy and Tajuddin (1985) using four moments.

When $R(X_I)$ denotes the rank of X_I in the combined samples, and the distance function is $\Delta_{IJ}=|R(X_I) - R(X_J)|^\nu$, the MRPP test statistic is denoted by δ_ν. Mielke, Berry, Brockwell and Williams (1981) consider $c_i=n_i/N$ and $\nu=1$ and 2, and use three moments to fit a Pearson type distribution. Using four moments, Tracy and Tajuddin (1986) obtain a closer Pearson type approximation, and compare the empirical powers of δ_1 and δ_2 for two equal sized subgroups from several symmetric populations. Tracy and Khan (1990) extend this study to the case of three equal sized subgroups.

Here, we do empirical studies for two and three equal sized subgroups from the symmetric kappa family of distributions, introduced by Mielke and Sen (1981),

$$f(x)=\frac{\lambda}{2}(\lambda + |x|^\lambda)^{-(\lambda+1)/\lambda}, \quad -\infty<x<\infty,$$

where the parameter $\lambda(> 0)$ determines the shape of the distribution. This is a versatile family — for small λ the distribution is heavy-tailed, and as λ increases, the distribution tends to a uniform distribution.

2 Empirical powers

We simulate 5000 samples each from symmetric kappa populations with $\lambda=0.3$, 0.5 and 5.0, and study the cases of $n_1=n_2=40$ and $n_1=n_2=n_3=30$, using four moments of δ_ν for $\nu = 1$ and 2. Since the variance of the symmetric kappa distribution does not exist for $\lambda \le 2$, we use the location shift $Q_p - Q_{.50}$, where Q_p is the pth quantile of the parent distribution; see Victoria (1986, p.47).

In the case of two equal sized subgroups, we shift the first half of the 80 randomly generated observations by $K\sigma$, where K is a constant and $\sigma = Q_{.75} - Q_{.50}$, and count the number of rejections to compute empirical powers. For three equal sized subgroups, we shift one-third of the 90 randomly generated observations by $K\sigma$, and another one-third by $2K\sigma$. Using nominal α levels as .001, .01, .05 and .10, we present the empirical powers of δ_1 and δ_2 for two equal sized subgroups in Table 1 and for three equal sized subgroups in Table 2. Power curves are also presented to give an overall view.

3 Conclusions

For both two and three equal sized subgroups, we find that the empirical powers of the statistic δ_1 are much higher than those of δ_2 for $\lambda = 0.3$. With $\lambda = 0.5$, the powers of δ_1 are still higher, but not by so much. With further increase to $\lambda = 5.0$, the empirical powers of δ_2 become higher than those of δ_1. Since the shape of the distribution here

TABLE 1. EMPIRICAL POWERS. TWO SUBGROUPS:
$n_1=n_2=40$

K	α	$\lambda=0.3$		$\lambda=0.5$		$\lambda=5.0$	
		δ_1	δ_2	δ_1	δ_2	δ_1	δ_2
0	.001	.0008	.0010	.0010	.0010	.0004	.0006
	.010	.0086	.0086	.0098	.0100	.0088	.0084
	.050	.0472	.0474	.0504	.0520	.0486	.0506
	.100	.0932	.0940	.0988	.1040	.0998	.1016
0.2	.001	.0324	.0124	.0098	.0060	.0036	.0038
	.010	.1294	.0616	.0522	.0328	.0246	.0272
	.050	.3084	.1686	.1560	.1074	.0900	.0982
	.100	.4554	.2604	.2490	.1768	.1574	.1692
0.4	.001	.0886	.0322	.0402	.0178	.0182	.0206
	.010	.2740	.1234	.1538	.0822	.0818	.0916
	.050	.5416	.2842	.3534	.2128	.2246	.2416
	.100	.6906	.3970	.4978	.3198	.3152	.3410
0.6	.001	.1512	.0584	.0912	.0392	.0652	.0740
	.010	.4078	.1838	.2902	.1466	.2186	.2390
	.050	.6798	.3776	.5450	.3296	.4184	.4558
	.100	.8094	.5004	.6770	.4480	.5526	.5928
0.8	.001	.2146	.0852	.1722	.0744	.1750	.2028
	.010	.5142	.2394	.4268	.2286	.4120	.4556
	.050	.7744	.4548	.6880	.4322	.6634	.7040
	.100	.8822	.5784	.8032	.5556	.7700	.8018
1.0	.001	.2806	.1122	.2598	.1102	.3598	.4066
	.010	.5970	.2868	.5482	.3068	.6594	.7012
	.050	.8362	.5164	.7918	.5288	.8452	.8722
	.100	.9250	.6390	.8860	.6430	.9094	.9298
1.2	.001	.3416	.1368	.3486	.1518	.5994	.6490
	.010	.6572	.3326	.6458	.3732	.8398	.8698
	.050	.8822	.5680	.8606	.6012	.9486	.9642
	.100	.9490	.6858	.9346	.7140	.9248	.9822
1.4	.001	.4006	.1614	.4326	.2038		
	.010	.7114	.3772	.7294	.4394		
	.050	.9134	.6156	.9084	.6716		
	.100	.9670	.7214	.9576	.7712		
1.6	.001	.4536	.1866	.5060	.2518		
	.010	.7556	.4166	.7894	.5004		
	.050	.9296	.6472	.9386	.7218		
	.100	.9784	.7532	.9770	.8152		
1.8	.001	.4982	.2120	.5760	.3018		
	.010	.7936	.4546	.8366	.5598		
	.050	.9468	.6790	.9578	.7682		
	.100	.9824	.7792	.9872	.8508		

TABLE 2. EMPIRICAL POWERS. THREE SUBGROUPS:
$n_1 = n_2 = n_3 = 30$

K	α	$\lambda=0.3$		$\lambda=0.5$		$\lambda=5.0$	
		δ_1	δ_2	δ_1	δ_2	δ_1	δ_2
0	.001	.0014	.0012	.0008	.0006	.0012	.0016
	.010	.0108	.0106	.0122	.0120	.0088	.0090
	.050	.0540	.0536	.0546	.0554	.0532	.0536
	.100	.1030	.1032	.1064	.1078	.1008	.1060
0.2	.001	.0624	.0200	.0182	.0096	.0068	.0084
	.010	.2108	.0818	.0950	.0476	.0412	.0468
	.050	.4476	.2180	.2424	.1526	.1398	.1520
	.100	.5966	.3158	.3672	.2388	.2242	.2484
0.4	.001	.1856	.0556	.1066	.0384	.0638	.0788
	.010	.4438	.1748	.2956	.1440	.2168	.2500
	.050	.7066	.3558	.5476	.3116	.4390	.4790
	.100	.8190	.4780	.6790	.4288	.5660	.6076
0.6	.001	.3072	.0956	.2374	.0936	.3002	.3488
	.010	.6094	.2550	.5188	.2536	.5876	.6350
	.050	.8358	.4720	.7582	.4700	.8038	.8366
	.100	.9160	.5934	.8552	.5948	.8802	.9010
0.8	.001	.4300	.1400	.3978	.1616	.8870	.7432
	.010	.7236	.3270	.6924	.3708	.8910	.9148
	.050	.9064	.5600	.8820	.6030	.9670	.9770
	.100	.9546	.6702	.9362	·.7122	.9832	.9892
1.0	.001	.5236	.1826	.5420	.2410	.9332	.9532
	.010	.7994	.3944	.8070	.4790	.9886	.9910
	.050	.9380	.6216	.9378	.7022	.9976	.9988
	.100	.9752	.7292	.9716	.8026	.9990	.9996
1.2	.001	.6016	.2184	.6622	.3160		
	.010	.8558	.4548	.8840	.5742		
	.050	.9600	.6748	.9672	.7768		
	.100	.9866	.7746	.9890	.8578		
1.4	.001	.6658	.2620	.7542	.3892		
	.010	.8894	.5030	.9236	.6528		
	.050	.9740	.7216	.9840	.8348		
	.100	.9910	.8122	.9940	.8968		
1.6	.001	.7194	.2946				
	.010	.9160	.5462				
	.050	.9830	.7560				
	.100	.9946	.8416				
1.8	.001	.7610	.3310				
	.010	.9328	.5896				
	.050	.9894	.7846				
	.100	.9964	.8626				

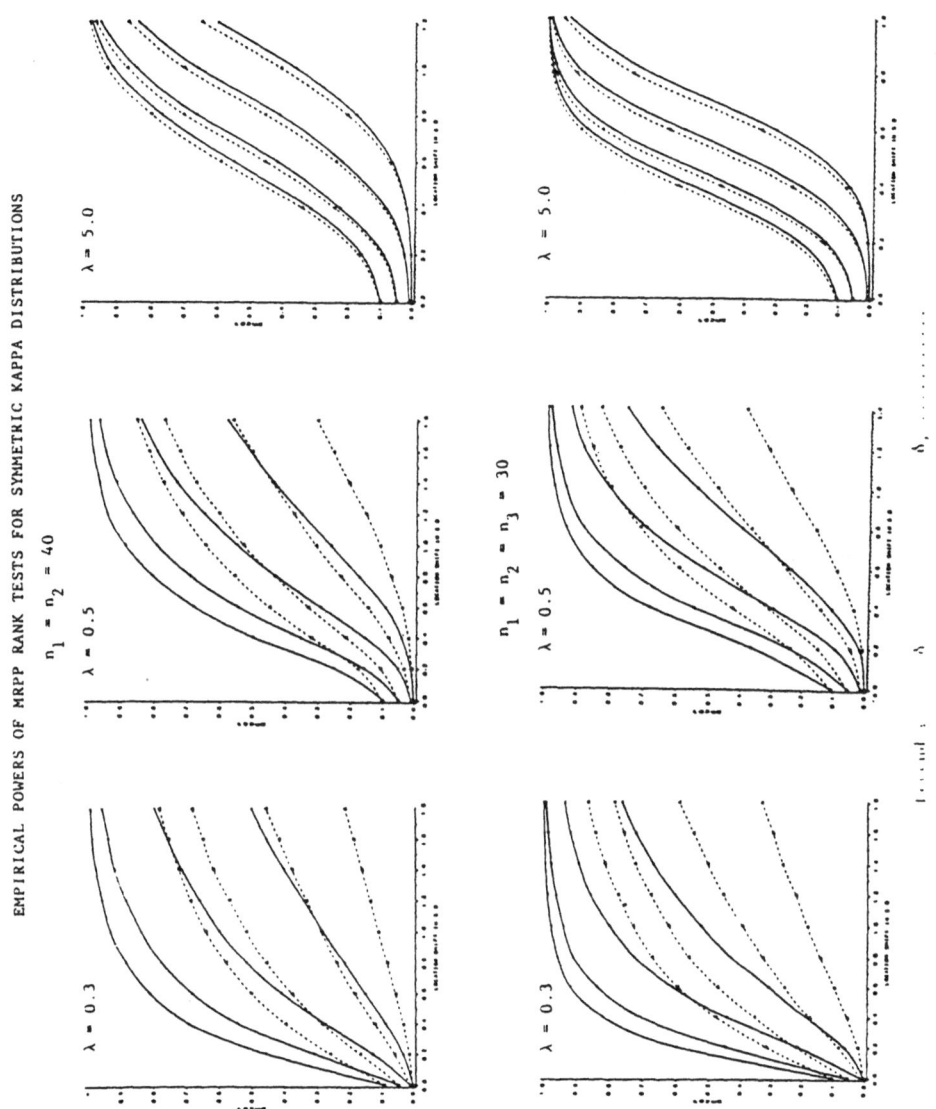

EMPIRICAL POWERS OF MRPP RANK TESTS FOR SYMMETRIC KAPPA DISTRIBUTIONS

approaches that of the uniform distribution, it is consistent with the results of Mielke, Berry, Brockwell and Williams (1981), Tracy and Tajuddin (1986) and Tracy and Khan (1990).

For $\lambda=0.3$, we note that the relative power performance of the statistic δ_1 is much better than that of δ_2, as compared to the cases of other heavy-tailed distributions such as Laplace and Cauchy, studied by Tracy and Tajuddin (1986) and Tracy and Khan (1990).

Acknowledgement

Partial support from NSERC Grant A-3111 is gratefuly acknowledged. Help received in preparing power plots from C. Kar Wong on an NSERC Undergraduate Summer Award is also acknowledged.

References

P.W. Mielke, K. J. Berry, P.J. Brockwell and J.S. Williams (1981). A class of non-parametric tests based on multiresponse permutation procedures. *Biometrika* **68**, 720-724.

P.W. Mielke, K.J. Berry and E.S. Johnson (1976). Multiresponse permutation procedures for *a priori* classifications. *Commun. Statist.-Theor. Meth.* **5**, 1409-1424.

P.W. Mielke and P.K. Sen (1981). On asymptotic non-normal null distributions for locally most powerful rank-test statistics. *Commun. Statist.-Theor. Meth.* **10**, 1079-1094.

D.S. Tracy and K.A. Khan (1990). Comparison of some MRPP and standard rank tests for three equal sized samples. *Commun. Statist.-Simula. Computa.* **19**, 315-333.

D.S. Tracy and I.H. Tajuddin (1985). Extended moment results for improving inferences based on MRPP. *Commun. Statist.-Theor. Meth.* **14**, 1485-1496.

D.S. Tracy and I.H. Tajuddin (1986). Empirical power comparisons of two MRPP rank tests. *Commun. Statist.-Theor. Meth.* **15**, 551-570.

J.S. Victoria (1986). *Power Comparison of Two MRPP Statistics Using Two Approximation Methods*. Ph.D. Dissertation, Colorado State University, Fort Collins, CO.

Part XXV

Applications

Dynamic Updating of Inter-Urban Traffic Network Models Using Conditional Independence Relationships

Joe Whittaker, Simon Garside, Karel Lindveld

Mathematics Department, Lancaster University and Hague Consulting Group

Abstract Traffic counts from conductance loops embedded in the road surface of motorway networks are now being wired to local and distant processors. The loops generate a stream of volumes and velocities, classified by vehicle type and lane, at a large number of monitoring stations on the network. Ways of exploiting this stochastic information, in particular for estimating current and predicting future travel times, are discussed.

1   Introduction

Many advanced industrial societies are feeling capacity constraints on their motorway networks in the form of increasing congestion and lengthening travel times. Extraction of additional capacity from the existing network by improved management is one of the few options available. Dynamic traffic models transform on-line roadside measurements of traffic flow into explicit assessments about the current and future state of an inter-urban network and thus provide transport management with a tool for monitoring, prediction and potentially control.

The electronics revolution has led to many new methods of transportation management. In particular conductance loops implanted in the road surface which take real time data from the passing traffic process are now centrally wired. This enables on line monitoring and management of the network, never previously possible, and naturally demands automatic methods of information processing. One methodological approach is dynamic state space modelling. The apparently large quantity of data (both in time and space) is mainly uninformative and is simplified by investigation of the conditional independence structure of the process. This structure in part reflects the given spatial topology of the network, and is incorporated in the state transition equations. The transition and observation equations lead to an associated Kalman filter which provides the relative discounting of past data in proportion to the underlying variability of the traffic process. Prediction ahead is an instance of Bayesian forecasting. Basic ideas of traffic flow, such as conservation of vehicles and the so-called fundamental diagram are built into the state transition equations. However some complications ensue because of non-linearities in speed-flow relationships and resulting non-linearity of the equations.

The paper describes project work underway at the University of Lancaster in the EC/DRIVE II initiative, and in cooperation with partners from the Hague Consulting Group, DVK (from the Dutch Ministry of Transport), CSST and ELASIS (part of the FIAT group), and from the Universities of Brussels and Napoli. The emphasis of this work is to provide practical working methods based on existing technology, but preliminary pilot studies have naturally raised some interesting statistical and theoretical questions.

Typical applications have networks with 50 or so on- off- ramps and 500 monitoring stations. The measurement process is based on a pair of embedded conductance loops ⊓⊓ in each lane of the motorway. As a vehicle axle passes over the loops, in principle 4 times are recorded and the vehicle passage, the vehicle speed and type can be recorded. Usually the information from the loop pair is processed at the station site with that from the other lanes to give just the flow and the velocity by 1 minute aggregates or averages. In fact some of the older stations give exponentially weighted moving averages of the measures. The newer stations, aggregate but also report standard deviations and other information such as occupancy and vehicle composition. The pre-processed data is then transmitted to the traffic control centre. Certain difficulties can occur, one is time synchronisation of measures from the different stations, another is missing observations from dead or malfunctioning loops.

The traffic variables (q, r, v) are flow, density and velocity. They vary in space and time (s, t) and satisfy the identity $q = rv$. As vehicles are conserved in a closed system an essential part of the dynamics of the process is an accounting equation that keeps track of the number of vehicles on the different links of the network. Traffic congestion leads to a reduction in velocity as expressed in the so-called 'fundamental traffic diagram' by a velocity - density function $v = V(r)$. Several such functions are described in Papageorgiou (1991) and one is

$$
v = \begin{cases}
a - br & \text{if} \quad r < r_{crit} \\
d(1/r - 1/r_{jam}) & \text{if} \quad r_{crit} < r < r_{jam} \\
v_{jam} & \text{otherwise.}
\end{cases}
\tag{1}
$$

The constants a, b, d, r_{crit}, r_{jam} and v_{jam} are here chosen to match empirical data. The free flow capacity of the link is reflected in the value of b. Congestion occurs when a critical density is reached. A striking feature of empirical versions of this diagram is the enormous variability in velocity in congested periods relative to free flow periods.

2   The state space model

The Kalman filter, Kalman (1960), is an important algorithmic device in the theory of automatic control, for example, Young (1984), and in economic time series modelling, for example, Harvey (1990).

Underlying the filter is the state space model, a multivariate stochastic process, for which the filter is an optimal Bayes solution to a prediction problem. The state space model supposes that at each discrete time point t, a (vector-valued) observation y_t, is related to x_t, the states, through an observation equation. The state process is an unobserved first order Markov process and is specified by a transition equation. On the basis of past observations $y^t = (y_1, y_2, \ldots, y_t)$ and the current observation y_{t+1} it is required to best predict x_{t+1}. It turns out that the filter depends on the observations only through the previous estimate of the state $\hat{x}_t$ and the current prediction error, and so allows a recursive computation.

The basis of the model are the transition and observation equations

$$\begin{aligned} x_{t+1} &= A_t x_t + B u_t + q_t, & \text{var}(q_t) &= Q \\ y_t &= C x_t + r_t, & \text{var}(r_t) &= R, \end{aligned} \qquad (2)$$

The structure of these equations is held in the conditional independence graph, or influence diagram,

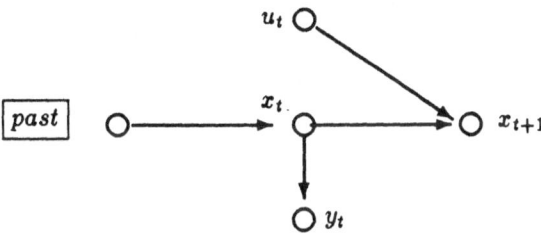

where it is the missing edges that indicate independences. See Dempster (1990) for the relationship between graphical modelling and Kalman filtering. The past represents past values of the variables. That there is no edge to x_{t+1} from the past indicates that enough information is held in the current state x_t to render past information redundant in predicting its evolution.

3 Networked traffic applications

The features of a state space model that make it appropriate for traffic network applications are that it is multivariate, stochastic - both inherently random and measurement error, and above all, from the manner in which structure may be imposed:

1. Variables exist at different levels: observed, latent, derived, so that link flow is observed, link density is latent, link travel time is derived, and on-ramp flow is exogenous.

2. Physical understanding, the traffic dynamics, is incorporated through the transition equation: density is predictable (but not observed) through an accounting equation; derived variables such as velocity are consequent on density via the fundamental traffic diagram; and flow (which is observed) is dependent on upstream and downstream links, and is a form of spatial interaction.

3. The temporal and spatial Markov structure of the physical process is mapped onto the Markov structure of the state space model; this both allows recursive processing and decomposition into subnetworks.

4. Flexible observation equations allow for the vagaries of the measurement process.

The one downside of the state space model is that it ignores the behavioural interpretation of traffic flow in terms of humans making choices of trip origin, destination and path, which at one level determine the observed traffic process.

The state space model under development consists of the following components: a latent state vector x_t dimension (1000x1) holds traffic variables (r, q) for all links on network, turning fractions at forks, and unknown parameters; an observation vector y_t (500x1) of measurements from loops at monitoring stations; a vector of exogenous input u_t (50x1) the on-ramp flows, from monitoring stations.

The lack of an edge between y_t and the past in the independence graph above indicates that given the state of the network x_t any further variation is due to independent measurement error. Importantly there is no arrow from the past or from x_t to the input, the on-ramp flows u_t. This assumes that traffic demand is not dependent on past state of the network (and of course this may be unrealistic).

3.1   The transition equations

The transition equations and the transition matrix A_t carries both the network topology (which are time invariant) and the traffic dynamics.

Network topology

The network is viewed as a set of stations, resultant links and blocks composed of links. A link is the section of motorway between two adjacent stations. Motorway networks are relatively simple because there are at most two upstream and two downstream links to any station.

A block then is a combination of links: either a single section, or the two sections that form a join, or the two sections that form a fork. An on-ramp is a join block, an off-ramp is a fork block; intersections are combinations of forks and joins. Roundabouts and cross-roads are prohibited. The whole motorway network may be decomposed into these 3 types of blocks.

If the state space vector holds just the density and the exit flow(s) for each block then the section and join blocks have dimension 2, while the fork blocks have dimension 3 as there are two exitting flows. The block adjacency graph in part determines the spatial structure of the influence diagram of the state space blocks.

Transitions: time varying traffic dynamics

The transition matrix $A_t = A(x_t)$ depends on traffic dynamics. Classically these follow differential equations analogous to those of fluid dynamics, and for a review see the articles in Papageorgiou (1991). There are three types of equations - a balance or accounting equation - a flow equation based on the identity $q = vr$, and - an equation determining velocity from the fundamental diagram $v = V(r)$ above.

There are three types of transitions to consider section to section, sections to join and fork to sections. The independence graph of a section to section transition is constructed by replicating the graph

time t

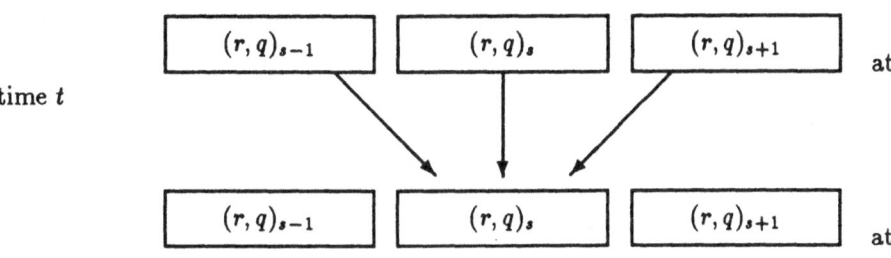

at

time $t + 1$

at

over s. There are no other directed arrows into block s.

The maths that lies behind these arrows is a balance equation of the form

$$r_{s,t+1} = r_{s,t} + k_s(q_{s-1,t} - q_{s,t})$$

and one of several possible flow equations. The simplest is

$$q_{s,t+1} = \min(v_{s,t}, v_{s+1,t})\, r_{s+1,t}.$$

The flow equations incorporate velocity, and so give a non-linear function of the state space, and reflect the physical reality of congestion.

Similar equations hold for sections to joins; fork to sections are slightly more complicated and need to introduce a state variable representing the turning fraction between the two downstream choices. The unknown parameters require estimation, which may be performed on-line from current measurements or off-line from an historical data base.

The observation equations for measured flow are straightforward, and graphically result in a single directed arrow from each traffic block to an independent observation y_s that holds the observed flow from the monitoring station exitting the block. Hence the C matrix is block diagonal. There are two observed flows in a fork block. The observation equation that incorporates measured velocity is slightly more involved but the graph is the same.

528

4   Concluding remarks

The consequence of having a transition matrix A_t which is a function of the state space and so of time is that the Kalman gains needs to be recomputed at each time point. Ways of simplifying and speeding up this calculation are under research.

We have outlined above a partial description of the monitoring and prediction system that is under development in the DYNA project. There are many issues not dealt with here, and specific topics of investigation include: self calibration for parameter estimation; forecasting network entry flows using historical data base methodology; assessment of system performance from diagnostic routines based on one step ahead forecasts; generalisation of traffic flow models at a point to encompass a whole network; assessment of non-linearities in the state transitions required to model traffic flow and their effect on longer term predictions; methods of handling large matrix multiplications and inversions; decomposition techniques for simplifying networks, especially cycles in which path choice plays a part and which correspond to non-decomposable graphical models.

References

Dempster, A.P. (1990). In Oliver, R.M. and Smith, J.Q. (eds) *Influence diagrams, belief nets and decision analysis.* Wiley: Chichester.

Harvey, A.C. (1989). *Forecasting, structural time series models and the Kalman filter.* C.U.P.: Cambridge.

Kalman, R.E. (1960). A new approach to linear filtering and prediction problems. *Journal of Basic Engineering, Transactions ASME. Series D* **82**, 35-45.

Papageorgiou, M. (1991). *Concise Encyclopedia of Traffic and Transportation Systems.* Pergamon Press: Oxford.

Whittaker, J. (1990). *Graphical Models in Applied Multivariate Statistics.* Wiley: Chichester.

Whittaker, J. and Garside, S. (1993). *State space models for dynamic traffic networks.* in Gunn, H. *DYNA-DRIVE II project V2036 Annual Project Review Report - Part A Section 2.* Submitted to EC R&D Program Telematic Systems in the Area of Transport.

Young, P. (1984). *Recursive Estimation and Time-Series Analysis.* Berlin: Springer-Verlag.

Part XXVI

Late Paper

Smoothing and Robust Wavelet Analysis

Andrew G. Bruce, David L. Donoho, H.Y. Gao, **R. Douglas Martin**

MathSoft, Inc., Seattle

ABSTRACT

In a series of papers, Donoho and Johnstone develop a powerful theory based on wavelets for extracting non-smooth signals from noisy data. Several nonlinear smoothing algorithms are presented which provide high performance for removing Gaussian noise from a wide range of spatially inhomogeneous signals. However, like other methods based on the linear wavelet transform, these algorithms are very sensitive to certain types of non-Gaussian noise, such as outliers. In this paper, we develop *outlier resistant* wavelet transforms. In these transforms, outliers and outlier patches are localized to just a few scales. By using the outlier resistant wavelet transforms, we improve upon the Donoho and Johnstone nonlinear signal extraction methods. The outlier resistant wavelet algorithms are included with the S+WAVELETS object-oriented toolkit for wavelet analysis.

1 INTRODUCTION

The introduction of wavelets in the late 1980's has spawned a flurry of research activity, exploring new techniques for analysis of data simultaneously in the time and frequency domains. Several new "wavelet-like" transforms have been developed, such as wavelet packets, local cosine bases, Wilson bases, and matching pursuits. Wavelets have proven valuable for a variety of statistical applications, such as the denoising of signals or estimation of spectral or probability densities.

The presence of outliers in data causes problems in traditional time series analysis techniques. Outliers can seriously distort the autocorrelation function, partial autocorrelation function, spectral density function, model identification, and parameter estimates for models. Outliers can also cause problems with methods based on the wavelet decomposition. Wavelets are a linear transformation of the data, and hence, outliers have unbounded influence on the wavelet coefficients.

---

[1]MathSoft, Inc., 1700 Westlake Ave. N, Suite 500, Seattle, WA 98109
[2]Department of Statistics, Stanford University, Stanford, CA 94305

In this paper, we review research into new robust wavelet decompositions which are designed for analysis of data which contains outliers. Based on these decompositions, we extend wavelet-based statistical algorithms to handle a broader class of problems. In particular, we focus on the robust "smoother-cleaner" wavelet decomposition. Smoother-cleaner wavelets are an adaptation of the pyramid algorithm in which outliers captured into robust residuals at different multiresolution levels. The algorithm is computationally very fast with $O(n)$ complexity.

The paper is organized as follows. Section 2 reviews the wavelet-based denoising procedure of Donoho and Johnstone. Section 3 motivates the need for new robust wavelet decompositions, and presents decompositions based on minimizing norms which are insensitive to outliers. These decompositions have nice theoretical properties but are computationally slow. Section 4 presents the robust smoother-cleaner wavelet algorithm. The algorithm is applied to simulated data and radar glint noise data. Finally, section 5 gives a discussion of related research. This includes research into other robust wavelet decompositions, and the development S+WAVELETS, an object-oriented toolkit for wavelet analysis.

2 DENOISING BY WAVELET SHRINKAGE

Suppose our data x_i are noisy samples from a function f:

$$x_i = f(i/n) + \epsilon_i \qquad (1)$$

where ϵ_i are iid $N(0, \sigma^2)$. We want to find an estimate $\hat{f}$ which minimizes the risk $R(\hat{f}, f) = E \parallel \hat{f} - f \parallel_2^2$. In a series of papers [6, 8, 7], Donoho and Johnstone propose a collection of related techniques which solve this problem. Their denoising procedure, which we refer to as WAVESHRINK, is based on a theoretically motivated nonlinear shrinkage of wavelet coefficients. The principle is that noise contributes to many coefficients but features contribute to only a few coefficients. Hence, by setting the smaller coefficients to zero in a statistically guided manner, we can nearly optimally eliminate noise while preserving the underlying signal.

The three steps in the WAVESHRINK algorithm are

[1] Apply the wavelet transform with J levels to the signal **X**, obtaining wavelet detail and smooth coefficients D(1), D(2), ..., D(J), S(J).

[2] Shrink the detail coefficients at the j finest scales to obtain new detail coefficients $\tilde{D}(1) = \delta_{\lambda_1}(D(1))$, ..., $\tilde{D}(j) = \delta_{\lambda_j}(D(j))$. A statistically

attractive form for the thresholding function is a *soft threshold*:

$$\delta_{\lambda_i}(x) = \begin{cases} 0 & \text{if } |x| \le \lambda_i \\ \text{sign}(x)(|x| - \lambda_i) & \text{if } |x| > \lambda_i \end{cases} \qquad (2)$$

Note that the threshold λ_i may vary from level to level.

[3] Apply the inverse wavelet transform to obtain the estimated smooth $\hat{X}$.

Theoretical results show that for certain choices of the λ_j, the WAVESHRINK estimate $\hat{f}_{w_s}$ can achieve nearly the *minimax* risk over a broad class of functions $\mathcal{F}$:

$$R(\hat{f}_{w_s}, f) \approx \inf_{\hat{f}} \sup_{f \in \mathcal{F}} R(\hat{f}, f) \qquad (3)$$

A consequence of this result is that the WAVESHRINK algorithm has a *locally adaptive bandwidth*. It has been shown to perform remarkably well on a broad range of spatially inhomogeneous signals. The smooth is completely automatic: no tuning constants are required (other than the choice of the wavelet filter and thresholding rule). The WAVESHRINK algorithm can be extended to other orthonormal bases as well, such as wavelet packets and local cosine bases[5].

3 ROBUST WAVELET DECOMPOSITIONS

The theory of Donoho and Johnstone demonstrates that wavelets provide a powerful framework for denoising data. However, this theory is based on the assumption that the noise ϵ_i is close to a Gaussian distribution. As a result, the WAVESHRINK algorithm is very sensitive to outliers.

Figure 1 compares WAVESHRINK for an artificial signal contaminated with Gaussian and non-Gaussian noise. Figure 1(a) displays the "jumpsine" signal: a sinusoid with a jump in the middle. The adjacent plot is the wavelet decomposition of the jumpsine signal. All of the large coefficients at the finer levels correspond to the jump. Figure 1(b) gives the signal plus Gaussian noise with the WAVESHRINK smooth. In this example, WAVESHRINK performs well and the smooth is very close to the original signal. This smooth is derived by inverting the "shrunken" wavelet decomposition, shown in Figure 1(b). By shrinking the coefficients, we are able to remove most of the noise while still maintain the underlying signal, including the level shift. The data in figure 1(c) is obtained by further corrupting the signal with non-Gaussian impulsive outlier noise. The outliers are patches of fixed magnitude but random sign and patch length. The resulting WAVESHRINK smooth is very sensitive to the impulsive noise. The problem is that outliers are treated as local features by the WAVESHRINK procedure. Hence, like the level shift,

534

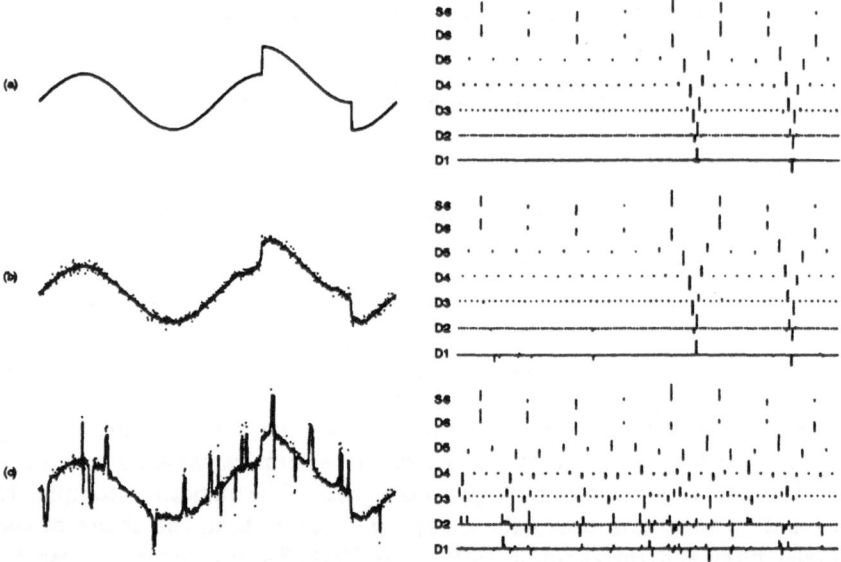

Figure 1: (a) The jumpsine signal (left plot) and its wavelet decomposition (right plot), (b) the signal contaminated by Gaussian noise (points), the WAVESHRINK smooth of the Gaussian contaminated data (solid line), and the wavelet decomposition corresponding to the WAVESHRINK smooth (c) The same as "(b)" except that the signal is also contaminated by impulsive patchy noise at random locations. While WAVESHRINK performs very well with Gaussian noise, WAVESHRINK is highly sensitive to outliers.

outliers are preserved (see the corresponding wavelet decomposition).

One aim of our research is to broaden the scope of situations for which WAVESHRINK and related procedures are useful. To achieve this goal, we are developing a suite of algorithms for producing multiresolution and wavelet decompositions designed for signals which have noise distributions F_ϵ of the form

$$F_\epsilon = (1 - \gamma)F + \gamma H \qquad (4)$$

F is the "core" model, H is a "long tailed" outlier producing distribution, and γ is the fraction of contamination. We consider a variety of contamination models, emphasizing those which generate outliers occuring in patches [14].

The classical wavelet transform produces a sequence of approximations $\hat{f}_\ell(t)$ which are the projections of a signal $f(t)$ onto the basis formed by the collection of scaling functions $\phi_{\ell,k}(t) = 2^{-\ell/2}\phi(2^{-\ell}t - k)$. These projections

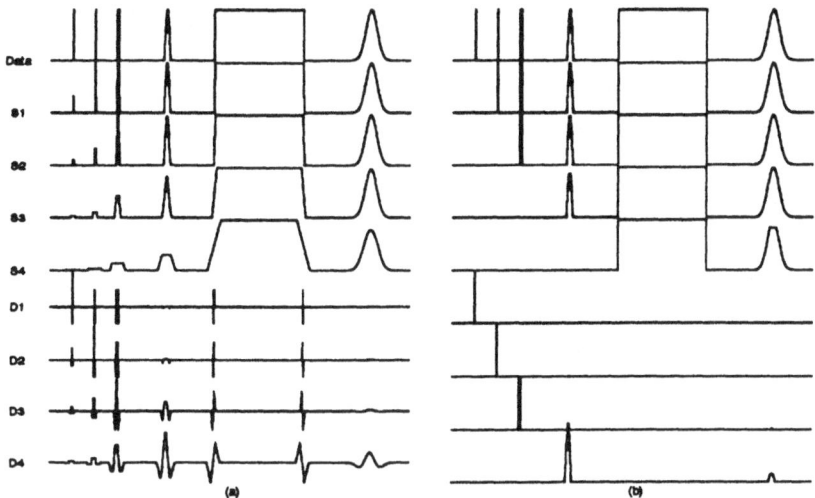

Figure 2: (a) L_2 multiresolution analysis using the Haar basis for an artificial signal consisting of outlier bursts, a smooth transient, a level shift, and smooth bump, and (b) L_1 multiresolution analysis for the same signal. The outlier bursts and transient are localized to the fine scales and concentrated in fewer coefficients for the L_1 decomposition.

minimize the L_2 norm

$$\| f(t) - \hat{f}_\ell(t) \|_2 \qquad (5)$$

The L_2 norm, however, is well known to be very sensitive to outliers. In this section, we consider decompositions obtained by minimizing norms which are robust towards noise generated from models such as (4).

3.1 L_1 Fitting and the Haar Basis

The L_1 norm is well known to be resistant towards outliers. As an example, we consider the analysis of an artificial signal using the Haar basis. For the Haar basis, the optimal L_2 and L_1 fits are given by the block mean and median respectively. Figure 2 compares the fits obtained by minimizing the L_2 norm (figure 2(a)) and L_1 norm (figure 2(b)). The original signal, given in the top plot, consists of three outlier bursts with different patch lengths, a smooth transient, a discontinuous level shift, and a smooth Gaussian kernel. The next four plots display the approximations $S(\ell) \equiv \hat{f}_\ell(t)$ for $\ell = 1, 2, 3, 4$. The final four plots display the differences $D(\ell) \equiv \hat{f}_{\ell-1}(t) - \hat{f}_\ell(t)$ for $\ell = 1, 2, 3, 4$. These differences are closely related to the "detail" coefficients in the classical wavelet transform. To ensure uniqueness of the L_1 fits, we use decimation by 3 in this examples.

This example illustrates two properties of L_2 and L_1 fits (in the decimation by 3 case):

P1: An outlier spike of length $\lfloor 3^\ell/2 \rfloor$ is isolated to levels $j = 0, 1, \ldots, \ell - 1$ for L_1 approximations $\hat{f}_\ell(t)$. By contrast, the outlier spikes are spread throughout the L_2 projections.

P2: The discontinuities and local transients are concentrated in fewer coefficients for the L_1 fits.

Hence, we can more easily remove outliers from the L_1 decomposition. Note also that the edges of the level shift are better preserved with the L_1 decomposition. The concentration of coefficients in the L_1 decomposition indicates that robust decompositions may have applications to data compression problems[4].

3.2 Smooth But Robust Fits

We can extend L_1 fitting to general wavelet bases and hence, obtain smoother projections. As an approximation, ℓ_1 fits can be used which are easily computed using using standard ℓ_1 regression techniques. However, L_1 fits are intrinsically non-smooth, and the so L_1 fits for general wavelet bases can still exhibit local roughness. We can obtain smooth but robust fits by using a hybrid loss function, such as

$$G(t) = \left\{ \begin{array}{ll} t^2 & \text{for } |t| \leq C \\ |t| & \text{for } |t| > C \end{array} \right. \tag{6}$$

Minimizing the norm defined by $G(t)$ retains the smoothness of L_2 projections and the robustness of L_1 fits.

To illustrate the difference between the different norms when using a smooth wavelet, we return to jumpsine example. Figure 3(a) plots the signal contaminated by non-Gaussian impulsive noise. Figure 3(b) gives the L_2 projection at level 4 (decimation of the original signal by 2^4) using the "least asymmetric" orthogonal wavelet with a filter of length 8 [3]. For illustrative purposes, a non-decimating shift invariant projection is performed [12]. Figures 3(c)-(d) give the corresponding fits based on minimizing the L_1 norm and the hybrid loss function respectively. The L_2 projection is significantly influenced by the outlier bursts. By contrast, L_1 and hybrid fits are relatively insensitive to the outliers. However, the hybrid fit is smoother and visually more appealing than the L_1 fit.

3.3 Computationally Unattractive

In general, the exact L_1 or hybrid approach is computationally too inefficient for practical use. Even in the Haar case, we do not retain the recursive

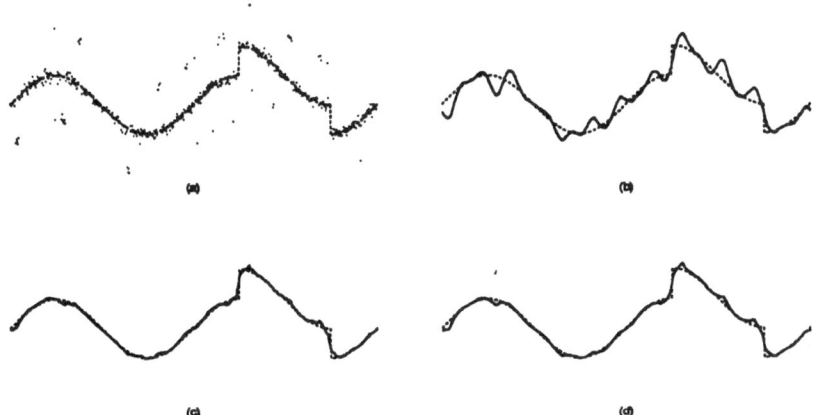

Figure 3: (a) The signal (line) contaminated by non-Gaussian noise (points), taken from in figure 1(c), (b) the L_2 projection at level 4 (decimation of the original signal by 2^4), (c) the L_1 fit at level 4, and (d) the hybrid fit at level 4 using the norm defined by (6). The hybrid analysis retains the smoothness of L_2 projection and the robustness of L_1 fit.

filtering pyramid which makes the wavelet approach so attractive. It is our aim to mimic the robustness properties of this approach without sacrificing the computation efficiency of the discrete wavelet transform.

4 ROBUST SMOOTHER-CLEANER WAVELETS

The goal of robust smoother/cleaner wavelets is to produce a fast wavelet decomposition which is robust towards outliers. Smoother-cleaner wavelets behave like the classical L_2 wavelet transform for Gaussian signals, but prevent outliers and outlier patches from leaking into the wavelet coefficients at coarse levels (like L_1 wavelets). However, in contrast to the L_1 wavelets, algorithm is very fast with computational complexity $O(n)$.

4.1 Basic Algorithm

The basic idea of robust smoother/cleaner wavelets is simple: the smooth coefficients are preprocessed with a fast and robust smoother/cleaner. The procedure is illustrated in figure 4. As usual, we start with a set of wavelet coefficients $S(0)$. Then, for each multiresolution level, the signal is decomposed into three components:

538

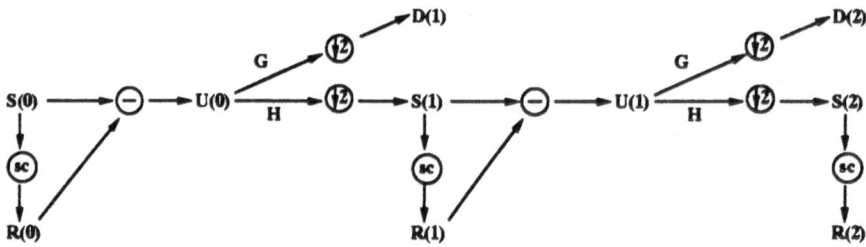

Figure 4: The robust smoother algorithm produces a pyramid decomposition with an extra component: the robust residual $R(\ell)$. For each multiresolution level, the low-pass coefficients $S(\ell)$ are first cleaned using a robust smoother cleaner, denoted by sc in the figure. The residuals are saved in the $R(\ell)$. The usual wavelet filters are then applied to the cleaned $S(\ell)$ to obtain $S(\ell+1)$ and $D(\ell+1)$.

1. A set of robust residuals $R(\ell-1)$, given by

$$R(\ell-1) = \delta_\lambda \left(S(\ell-1) - \hat{S}(\ell-1) \right)$$

   where δ_λ is the thresholder function (2) and $\hat{S}(\ell-1)$ is a robust smooth of $S(\ell-1)$ (e.g., running medians of 5). The threshold λ is chosen so that most of the robust residuals are zero.

2. The smooth wavelet coefficients $S(\ell)$ obtained by applying the usual low-pass/decimation wavelet filter H to the cleaned smooth coefficients $U(\ell-1) = S(\ell-1) - R(\ell-1)$.

3. The detail wavelet coefficients $D(\ell)$ obtained by applying the usual high-pass/decimation wavelet filter G to $U(\ell-1)$.

4.1.1 Choice of Wavelet Filters

The low-pass decomposition filters should be short in order to avoid leakage of outlier patches to the smooth coefficients. In general, the longer the low-pass filter, the more an outlier patch tends to get smeared when going from fine to coarse levels. The smearing is undesirable since it then is difficult to isolate and identify the outlier patch (as in the L_2 case). On the other hand, it is desirable to have longer filters to ensure sufficient smoothness with the underlying basis functions.

The "b-spline" biorthogonal wavelets[3] is one class of filters which satisfy both requirements: short filters can be used for decomposition while longer filters for reconstruction.

4.1.2 Choice of Robust Filter

The robust filter should be simple, computationally fast to compute, and have a very high breakdown point. Median filters are an attractive choice, and enjoy extensive usage in the engineering community. The width of the robust filter L should be as small as possible to provide minimal distortion of the underlying signal. However, L must be sufficiently big to prevent outlier patches from getting smeared in coarser scales.

In theory, for a low-pass wavelet filter of length M, smearing is prevented by using median filters of length $L \geq 2M + 1$. In practice, we find that using median filters of length 5 or 7 is usually sufficient to avoid smearing for most types of wavelets.

4.1.3 Setting the Robust Residual Threshold

The threshold λ determines the number of non-zero robust residuals. Setting λ too big will result in leakage of outliers into the signal and setting λ too small will cause distortion of the signal. We set λ so that an average of $100 \times p\%$ non-zero robust residuals remain after thresholding in the Gaussian case. The tuning parameter p is set to some small value (e.g., .01). A table for λ is obtained by simulation based on the Gaussian model. This value of λ is quite insensitive to the stochastic characteristics of the underlying signal.

4.2 Key Properties

- In the Gaussian noise case, the robust smoother-cleaner wavelet transform produces essentially the same decomposition as the classical wavelet transform. By design, only a small number of robust residuals are detected, and these will be small in magnitude (by virtue of the soft shrinkage function).

- If the data contains outliers and outlier patches, then the decomposition retains the dyadic equivalent of property **P1** for the L_1 decomposition of section 2: outliers patches of length $2^\ell + 2$ are isolated to wavelet coefficients at levels $j \leq \ell$.

- For outlier models of the type (4), it can be shown that WAVESHRINK, when applied appropriately to the robust smoother-clean wavelet decomposition (see below), can achieve a near optimality property similar to (3).

To illustrate the smoother-cleaner wavelet decomposition, we return to the jumpsine example of section 3. Figure 5(a) displays the smoother-cleaner wavelet decomposition for the outlier contaminated signal. In this example, we have used the Haar basis and median filter of length 5. The robust residuals correspond to the outlier bursts. Generally, the residuals R0 correspond

540

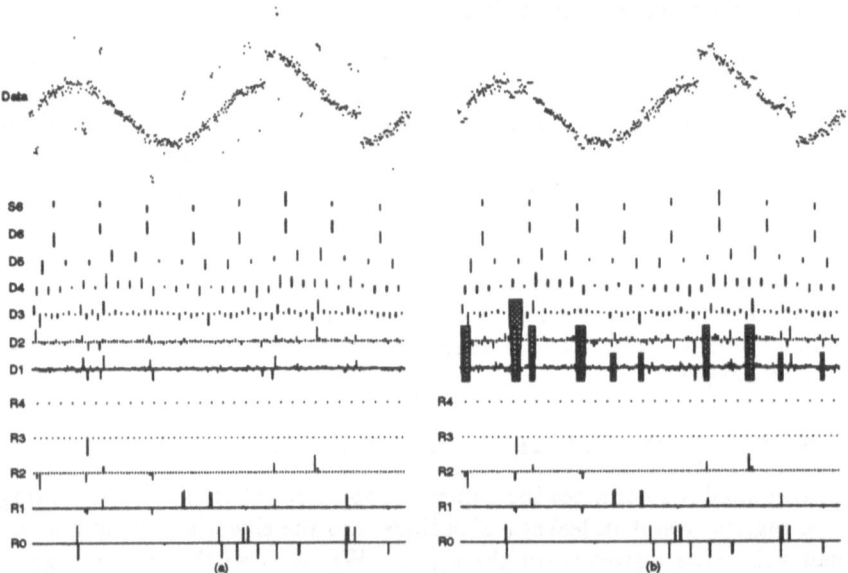

Figure 5: (a) Robust smoother-cleaner decomposition of the outlier contaminated jumpsine signal and (b) zero cones applied to the robust smoother decomposition in "(a)".

to isolated outliers or pairs of outliers and the residuals R1-R3 correspond to longer outlier patches. Note that some longer patches are captured by pairs of residuals in R1 while other patches are captured by single (large) residuals in R2 or even R3. The difference in the way in which outlier patches are represented in the robust residuals is due to decimation. Relative to the classical wavelet transform (see figure 1(c)), the robust smoother-cleaner wavelets have less leakage of the outlier patches into the coarse detail coefficients: compare the D3 and D4 coefficients.

The differences are even more evident when we look compare the robust and classical multiresolution analyses. Figure 6(a) gives a sequence of successively coarser estimates of the signal based on reconstructing from the $S1$, $S2$, ... classical wavelet coefficients. This is equivalent to annihilating all detail coefficients at scales $D1$, $(D1, D2)$, $(D1, D2, D3)$, ... Figure 6(b) gives the analogous sequence based on the robust smoother-cleaner wavelet coefficients. The robust reconstructions are much less sensitive to the outlier bursts. The $S4$ robust fit closely mimics the L_1 and Huber fits of figure 3.

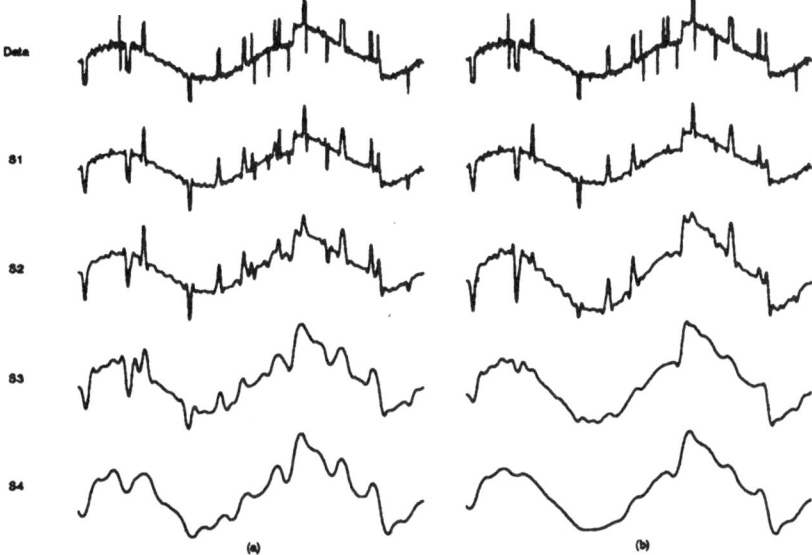

Figure 6: (a) Linear shrinkage applied to successively coarser levels for the discrete wavelet transform of the outlier corrupted jumpsine signal, and (b) linear shrinkage applied to robust smoother cleaner wavelet transform of the same data. The robust smoother-cleaner wavelet decomposition is superior to the classical wavelet transform for denoising by linear shrinkage.

4.3 Combining with Waveshrink

Non-linear shrinkage, such as that used by the WAVESHRINK procedure, can outperform linear shrinkage when the signal contains local features which we want to preserve in the finest detail wavelet coefficients. For example, note how the jump in the signal becomes blurred at the coarser scales in figure 6. By contrast, the WAVESHRINK estimate for the same signal contaminated only by Gaussian data preserves the jump: see figure 1(b). In this section, we discuss how the WAVESHRINK algorithm can be applied to data with outliers by using the robust smoother-cleaner basis.

The simplest procedure is to discard the robust residuals and to use WAVESHRINK on the wavelet coefficients. If the data contains only isolated or pairs of outliers, this procedure will generally work. However, if the data contains longer bursts of outliers, then this procedure is likely to breakdown. While the robust smoother-cleaner prevents outliers from leaking into the coarse scale detail and smooth coefficients, it does not prevent outliers from patches leaking into fine scale detail coefficients. In figure 5(a), we see that the largest D1 and D2 coefficients are associated with outlier patches. Hence, applying WAVESHRINK to the decomposition of figure 5(a) will result in some leakage of the noise into the signal.

To get around this problem, we provide two solutions: selective annihilation of coefficients using "zero cones" and a "clean and repeat" procedure. These are discussed below.

4.3.1   Zero Cones

The basic principles behind zero cones are that

- Every patch of outliers will eventually be detected by some robust residual.

- An influence cone can be constructed indicating which detail coefficients at levels ℓ, $\ell - 1$, ..., 1 are computed from the outlier patch associated with a robust residual at level ℓ

By shrinking all coefficients in the influence cone to zero (or to a suitable threshold), we can ensure that we bound the influence of any large detail coefficients associated with an outlier patch. We denote this as the "zero cone" procedure, since all coefficients are annihilated in a specified cone. In practice, to avoid artifacts caused by over-shrinking, we use zero cones for only moderate or large robust residuals.

Figure 5(b) displays the result of applying the zero cone procedure to the smoother-cleaner wavelet decomposition. The data at top is the result of reconstructing from the zero cone wavelet decomposition (without the robust residuals). The influence of the outlier patches has been almost entirely removed. The zero cones are superimposed on the plot of the decomposition. The largest detail coefficients at levels D1 and D2 have been set to zero by the zero cones: compare with figure 5(a). The remaining large coefficients at these levels correspond to the level shift.

The smoother-cleaner wavelets combined with the zero-cone procedure achieves our goal: large detail coefficients associated with outlier patches are annihilated but those associated with features are preserved. The result of applying the WAVESHRINK procedure to the zero-cone wavelets in this example is given in figure 7(a). The estimated signal (solid blocky line) faithfully tracks the true signal (dashed line), preserving the discontinuity. The outlier patches result in minimal leakage into the signal. The estimated signal is blocky since the Haar basis is used.

We remark that the zero cone procedure is especially attractive when combined with the power of an modern graphics workstation. Using the mouse, the user can interactively select cones which correspond to suspected

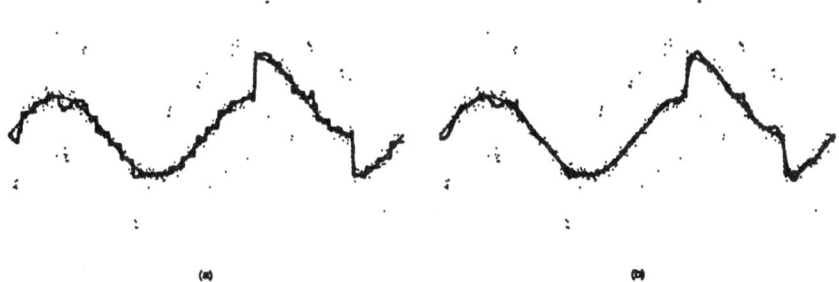

Figure 7: (a) The estimated signal obtained by the WAVESHRINK algorithm applied to the smoother-cleaner wavelet decomposition combined with the zero cone procedure and (b) the estimated signal using WAVESHRINKapplied to decomposition obtained from the clean and repeat procedure.

outlier patches. In this context, zero cones can be applied both to the robust smoother-cleaner transform as well as the classical wavelet transform. See section 5 for further discussion of software.

4.3.2   Clean and Repeat

While zero cones are intuitive and computationally very fast, they have the drawback that the cones can get very wide for general wavelet bases. In addition, zero cones require careful tracking of the indices taking into account the various possible configurations which result from decimation. A very simple alternative to the zero cones procedure which is almost as fast and empirically produces as good or better results is the "clean and repeat" procedure:

[0] Initialize with $j = 0$ and $\hat{X}_0(t) = X(t)$.

[1] Apply the robust smoother-cleaner wavelet decomposition to a data sequence $\hat{X}_j(t)$. If the number and magnitude of the robust residuals is sufficiently small, then quit.

[2] Reconstruct without the robust residuals to obtain a clean data sequence $\hat{X}_{j+1}(t)$.
Set $j = j + 1$ and go to step 1.

The basic idea is that the repeated applications of the the robust smoother-cleaner will capture any outliers which leak into the detail coefficients in previous applications. In practice, only *two passes* of the robust smoother-cleaner operations are necessary to clean the data. The resulting decomposition contains the usual wavelet coefficients plus $j-1$ sets of robust residuals.

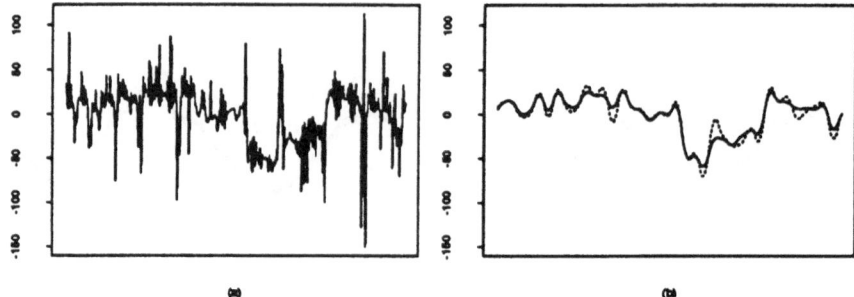

Figure 8: (a) Radar glint noise data in degrees, and (b) denoising by linear shrinkage of wavelets (dashed line) compared with denoising by WAVESHRINK combined with robust smoother-cleaner wavelets obtained by the clean and repeat procedure (solid line).

Figure 7(b) shows the result of applying WAVESHRINK to the "clean and repeat" decomposition of the outlier contaminated jumpsine data. The "b-spline" biorthogonal wavelet $\psi_{1,5}[3]$ is used for this example. The smooth is very similar to the estimate obtained by the zero cone procedure. The main difference is that we have used a smooth basis function instead of the Haar basis.

4.4   Application: Denoising Radar Glint Noise Data

We now apply the robust denoising procedures to radar glint noise data. The original noisy signal, which is the angle of the target in degrees, is displayed in Figure 8(a) The true signal is a low-frequency oscillation about $0°$. The signal contains a number of glint spikes, causing the apparent signal to be different from the true signal by as much as $150°$.

Figure 8(b) compares denoising with linear shrinkage of wavelets (dashed line) to denoising with WAVESHRINK combined with robust smoother-cleaner wavelets obtained by the clean and repeat procedure (solid line). The linear shrinkage is based on annihilating all detail coefficients of the classical wavelet transform at levels $\ell = 1, 2, 3, 4$. While linear shrinkage estimate is smooth, it is still somewhat sensitive to the glint spikes. By contrast, the clean and repeat procedure is quite resistant to the glint spikes but effectively tracks the sudden changes in the series.

5 DISCUSSION

Our research was motivated by a problem central in time series analysis: how to extract non-stationary signals which may have abrupt changes, such as level shifts, in the presence of impulsive outlier noise. A variety of techniques have been employed to deal with the problem, such as robust Kalman filtering[10, 15] and iterative outlier identification[2]. Our research indicates that a wavelet approach is an attractive alternative, offering a very fast algorithm with good theoretical properties.

Wavelets are not an appropriate basis for analysis of all types of signals. Researchers have offered various alternative bases, such as wavelet packets and local cosine bases. In this paper, we have presented some variations of wavelet analysis for data which contains impulsive outlier noise. See below for a discussion of other related research efforts.

A rich software environment is needed to support the rapid proliferation of wavelet-like techniques for analyzing data require. In a complimentary research project, we are developing S+WAVELETS, an object-oriented toolkit for wavelet analysis. The robust algorithms discussed in this paper are embedded in this toolkit. S+WAVELETS is briefly discussed below.

5.1 Related Research

Robust multiresolution decompositions based on median filtering have been developed elsewhere[4, 11] and applied to analysis of mammograms[16]. We are developing other new algorithms for robust wavelet analysis. In one approach, we develop a nonlinear triadic refinement scheme in which the wavelet coefficients are possibly nonlinear combinations of finitely many block medians at a given scale. These wavelets are nonlinear cousins of the Deslaurier-Dubuc[9] interpolation scheme. In another approach, we combine approximate conditional mean smoothers[13] with wavelets implemented using IIR filters.

5.2 S+Wavelets Object-Oriented Toolkit

S+WAVELETS is an extensible *object-oriented language* for wavelet analysis. S+WAVELETS is based on the S language for data analysis, graphics, and statistics [1]. Classes of objects are represented by the nodes. The arcs represent conceptual links between objects. The toolkit offers an array of basic building blocks, including waveform "atoms" and "crystals", wavelet filters, and time frequency dictionaries. These low-level objects are used to construct the higher-level objects, such as a wavelet transform or multiresolution analysis. The high-level objects are organized into a class hierarchy, utilizing inheritance for sharing behavior. In addition to the new robust algorithms, S+WAVELETS includes the discrete wavelet transform, wavelet packets and local cosine bases, non-decimating wavelets, a variety of graphical displays, and careful treatment of boundary related issues.

6 ACKNOWLEDGEMENTS

This research was supported by the Office of Naval Research.

7 REFERENCES

[1] R. A. Becker, J. C. Chambers, and A. R. Wilks. *The New S Language: An Programming Environment for Data Analysis and Graphics.* Wadsworth, 1988.

[2] I. Chang, G. C. Tiao, and C. Chen. Estimation of time series parameters in the presence of outliers. *Technometrics*, 30:193–204, 1988.

[3] I. Daubechies. *Ten lectures on wavelets.* Society for industrial and applied mathematics, Philadelphia, PA, 1992.

[4] Ronald A. DeVore, Bjorn Jawerth, and Bradly J. Lucier. Image compression through wavelet transform coding. *IEEE Transactions on Information Theory*, 38(2):719–746, 1992.

[5] David L. Donoho. Nonlinear wavelet methods for recovery of signals, densities, and spectra from indirect and noisy data. In Ingrid Daubechies, editor, *Proceedings of the Symposia in Applied Mathematics*. American Mathematical Society, 1993.

[6] David L. Donoho and Iain M. Johnstone. Adapting to unknown smoothness via wavelet shrinkage. Technical report, Department of Statistics, Stanford University, 1992.

[7] David L. Donoho and Iain M. Johnstone. Ideal spatial adaptation via wavelet shrinkage. Technical report, Stanford University, 1992.

[8] David L. Donoho and Iain M. Johnstone. Minimax estimation via wavelet shrinkage. Technical Report 402, Stanford University, 1992.

[9] S. Dubuc. Interpolation through an iterative scheme. *J. Math. Anal. and Appl.*, 114:185–204, 1986.

[10] P. J. Harrison and C. F. Stevens. Bayesian forecasting. *Journal of the Royal Statistical Society, Series B*, 38:205–247, 1976.

[11] H. Longbotham. A class of robust nonlinear filters for signal decomposition and filtering utilizing the Haar basis. In *ICASSP-92*, volume 4. IEEE Signal Processing Society, March 1992.

[12] Stéphane Mallat and Sifen Zhong. Characterization of signals from multiscale edges. *IEEE Transactions on Pattern Analysis and Machine Intelligence*, 14(7):710–732, 1992.

[13] R. D. Martin. Approximate conditional-mean type smoothers and interpolators. In T. Basser and M. Rosenblatt, editors, *Smoothing Techniques for Curve Estimation*, pages 147–176. Academic Press, New York, 1979.

[14] R. D. Martin and V. J. Yohai. Influence curves for time series. *The Annals of Statistics*, 11:781–818, 1986.

[15] C. J. Masreliez and R. D. Martin. Robust Bayesian estimation for the linear model and robustifying the Kalman filter. *IEEE Transactions on Automatic Control*, AC-22:361–371, 1977.

[16] Walter B. Richardson. Nonlinear filtering and multiscale texture discrimination for mammograms. Technical report, University of Texas, San Antonio, TX 78249, 1993.

Author Index

Address List of Authors

Achcar, Jorge Alberto
  Universidade de São Paulo, ICMSC
  S.P.
  C.P. 668
  13560-970 São Carlos
  BRAZIL

Adèr, Herman J.
  Faculty of Medicine, Dept. of
  Epidemiology and Biostatistics,
  Vrije Universiteit Amsterdam
  v.d. Boechorststr. 7
  1081BT Amsterdam
  THE NETHERLANDS

Altenburg, Hans-Peter
  University of Heidelberg, Fakultät
  für Klinische Medizin Mannheim,
  Med. Statistik, Biomathematik und
  Informationsverarbeitung
  Theodor-Kutzer-Ufer
  D-68135 Mannheim
  GERMANY

Atkinson, Anthony C.
  Department of Statistics, London
  School of Economics
  Houghton Street
  London WC2A 2AE
  GREAT BRITAIN

Bacha, Mostafa
  INRIA-Rocquencourt, Domaine de
  Voluceau
  BP 105
  F-78153 Le Chesnay Cedex
  FRANCE

Baczkowski, Andrew J.
  Department of Statistics, University
  of Leeds
  Leeds LS2 9ST
  GREAT BRITAIN

Bäumer, Hans-Peter
  HRZ–Angewandte Statistik, Carl
  von Ossietzky Universität Oldenburg
  Postfach 2503
  D-26015 Oldenburg
  GERMANY

Benner, Axel
  Biostatistik, Deutsches
  Krebsforschungszentrum, Heidelberg
  Postfach 101949
  D-69009 Heidelberg
  GERMANY

Besse, Philippe C.
  Laboratoire de Statistique et
  Probabilités, U.A. CNRS 745,
  Université Paul Sabatier
  F-31062 Toulouse cedex
  FRANCE

Bezenchek, Antonia
  INFN - Sezione Sanità, Roma
  viale Regina Elena 299
  I-00161 Roma
  ITALY

Celeux, Gilles
  Centre Hospitalier Universitaire de
  Grenoble, Service S.I.I.M.
  B.P. 217
  F-38043 Grenoble cedex 9
  FRANCE

Choukair, Lamia
  Dept. INF - ENST, Télécom-Paris
  46, rue Barrault
  F-75634 Paris CEDEX 13
  FRANCE

Ciampi, Antonio
Department of Epidemiology &
Biostatistics, McGill University,
Montreal and McGill
University/Montreal Children's
Hospital Research Institute
4060 Ste.-Catherine West
Montréal
CANADA Que H3Z 2Z3

Cook, Dianne
Department of Statistics, Iowa State
University
323 Snedecor Hall
Ames
USA IA 50011

Croux, Christophe
Department of Mathematics and
Computer Science, University of
Antwerp (U.I.A.)
Universiteitsplein 1
B-2610 Wilrijk
BELGIUM

Darius, Paul L.
Katholieke Universiteit Leuven,
Laboratory for Statistics and
Experimental Design
Kardinaal Mercierlaan 92
B-3001 Heverlee
BELGIUM

Durand, Jean-François
ENSAM-INRA-UM II, Unité de
Biométrie, Montpellier
9, place Pierre Viala
F-34060 Montpellier
FRANCE

Edler, Lutz
Biostatistik, Deutsches
Krebsforschungszentrum,
Heidelberg, Abt. 0820
Im Neuenheimer Feld 280
D-69009 Heidelberg
GERMANY

Ekblom, Håkan
Department of Mathematics, Luleå
University
S-951 87 Luleå
SWEDEN

Falguerolles, Antoine de
Laboratoire de Statistique et
Probabilités, U.A. CNRS D0745,
Université Paul Sabatier
F-31062 Toulouse cedex
FRANCE

Fedorov, Valery
University of Minnesota, Dept. of
Applied Statistics
1994 Buford Ave
St.Paul
USA MN 55108

Frigessi, Arnoldo
Laboratorio di Statistica, Università
di Venezia
Cà Foscari
I-30123 Venezia
ITALY

Genz, Alan
Department of Mathematics,
Washington State University
Pullman
USA WA 99164-3113

Giani, Guido
Abteilung Biometrie und
Epidemiologie,
Diabetes-Forschungsinstitut an der
Universität Düsseldorf
Auf'm Hennekamp 65
D-40225 Düsseldorf
GERMANY

Gilchrist, Robert
University of North London
Holloway Rd.
London N7 8DB
GREAT BRITAIN

Gill, Christine A.
Department of Statistics, University
of Leeds
Leeds LS2 9JT
GREAT BRITAIN

Hartmann, Peter
Institute for Statistics and
Econometrics, University of Basel
Petersgraben 51
CH-4051 Basel
SWITZERLAND

Hauser, Michael A.
Department of Statistics, University
of Economics and Business
Administration, Vienna
Augasse 2-6
A-1090 Vienna
AUSTRIA

Hitz, Martin
Department of Statistics, Operations
Research and Computer Methods,
University of Vienna
Universitätsstr.5
A-1010 Vienna
AUSTRIA

Hörmann, Wolfgang
Department of Statistics, University
of Economics and Business
Administration
Augasse 2-6
A-1090 Vienna
AUSTRIA

Hornik, Kurt
Institut für Statistik und
Wahrscheinlichkeitstheorie,
Technische Universität Wien
Wiedner Hauptstr. 8-10/1071
A-1040 Vienna
AUSTRIA

Huber, Peter J.
Universität Bayreuth, Mathematik 7
Universitätsstr. 30
D-95447 Bayreuth
GERMANY

Hudec, Marcus
Department of Statistics, Operations
Research and Computer Methods,
University of Vienna
Universitätsstr. 5
A-1010 Vienna
AUSTRIA

Hušková, Marie
Department of Statistics, Charles
University
Sokolovská 83
CZ-186 00 Praha
CZECH REPUBLIC

Jaupi, Luan
Département de Mathématiques;
Conservatoire National des Arts et
Métiers, Paris
292, rue Saint-Martin
F-75003 Paris
FRANCE

Joanes, Derrick N.
Department of Statistics, University
of Leeds
Leeds LS2 9JT
GREAT BRITAIN

Katkovnik, Vladimir Y.
Department of Statistics, University
of South Africa
P.O. Box 392
0001 Pretoria
SOUTH AFRICA

Kropf, Siegfried
Institut für Biometrie und
Medizinische Informatik,
Otto-von-Guericke-Universität
Magdeburg
Leipziger Str. 44
D-39120 Magdeburg
GERMANY

Lenz, Hans-Joachim
Dept. of Statistics and
Econometrics, Free University of
Berlin
Garystr. 21
D-14195 Berlin
GERMANY

Low, Janice Lorraine
Mathematics Department,
University of Southampton
Highfield
Southampton S09 5NH
GREAT BRITAIN

Markus, Monica
Department of Data Theory, Leiden
University
P.O. Box 9555
2300 AK Leiden
THE NETHERLANDS

Martin, R. Douglas
MathSoft, Inc., Seattle
1700 Westlake Ave. N
Suite 500
Seattle
USA WA 98195

Mazur, Viacheslav
Firm "BNK", Moscow
Gvardeyskaya ul. 4-77
Moscow
RUSSIA

Militký, Jiří
Department of Textile Materials,
Textile Faculty, Technical University
Liberec
CZ-461 17 Liberec
CZECH REPUBLIC

Mucha, Hans-Joachim
IAAS, Berlin
Mohrenstr. 39
D-10117 Berlin
GERMANY

Müller, Christine
Freie Universität Berlin, Fachbereich
Mathematik und Informatik, WE 1
Arnimallee 2-6
D-14195 Berlin
GERMANY

Murphy, Brian P.
Department of Information
Management & Marketing,
University of Western Australia
Nedlands
WA 6009 AUSTRALIA

O'Brien, Timothy E.
Department of Statistics,
Washington State University
Pullman
USA WA 99164-3144

Payne, Roger W.
Rothamsted Experimental Station,
AFRC Institute of Arable Crops
Research
Harpenden
Hertfordshire AL5 2JQ
GREAT BRITAIN

Pesarin, Fortunato
Dipartimento di Scienze Statistiche,
University of Padua
Via S. Francesco 33
I-35121 Padua
ITALY

Pfeiffer, Karl-Peter
  Ludwig Boltzmann-Institut für
  Epidemiologie &
  Gesundheitssystemforschung, Graz
  Harrachg. 21/5
  A-8010 Graz
  AUSTRIA

Prat Bartes, Albert
  Universidad Politécnica de Cataluña,
  Departamento de Estadística e
  Investigación Operativa
  Av. Diagonal 647
  planta 6.ª
  E-08028 Barcelona
  SPAIN

Prescott, Philip
  Department of Mathematics,
  University of Southampton
  Highfield
  Southampton SO9 5NH
  GREAT BRITAIN

Redfern, Edwin J.
  Department of Statistics, Leeds
  University
  Leeds LS2 9JT
  GREAT BRITAIN

Richardson, Thomas
  Philosophy Department,
  Carnegie-Mellon University
  Baker Hall 135
  Pittsburgh
  USA PA 15213

Rigby, Robert A.
  School of Mathematical Sciences,
  University of North London
  166-220 Holloway Road
  London N7 8DB
  GREAT BRITAIN

Ringrose, Trevor J.
  Department of Mathematical
  Sciences, University of Aberdeen
  The Edward Wright Building
  Dunbar Street
  Old Aberdeen AB9 2TY
  GREAT BRITAIN

Ross, Gavin J.S.
  Statistics Department, IACR
  Rothamsted Experimental Station
  Herts
  Harpenden AL5 2JQ
  GREAT BRITAIN

Roure, David C. De
  Department of Electronics and
  Computer Science & Department of
  Social Statistics, University of
  Southampton
  Highfield
  Southampton SO17 1BJ
  GREAT BRITAIN

Schimek, Michael G.
  Medical Biometrics Group,
  University of Graz Medical Schools
  Auenbruggerplatz 30/IV
  A-8036 Graz
  AUSTRIA

Schmidtke, J.
  Biorat GmbH, Rostock
  Joachim-Jungius-Str. 9
  D-18059 Rostock
  GERMANY

Siciliano, Roberta
  Dipartimento di Matematica e
  Statistica, Università di Napoli
  Federico II
  Via Cintia - Monte S. Angelo
  I-80126 Napoli
  ITALY

Smith, Peter W.F.
  Department of Social Statistics,
  University of Southampton
  Southampton SO17 1BJ
  GREAT BRITAIN

Snyman, J. Louise J.
  Department of Statistics,
  Potchefstroom University
  2520, Private Bag X6001
  Potchefstroom
  SOUTH AFRICA

Stasinopoulos, Mikis D.
  School of Mathematical Sciences,
  University of North London
  166-220 Holloway Road
  London N7 8DB
  GREAT BRITAIN

Tanaka, Yutaka
  Department of Statistics, Okayama
  University
  Tsushima
  Okayama
  700 JAPAN

Tracy, Derrick S.
  Department of Mathematics and
  Statistics, University of Windsor
  Ontario
  401 Sunset
  Windsor
  CANADA N9B 3P4

Whittaker, Joe
  Mathematics Department, Lancaster
  University and Hague Consulting
  Group
  Lancaster LA1 4YF
  GREAT BRITAIN

COMPSTAT 1994

11th Symposium on
Computational Statistics
22 – 26 August 1994
Vienna, Austria

The European Regional Section (ERS) of the International Association for Statistical Computing (IASC), a Section of the International Statistical Institute (ISI).

Y. Dodge, University of Neuchâtel, Switzerland
J. Whittaker, Lancaster University, UK (Eds.)

Computational Statistics

Proceedings of the 10th Symposium on
Computational Statistics.
COMPSTAT. Neuchâtel, Switzerland, August 1992

Volume 1

1992. XVI, 578 pp. 102 figs. Hardcover DM 198,–
ISBN 3-7908-0634-X

Volume 2

1992. X, 440 pp. 97 figs. Hardcover DM 148,–.
ISBN 3-7908-0640-4

The papers assembled in this book were presented at the biannual Symposium of the International Association for Statistical Computing in Neuchâtel, Switzerland, in August 1992. This congress maintaines the tradition of providing a forum for the open discussion of progress made in computer oriented statistics and the dissemination of new ideas throughout the statistical community. The papers are published in two volumes according to the emphasis of the topics: **volume 1** gives a slight leaning towards statistics and modelling, while **volume 2** concentrates on computation.

Volume 1 brings together a wide range of topics and perspectives in the field of statistics. It contains invited and contributed papers that are grouped for the ease of orientation in ten parts: Statistical Modelling; Multivariate Analysis; Classification and Discrimination; Symbolic Data; Graphical Models; Time Series Models; Nonlinear Regression; Robust and Smoothing Techniques; Industrial Applications: Pharmaceutics and Quality Control; Bayesian Statistics.

Volume 2 discusses the following topics and perspectives grouped in eight parts: Programming Environments; Computational Inference; Package Developments; Experimental Design; Image Processing and Neural Networks; Meta Data; Survey Design;
Data Base.

Please order through your bookseller or from
Physica-Verlag, c/o Springer-Verlag GmbH & Co.KG,
Auftragsbearbeitung, Postfach 31 13 40,
D-10643 Berlin, Germany.

Computational Statistics

ISSN 0943–4062

Editors: W. Härdle, Berlin, Germany; D.W. Scott, Houston, TX, USA
Editorial Assistant: S. Klinke, Berlin, Germany
Associate Editors: L. Bauwens, J. Beirlant, P. Besse, J. Booth, B.B. Brown,
D.B. Carr, D. Girard, Ch. Jennison, H. Läuter, K.Ch. Li,
E. Mammen, J.S. Marron, R. Ostermann, B.U. Park,
W. Polasek, W.M. Sallas, M. Schimek, C.D. Sutton,
G.R. Terrel, T.M. Therneau, M. Veall, A. Wegmann, A.R. Wilks
Book Review Editors: H. Läuter, Potsdam, Germany; M. Schimek, Graz, Austria
Software Developers: R. Buhler, M. Clarke, B. Ford, M. Norusis, R. Payne,
J.P. Sall, R. Tomassone
Software Review Editor: R. Ostermann, Siegen, Germany

Computational Statistics (CompStat) is an **international journal** which promotes
the publication of applications and methodological research in the field of
Computational Statistics. The focus of papers in **CompStat** is on the contribution
to and influence of computing on statistics and vice versa. The journal provides
a forum for computer scientists, mathematicians, and statisticians in a variety of
fields of statistics such as biometrics, econometrics, data analysis, graphics,
simulation, algorithms, knowledge based systems, and Bayesian computing.
CompStat publishes hardware, software plus package reports as well as book reviews

Fields of interest: Statistics, computer sciences
Abstracted/Indexed in: Current Index to Statistics, Journal Contents in Quantitative
Methods, Mathematical Reviews, Statistical Theory & Method Abstracts,
Zentralblatt für Mathematik

Subscription information 1994:
Vol. 9 (4 issues) DM 248,00 suggested list price,
plus carriage charges

Please order through your bookseller or from **Physica-Verlag**, c/o Springer-Verlag
GmbH & Co.KG, Auftragsbearbeitung, Postfach 31 13 40, D-10643 Berlin, Germany